SNOWTHROW
Service Manual

(3rd Edition)

Aircap Snow Champ
Atlas
Bolens
Craftsman
Crary Bear Cat
Cub Cadet
John Deere
Deutz-Allis
Ford
Gilson
Homelite/Jacobsen
Honda

International Harvester
Lawn-Boy
MTD
Montgomery Ward
Noma
J.C. Penney
Snapper
Toro
Troy-Bilt
White
Yard-Man

PUBLISHED BY:
INTERTEC PUBLISHING
P.O. Box 12901, Overland Park, KS 66282-2901
Phone: 800-262-1954 • Fax: 800-633-6219

Cover photo courtesy of:
Homelite Division of Textron, Inc.
14401 Carowinds Blvd.,
P.O. Box 7047
Charlotte, NC 28241

©Copyright 1990 by Intertec Publishing   Printed in the United States of America.
Library of Congress Catalog Card Number 90-055417

CONTENTS

CHASSIS SERVICE SECTION

CONTENTS (CONT.)

CHASSIS SERVICE SECTION (CONT.)

GENERAL

ENGINE SERVICE SECTION

GEAR TRANSMISSION SERVICE SECTION

DUAL DIMENSIONS

This service manual provides specifications in both the U.S. Customary and Metric (SI) systems of measurement. The first specification is given in the measuring system perceived by us to be the preferred system when servicing a particular component, while the second specification (given in parenthesis) is the converted measurement. For instance, a specification of "0.011 inch (0.28 mm)" would indicate that we feel the preferred measurement, in this instance, is the U.S. system of measurement and the metric equivalent of 0.011 inch is 0.28 mm.

AIRCAP SNOW CHAMP

AIRCAP INDUSTRIES INC.
P.O. Box 2120
Tupelo, MS 38803-2120

Model	Engine Make	Engine Model	Self-Propelled	No. of Stages	Scoop Width
8451	Tecumseh	HSSK50	Yes	2	20 in. (508 mm)
8461	Tecumseh	HSSK50	Yes	2	24 in. (610 mm)
8551	Tecumseh	HSSK50	Yes	2	24 in. (610 mm)
8481	Tecumseh	HMSK80	Yes	2	26 in. (660 mm)

LUBRICATION

ENGINE

Refer to Tecumseh 4-stroke engine section for engine lubrication requirements. Recommended fuel is regular or unleaded regular gasoline.

DRIVE MECHANISM

Lubricate drive chains, including chains in track drives, after every 25 hours of operation or at least once each season using a good-quality chain oil. Auger shaft bushings and traction drive shaft bushings should be oiled after every 25 hours of operation with SAE 30 motor oil. Lubricate all control pivot and slide points for smooth operation. Once each season or when necessary to replace auger flight shear bolts, remove shear bolts and squirt oil in hole and at ends of auger flights, then turn auger flight to distribute oil. Remove wheels once each season and grease axle shafts. On track drive models, keep weight transfer linkage points greased with a clinging lubricant such as "Lubriplate."

AUGER GEARBOX

The auger gearbox is factory lubricated and should not require further lubrication unless disassembled for overhaul. Recommended lubricant is 10 ounces (295 mL) of Shell Alvania grease EPR00.

MAINTENANCE

Inspect belts and pulleys and renew belts if excessively worn, frayed or burned. Renew pulleys if belt grooves are worn or damaged, or if bearings in idler pulleys are loose or rough. Check drive chains and sprockets and renew if excessively worn or damaged. On track drive models, check track adjustment.

ADJUSTMENTS

AUGER HEIGHT

Auger height can be adjusted by loosening skid shoe mounting bolts and repositioning the skid shoes. Auger height should be at lowest setting (skid shoes raised) when operating over smooth, paved surfaces and toward the highest setting (skid shoes lowered) when operating over rough surfaces. Be sure skid shoes are adjusted equally.

Fig. AC1 — Exploded view of handle assembly and controls.

1. Drive control lever
2. Auger control lever
3. Auger control rod
4. Drive control rod
5. Shift control lever
6. Shift control rod
7. Chute crank
8. Drive clutch spring
9. Adjusting nut
10. Shoulder bolt
11. Drive clutch bracket
12. Drive clutch rod
13. Drive spring
14. Ferrule
15. Shift bracket
16. Ferrule
17. Auger clutch bracket
18. Drive pulley support
19. Frame cover
20. Auger spring

SHIFT ROD

Disconnect shift control rod (6—Fig. AC1) from shift control lever (5) and place both the transmission and shift lever in neutral position. Turn shift control rod in or out of ferrule (14) at bottom of rod so upper end can be inserted in hole in shift lever. Secure rod in lever with cotter pin.

DRIVE CLUTCH

Drive clutch spring (8—Fig. AC1) at lower end of drive control rod (4) should be loose with clutch disengaged. If not, unhook spring from drive clutch bracket

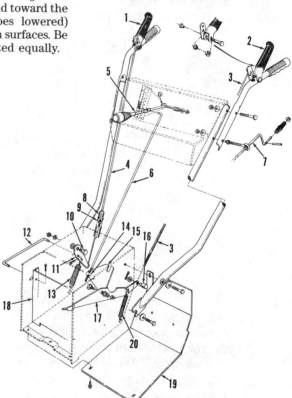

(11). Turn adjusting nut (9) so hook of spring is aligned with center of hole in drive clutch bracket then hook spring into bracket; the spring should be loose. Depress drive control lever against handle grip; the clutch should engage and move the snowthrower without belt slippage.

AUGER CLUTCH

To check adjustment, remove belt cover and hold auger control lever (2—Fig. AC1) against handle. The outer end of the slot in the end of auger brake linkage should be against the spacer on the auger belt idler as shown in Fig. AC3. If not, disconnect auger control rod ferrule (Fig. AC3) from auger clutch bracket and turn ferrule until proper adjustment is obtained with ferrule reconnected to auger clutch bracket.

TRACK DRIVE

Drive track tension is properly adjusted when the track can be lifted approximately 1/2 inch (13 mm) at midpoint between track rollers. Adjustment is provided by adjuster cam (21—Fig. AC6) on each side of left and right track assembly. If necessary to adjust track tension, loosen the cap screws retaining the adjuster cams on each side of track. Turn the adjuster cams equally with a screwdriver placed between tab on adjuster cam and heavy washer, then tighten the cap screws.

OVERHAUL

ENGINE

Engine make and model numbers are listed at beginning of this section. Refer to Tecumseh 4-stroke engine section for engine overhaul information.

DRIVE BELTS

Remove the deflector chute crank at chute and remove upper belt cover. Remove the two shoulder bolt belt guides at engine crankshaft pulley. On track drive Model 8551, place weight transfer lever in "Packed Snow" position. Remove the two top bolts attaching auger housing to snowthrower frame assembly and loosen but do not remove the two bottom bolts. Lift up on auger belt to pull auger housing off of snowthrower frame and separate auger housing from the main unit.

To remove auger drive belt, remove the four shoulder bolt belt guides at auger pulley. The auger belt can now be removed from the bottom pulley. To remove traction drive belt, disconnect drive idler spring (8—Fig. AC2) at engine plate and remove belt from engine pulley and bottom drive pulley. Install new belt(s) and reassemble by reversing disassembly procedure. Be sure pin on end of auger clutch bracket engages slot in auger brake bracket (see Fig. AC4) before installing auger housing mounting bolts. Adjust auger and drive clutches as outlined in preceding paragraphs.

FRICTION WHEEL

To renew friction wheel, proceed as follows: Move shift lever to reverse position. Tip snowthrower up so it rests on front of auger housing and remove bottom cover, then remove friction wheel (38—Fig. AC5) from wheel adapter (40). Place new wheel with cup side away from adapter and install retaining bolts finger tight. Turn the wheel to be sure it does not wobble, then securely tighten retaining bolts. Replace bottom cover.

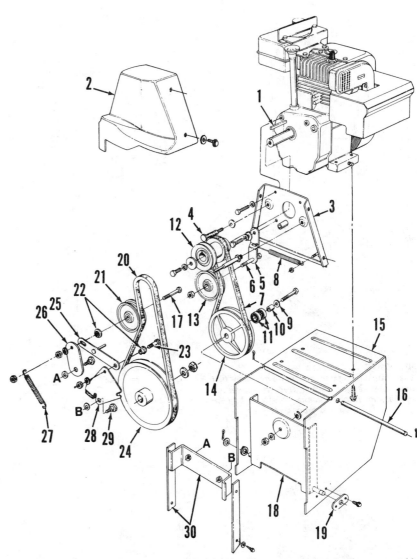

Fig. AC2—Exploded view of drive belt and auger belt drive components and engine/frame assembly. Back side of drive pulley (14) has drive surface for friction wheel; depressing drive control lever moves pulley against friction wheel (38—Fig. AC5).

1. Engine pulley key	8. Drive idler spring	16. Sliding bracket rod	23. Brake pivot bolt
2. Belt cover	9. Belleville washer	17. Auger idler bolt	24. Auger pulley
3. Engine plate	10. Spacer	18. Drive pulley support	25. Brake linkage
4. Shoulder bolt belt guide	11. Ball bearings	19. Support axle bracket	26. Auger idler bracket
5. Drive clutch idler bracket	12. Engine pulley	20. Auger belt	27. Auger idler spring
6. Idler bolt	13. Drive belt idler	21. Auger belt idler	28. Brake bracket
7. Drive belt	14. Drive pulley	22. Shoulder spacers	29. Shoulder bolt
	15. Main frame		30. Blower housing parts

TRACK DRIVE

To renew drive track or chain inside track assembly on track drive models, block up snowthrower frame so track assemblies are off the ground. Loosen the cap screws retaining track adjuster cams (21—Fig. AC6) at each side of assembly and turn adjusters so all track tension is removed. Roll the track off of track rollers and side supports. Remove roller chain (14) by disconnecting master link. Reassemble by reversing disassembly procedure, making sure to install track with cleats pointing toward front of snowthrower. Adjust track tension as outlined in adjustment paragraph.

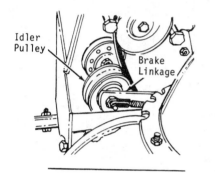

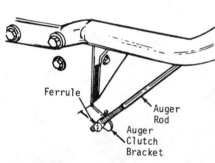

Fig. AC3—End of slot in brake linkage (25—Fig. AC2) should be against spacer on the auger belt idler when auger control lever is depressed against snowthrower handle. Refer to text for adjustment procedure.

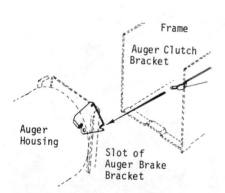

Fig. AC4—Be sure pin on end of auger clutch bracket enters slot in auger brake bracket when reassembling snowthrower.

Fig. AC5—Exploded view of drive components; also see Fig. AC2. Notch in shift linkage bracket (15) engages pin on sliding bracket (44). Sliding bracket is supported by sliding bracket rod (16). On all models except Model 8481, only one flat washer (34) is used at right end of axle shaft (53).

6. Shift rod
14. Ferrule
15. Shift linkage bracket
16. Sliding bracket rod
31. Belleville washer
32. Flat washer
33. Flange bearing
33A. Flange bearing
34. Flat washer(s)
35. Spacer
36. Sprocket
37. Hi-pro key
38. Friction wheel
39. Hex shaft
40. Friction wheel adapter
41. Snap ring
42. Flat washer
43. Bearing
44. Sliding bracket
45. Flat washer
46. Roller chain
47. Sprocket
48. Snap ring
49. Sprocket
50. Flat washer
51. Bronze bearing
52. Flat washer
53. Sprocket & axle assy. (except Model 8481)
54. Sprocket (Model 8481)
54A. Axle (Model 8481)
55. Roller chain
56. Shoulder bolt
57. Spacer (Model 8481)
58. Flat washer (Model 8481)
59. Wheel assy. (Models 8461 & 8481)
60. Sleeve bearing (Models 8461 & 8481)
61. Snap ring
62. Klick pin
63. Wheel assy. (Model 8451)

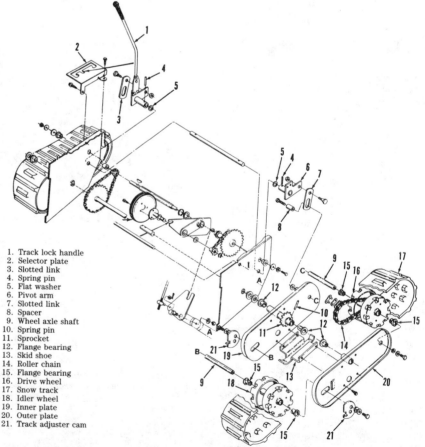

1. Track lock handle
2. Selector plate
3. Slotted link
4. Spring pin
5. Flat washer
6. Pivot arm
7. Slotted link
8. Spacer
9. Wheel axle shaft
10. Spring pin
11. Sprocket
12. Flange bearing
13. Skid shoe
14. Roller chain
15. Flange bearing
16. Drive wheel
17. Snow track
18. Idler wheel
19. Inner plate
20. Outer plate
21. Track adjuster cam

Fig. AC6—Exploded view of track drive assembly for Model 8551. Drive components are similar to wheel drive models shown in Fig. AC5. Track lock handle changes auger down force and has three positions: "Transport," "Normal Snow" and "Packed Snow."

SERVICE MANUAL

AUGER GEARBOX

To remove auger gearbox, follow procedure outlined in preceding paragraph for removal of auger drive belt. Then, remove auger drive pulley (7—Fig. AC7) from rear of gearbox shaft and unbolt auger bearing housings (37) at each end of auger housing. Remove the gearbox and auger flight assembly from front of auger housing, then remove impeller (18) from input shaft and auger flights from driveshaft. Note the position of washers and spacers as you are disassembling the unit.

To disassemble gearbox, remove the screws retaining upper gear housing half (19). Note thrust washer and spacer positions as you disassemble gearbox. Clean and inspect all parts and renew any worn or damaged parts. Lubricate all parts prior to reassembly and pack gearbox with 10 ounces (295 mL) of Shell Alvania Grease EPR00.

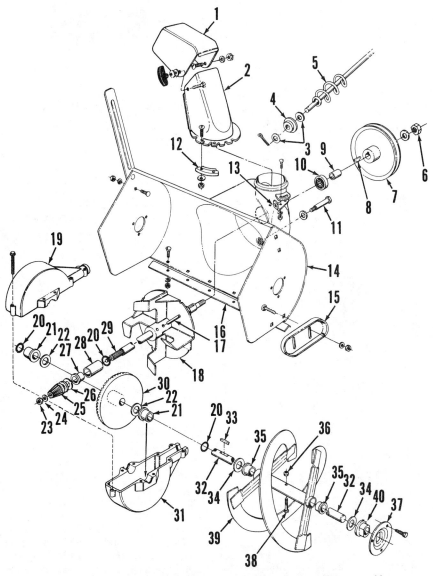

Fig. AC7—Exploded view of auger housing, gearbox and auger flight assembly.

1. Upper chute
2. Lower chute
3. Flat washers
4. Bushing
5. Chute crank
6. Pulley nut
7. Auger drive pulley
8. Hi-pro key
9. Spacer
10. Ball bearing
11. Shoulder bolt
12. Chute flange keeper
13. Chute crank bracket
14. Auger housing
15. Skid shoes
16. Scraper bar (shave plate)
17. Spring pin
18. Impeller (blower fan)
19. Upper gear housing half
20. "O" ring
21. Plastic flange bearing
22. Flat washer
23. Hex nut
24. Flat washer
25. Pinion gear
26. Flat washer(s)
27. Flange bearing
28. Sleeve bearing
29. Impeller (blower) axle
30. Bevel gear
31. Lower gear housing half
32. Auger flight (spiral) axle
33. Key (Hi-pro or square)
34. Flat washer
35. Flange bearing
36. Shear bolt nut
37. Bearing housing
38. Shear bolt
39. Auger flight (spiral)
40. Flange bearing

ATLAS

POWER EQUIPMENT CO.
P.O. Box 70
Harvard, IL 60033

Model	Engine Make	Engine Model	Self-Propelled?	No. of Stages	Scoop Width
SN320	B&S	80000	No	1	20 in. (508 mm)
SN520	B&S	130000	Yes	1	20 in. (508 mm)
ESN524	B&S	130000	Yes	1	24 in. (610 mm)
15-3200	B&S	80000	No	1	20 in. (508 mm)
15-3210	B&S	80000	No	1	20 in. (508 mm)
15-5200	B&S	130000	Yes	1	20 in. (508 mm)
15-5210	B&S	130000	Yes	1	20 in. (508 mm)
15-5210A	B&S	130000	Yes	1	20 in. (508 mm)
15-5211	B&S	130000	Yes	1	20 in. (508 mm)
15-5220	B&S	130000	Yes	2	22 in. (559 mm)
15-5240	B&S	130000	Yes	1	24 in. (610 mm)
15-5240A	B&S	130000	Yes	1	24 in. (610 mm)
15-8260	B&S	190000	Yes	2	26 in. (660 mm)

LUBRICATION

ENGINE

Refer to Briggs & Stratton section for engine lubrication requirements. If unit is to be stored over thirty days, drain fuel tank and run engine until all fuel is consumed. Pour 1 ounce (29.5 mL) of engine oil through spark plug hole, then crank engine to lubricate cylinder and piston.

All models except SN320, 15-3200, 15-3210 and 15-8260 are equipped with a reduction gearbox mounted on pto side of engine (see Fig. AT1). Remove oil level plug (G) and check oil level after every 100 hours of operation. Oil level should be even with bottom of plug hole. Remove oil fill plug (P) and fill gearbox with same type oil recommended for engine. Do not interchange oil level and oil fill plugs as oil fill plug has a vent hole.

DRIVE MECHANISM
All Models Except
Model 15-8260

Lubricate drive chains(s) with a suitable chain lubricant before and after snowthrower operation. Lubricate axle bushings after each use of snowthrower. Sprocket shaft bushing (9—Fig. AT2 or AT3) on all models except SN320 and 15-3200 and auger bushings on all models should be lubricated periodically through grease fittings in bushing housing (9) and auger. Lubricate control cables and cable and rod pivot points as required for smooth operation.

Model 15-8260

Lubricate drive chains with a suitable chain lubricant before and after snowthrower operation. After every 15

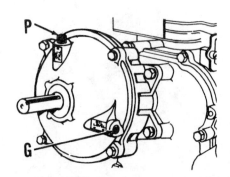

Fig. AT1—View showing location of fill plug (P) and oil level plug (G) on gear reduction unit of models so equipped.

hours of operation, remove rear panel and lubricate bushings, shafts and control rod pivot points. Note grease fitting on friction drive bearing housing (8—Fig. AT5). After every 15 hours lubricate auger shafts through grease fittings in augers; augers should also be lubricated if a shear pin has broken.

Gear box lubricant should be checked at least once a year. Oil level should be even with bottom of fill plug (14-Fig. AT12) with unit on level surface. Fill gear box with Atlas oil part No. 129-8 or SAE 90 gear oil.

MAINTENANCE

Check condition of control cables and renew if frayed or worn. Drive chains which have stretched excessively thereby preventing chain adjustment should be renewed. Renew worn or damaged chain sprockets. Inspect drive belt on models so equipped and renew belt if frayed, cracked, burnt or otherwise damaged. Inspect belt pulley groove and renew pulley if damaged or excessively worn.

A brass shear pin (42—Fig. AT2) secures the output sprocket (43) to engine on Models SN320, SN520, ESN524 and 15-3200. Should shear pin break

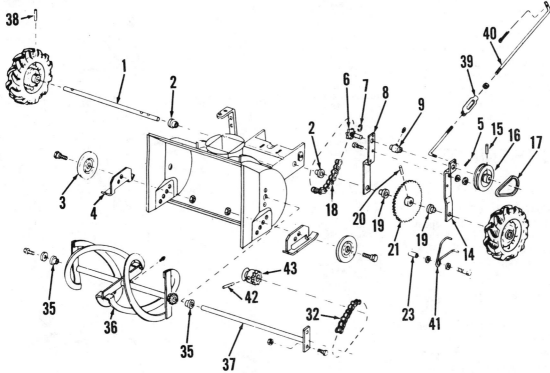

Fig. AT2—Exploded view of drive mechanism on Models SN520 and ESN524. Auger drive is similar on Models SN320, 15-3200 and 15-3210. Refer to Fig. AT3 for parts identification except for: 39. Turnbuckle; 40. Clutch rod; 41. Belt guard; 42. Shear pin; 43. Drive sprocket.

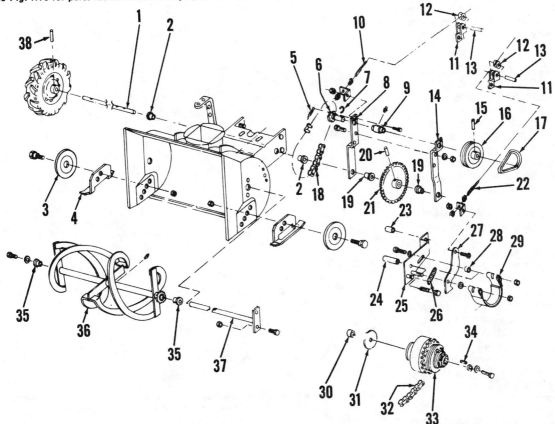

Fig. AT3—Exploded view of drive mechanism on Models 15-5210A, 15-5211 and 15-5240A. Models 15-5200, 15-5210 and 15-5240 are similar but use a clutch rod in place of clutch cable (10).

1. Axle	8. Inner Bracket	14. Outer bracket	20. Pin	26. Spring	32. Chain
2. Bushings	9. Bushing & housing	15. Pin	21. Sprocket	27. Lever	33. Drive clutch assy.
3. Runner wheel	10. Clutch cable	16. Belt pulley	22. Auger control cable	28. Roller	34. Key
4. Runner	11. Pulley bracket	17. Belt	23. Spacer	29. Brake band	35. Bushings
5. Spring	12. Cable pulley	18. Chain	24. Spacer	30. Spacer	36. Auger
6. Sprocket & shaft	13. Pin	19. Bushings	25. Bracket	31. Washer	37. Auger shaft
7. Snap ring					38. Pin

due to auger stoppage, disconnect chain, remove output sprocket and drive out shear pin.

Model 15-8260 is equipped with shear bolts (7—Fig. AT6) which secure augers to shafts. Do not replace shear bolts with standard bolts as gearbox may be damaged. Lubricate auger through grease fittings after shear bolt installation.

ADJUSTMENT

All Models Except 15-8260

HEIGHT ADJUSTMENT. Height of scoop front edge is adjusted by repositioning scoop runners (4—Fig. AT2 or AT3). Normal height setting is 1/8 to 5/8-inch (3.2-15.9 mm) for use over a smooth surface.

AUGER CHAIN ADJUSTMENT. Auger chain tension is determined by engine position. To adjust auger chain tension on Models SN320, SN520, 15-3200 and 15-5200, loosen engine mounting bolts and move engine rearward to tighten chain. On all other models, loosen bolts securing engine mount plate and move engine with plate rearward. Chain should be taut without binding during operation. Retighten bolts.

AXLE DRIVE CHAIN. To adjust axle drive chain, drive out roll pin (15—Fig. AT2 or AT3) and remove belt pulley (16). Unscrew bolts and separate brackets (8 and 14). Remove bushing housing (9) and note octagonal end of housing which fits in inner bracket (8) hole. Reposition bushing housing (9) in inner bracket (8) to adjust chain tension. Reassemble components.

BELT TENSION ADJUSTMENT. The drive belt on self-propelled models must be adjusted so belt does not slip when belt pulley engages belt but so belt will not grab when pulley is disengaged from belt. Belt pulley position is controlled either by a control rod or by a control cable. On models equipped with a control rod, rotate turnbuckle (39—Fig. AT2) to obtain desired belt tension in both engaged and disengaged positions. On models equipped with a control cable, turn adjusting nuts at cable end (10-Fig. AT3) to adjust belt tension.

DRIVE CLUTCH ADJUSTMENT. Models 15-5200, 15-5210, 15-5210A, 15-5211, 15-5240 and 15-5240A are equipped with a cable controlled auger drive clutch. Pulling the control cable releases brake band (29-Fig. AT3) and allows clutch to engage and turn drive chain sprocket. Turn nuts on end of

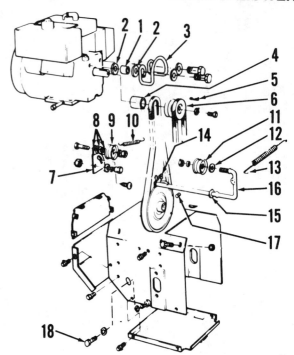

Fig. AT4—Exploded view of engine and associated components on Model 15-8260.

1. Spacer
2. Washer
3. Belt guard
4. Spacer
5. Key
6. Belt pulleys
7. Idler pulley bracket
8. Washers
9. Idler pulley
10. Spring
11. Idler pulley
12. Washer
13. Spring
14. Auger clutch lever
15. Washer
16. Idler pulley rod
17. Set screw
18. Stop bolt

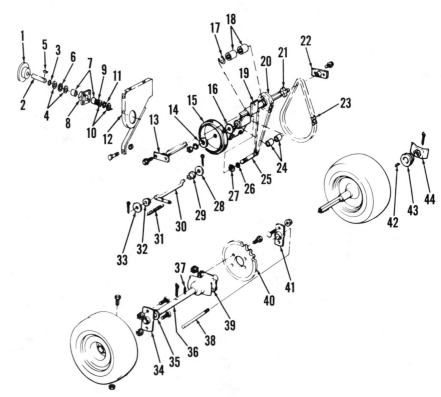

Fig. AT5—Exploded view of drive mechanism on Model 15-8260. Tightening friction nut (44) will reduce differential action and lock wheels together.

1. Friction disc	12. Plate	23. Chain	34. Bearing & housing
2. Shaft	13. Bearing & housing	24. Bearings	35. Washer
3. Seal	14. Washer	25. Shaft & sprocket	36. Axle shaft
4. Thrust washers	15. Friction wheel	26. Snap ring	37. Pin
5. Key	16. Spacer	27. Washer	38. Differential lock shaft
6. Thrust bearing	17. Snap ring	28. Washer	39. Differential
7. Needle bearings	18. Bearings	29. Bushing	40. Sprocket
8. Bearing housing	19. Control arm	30. Clutch lever	41. Bearing & housing
9. Seal	20. Sprocket	31. Spring	42. Key
10. Washers	21. Sprocket	32. Bushing	43. Friction hub
11. Snap ring	22. Bearing & housing	33. Washer	44. Nut

cable (22) so there is a small amount of play in cable with control lever released.

Model 15-8260

HEIGHT ADJUSTMENT. Loosen six bolts securing scraper bar to front of scoop and position bar so it is 1/8 to 1/4-inch below scoop. Height of scraper bar is adjusted by loosening runner bolts and repositioning runners. Scraper bar should be 1/8 to 5/8-inch (3.2-15.9

mm) above a smooth surface; increase height setting for irregular surfaces.

AUGER BELT ADJUSTMENT. When auger control lever is depressed, idler (11—Fig. AT4) is forced against drive belt to transfer power to auger. Adjust contact of idler with belt by rotating turnbuckle (4—Fig. AT7) on control rod. Belt should turn without slipping when control lever is depressed but disengage when lever is released. Be sure auger does not turn

with control lever released.

ADJUST SPEED CONTROL. Forward speed of unit is determined by contact point of friction wheel (15—Fig. AT5) on drive disc (1). With speed control lever in neutral (N) position, friction wheel should be centered on drive disc. To adjust speed control, remove rear panel and with speed control lever in neutral note location of friction wheel on drive disc. If adjustment is required, rotate turnbuckle (9—Fig. AT7) in speed control rod.

FRICTION DRIVE CLUTCH ADJUSTMENT. With speed control lever in number "1" position and clutch control rod disconnected from handlebar control lever, clearance between friction wheel (15—Fig. AT5) and drive disc (1) should be 0.090-0.120 inch (2.3-3.0 mm). Adjust clearance by loosening then repositioning stop bolt (18–Fig. AT8) in slot. Tighten stop bolt and connect clutch control rod to lever.

Adjust length of clutch control rod (6—Fig. AT7) by rotating turnbuckle so upper cotter pin hole (H—Fig. AT8) in rod is 1/4 to 3/8-inch (6.4-9.5 mm) above clutch lever (30) when clutch control lever on handlebar is fully compressed. See Fig. AT8.

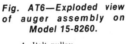

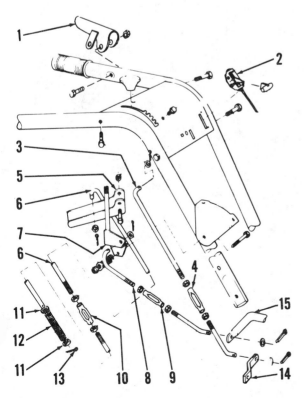

Fig. AT6—Exploded view of auger assembly on Model 15-8260.

1. Belt pulley
2. Set screw
3. Bearing & housing
4. Bearing & housing
5. Washer
6. Right auger
7. Shear bolt
8. Left auger
9. Gearbox
10. Scrapper bar
11. Runner

Fig. AT7—Model 15-8260 handlebar assembly.

1. Auger control lever
2. Throttle control
3. Auger control rod
4. Turnbuckle
5. Clutch control lever
6. Clutch control rod
7. Speed control lever
8. Speed control rod
9. Turnbuckle
10. Turnbuckle
11. Washers
12. Spring
13. Cotter pin
14. Auger clutch lever
15. Speed control arm

OVERHAUL

ENGINE

Engine make and model number are listed at beginning of Atlas section. Refer to Briggs & Stratton engine section for engine overhaul.

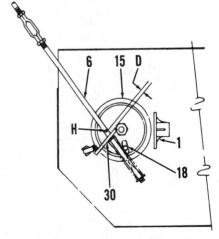

Fig. AT8 — Distance (D) between upper cotter pin hole (H) and clutch lever (30) should be 1/4 to 3/8-inch (6.4-9.5 mm) when clutch lever on handlebar is fully compressed.

1. Friction disc
6. Clutch control rod
15. Friction wheel
18. Stop bolt
30. Clutch lever

REDUCTION GEARBOX

All models except SN320, 15-3200, 15-3210 and 15-8260 are equipped with a reduction gearbox mounted to pto side of engine and shown in Fig. AT9. To overhaul gearbox, remove drive clutch then unbolt gearbox from engine. Drain lubricant from gearbox. Unscrew cover (2) screws and remove cover and gear assembly from housing (6). Inspect gears and seals and renew if necessary. Reassemble gearbox by reversing disassembly procedure. Refill gearbox as outlined in LUBRICATION section.

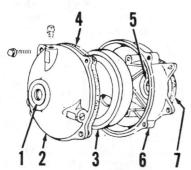

Fig. AT9—Exploded view of reduction gearbox used on all models except SN320, 15-3200, 15-3210 and 15-8260.

1. Seal
2. Cover
3. Gear assy.
4. Gasket
5. Seal
6. Housing
7. Gasket

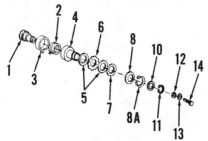

Fig. AT10—Exploded view of drive clutch used on Models 15-5200, 15-5210 and 15-5240. Refer to Fig. AT11 for parts identification except for: 8A. Snap ring.

DRIVE CLUTCH

Models 15-5200, 15-5210 and 15-5240 are equipped with the drive clutch assembly shown in Fig. AT10. The drive clutch is mounted on the reduction gearbox output shaft and is not adjustable. Clutch discs (5) may be renewed if clutch slippage is evident. Extract and renew spring (2) if broken.

Models 15-5210A, 15-5211 and 15-5240A are equipped with the drive

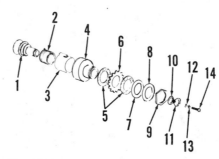

Fig AT11—Exploded view of drive clutch used on Models 15-5210A, 15-5211 and 15-5240A.

1. Inner hub
2. Spring
3. Clutch sleeve
4. Outer hub
5. Clutch discs
6. Sprocket
7. Washer
8. Belleville washer
9. Nut
10. Spacer
11. Snap ring
12. Washer
13. Lockwasher
14. Cap screw

Fig. AT12—View of auger gearbox used on Model 15-8260.

1. Auger & shaft
2. Seal
3. Needle bearing
4. Side cover
5. Gasket
6. Bearing
7. "O" ring
8. Spacer
9. Gear
10. Pin
11. Washer
12. "O" ring
13. Housing
14. Plug
15. Thrust washers
16. Pin
17. Worm gear
18. Needle bearing
19. Bearing
20. Auger shaft

clutch assembly shown in Fig. AT11. If clutch slips, disassemble unit as follows: Remove side cover, disconnect auger drive chain, unscrew retaining screw (14) and remove drive clutch assembly. Remove snap ring (11) and using a counter-clockwise twisting motion remove inner hub (1) and spring (2). Renew spring if broken. Unscrew nut (9) which is retained with Loctite and remove washers and sprocket. Note condition of fiber clutch discs (5) and renew discs if broken or excessively worn—discs are approximately 1/8-inch (3.2 mm) thick when new. Reverse disassembly procedure when assembling clutch. Install Belleville washer (8) with convex side against nut (9). Apply Loctite to nut threads and tighten nut to 23 ft.-lbs (31.2 N·m). Install spring (2) with outer end nearest pulley on inner hub (1) and in clutch sleeve (3) groove.

AUGER GEARBOX

Model 15-8260 is equipped with the worm drive gearbox shown in Fig. AT12. Disassembly of unit is evident after removal and referral to Fig. AT12. Inspect components and renew any which are damaged or excessively worn.

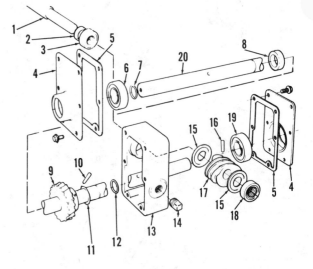

ATLAS

Model	Engine Make	Engine Model	Self-Propelled?	No. of Stages	Scoop Width
3222	B&S	32032	No	1	22 in. (559 mm)

NOTE: Model number will have letter prefix and suffix such as "A3222ESH." "A" prefix indicates Atlas brand, "C" indicates Chieftain brand, and "SC" prefix indicates Snow Chief brand. "ES" suffix indicates model is equipped with electric start. The last suffix letter (in addition too or without suffix "ES") changes with model year (B,...G, H).

LUBRICATION

ENGINE

Refer to Briggs & Stratton 2-stroke engine section for engine information. Oil is mixed with fuel to lubricate the 2-stroke engine. Recommended fuel:oil mixture ratio after initial break-in period is 32:1. Recommended fuel is lead free non-premium gasoline. Recommended oil is a BIA certified 2-cycle oil rated TC-W.

DRIVE MECHANISM

The auger bearings are sealed ball bearing type and do not require lubrication. Lubricate belt idler bracket pivot point lightly, being careful not to allow lubricant on belt or pulleys. Wheel bearings are plastic. Removing the hub cap and placing a drop or two of oil on the axle will help the snowthrower roll easier.

MAINTENANCE

Inspect belt and pulleys. Renew belt if frayed, cracked, burned or excessively worn. Renew idler pulley if bearing is worn or pulley turns rough. Check clutch cable for binding or stiff operation and be sure idler bracket moves smoothly. Renew auger assembly (45—Fig. AT15) as a unit if auger flights are excessively worn or damaged or return the assembly to an authorized service center for repairs.

ADJUSTMENTS

AUGER HEIGHT

There are no auger height adjustments on this snowthrower.

CLUTCH CABLE

If belt slips under load, unbolt cable bracket (4—Fig. AT15) from snowthrower handle and move the bracket down the handle one notch of the bracket. Tighten bolt securely and recheck clutch adjustment. Auger should not rotate with clutch lever released. Install new belt if moving the cable end bracket to its farthest downward position does not provide sufficient belt tension.

OVERHAUL

ENGINE

Engine make and model number are listed at beginning of this section. Refer to Briggs & Stratton 2-stroke engine section for engine overhaul information.

AUGER ASSEMBLY

Refer to exploded view in Fig. AT15. Remove drive belt cover (24) and belt (23), then remove snap ring (25) and auger pulley (26) from auger shaft. Remove snap ring (40) at outer side of right auger bearing (42). Loosen both bearings and slide auger to left to clear right side bearing and remove auger from housing. Remove the screws retaining auger assembly to auger shaft and remove the shaft from auger. Parts for the auger assembly are available only from factory authorized service centers and must be renewed as a unit in the field or returned to a factory service center for repair. Auger bearing retainers (41 & 44) are pop-riveted on auger housing

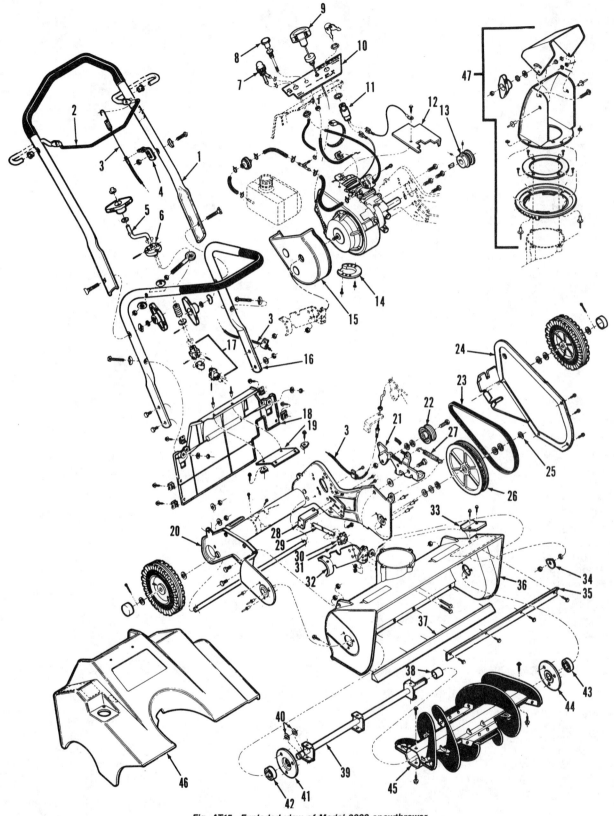

Fig. AT15—Exploded view of Model 3222 snowthrower.

1. Upper handle
2. Auger control bail
3. Auger cable & spring
4. Cable bracket
5. Chute crank
6. Chute crank detent
7. Fuel primer
8. Choke control
9. Engine start grip
10. Panel decal

11. Ignition switch
12. Engine heat shield
13. Engine pulley
14. Muffler deflector
15. Engine blower shield
16. Lower handle
17. Chute crank U-joint
18. Back panel
19. Panel support
20. Frame

21. Idler arm
22. Idler pulley
23. Drive belt
24. Belt cover
25. Snap ring
26. Auger pulley
27. Spring
28. Auger cable bracket
29. Chute sprocket shaft

30. Chute sprocket
31. Roll pin
32. Engine support
33. Tab
34. Backup washer
35. Auger housing baffle
36. Auger housing
37. Scraper bar
38. Spacer

39. Auger shaft
40. Snap rings
41. Bearing retainer
42. Auger bearing, R.H.
43. Auger bearing, L.H.
44. Bearing retainer
45. Auger assy.
46. Engine cover
47. Chute assy.

and can be renewed by cutting rivets, then the new parts riveted in place.

DRIVE BELT

The drive belt can be renewed after removing the belt cover from left side of snowthrower. Push the idler pulley slightly downward to release front of idler bracket from auger pulley and belt. Then, remove the belt from engine and auger pulleys. With belt removed, check condition of pulleys and be sure idler bracket moves freely on pivot. Reverse removal procedure to install new belt. Refer to Fig. AT16 for belt routing diagram. It may be necessary to adjust position of clutch cable bracket on snowthrower handle as outlined in previous CLUTCH CABLE paragraph.

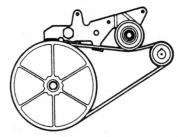

Fig. AT16—View showing routing of drive belt.

ATLAS

Model	Engine Make	Engine Model	Self-Propelled?	No. of Stages	Scoop Width
3216E	Tecumseh	AH600	No	1	21 in. (533 mm)

NOTE: Model number will have letter prefix and suffix such as "SC3216ES." "A" prefix indicates Atlas brand, "C" indicates Chieftain brand, and "SC" prefix indicates Snow Chief brand. "ES" suffix indicates model is equipped with electric start. The last suffix letter (in addition to or without suffix "ES") changes with model year (B,...G, H).

LUBRICATION

ENGINE

Refer to Tecumseh 2-stroke engine section for engine lubrication information. Recommended fuel is regular or unleaded gasoline. After every 25 hours of operation, apply a small amount of SAE 30 engine oil on choke knob (24—Fig. AT21) at control panel and work knob up and down to distribute oil.

DRIVE MECHANISM

Drive belt idler pivot arm should be lubricated at pivot stud (12—Fig. AT23) with SAE 30 engine oil after every 25 hours of operation. To gain access to pivot arm, remove belt cover from left side of snowthrower housing. Be careful not to get any oil on drive belt or pulleys.

MAINTENANCE

Inspect drive belt and renew if frayed, cracked, burned or excessively worn. Pulleys should be renewed if belt groove is excessively worn or pulley is otherwise damaged. Check auger clutch cable; lubricate or renew cable if binding. Check belt idler pulley for bearing being loose or rough and renew idler pulley assembly if defect is noted. Check the rubber impeller paddles and renew if excessively worn or damaged.

ADJUSTMENTS

IMPELLER CLUTCH

If impeller fails to turn or belt slips under load, check to be sure idler arm is

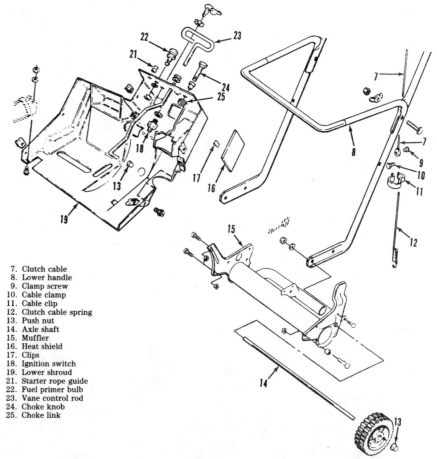

7. Clutch cable
8. Lower handle
9. Clamp screw
10. Cable clamp
11. Cable clip
12. Clutch cable spring
13. Push nut
14. Axle shaft
15. Muffler
16. Heat shield
17. Clips
18. Ignition switch
19. Lower shroud
21. Starter rope guide
22. Fuel primer bulb
23. Vane control rod
24. Choke knob
25. Choke link

Fig. AT21—Exploded view of handle and controls. Heat shield (16) fastens to inside wall of lower shroud with clips (17). Choke knob (24) screws into coil end of choke link (25).

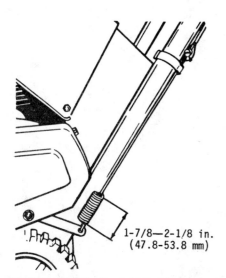

1-7/8—2-1/8 in. (47.8-53.8 mm)

Fig. AT20—With impeller clutch lever in engaged position, length of clutch cable spring should be as shown.

working freely on pivot stud (12—Fig. AT23). Check idler arm return spring (13) and clutch cable spring (12—Fig. AT21). With control lever engaged against handle, check length of spring as shown in Fig. AT20. If spring length is not between measurements shown, release control lever, loosen clamp screw (9—Fig. AT21) holding control cable (7) to spring (12) and lengthen or shorten spring as necessary.

BELT GUIDES

With impeller clutch lever in engaged position, gap between belt guides (9 & 14—Fig. AT23) and drive belt should be 1/16 inch (1.6 mm). If gap is not correct, loosen guide retaining nuts, reset gap and tighten nuts securely.

OVERHAUL

ENGINE

Refer to Tecumseh 2-stroke engine section for engine overhaul information.

DRIVE MECHANISM

Overhaul of basic unit should be evident after inspection of unit and reference to Fig. AT23. Impeller pulley (19) screws onto impeller shaft (23—Fig. AT22) with left-hand threads. If engine or impeller pulley has been removed, place straightedge on pulleys to check alignment. Move engine crankshaft pulley, if necessary, to bring alignment to within 0.030 inch (0.76 mm).

DRIVE BELT

To renew drive belt, remove cover from left side of snowthrower housing and, using pliers, pull fixed belt guide pin (21—Fig. AT23) from housing (8—Fig. AT22). Loosen belt guides (9 & 14—Fig. AT23). Remove belt from under idler pulley and from engine pulley, then lift belt from impeller pulley. Install new belt and tap fixed guide pin back into auger housing. Be sure that groove in pin is away from belt. Adjust engine pulley belt guides as outlined in adjustment paragraph.

IMPELLER PADDLES

Impeller paddles (22—Fig. AT22) can be renewed after removing the row of cap screws retaining each paddle in the impeller hub assembly, then pulling paddle from hub. Paddles must be replaced in pairs to avoid an unbalanced condition in impeller. Be sure that curve (cup) in paddle faces direction of impeller rotation.

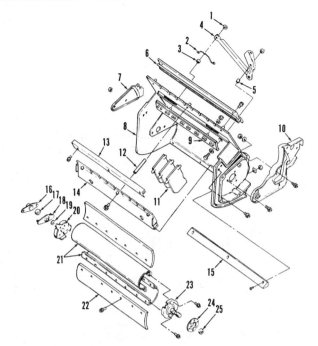

Fig. AT22—Exploded view of impeller and discharge vane assembly.

1. Self-thread nut
2. Detent spring
3. Bushing
4. Vane control lever
5. Bushing
6. Rear vane retainer
7. Right bracket
8. Impeller housing
9. Vane adjustment bar
10. Left bracket
11. Vanes
12. Spacer tube
13. Front vane retainer
14. Front vane mount
15. Scraper bar
16. Bearing retainer
17. Bearing
18. Bearing flange
19. Flat washer
20. Right impeller shaft
21. Impeller drum
22. Impeller paddle
23. Left impeller shaft
24. Bearing retainer
25. Bearing

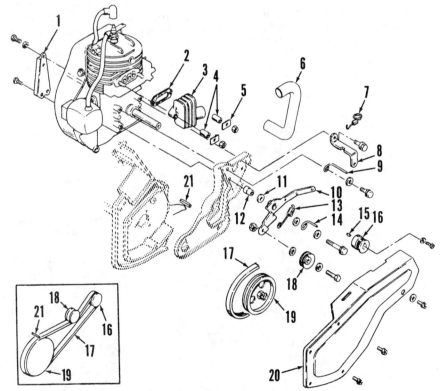

Fig. AT23—Exploded view showing engine and impeller drive mechanism. Spring (7) hooks in notch of belt cover bracket (8) and around exhaust pipe (6) to hold pipe in position.

1. Engine bracket
2. Manifold gasket
3. Exhaust manifold
4. Spacers
5. Nut retainer tab
6. Exhaust pipe
7. Pipe retainer spring
8. Belt cover bracket
9. Belt guide
10. Belt idler arm
11. Flat washer
12. Pivot stud
13. Idler arm return spring
14. Belt guide
15. Woodruff key
16. Engine pulley
17. Impeller belt
18. Idler pulley
19. Impeller pulley
20. Belt cover
21. Belt guide pin

ATLAS

Model	Engine Make	Engine Model	Self-Propelled?	No. of Stages	Scoop Width
420T	Tecumseh	HS40	Yes	2	20 in. (508 mm)

NOTE: Model numbers will have letter prefix and suffix such as "A420TH." "A" prefix indicates Atlas brand, "C" indicates Chieftain brand, and "SC" prefix indicates Snow Chief brand. "T" suffix indicates model is equipped with track drive. The last suffix letter (in addition to suffix "T") changes with model year (E, F, G, H).

LUBRICATION

ENGINE

Refer to Tecumseh 4-stroke engine section for engine lubrication information. Recommended fuel is regular or unleaded regular gasoline.

DRIVE MECHANISM

Recommended lubrication period is after every 10 hours of operation and before storing unit at end of season. Oil the drive chains and sprockets with SAE 10W-30 oil. The hex shaft assembly in the variable speed friction drive requires no lubrication except to wipe the hex shaft and gears with SAE 10W-30 motor oil at end of season to prevent rusting. The hex shaft bearings are of the sealed type. Be careful not to get any grease or oil on the belts, disc drive plate or friction wheel; clean these parts thoroughly if oil or grease comes in contact with these parts.

Lubricate all moving part pivot points with SAE 10W-30 oil. Lubricate weight transfer plate and pivot points with clinging type grease such as "Lubriplate."

The auger flights do not have a grease fitting; before storage at end of season, remove the auger shear bolts and squirt oil into shear bolt holes. Turn the auger to distribute the oil throughout length of auger shaft.

AUGER GEARBOX

The auger gearbox is factory lubricated for life and unless grease leaks out, lubrication should not be necessary unless gearbox is being overhauled. Some models have grease plug (33—Fig. AT33) that can be removed. Use a wire as a dipstick to check for presence of grease.

ADJUSTMENTS

AUGER HEIGHT

Lower the adjusting skids to raise auger scraper bar on rough surfaces. On paved surfaces, the skids can be raised to bring the auger assembly down. Keep in mind that the scraper bar below the auger should be 1/8 inch (3.2 mm) above

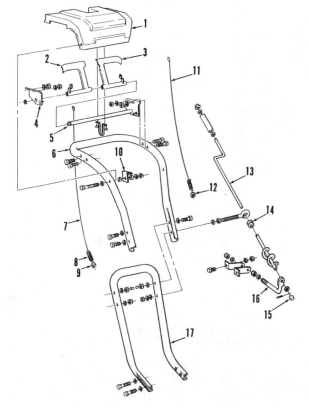

Fig. AT30—Exploded view of control and handle assembly.

1. Control panel
2. Auger clutch lever
3. Drive clutch lever
4. Lever rod bracket
5. Rod & bracket assy.
6. Upper handle
7. Auger cable
8. Auger clutch spring
9. Locknut
10. Panel bracket
11. Drive clutch cable & spring
12. Locknut
13. Chute crank assy.
14. Grommet
15. Plastic cap
16. Chute crank bracket
17. Lower handle

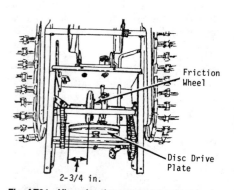

Fig. AT31—View showing measurement to check friction wheel adjustment.

sidewalk or other paved area to be cleaned. This clearance can be obtained by placing spacers (the extra shear pins stored under snowthrower handle can be used for spacers) under the skids so they contact paved surface. To compensate for wear, the scraper bar position is adjustable, and the bar can be reversed or replaced. Be sure the bar is adjusted parallel with the working surface, and for proper clearance, above skid shoes.

DRIVE BELTS

The traction drive belt is spring loaded and does not require adjustment; check control cable adjustment or replace belt if it is slipping. The auger drive belt can be adjusted after removing belt cover, then proceed as follows: Loosen nut on auger belt idler (16—Fig. AT32) and move idler in slot of idler arm (14) toward belt about 1/8 inch (3 mm). Tighten nut, engage auger drive clutch lever and check belt adjustment; belt should deflect about 1/2 inch (12.7 mm) with moderate pressure. Reset idler pulley until proper belt tension is obtained with drive engaged. The belt guides at engine pulley should be adjusted to clear belt 3/32 inch (2.4 mm) with auger drive clutch engaged.

CONTROL CABLES

Refer to Fig. AT30 and with auger cable (7) "z" fitting disconnected from auger clutch lever (2), move lever forward to contact plastic bumper and pull upward on cable; the center of the "z" fitting should be between the top and the center of the hole in the lever. If not, slide cable down through auger clutch spring (8) and turn locknut (9) to obtain proper adjustment. Then, pull cable up through spring and reconnect "z" fitting on cable in hole of lever. Repeat the procedure for drive clutch cable (11) adjustment.

FRICTION WHEEL

Remove bottom panel from snowthrower. Place shifter lever in first (Position 1) gear and check position of friction wheel on disc drive plate. Distance from left outer side of disc drive plate to right side of friction wheel should be 2-3/4 inches (70 mm) as shown in Fig. AT31. If necessary to adjust position of friction wheel, loosen bolts retaining upper speed shift control lever (29—Fig. AT32) to shift lever adapter (32). Then move friction wheel to proper position and tighten lever bolts.

TRACK DRIVE

To check track tension, pull up gently on center of track at midpoint between track rollers. Distance between track and top of track support frame should then be no more than 1-1/4 inches (32 mm). To adjust track, loosen jam nuts (3—Fig. AT34) on adjusting bolt (11) at rear end of track frame (4) and tighten or loosen adjusting bolt as necessary. Adjust opposite track in same manner. Be sure adjusting bolt jam nuts are securely tightened.

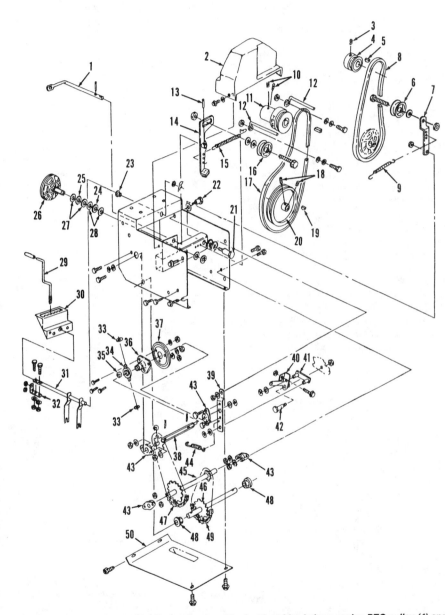

Fig. AT32—Exploded view of drive mechanism. The traction drive is from engine PTO pulley (4) and the auger is driven from engine crankshaft pulley (11). Drive disc (26) and shaft are separate items on some models.

1. Auger clutch lever	15. Auger brake arm spring	39. Drive clutch lever
2. Belt cover	16. Auger belt idler	40. Drive lever assy.
3. Set screw	17. Auger belt	41. Drive lever link
4. Engine PTO pulley	18. Set screws	42. Shoulder bolt
5. Woodruff key	19. Woodruff key	43. Self-centering bearing
6. Drive idler pulley	20. Auger drive pulley	44. Return spring
7. Drive idler arm	21. Shoulder bolt	45. Jackshaft assy.
8. Drive belt	22. Shoulder bolt	46. Wheel axle & sprocket
9. Idler arm spring	23. Plastic bushing	47. Roller chain
10. Set screws	24. Thrust washer	48. Axle bearing
11. Engine crankshaft pulley	25. Needle thrust bearing	49. Roller chain
12. Belt guide	26. Drive disc	50. Bottom cover panel
13. Roll pin	27. Thrust washers	
14. Auger brake & idler arm	28. Needle thrust bearings	
	29. Shift lever	
	30. Shift bracket	
	31. Shift yoke	
	32. Shift lever/yoke adapter	
	33. Flat washers	
	34. Trunion bearing	
	35. Snap ring	
	36. Friction wheel hub	
	37. Friction wheel	
	38. Hex shaft	

OVERHAUL

ENGINE

Engine make and model numbers are listed at beginning of section. Refer to Tecumseh engine section for engine overhaul.

AUGER DRIVE BELT

To remove auger drive belt, remove belt cover (2—Fig. AT32) and loosen the right-hand belt guide (12) and move it away from drive pulley. Loosen auger belt idler (16) and move it away from the belt. Engage auger drive clutch lever to move brake away from belt and slip the belt out from between auger pulley and brake. Remove the top two bolts securing auger housing to motor mount frame. Place weight transfer system pedal in transport position (latched in top notch). Slightly loosen bottom two bolts attaching auger housing to motor mount frame. Separate auger housing and motor mount frame by hinging on bottom two bolts. Remove belt from engine auger drive pulley. Install new belt by reversing removal procedure. Place weight transfer system in lowest notch in bracket to pivot auger housing and motor mount frame back into position. Adjust belt as outlined in preceding paragraph.

TRACTION DRIVE BELT

To remove traction drive belt, first remove upper belt cover, pull traction idler pulley back and slip belt past idler pulley, then remove belt from engine pulley (It may be necessary to loosen and remove engine auger drive pulley for clearance to remove belt). Place speed selector in sixth gear position and remove belt from between rubber drive disc and the combination drive pulley and drive disc plate. Install new belt by reversing removal procedure. If engine auger drive pulley was removed, be sure auger drive pulleys are aligned and drive key is flush with end of engine shaft before tightening engine pulley set screw. Adjust belts as outlined in belt adjustment paragraph.

AUGER SHEAR BOLT

The augers are driven from the auger shaft through special bolts (37—Fig. AT33) that are designed to break to protect the machine if an object becomes lodged in the auger housing. Only origi-

nal equipment replacement shear bolts should be used for safety reasons. Before installing new bolt, squirt oil between auger flight and auger shaft and turn auger flight to distribute oil throughout the full length of shaft.

AUGER ASSEMBLY

To remove auger, first remove auger drive belt as outlined in previous paragraph. Remove auger drive pulley and

Fig. AT33—Exploded view of auger housing, gearbox assembly and augers.

1. Deflector chute
2. Hand guard (not all models)
3. Chute extension
4. Mounting flange
5. Flange retainer
6. Impeller shaft bearing
7. Auger housing
8. Left side plate
9. Right side plate
10. Auger shaft bearing
11. Skid shoe
12. Scraper bar
13. Oil seal
14. Bushing
15. Thrust washers
16. Needle bearing
17. Roll pin
18. Thrust collar
19. Impeller shaft
20. Right case half
21. Oil seal
22. Bushing
23. Thrust washer
24. Woodruff key
25. Worm gear
26. Bushing
27. Gasket
28. Auger shaft
29. Thrust washer
30. Bushing
31. Oil seal
32. Left case half
33. Grease plug
34. Impeller bolts
35. Impeller assy.
36. Locknuts
37. Auger shear bolts
38. Left auger assy.
39. Right auger assy.

impeller shaft bearing (6—Fig. AT33) from rear of housing, unbolt auger shaft bearings (10) at each side of housing and remove the auger and gearbox assembly. Remove the auger drive shear bolts (37) and slide the auger flights from auger shaft. Remove impeller bolts (34) and pull impeller from shaft. Remove the bolts holding the case halves (20 & 32) together and disassemble gearbox. Clean and inspect all parts and renew any worn or damaged parts. Coat the parts with clean grease and pack the

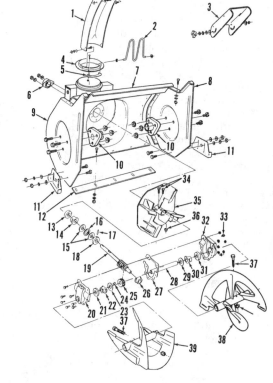

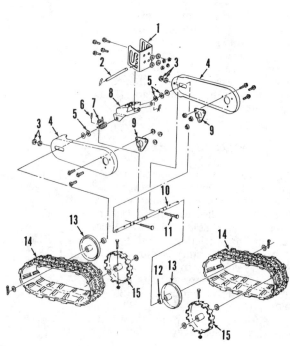

Fig. AT34—Exploded view of track drive assembly. Track drive wheels (15) mount on axle shaft (46—Fig. AT32). Foot pedal (8) operates weight transfer system.

1. Mounting bracket
2. Pivot pin
3. Jam nuts
4. Track frame
5. Snap ("E") rings
6. Roll pin
7. Foot pedal spring
8. Foot pedal
9. Frame support bearing
10. Idler axle shaft
11. Track adjusting bolt
12. Spacer
13. Track idler wheel
14. Rubber track
15. Track drive wheel

gearbox with approximately 3 ounces (85 grams) grease on assembly. Some recommended greases are Benalene #372, Shell Darina 1, Texaco Thermatex EP1, and Mobiltem 78.

FRICTION DRIVE WHEEL

To replace friction wheel, proceed as follows: Stand snowthrower up on auger housing end and remove bottom cover panel (50—Fig. AT32). Unbolt bearings (43) from engine frame at each end of hex shaft (38). Slide the hex shaft with friction wheel, trunnion bearing and end bearings out of snowthrower. Be careful not to lose flat washers (33) from studs on trunnion bearing (34). Unbolt friction wheel (37) from hub and install new wheel (cup side away from hub) by reversing removal procedure. Check friction wheel adjustment as outlined in preceding paragraph before reinstalling bottom panel.

TRACK ASSEMBLY

Refer to exploded view of track assembly in Fig. AT34. Disassembly and overhaul is evident after inspection of unit and reference to exploded view. Readjust track as outlined in previous paragraph.

ATLAS

Model	Engine Make	Engine Model	Self-Propelled?	No. of Stages	Scoop Width
5230	Tecumseh	HS50	Yes	2	23 in. (584 mm)
5250	Tecumseh	HS50	Yes	2	25 in. (635 mm)
825T	Tecumseh	HS80	Yes	2	25 in. (635 mm)
8250	Tecumseh	HS80	Yes	2	25 in. (635 mm)

NOTE: Model numbers will have letter prefix and suffix such as "A825TH." "A" prefix indicates Atlas brand, "C" indicates Chieftain brand, and "SC" prefix indicates Snow Chief brand. "T" suffix indicates model is equipped with track drive. The suffix letter (in addition to or without suffix "T") changes with model year (E, F, G, H).

LUBRICATION

ENGINE

Atlas recommends use of SAE 10W-30 motor oil, or when temperatures are consistently below 20° F (minus 7° C), use SAE 5W-30 motor oil. Change engine oil after every 25 hours of operation. Refer also to Tecumseh engine section.

DRIVE MECHANISM

Recommended lubrication period is after every 10 hours of operation and before storing unit at end of season. Oil the drive chains and sprockets with SAE 10W-30 oil. The hex shaft assembly in the variable speed friction drive requires no lubrication except to wipe the hex shaft and gears with SAE 10W-30 motor oil at end of season to prevent rusting. The hex shaft bearings are of the sealed type. Be careful not to get any grease or oil on the disc drive plate or friction wheel; clean these parts thoroughly if oil or grease comes in contact.

Lubricate all moving part pivot points with SAE 10W-30 oil. On track drive models, lubricate weight transfer plate and pivot points with clinging type grease such as "Lubriplate."

Using a hand-operated grease gun, lubricate the auger shaft zerk fittings after every 10 hours of operation.

AUGER GEARBOX

Once each season, remove the grease level plug (17—Fig. AT45) and, if grease is not visible, use a wire as a dipstick to check for grease in gearbox. Recommended lubricants are Benalene #372, Shell Darina 1, Texaco Thermatex EP1 and Mobiltem 78.

ADJUSTMENTS

AUGER HEIGHT

Lower the adjusting skids to raise auger scraper bar on rough surfaces.

NOTE: On track drive models, the weight transfer system will not operate with skid plates adjusted for maximum ground clearance.

On paved surfaces, the skids can be raised to bring the auger assembly down. Keep in mind that the scraper bar below the auger should be 1/8 inch (3.2 mm) above sidewalk or other paved area to be cleaned. This clearance can be obtained by placing spacers under the scraper bar, then adjusting the skids so they contact paved surface. To compensate for wear, the scraper bar position is adjustable, and the bar can be reversed or renewed. Be sure the bar is adjusted parallel with the working surface and for proper clearance, above skid shoes.

DRIVE BELTS

The traction drive belt is spring loaded and does not require adjustment; check control cable adjustment or renew belt if it is slipping. The auger drive belt can be adjusted after removing belt cover, then proceed as follows: Loosen nut on idler pulley (18—Fig. AT41) and move idler in idler arm slot toward belt about 1/8 inch (3 mm). Tighten nut, engage auger drive clutch lever and check belt adjustment; belt should deflect about 1/2 inch (12.7 mm) with moderate pressure. Reset idler pulley until proper belt tension is obtained with drive engaged. Belt guides (11 and 12) should be adjusted to clear belt 3/32 inch (2.4 mm) with auger drive clutch engaged.

CONTROL CABLES

Refer to Fig. AT40 and with drive clutch cable (23) "z" fitting disconnected from control lever (3), move lever forward to contact plastic bumper on handle; the center of the "z" fitting should be between the top and the center of the hole in the lever. If not, slide cable down through drive cable spring (24), hold square part of cable end from turning and adjust locknut (26) to obtain correct cable length. Then, pull cable up through spring and reconnect "z" fitting on cable to hole in lever. Repeat adjustment procedure for auger clutch cable (18).

FRICTION WHEEL

Remove bottom panel from snowthrower. Place speed control lever in first gear and check position of friction wheel on disc drive plate. The distance should be 3-3/8 inches (86 mm) as shown in Fig. AT43. If adjustment is necessary, loosen jam nut (12—Fig. AT40) on speed select rod. Remove ball joint (15) from speed control lower lever (16) and turn adapter (13) to obtain correct friction wheel position. Reinstall ball joint, tighten jam nut and reinstall bottom panel.

TRACKS

Refer to Fig. AT46 and loosen cap screw (13) on each side of track assembly. Rotate adjusting cam (14) equally on each side of assembly so deflected distance between top of track plate (12) and inside of track is not greater than 2 inches (50.8 mm). Uneven adjustment of adjusting cams will result in a twist in the rubber track. Repeat adjustment for opposite track assembly.

OVERHAUL

ENGINE

Engine make and model numbers are listed at beginning of section. Refer to Tecumseh engine section for engine overhaul.

AUGER DRIVE BELT

To remove auger drive belt, remove belt drive cover and loosen belt guides (11 & 12—Fig. AT41) and move them away from engine crankshaft pulley (24). Loosen auger belt idler pulley (18) and move it away from the belt. Engage auger drive clutch lever to move brake away from belt and slip the belt off au-ger pulley. Remove belt from engine drive pulley. Reverse removal procedure to install new belt and adjust belt and belt guides as outlined in a preceding paragraph.

TRACTION DRIVE BELT

To remove traction drive belt, remove belt cover and loosen the left-hand belt guide (11—Fig. AT41) and move guide away from engine PTO pulley. Pull spring loaded traction idler pulley back and slip belt past idler pulley. Remove belt from engine pulley and up between the auger and traction drive pulleys. Reverse removal procedure to install new belt and adjust belt and belt guides as outlined in a preceding paragraph.

AUGER SHEAR BOLT

The augers are driven via the auger shaft through special shear bolts (16—Fig. AT44) that are designed to shear to protect the machine if an object becomes lodged in the auger housing. Only original equipment replacement shear bolts should be used for safety reasons. When replacing shear bolt(s), be sure to install new shear bolt spacer (17) and, using a hand-operated grease gun, lubricate auger shaft grease fittings.

AUGER ASSEMBLY

To remove auger, remove auger drive belt, disconnect the chute crank from auger housing and separate the auger housing from main (motor mount)

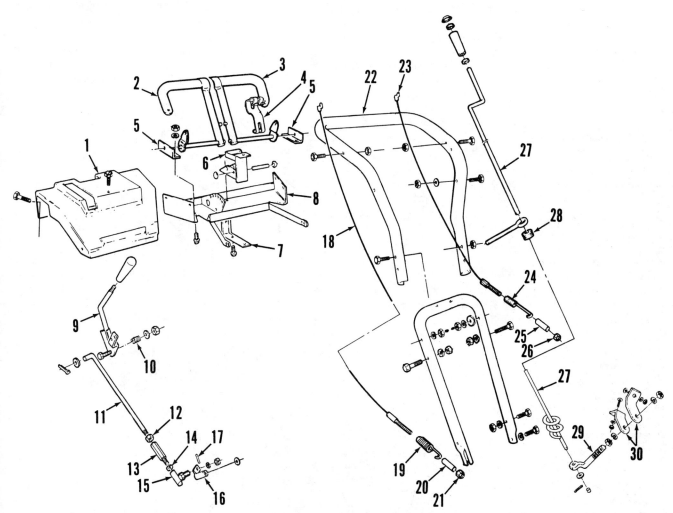

Fig. AT40—Exploded view of handle and controls. Some models have different design handles and control levers, but control linkage parts are basically same as shown.

. Console	8. Console bracket	14. Lockwasher
2. Auger clutch lever	9. Speed control lever	15. Ball joint
3. Drive clutch lever	10. Friction spring	16. Speed control
4. Cam lock	11. Speed control rod	lower lever
5. Handle bracket	12. Jam nut	17. Roll pin
6. Light bracket	13. Adapter	18. Auger clutch cable
7. Console support		

19. Auger cable spring	25. Spacer
20. Spacer	26. Locknut
21. Locknut	27. Chute crank
22. Upper handle	28. Grommet
23. Drive clutch cable	29. Chute crank bracket
24. Drive cable spring	30. Chute crank brackets

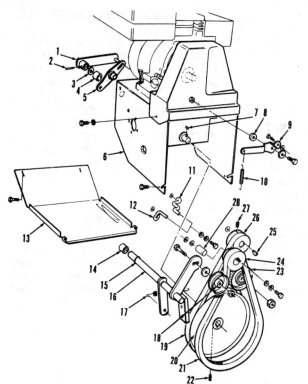

frame. Remove auger drive pulley (1—Fig. AT44) and impeller shaft bearing (2) from rear of housing. Unbolt auger shaft bushings (14) at each side of housing and remove the assembly. Remove the auger drive shear bolts (16) and remove the auger flights from auger shaft.

Fig. AT41—Exploded view of drive belt mechanism. Refer to Fig. AT42 for friction drive exploded view.

1. Auger clutch lever
2. Roll pin
3. Flat washer
4. Plastic bushing
5. Auger cable lever
6. Main (motor mount) frame
7. Hi-pro key
8. Spacer
9. Drive idler bracket
10. Drive idler spring
11. Belt guide
12. Belt guide
13. Bottom panel
14. Plastic bushing
15. Flat washer
16. Auger clutch rod
17. Auger idler return spring
18. Auger belt idler pulley
19. Drive belt
20. Drive pulley
21. Auger belt
22. Set screw
23. Drive belt idler
24. Engine crankshaft pulley
25. Woodruff key
26. Engine PTO pulley
27. Set screw
28. Spacer

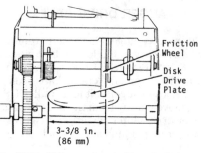

Fig. AT43—With speed control lever in first gear, friction wheel should be positioned as shown.

Fig. AT42—Exploded view of disc drive mechanism.

1. Bearing plate
2. Bearing
3. Flat washer
4. Hex shaft
5. Roll pins
6. Locknut
7. Speed control yoke
8. Plastic bushing
9. Washer
10. Snap ring
11. Washer
12. Trunnion bearing
13. Friction wheel hub
14. Friction wheel
15. Pinion gear
16. Flat washer
17. Bearing
18. Bearing plate
19. Spacers
20. Traction pivot
21. Push nut
22. Bushing
23. Flat washer
24. Cotter pin
25. Traction shaft
26. Trunnion bearing
27. Traction spring bracket
28. Flat washer
29. Traction spring
30. Spring pin
31. Traction rod
32. Drive disc
33. Return spring
34. Hi-pro key
35. Washer
36. Drive disc shaft
37. Snap ring
38. Thrust bearing
39. Needle bearings
40. Washer
41. Axle bearing
42. Bolt
43. Driven gear
44. Axle
45. Grip rings
46. Klick pin

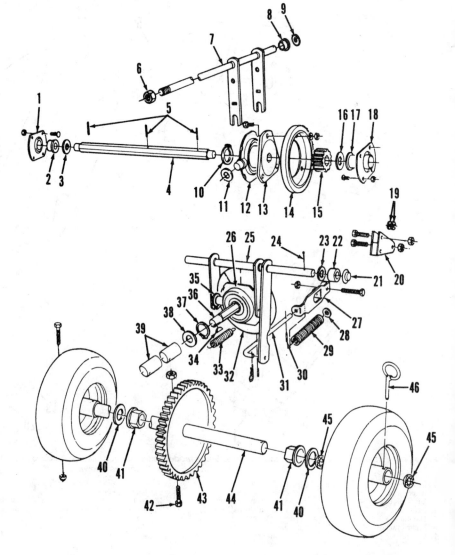

With auger flights removed, refer to Fig. AT45 for exploded view of auger drive gearbox. To disassemble, remove the cap screws holding the gearbox housing halves (16 & 22) and separate the unit. Remove impeller shaft bushing (14), snap ring (13) and thrust washer (12) from impeller shaft. Press shaft from worm gear (11) and remove needle thrust bearing (8) and races (7 & 9), rear impeller shaft bushing (6) and "O" ring (5) from impeller shaft. Clean and inspect all parts and renew any excessively worn or damaged parts. To install auger shaft bushings (18), press bushings into case from inside and flush with inside of casting. Install new seals (15) in gearbox halves.

When reassembling, coat all parts with grease and pack gearbox with a maximum of 3-1/4 ounces (96 mL) of one of the following greases: Benalene #372,

Shell Darina 1, Texaco Thermatex EP1, or Mobiltem 78.

FRICTION DRIVE

To renew friction wheel, proceed as follows: Stand snowthrower up on auger housing end and remove bottom panel (13—Fig. AT41). On Model 825T, disconnect track connecting rod (2—Fig. AT46) at right track plate (1) and rotate right track assembly parallel to ground to gain clearance to hex shaft end bearing. Remove the three bolts securing friction wheel (14—Fig. AT42) to hub (13) and move speed control lever to first speed position. Loosen but do not remove the four nuts securing right hex shaft bearing plate (1). Move speed control lever to sixth speed position. Place

tape over the four hex shaft bearing plate bolts inside main (motor mount) frame and remove the previously loosened retaining nuts. Remove bearing plate and slide hex shaft to right until friction wheel can be removed from end of hex shaft. Install replacement friction wheel retaining bolts loosely, then complete reassembly in reverse order of disassembly. Check friction wheel adjustment as outlined in a preceding paragraph.

TRACK ASSEMBLY

Disassembly and overhaul is evident after inspection of unit and reference to Fig. AT46. Readjust track after servicing as outlined in a previous paragraph.

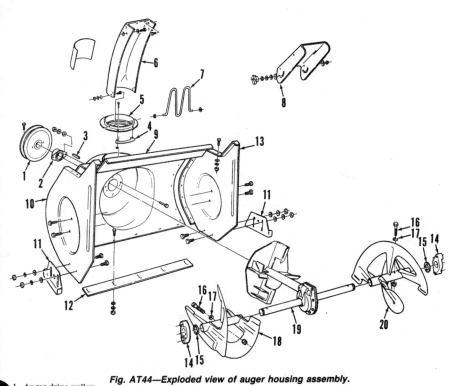

Fig. AT44—Exploded view of auger housing assembly.

1. Auger drive pulley	7. Hand guard	11. Skid shoe	16. Shear bolt
2. Impeller shaft bearing	8. Upper chute	12. Scraper bar	17. Shear bolt spacer
3. Square key	extension	13. L.H. side plate	18. R.H. auger flight
4. Chute retainer	9. Center auger housing	14. Auger shaft bushing	19. Gearbox assy.
5. Chute flange	10. R.H. side plate	15. Spacer washer	20. L.H. auger flight
6. Bottom chute			
extension			

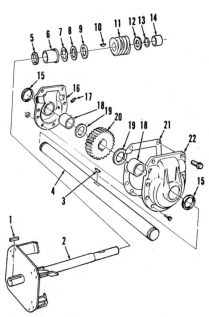

Fig. AT45—Exploded view of auger gearbox. Impeller is an integral part of impeller shaft (2).

1. Square key	11. Worm gear
2. Impeller & shaft assy.	12. Thrust washer
3. Woodruff key	13. Snap ring
4. Auger shaft	14. Impeller shaft bushing
5. Square cut "O" ring	15. Oil seals
6. Impeller shaft bushing	16. L.H. housing half
7. Thrust bearing race	17. Grease plug
8. Needle thrust bearing	18. Auger shaft bushing
9. Thrust bearing race	19. Thrust washer
10. Woodruff key	20. Gear
	21. Gasket
	22. R.H. housing half

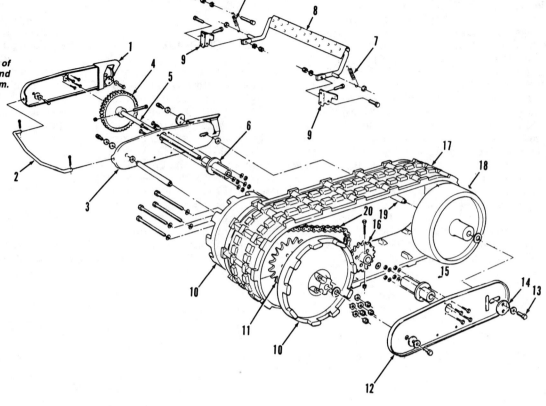

Fig. AT46—Exploded view of left track assembly and weight transfer mechanism.

1. Track plate, inner R.H.
2. Connecting rod
3. Track plate, inner L.H.
4. Axle drive gear
5. Axle shaft
6. Inner axle hub
7. Pedal spring
8. Weight transfer pedal
9. Pedal bracket
10. Track drive wheel
11. Track drive sprocket
12. Track plate, outer L.H.
13. Cap screw
14. Adjusting cam
15. Outer axle hub
16. Axle sprocket
17. Rubber track
18. Track idler
19. Track idler shaft
20. Track drive chain

ATLAS

Model	Engine Make	Engine Model	Self-Propelled?	No. of Stages	Scoop Width
8280	Tecumseh	HMSK80	Yes	2	28 in. (711 mm)
1032	Tecumseh	HMSK100	Yes	2	32 in. (813 mm)

LUBRICATION

ENGINE

Refer to Tecumseh engine section for engine lubrication information. Recommended fuel is regular or low-lead gasoline.

DRIVE MECHANISM

Recommended lubrication period is after every 10 hours of operation and before storing unit at end of season. Oil the drive chains and sprockets with SAE 10W-30 oil. Using a hand grease gun, lubricate grease fitting (4—Fig. AT52) on the spindle assembly and grease fittings (23) on speed control bracket (13) once each season. Be careful not to get any grease or oil on the friction wheel drive plate or friction wheel; clean these parts thoroughly if oil or grease comes in contact. Remove the drive wheels and lubricate axle shafts with any automotive type grease at least once a year and prior to storage. Lubricate all moving part pivot points with SAE 10W-30 oil.

Using a hand operated grease gun, lubricate the auger flight grease fittings after every 10 hours of operation and each time an auger shear bolt is replaced.

AUGER GEARBOX

Once each season, remove grease plug (18—Fig. AT56) and, if grease is not visible, use a wire as a dipstick to check for grease in gearbox. If gearbox is dry, add recommended grease such as Benalene #372, Shell Darina 1, Texaco Thermatex EP1 and Mobiltem 78.

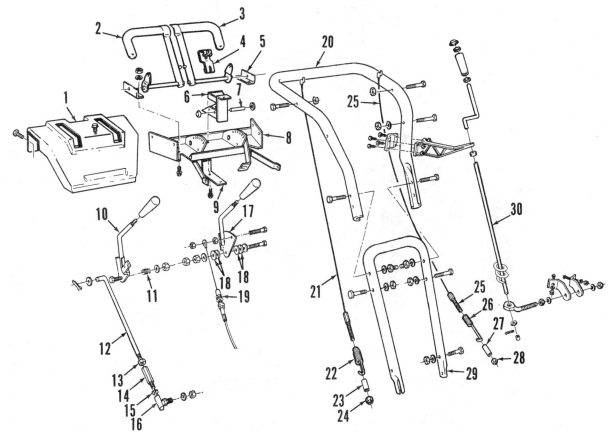

Fig. AT50—Exploded view of handle and controls. Some models have different design handles and control levers and do not have remote chute deflector lever (17), but other controls are basically similar to those in illustration.

1. Console	8. Console bracket	14. Adapter	19. Remote deflector cable	24. Locknut
2. Auger clutch lever	9. Console support	15. Lockwasher	20. Upper handle	25. Traction drive cable
3. Drive clutch lever	10. Speed control lever	16. Ball joint	21. Auger clutch cable	26. Drive clutch spring
4. Cam lock	11. Friction spring	17. Remote chute deflector lever	22. Auger clutch spring	27. Spacer sleeve
5. Lever brackets	12. Speed control rod	18. Belleville washers	23. Spacer sleeve	28. Locknut
6. Light bracket	13. Jam nut			29. Lower handle
7. Latch pin				30. Chute crank

ADJUSTMENTS

AUGER HEIGHT

Lower the adjusting skids to raise auger scraper bar on rough surfaces. On paved surfaces, the skids can be raised to bring the auger assembly down. Keep in mind that the scraper bar below the auger should be 1/8 inch (3 mm) above sidewalk or other paved area to be cleaned. This clearance can be obtained by placing spacers under the scraper bar, then adjusting the skids so they contact paved surface.

To compensate for wear, the scraper bar position is adjustable, and the bar can be reversed for longer wear or replaced when both edges are worn. Be sure the bar is adjusted parallel with the working surface and for proper clearance above skid shoes.

DRIVE BELTS

The traction drive belt is spring loaded and does not require adjustment; replace belt if it is slipping. The auger drive belt can be adjusted after removing belt cover, then proceed as follows: Loosen nut on auger idler pulley (17—Fig. AT51) and move idler in slot of idler bracket (16) toward belt about 1/8 inch (3 mm). Tighten nut, engage auger drive clutch lever and check belt adjustment; belt should deflect about 1/2 inch (12.7 mm) with moderate pressure. Reset idler pulley until proper belt tension is obtained with drive engaged.

CONTROL CABLES

Refer to Fig. AT50 and with "z" fitting on upper end of auger clutch cable (21) disconnected from lever (2), move lever forward to contact plastic bumper; the center of the "z" fitting should

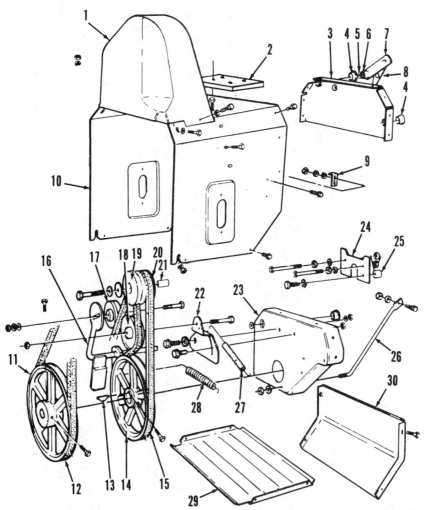

Fig. AT51—Exploded view of belt drive mechanism. Refer to Fig. AT52 for friction wheel drive components.

1. Belt cover
2. Engine riser plate
3. Rear main frame plate
4. Plastic bushings
5. Flat washer
6. Roll pin
7. Auger clutch lever
8. Clutch cable lever
9. Cable adjustment bracket
10. Main (motor mount) frame

11. Auger pulley
12. Auger belt
13. Drive pulley key
14. Traction drive pulley
15. Traction drive belt
16. Auger idler bracket & brake
17. Auger idler pulley
18. Drive idler pulley
19. Auger drive pulley
20. Traction drive pulley

21. Engine pulley spacer
22. Drive idler bracket
23. Drive spindle plate
24. Belt guide plate
25. Spacer
26. Plate support rod
27. Drive idler spring
28. Auger brake spring
29. Bottom cover panel
30. Rear cover panel

be between the top and the center of the hole in the lever. If not, push cable through auger clutch spring (22) and adjust locknut (24) on lower end of cable to align "z" fitting with hole in lever. Then, pull cable up through springs and reconnect "z" fitting to lever. Repeat this procedure to adjust traction drive cable (25).

FRICTION WHEEL

Remove rear cover panel (30—Fig. AT51) and bottom cover panel (29) from main (motor mount frame). Place speed control lever in first (1) speed position and check location of friction wheel on friction wheel drive plate as shown in Fig. AT53. The left side of friction wheel should be 2-5/8 inches (66.7 mm) from left outer edge of friction wheel drive plate. If necessary to adjust position of friction wheel, refer to Fig. AT50 and loosen jam nut (13) on speed control rod (12). Remove ball joint (16) from shifter bracket and turn adapter (14) to obtain correct friction wheel position. Reinstall ball joint, tighten jam nut and reinstall panels.

OVERHAUL

ENGINE

Engine make and model numbers are listed at beginning of section. Refer to Tecumseh engine section for engine overhaul.

DRIVE BELTS

To remove auger belt, remove upper (belt) cover and remove auger drive engine pulley and belt from engine; this allows removal of belt from pulley without removing belt guides. Disconnect chute crank mechanism at chute. Remove the two top bolts and loosen the bottom two bolts holding auger housing to engine mount frame. Separate auger housing from engine mount by pivoting on bottom bolts. Bend belt retainer tabs (Fig. AT54) away from auger pulley only as necessary to remove belt. Install new belt on auger pulley and while holding it tightly in pulley groove, bend retainer tabs back into position leaving 1/16 to 1/8 inch (1.5-3 mm) clearance between tabs and belt. Pull up on belt, engage auger clutch and swing engine mount frame back into position. Replace two top housing bolts and tighten the two lower bolts. Install auger drive belt on engine pulley and reinstall pulley on engine shaft. Adjust auger belt as outlined in a preceding paragraph.

To remove traction drive belt, remove auger belt as in preceding paragraph. Pull spring loaded traction idler pulley back and slip belt past idler pulley and off of engine pulley. It may be necessary to remove auger brake assembly to remove belt from traction drive pulley. Install new belt and reassemble in reverse of disassembly procedure. Adjust belts as outlined in a preceding paragraph.

Fig. AT52—Exploded view of friction wheel drive mechanism. Friction wheel spindle housing (3) mounts in spindle plate (23—Fig. AT51).

1. Seal
2. Needle roller bearing
3. Spindle housing
4. Grease fitting
5. Thrust bearing races
6. Needle thrust bearing
7. Friction wheel drive plate & spindle
8. Thrust washer
9. Snap ring
10. Hex shaft drive sprocket
11. Roller chain
12. Hex shaft sliding bearing
13. Speed control bracket
14. Speed control yoke
15. Snap ring
16. Thrust washer
17. Friction wheel hub
18. Friction wheel
19. Hex shaft end bearing
20. Square key
21. Gear & shaft assy.
22. Bearing
23. Grease fitting
24. Push on end cap
25. Flat washer
26. Bushing
27. Hex shaft end bearing
28. Snap ring
29. Hex shaft
30. Axle shaft
31. Axle drive chain
32. Axle bearing bracket
33. Thrust washer
34. Grip type retainer
35. Klick pin
36. Bolt
37. Yoke drive
38. Yoke return spring
39. Cotter pin

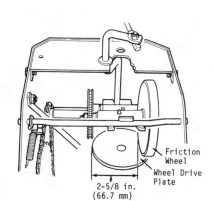

Fig. AT53—With shift (speed control) lever in first speed position, friction wheel position should be as shown.

2-5/8 in. (66.7 mm)
Friction Wheel
Wheel Drive Plate

Fig. AT54—Bend auger belt retainer tabs back to remove auger belt. On reassembly, bend tabs to clear belt as shown.

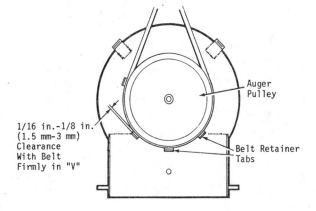

Auger Pulley

1/16 in.-1/8 in. (1.5 mm-3 mm) Clearance With Belt Firmly in "V"

Belt Retainer Tabs

AUGER SHEAR BOLT

The augers are driven by the auger shaft via shear bolts (15—Fig. AT55) that are designed to shear to protect the machine if an object becomes lodged in the auger housing. Only original equipment replacement shear bolts should be used for safety reasons. When replacing shear bolt(s), be sure to install new shear bolt spacer (14) and, using a hand operated grease gun, lubricate auger shaft grease fittings.

AUGER ASSEMBLY

To remove auger, first remove auger drive belt as previously outlined and refer to Fig. AT55. Remove auger drive pulley (1) and impeller shaft bearing (2) from rear of housing. Unbolt auger shaft bearings (12) at each side of housing and remove the assembly. Remove the auger drive shear bolts (15) and pull the auger flights from auger shaft.

With auger flights removed, refer to Fig. AT56 for exploded view of auger drive gearbox. To disassemble, remove the cap screws holding gearbox housing halves (17 & 23) and separate the unit. Remove impeller shaft bushing (15),

snap ring (14) and thrust washer (13) from impeller shaft. Press shaft from worm gear (12) and remove needle thrust bearing (9) and races (8 & 10), impeller shaft bearing (7) and "O" ring (6) from impeller shaft. If necessary, impeller fan (2) can be removed from the impeller shaft. Clean and inspect all parts and renew any excessively worn or damaged parts. To install auger shaft bushings (19), press bushings into case from inside until flush with inside of casting. Install new seals (16) in gearbox halves.

When reassembling, coat all parts with grease and pack gearbox with a maximum of 3-1/4 ounces (92 grams) of one of the following greases: "Benalene #372, Shell Darina 1, Texaco Thermatex EP1, or Mobiltem 78.

FRICTION DRIVE

To replace friction wheel, stand snowthrower up on auger housing end and remove rear panel and bottom panel. Hold hex shaft from turning with a 11/16-inch open-end wrench and remove the nut and washers holding friction wheel (18—Fig. AT52) and hub (17) assembly on shaft. Remove the friction

wheel and hub assembly, install new friction wheel, cup side away from hub, and reassemble by reversing disassembly procedure. Be sure that square key (20) is in place in friction wheel hub and that thrust washer (16) is completely on the shoulder of the shaft, not trapped between the shoulder and threaded part of shaft.

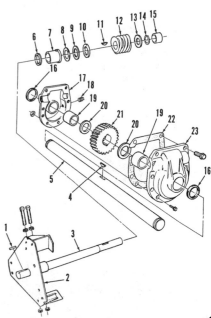

Fig. AT56—Exploded view of auger gearbox and impeller assembly.

1. Square key	
2. Impeller fan	13. Thrust washer
3. Impeller shaft	14. Snap ring
4. Woodruff key	15. Impeller shaft bearing
5. Auger shaft	16. Auger shaft seal
6. Square cut "O" ring	17. L.H. housing half
7. Impeller shaft bearing	18. Grease plug
8. Thrust bearing race	19. Auger shaft bushing
9. Needle thrust bearing	20. Thrust washer
10. Thrust bearing race	21. Auger shaft gear
11. Key	22. Gasket
12. Worm gear	23. R.H. housing half

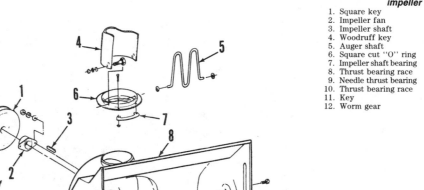

Fig. AT55—Exploded view of auger housing assembly. Refer to Fig. AT56 for exploded view of auger gearbox assembly (18).

1. Auger drive pulley
2. Impeller shaft bearing
3. Square key
4. Chute
5. Hand guard
6. Chute flange
7. Chute retainer
8. Auger housing
9. Skid shoes
10. Scraper plate
11. Spacer plates
12. Auger shaft bearing
13. Thrust washer
14. Shear bolt spacer
15. Shear bolt
16. Auger flight, R.H.
17. Grease fitting
18. Gearbox assy.
19. Gearbox brace
20. Auger flight, L.H.

BOLENS

BOLENS INTERNATIONAL
215 South Park Street
Port Washington, Wisc. 53074

Model	Engine Make	Engine Model	Self-Propelled?	No. of Stages	Scoop Width
524	Tecumseh	H50	Yes	2	24 in. (610 mm)
625	Tecumseh	HM80	Yes	2	24 in. (610 mm)
724	Tecumseh	H70	Yes	2	24 in. (610 mm)
726	Tecumseh	H70	Yes	2	26 in. (660 mm)
826	Tecumseh	HM80	Yes	2	26 in. (660 mm)
832	Tecumseh	HM80	Yes	2	32 in. (813 mm)
1032	Tecumseh	HM100	Yes	2	32 in. (813 mm)

LUBRICATION

ENGINE

Refer to Tecumseh engine section for engine lubrication requirements. Recommended fuel is regular or low lead gasoline.

Periodically apply oil to engine control cable to prevent binding and corrosion.

DRIVE MECHANISM

Lubricate drive chain and all linkage pivot points before and after snowthrower use. After every 25 hours of operation, remove bottom panel of main frame and using good quality multi-purpose grease, lubricate gears and hex and sliding fork shafts. After each season, lubricate auger shaft by forcing grease through grease fittings in augers.

Auger gearbox oil level should be checked after each season. With unit on level ground, oil level in gearbox should reach lower edge of oil fill plug hole. Fill gearbox with SAE 90 gear oil.

MAINTENANCE

Inspect drive belts and renew if frayed, cracked, burnt or otherwise damaged. Renew belt pulley if groove is excessively worn or damaged. Inspect drive chain and sprockets and renew chain if stretched or sprockets if worn.

Augers are secured to auger shaft by shear bolts. If a shear bolt fails, drive out remainder of old bolt and install new shear bolt. DO NOT install standard bolts in place of shear bolts.

Recommended tire pressure is 20 psi (138 kPa) for Models 524, 625, 724 and 726 and 8-12 psi (56-82 kPa) for Models 826, 832 and 1032.

ADJUSTMENT

HEIGHT

To adjust auger height, loosen runner mounting bolts and position runner at desired height. Recommended height of scraper bar above smooth surface is 1/8-inch (3.2 mm). Height should be increased when operating on irregular surface.

AUGER CONTROL ROD

With auger control lever in "ON" position, spring (18—Fig. BN2) on lower end of control rod should compress approximately 3/8-inch (9.5 mm). To adjust spring compression, detach pivot block (20), loosen locknut at upper end of control rod and turn control rod in upper pivot block. Reattach lower pivot block (20) and check adjustment.

AUGER BRAKE

On all models except 726 and 832, the auger drive belt is held by a brake arm (35—Fig. BN3) when auger control lever is in "OFF" position. Brake is properly adjusted if there is 1/4 to 1/2-inch (6.4-12.7 mm) gap between rear of control lever and rear of lever slot when in "OFF". To adjust gap, turn adjusting nut (38) on idler pulley arm.

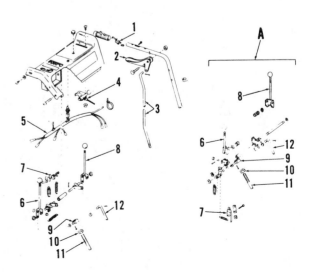

Fig. BN1—Exploded view of control assembly. Early models use speed and auger control assembly shown in bracket "A".

1. Safety switch
2. Clutch lever
3. Clutch rod
4. Throttle control
5. Wiring harness
6. Auger clutch lever
7. Safety switch
8. Speed control lever
9. Pivot block
10. Locknut
11. Auger clutch rod
12. Speed control rod

Bolens

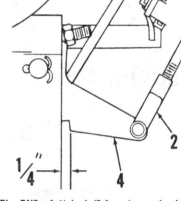

11. Auger clutch rod
17. Pins
18. Spring
19. Washers
20. Pivot block
21. Washer
22. Control arm
23. Pin
25. Washer
26. Belleville washer
27. Bushing
28. Main frame
29. Bushings
31. Spring
32. Engine drive belt idler pulley
33. Idler arm
34. Spring
35. Brake arm
36. Auger belt idler pulley
37. Idler arm & shaft
38. Nut

ENGINE

Engine make and model are listed at beginning of Bolens section. Refer to Tecumseh engine section for engine overhaul.

DIFFERENTIAL

Models 625, 726, 832 and Model 826 prior to 1978 are equipped with variable slip differential shown in Fig. BN6. Left axle (35) is keyed to differential shaft (28) which is driven by left differential side gear pin (27). Right axle (23) is driven by right differential side

SPEED CONTROL ROD

Remove cotter pin in coupler (14—Fig. BN4), push rod end (16) down into main frame as far as possible and position speed control lever in number "4" notch. Rotate rod end (16) so cotter pin hole in coupler (14) aligns with hole in upper control rod (12) and install cotter pin.

FRICTION DRIVE CLUTCH

With clutch control lever squeezed, there should be ¼-inch (6.4 mm) gap between notch in clutch lever (4–Fig. BN5) and rear of main frame. To adjust gap size, disconnect and rotate rod clevis (2).

Remove bottom plate and check clearance between friction wheel (25–Fig. BN4) and friction disc (1–Fig. BN6). Clearance should be 0.015-0.030 inch (0.4-0.8 mm) with clutch control lever released and clutch control rod properly adjusted.

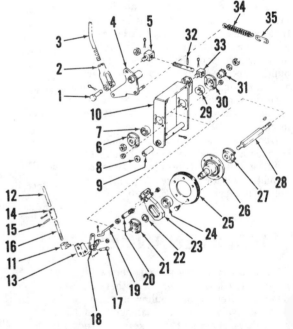

Fig. BN5—A ¼-inch (6.4 mm) gap should exist between notch in clutch lever (4) and rear of main frame with clutch lever fully squeezed. Turn clevis (2) for adjustment.

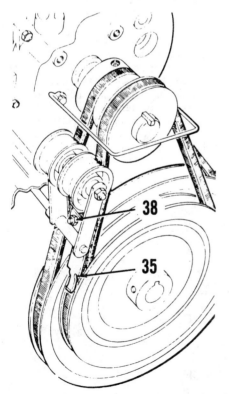

Fig. BN3—Adjust position of brake arm (35), on models so equipped, by turning adjusting nut (38). See text.

Fig. BN4—Exploded view of friction wheel carriage. Friction disc (1—Fig. BN6) drives friction wheel (25). Drive sprocket (31) transmits power through intermediate sprocket and gear (13—Fig. BN6) to rear axle gear.

1. Pin
2. Clevis
3. Clutch control rod
4. Clutch lever
5. Pivot block
6. Bearing retainer
7. Bearing
8. Washer
9. Spacer
10. Carriage
11. Pivot block
12. Speed control rod
13. Bracket
14. Coupler
15. Pin
16. Rod end
17. Pin
18. Lever
19. Link
20. Ball joint
21. Bearing retainer
22. Snap ring
23. Speed control fork
24. Thrust bearing
25. Friction wheel
26. Hub
27. Bearing retainer
28. Hex shaft
29. Bearing
30. Bearing retainer
31. Sprocket
32. Rod
33. Pivot block
34. Spring
35. Link

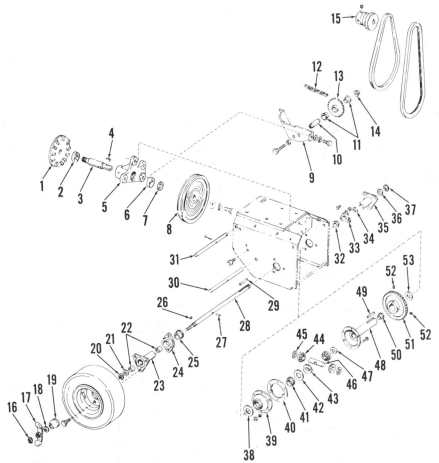

Fig. BN6—Exploded view of friction disc assembly used on all models and axle and differential used on Models 625, 726, 832 and Model 826 prior to 1978. Refer to Fig. BN7 for axle and differential used on other models.

1. Friction disc	15. Engine pulley	28. Differential shaft	41. Bushing
2. Bearing	16. Snap ring	29. Key	42. Washer
3. Shaft	17. Wing nut	30. Shaft	43. Side gears
4. Key	18. Spring washer	31. Shaft	44. Pinion gears
5. Housing	19. Friction block	32. Bushing	45. Thrust washers
6. Bearing	20. Snap ring	33. Bushing retainer	46. Pinion shaft
7. Washer	21. Washer	34. Washer	47. Washer
8. Pulley	22. Bushing	35. Left axle	48. Differential housing
9. Support	23. Right axle	36. Washer	& shaft
10. Shaft	24. Bushing retainer	37. Snap ring	49. Key
11. Bushings	25. Bushing	38. Washer	50. Bushing
12. Chain	26. Key	39. Differential housing	51. Gear
13. Sprocket & gear	27. Pin	40. Plate	52. Set screws
14. Nut			53. Washer

gear. Wing nut (17) is attached to right end of differential shaft (28) and tightening wing nut forces friction block (19) against right axle face thereby decreasing or eliminating differential action. Wing nut should only be tightened by hand and never with a tool.

Model 826 after 1977 and Model 1032 are equipped with differential shown in Fig. BN7. Differential shaft (13) extends through differential and inserting pin (14) through differential shaft end and left axle (17) will lock differential.

Overhaul of differential is evident after inspection of unit and referral to Fig. BN6 or BN7. Note differential on Model 826 after 1977 and Model 1032 is available only as a unit assembly. Note

Fig. BN7—Exploded view of axle and differential assembly used on Model 826 after 1977 and Model 1032.

1. Right axle
2. Bushing retainer
3. Washer
4. Washer
5. Differential
6. Spacers
7. Gear
8. Bushing
9. Washer
10. Bushing
11. Bushing
12. Pin
13. Differential shaft
14. Pin
15. Washer
16. Snap ring
17. Left axle

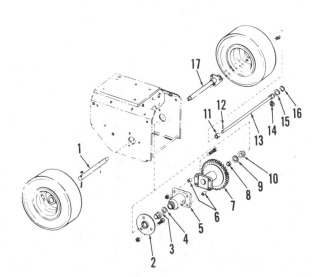

differential side gears (43—Fig. BN6) on Models 625, 726, 832 and Model 826 prior to 1978, have slots that align with differential shaft drive pin (27) and drive lug on right axle (23).

AUGER GEARBOX

All models are equipped with the auger gearbox shown in Fig. BN9. Overhaul of auger gearbox is evident after inspection of unit and referral to Fig. BN9. Clean components and renew any which are damaged or excessively worn.

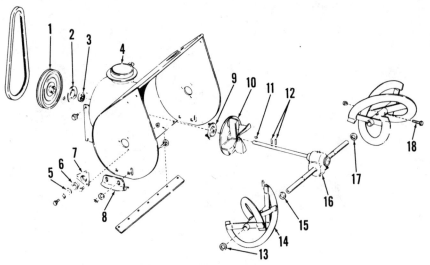

Fig. BN8—Drawing of auger assembly.

1. Pulley	6. Bushing	10. Impeller	14. Auger
2. Bearing retainer	7. Bushing retainer	11. Key	15. Washer
3. Bearing	8. Runner	12. Pins	16. Auger gearbox
4. Auger frame	9. Bearing retainer	13. Washer	17. Washer
5. Washer			18. Shear bolt

Fig. BN9—Exploded view of auger gearbox.

1. Seal
2. Needle bearing
3. Thrust bearing
4. Thrust washers
5. Snap ring
6. Worm gear & shaft
7. Gearbox
8. Cover
9. Gasket
10. Seal
11. Bushing
12. Thrust washer
13. Gear
14. Plug
15. Washer
16. Snap ring
17. Thrust washer
18. Spacer
19. Needle bearing
20. Snap ring

21. Plug
22. Thrust washer
23. Bushing

24. Seal
25. Shaft
26. Key

BOLENS

Model	Engine Make	Engine Model	Self-Propelled?	No. of Stages	Scoop Width
225, 225E	Tecumseh	AH520	No	1	20 in. (508 mm)
350, 350E	Tecumseh	H35	No	1	20 in. (508 mm)
500, 500E	Tecumseh	HS50	No	1	24 in. (610 mm)

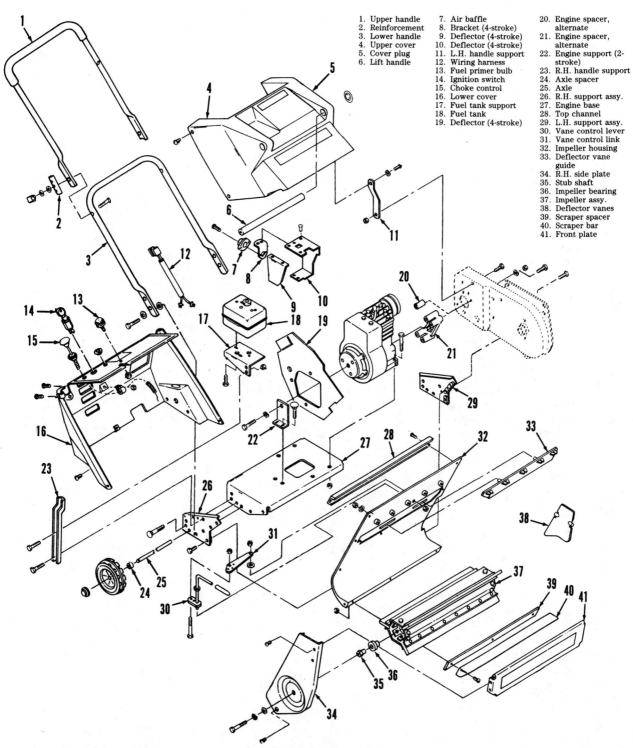

1. Upper handle
2. Reinforcement
3. Lower handle
4. Upper cover
5. Cover plug
6. Lift handle
7. Air baffle
8. Bracket (4-stroke)
9. Deflector (4-stroke)
10. Deflector (4-stroke)
11. L.H. handle support
12. Wiring harness
13. Fuel primer bulb
14. Ignition switch
15. Choke control
16. Lower cover
17. Fuel tank support
18. Fuel tank
19. Deflector (4-stroke)
20. Engine spacer, alternate
21. Engine spacer, alternate
22. Engine support (2-stroke)
23. R.H. handle support
24. Axle spacer
25. Axle
26. R.H. support assy.
27. Engine base
28. Top channel
29. L.H. support assy.
30. Vane control lever
31. Vane control link
32. Impeller housing
33. Deflector vane guide
34. R.H. side plate
35. Stub shaft
36. Impeller bearing
37. Impeller assy.
38. Deflector vanes
39. Scraper spacer
40. Scraper bar
41. Front plate

Fig. BN12—Exploded view of handle, frame and impeller housing. Refer to Fig. BN13 for exploded view of impeller assembly (37).

LUBRICATION

ENGINE

For engine lubrication information, refer to Tecumseh 2-stroke engine section for Models 225 and 225E, and to Tecumseh 4-stroke engine section for all other models. Recommended fuel is unleaded or leaded regular gasoline.

DRIVE MECHANISM

Using lightweight motor oil, lubricate belt drive idler arm pivot, vane control mechanism, impeller bearings at each end of impeller and the wheel bearings. Be careful not to get any oil on drive belt and pulleys. After each season, wipe oil or grease on all bare metal surfaces.

ADJUSTMENTS

As the belt drive idler is spring loaded, no adjustment is necessary. If drive belt slips, check to see that idler bracket spring (6—Fig. BN13) is not stretched or broken and that idler arm (11) pivots freely.

OVERHAUL

ENGINE

Refer to appropriate Tecumseh 2-stroke or 4-stroke engine section for engine overhaul information.

DRIVE BELT

The drive belt can be replaced after removing drive cover (13—Fig. BN13) at left side of impeller housing. Release tension on belt by either prying up on idler bracket or removing the tension spring, then remove and replace belt. Lubricate idler bracket pivot bushing (10) with light engine oil and be sure arm is working freely before installing new belt.

IMPELLER PADDLES

If paddles (3—Fig. BN13) are worn or damaged, a complete set of three new paddles should be installed. However, remove and replace only one paddle at a time by removing blade screws (4) retaining paddle between impeller frames (1).

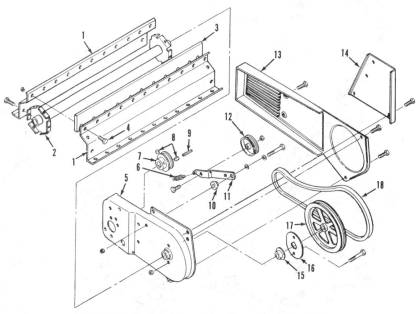

Fig. BN13—Exploded view of impeller assembly and belt drive.

1. Impeller frame (3)	7. Engine pulley	13. Belt drive cover
2. Impeller tube assy.	8. Set screws	14. L.H. side plate
3. Impeller paddle (3)	9. Pulley key	15. Flange bushing
4. Blade screw (27)	10. Pivot bushing	16. Bushing retainer
5. Drive support	11. Idler arm	17. Impeller pulley
6. Idler bracket spring	12. Idler pulley	18. Drive belt

BOLENS

Model	Engine Make	Engine Model	Self-Propelled?	No. of Stages	Scoop Width
322	B&S	62032	No	1	22 in. (559 mm)

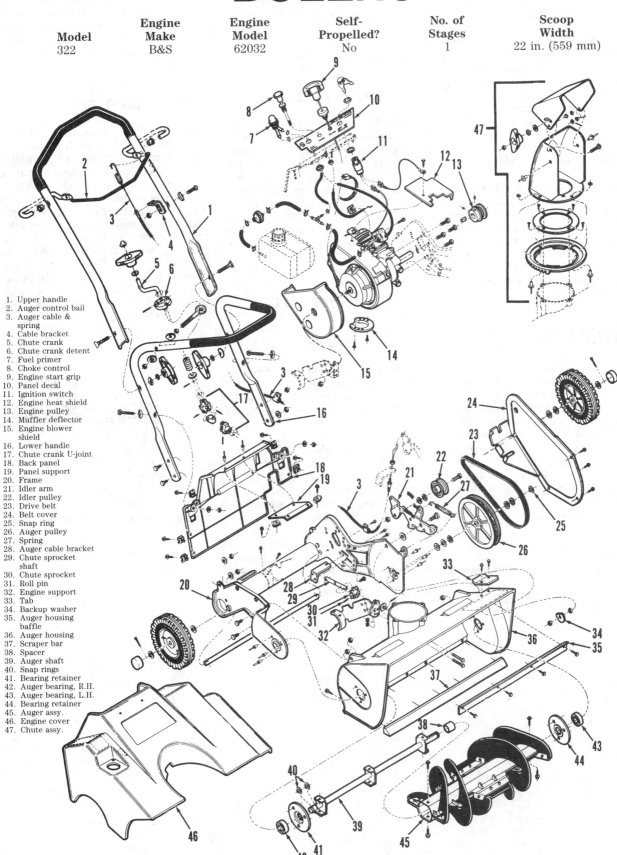

1. Upper handle
2. Auger control bail
3. Auger cable & spring
4. Cable bracket
5. Chute crank
6. Chute crank detent
7. Fuel primer
8. Choke control
9. Engine start grip
10. Panel decal
11. Ignition switch
12. Engine heat shield
13. Engine pulley
14. Muffler deflector
15. Engine blower shield
16. Lower handle
17. Chute crank U-joint
18. Back panel
19. Panel support
20. Frame
21. Idler arm
22. Idler pulley
23. Drive belt
24. Belt cover
25. Snap ring
26. Auger pulley
27. Spring
28. Auger cable bracket
29. Chute sprocket shaft
30. Chute sprocket
31. Roll pin
32. Engine support
33. Tab
34. Backup washer
35. Auger housing baffle
36. Auger housing
37. Scraper bar
38. Spacer
39. Auger shaft
40. Snap rings
41. Bearing retainer
42. Auger bearing, R.H.
43. Auger bearing, L.H.
44. Bearing retainer
45. Auger assy.
46. Engine cover
47. Chute assy.

Fig. BN15—Exploded view of Model 322 snowthrower.

LUBRICATION

ENGINE

Refer to Briggs & Stratton 2-stroke engine section for engine lubrication information. Oil is mixed with fuel to lubricate the 2-stroke engine. Recommended fuel:oil mixture ratio after initial break-in period is 32:1. Recommended oil is a BIA certified 2-cycle oil rated TC-W.

DRIVE MECHANISM

The collector/auger bearings are sealed ball bearing type and do not require lubrication. Lubricate belt idler bracket pivot point lightly, being careful not to allow lubricant on belt or pulleys. Wheel bearings are plastic. Removing the hub cap and placing a drop or two of oil on the axle will help the snowthrower roll easier.

MAINTENANCE

Inspect belt and pulleys. Renew belt if frayed, cracked, burned or excessively worn. Renew idler pulley if bearing is worn or pulley turns rough. Check clutch cable for binding or stiff operation and be sure idler bracket moves smoothly. Renew auger assembly (45—Fig. BN15) as a unit if auger flights are excessively worn or damaged or return the assembly to an authorized service center for repairs.

ADJUSTMENTS

AUGER HEIGHT

There are no auger height adjustments on this snowthrower.

CLUTCH CABLE

If belt slips under load, unbolt cable bracket (4—Fig. BN15) from snowthrower handle and move the bracket down the handle one notch in the bracket. Tighten bolt securely and recheck clutch adjustment. Auger should not rotate with clutch lever released. Install new belt if moving the cable end bracket to its farthest downward position does not provide sufficient belt tension.

OVERHAUL

ENGINE

Engine make and model numbers are listed at beginning of this section. Refer to Briggs & Stratton 2-stroke engine section for engine overhaul information.

AUGER ASSEMBLY

Refer to exploded view in Fig. BN15. After removing drive pulley and bearings, the auger assembly can be withdrawn from collector housing. Remove the screws retaining auger assembly to auger shaft and remove the shaft from auger. Parts for the auger assembly are available only from factory authorized service centers and must be renewed as a unit in the field or return the auger assembly to the factory service center for repair.

DRIVE BELT

The drive belt can be renewed after removing the belt cover from left side of snowthrower. Push the idler pulley slightly downward to release front of idler bracket from auger pulley and belt. Then, remove the belt from engine and auger pulleys. With belt removed, check condition of pulleys and be sure idler bracket moves freely on pivot. Reverse removal procedure to install new belt. Refer to Fig. BN16 for belt routing diagram. It may be necessary to adjust position of clutch cable bracket on snowthrower handle as outlined in previous CLUTCH CABLE paragraph.

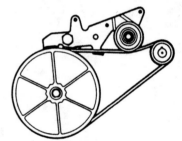

Fig. BN16—View showing routing of drive belt.

BOLENS

Model	Engine Make	Engine Model	Self-Propelled?	No. of Stages	Scoop Width
5210	Tecumseh	HSSK50	Yes	2	22 in. (559 mm)

LUBRICATION

ENGINE

Refer to Tecumseh 4-stroke engine section for engine lubrication information. Recommended fuel is unleaded or leaded regular grade gasoline.

DRIVE MECHANISM

Auger gearbox oil level should be maintained level with plug opening (23—Fig. BN25) in gearbox with SAE 90 gear lubricant. After each use, lubricate all handlebar controls, lever pivot points and auger shaft bearings with SAE 30 motor oil; silicone spray may be used on controls instead of motor oil. Lubricate traction drive chain after every 50 hours of operation with a light coat of oil. After every 50 hours of operation, remove drive wheels and apply a light coat of multipurpose grease on axle shaft bearings. Lubricate chute worm gear and toothed part of chute flange with multipurpose grease. The gear type transmission is lubricated at the factory and does not require lubrication except during overhaul; refer to Peerless transmission section for information.

MAINTENANCE

Inspect belts and pulleys and renew belts if frayed, cracked, burned or excessively worn. Check pulley grooves for excessive wear or damage and renew if defect is found. Check idler pulley bearings and renew idler pulley assembly if bearings are loose or rough. Inspect traction drive chain and sprockets and renew if excessively worn or damaged.

The auger flights are driven by the auger gearbox shaft via auger shear bolts (18—Fig. BN24) which may shear if the auger encounters a solid obstruction. If this happens, turn auger so holes in auger tube and gearbox shaft are aligned. Drive out remainder of old bolt and install a new factory replacement shear bolt.

ADJUSTMENTS

AUGER HEIGHT

The auger height may be adjusted by loosening skid shoe mounting bolts and repositioning the skid shoes. Auger should be in lowest position (skid shoes raised on housing) when operating over smooth surfaces, and in highest position when operating over rough surfaces.

SHIFT CONTROL

The shift lever is mounted directly to the transmission shift rod and fork assembly and adjustment is not required.

DRIVE CABLE

To check drive cable adjustment, first remove belt cover at front of engine. Refer to Fig. BN20 and hold the idler pulley against belt so belt is tight. Loosen idler stop screw nut (J) and adjust drive idler stop (I) so end of stop is 3/8 inch (9.5 mm) away from main frame, then tighten nut. With stop adjusted, all slack should be out of drive cable with control lever released. If cable is too tight or loose, adjust cable housing (K—Fig. BN21) position at bracket on handlebar to obtain desired cable tension. Then, with spark plug removed from engine, pull engine recoil starter while watching engine drive pulley. The belt should not move as engine is turned; if it does, decrease the gap between idler stop and frame below the distance previously set.

AUGER CABLE

To check auger control cable adjustment, remove belt cover and push auger belt idler against belt. While holding idler in this position, check to see that hole in brake arm adjustment indicator (U—Fig. BN22) is flush with main frame. If not, loosen the two screws (V) and move brake arm (W) sideways as

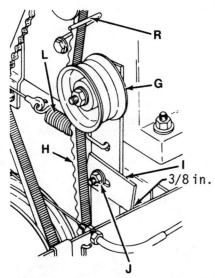

Fig. BN20—View showing adjustment of drive idler stop.

G. Drive belt idler
H. Traction drive belt
I. Drive idler stop
J. Stop screw nut
R. Belt guide

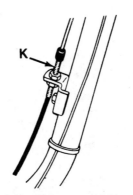

Fig. BN21—Cables (K) for auger and drive controls are adjusted at bracket on snowthrower handle. Auger cable bracket is on right handle and drive cable on left handle.

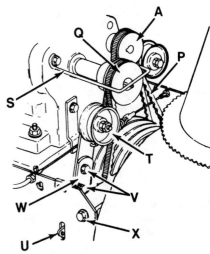

Fig. BN22—View showing adjustment of auger drive brake arm.

A. Engine PTO pulley
P. Auger drive belt
Q. Engine crankshaft pulley
S. Belt guide
T. Auger belt idler
U. Brake arm adjustment indicator
V. Brake arm screws
W. Brake arm
X. Auger housing bolt

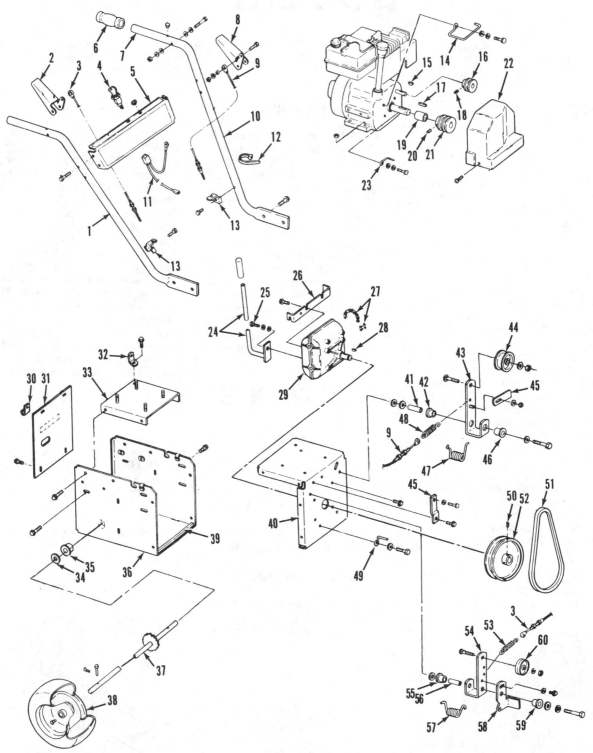

Fig. BN23—Exploded view showing handles, controls and drive system. Letters following callouts indicate same parts shown in Figs. BN20, BN21 and BN22.

1. Right handle
2. Auger control lever
3. Auger control cable (K)
4. Ignition switch
5. Console panel
6. Handle grips
7. Lever bumpers
8. Drive control lever
9. Drive control cable (K)
10. Left handle
11. Ignition cable assy.
12. Wiring tie strap

13. Cable brackets
14. Belt guide (S)
15. Woodruff key
16. Engine PTO pulley (A)
17. Square key
18. Set screw
19. Spacer
20. Set screw
21. Engine pulley (Q)
22. Belt cover
23. Belt guide (R)
24. Shift lever
25. Shift lever bolt

26. Reinforcement bracket
27. Drive chain & connector
28. Woodruff key
29. Peerless transmission
30. Auger cable clamp
31. Rear cover
32. Auger crank support
33. Motor mount plate
34. Flat washer

35. Axle flange bushing
36. Main frame
37. Drive axle assy.
38. Drive wheel
39. Rubber pad
40. Frame cover
41. Spacer
42. Bushing
43. Drive idler arm
44. Drive belt idler (G)
45. Drive idler stop (I)
46. Bushing
47. Torsion spring

48. Drive cable spring
49. Belt guide
50. Set screw
51. Drive belt (H)
52. Transmission pulley
53. Auger cable spring
54. Auger idler bracket
55. Bushing
56. Spacer
57. Torsion spring
58. Auger brake (W)
59. Bushing
60. Auger belt idler (T)

necessary, then tighten the screws and recheck position of brake arm hole. When this adjustment is completed, check auger control cable at control lever; all slack should be out of cable with control lever released. If not, adjust position of control cable housing (K—Fig. BN21) at bracket on handlebar.

OVERHAUL

ENGINE

Engine make and model numbers are listed at beginning of this section. Refer to Tecumseh 4-stroke engine section for engine overhaul information.

DRIVE BELTS

To remove either drive belt, remove belt cover, unbolt chute worm gear bracket and remove the worm gear assembly. Remove the two auger housing bolts (X—Fig. BN22), one at each side of frame. Pivot rear half of snowthrower back and rest handlebars on solid support. Loosen bolt holding belt guide (S) and remove auger drive belt (P) from engine crankshaft pulley (Q). Loosen two belt guides located on back of auger housing below the auger pulley, turn guides out of the way and remove belt from auger pulley.

The traction drive belt can be removed after removing auger belt from engine pulley. Loosen belt guide (R—Fig. BN20) on engine and the belt guides on front of main frame chassis below the transmission pulley. Take the drive belt off of transmission pulley, then remove it from engine auxiliary PTO pulley.

Reassemble by reversing disassembly procedure and adjust drive and auger controls as outlined in adjustment sections. When remounting chute worm gear, position gear mounting bracket so gear turns freely after tightening mounting bolt.

AUGER GEARBOX

To remove auger gearbox, first remove auger belt and drive belt as outlined in preceding paragraph. Completely separate auger housing from snowthrower frame. Remove set screw (1—Fig. BN24) from auger pulley (2), then remove auger pulley and Woodruff key (13) from rear end of impeller shaft. It may be necessary to heat auger pulley hub as thread locking compound is used on installation. Remove impeller shaft bearing retainer (3), shims (5) and ball bear-

ing (4). Remove auger bearings (20) from ends of auger shaft and bearing supports (6) from housing, then withdraw auger and blower fan assembly from housing. Remove shear bolts (18) and slide augers from auger shaft. Remove roll pin (14) from impeller and remove impeller from rear of auger gearbox input shaft. When reinstalling auger assembly, apply thread locking compound when mounting auger pulley on impeller shaft.

To disassemble gearbox, remove cover (1—Fig. BN25) and withdraw auger shaft and worm gear (6) as an assembly. If necessary to renew auger shaft (8) or gear, press shaft from gear and remove Woodruff key (9). Remove expansion plug (22) and internal snap ring (21)

from front of gearbox. Impeller shaft (15) with spacer (19), needle bearing (20), thrust race (18), thrust washer (17) and spacer (16) can then be removed out front opening of gearbox case. Removal of flange bushings (4), needle bearings (11 and 20) and seals (3 and 10) can now be accomplished. When installing new needle bearings, press only on lettered end of bearing cage. On reassembly, refill gearbox to level plug opening with SAE 90 gear lubricant.

TRANSMISSION

With transmission removed, refer to Peerless transmission section of this manual for overhaul information.

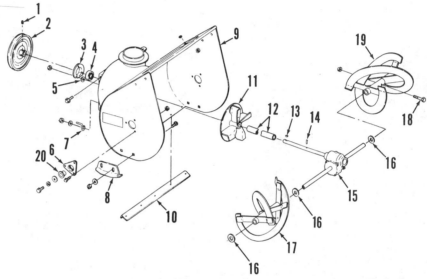

Fig. BN24—Exploded view of auger housing assembly. Exploded view of auger gearbox is shown in Fig. BN25.

1. Set screw	6. Bearing support	11. Impeller	16. Special washer
2. Auger pulley	7. Belt guide	12. Sleeve	17. R.H. auger
3. Bearing retainer	8. Skid shoes	13. Woodruff key	18. Auger shear bolt
4. Ball bearing	9. Auger housing	14. Roll pin	19. L.H. auger
5. Shim	10. Scraper blade	15. Gearbox assy.	20. Auger bearing

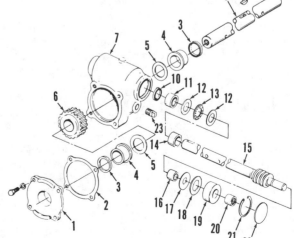

Fig. BN25—Exploded view of auger gearbox assembly.

1. Gearbox cover
2. Cover gasket
3. Oil seal
4. Flange bushing
5. Thrust bearing race
6. Worm gear
7. Gearbox case
8. Auger shaft
9. Woodruff key
10. Oil seal
11. Needle bearing
12. Thrust bearing race
13. Thrust bearing
14. Spacer
15. Impeller shaft & gear
16. Spacer
17. Thrust washer
18. Thrust race
19. Bearing spacer/retainer
20. Needle bearing
21. Internal snap ring
22. Expansion plug
23. Oil plug

BOLENS

Model	Engine Make	Engine Model	Self-Propelled?	No. of Stages	Scoop Width
624	Tecumseh	HSK60	Yes	2	24 in. (610 mm)
824	Tecumseh	HMSK80	Yes	2	24 in. (610 mm)
824A	Tecumseh	HMSK80	Yes	2	24 in. (610 mm)
1026	Tecumseh	HMSK100	Yes	2	26 in. (660 mm)
1032*	Tecumseh	HMSK100	Yes	2	32 in. (813 mm)

*Model 1032 is later production and different from Model 1032 included in previous Bolens section.

LUBRICATION

ENGINE

Refer to Tecumseh 4-stroke engine section for engine lubrication information. Recommended fuel is leaded or unleaded regular gasoline.

DRIVE MECHANISM

After each use, use clean engine oil or silicone spray to lubricate handlebar control lever pivot points; be careful not to get oil or spray on handles. Apply a few drops of oil on the pivot points of engine control lever and speed control lever. After every 25 hours of operation, check oil level at plug (23—Fig. BN38) in auger gearbox and add SAE 90 weight gear oil as necessary to bring oil level to bottom of plug opening. Use a hand-operated grease gun filled with multipurpose grease to lubricate the four auger flight grease fittings. Clean any dirt and old grease from the chute worm gear and tooth portion of chute flange and apply new multipurpose grease. Remove left drive wheel and grease axle shaft. Remove bottom panel of snowthrower, then lubricate hex fork shaft with multipurpose grease and apply light motor oil on drive chains.

MAINTENANCE

Inspect drive belts and renew if frayed, cracked, burned or excessively worn. Renew belt pulleys if pulley groove is worn or damaged. Idler pulleys should be renewed if bearing is rough or loose. Inspect drive chains and sprockets and renew if worn or if chains are stretched.

Augers are driven from auger shaft by shear bolts (19—Fig. BN37) which will fracture if a solid obstruction is encountered. If shear bolt fails, drive out any remaining part of bolt and install new factory replacement bolt. Before installing new bolt, use a hand-operated grease gun to lubricate auger shaft grease fittings and be sure auger turns freely on shaft. Do not use substitute shear bolts.

Recommended tire air pressure is 8-12 psi (56-82 kPa). Be sure both tires are inflated to same air pressure.

ADJUSTMENTS

AUGER HEIGHT

Both the skid shoes and scraper plate are adjustable. With auger height at lowest setting (skid shoes raised), scraper bar should be adjusted to height of 1/8 inch (3.2 mm) from flat, smooth surface. The skid shoes should be lowered (auger housing raised) when operating over rough, irregular surface.

SPEED CONTROL ROD

To adjust speed control rod, place control lever in fifth speed position on console, then loosen locknut (34—Fig. BN30) and back nut off as far as possible toward head of adjusting bolt (33). Turn locating screw in until locknut contacts main frame housing. Disconnect control rod pivot (10—Fig. BN31) from speed control lever and place lever in No. 1 position. While holding control rod up, the control rod pivot should just fit into hole in lever. If not, loosen jam nut (11) and turn control rod pivot up or down on speed control rod (28) until it will fit into lever when pulling up on

rod. Reinstall control rod pivot into lever and tighten control rod jam nut. Unscrew adjusting bolt (33—Fig. BN30) out of housing approximately 1/2 inch (12.7 mm) and tighten locknut (34). There should be no interference to movement of the speed control lever throughout its range of travel; if interference is noted, check drive disc clearance as outlined in following paragraph.

DRIVE DISC CLEARANCE

Remove main frame cover (35—Fig BN30) from main frame to provide access to drive disc. When drive clutch lever is released, there should be from 0.060 to 0.125 inch (1.5-3.2 mm) clearance between rubber friction drive

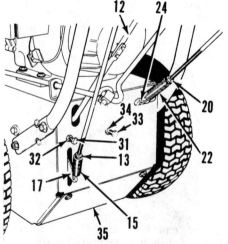

Fig. BN30—View showing control linkage adjusting points at rear of snowthrower.

12. Speed control rod	
13. Jam nuts	
15. Drive control spring	24. Auger control arm
17. Drive control arm	31. Adjusting bolt
20. Jam nuts	32. Locknut
22. Auger clutch	33. Adjusting bolt
control rod spring	34. Locknut
	35. Main frame cover

Information Courtesy of Garden Way Incorporated, Troy, New York wheel (FW—Fig. BN33) and friction disc (FD). If adjustment is necessary, loosen locknut (32—Fig. BN30) and turn adjusting bolt (31) to obtain correct clearance, then tighten locknut.

DRIVE BELT TENSION

To check adjustment, refer to Fig. BN30 and measure length of drive control spring (15) with drive clutch lever released, then measure length of spring again with lever held against handlebar grip. The difference in measurements should be 1/2 inch (12.7 mm). If not, unhook spring from drive control arm (17); it may be necessary to remove hand lever pivot pin to gain slack necessary to unhook spring. Loosen jam nuts (13) and turn the nut inside spring up to increase spring measurement or down to decrease measurement. Each 1-1/4 turn of nut will equal approximately 1/16 inch (1.6 mm). Tighten jam nuts, reconnect spring to drive control arm and recheck adjustment.

AUGER BELT TENSION

To check adjustment, measure length of auger clutch control rod spring (22—Fig. BN30) with auger control lever released. Then, hold lever against handlebar grip and measure spring in extended length. The difference between the two measurements should be 1-9/16 inches (40.3 mm). If not, unhook spring from auger control arm (24); it may be necessary to remove handlebar lever pivot bolt to provide slack needed to unhook the spring. Loosen jam nuts (20) and turn hex nut inside spring up the control rod to increase spring extension or down to decrease measurement. One turn of the nut will change spring extension about 1/16 inch (1.6 mm). After adjusting, tighten jam nuts, then reconnect spring to lever and recheck adjustment.

AUGER BRAKE ARM

To check brake arm adjustment, first remove belt cover at front of engine. Hold auger clutch control lever against handlebar grip and measure gap between brake arm (46—Fig. BN32) and auger drive belt. The gap should measure between 7/16 and 1/2 inch (11.1-12.7 mm). If gap is not correct, loosen jam nuts (47 and 50) and position screw (51) to obtain desired gap, then tighten jam nuts. The adjusting screw can be held with an Allen wrench while turning jam nuts.

LIMITED SLIP DIFFERENTIAL

Remove dust cap (54—Fig. BN36) from right end of drive axle and loosen jam nuts (55). Remove outer nut, turn inner jam nut in finger tight and then using a wrench, tighten inner jam nut 1-1/4 additional turns. Hold inner jam nut in this position with a wrench, then install and tighten outer jam nut.

OVERHAUL

ENGINE

Engine make and model numbers are listed at beginning of this section. Refer to Tecumseh 4-stroke engine section for engine overhaul information.

TRACTION DRIVE BELT

Remove belt cover at front of engine and remove the belt guide (G—Fig. BN35) from engine. Take auger drive belt off of engine pulley. Release tension from drive belt idler (JJ) and take drive belt off of lower (driven) pulley. Then move lower part of belt up the gap between the auger and drive pulleys and

remove belt from engine pulley. Install new belt by reversing removal procedure and check drive belt adjustment as outlined in adjustment paragraph.

AUGER DRIVE BELT

Remove belt cover at front of engine and remove belt guide (G—Fig. BN35) from engine. Take auger drive belt off of engine pulley, then remove it from lower pulley; it may be necessary to pull auger brake arm (R) outward to have room for removing belt. Models 824A, 1026 and 1032 have dual auger drive belts. Pull belt up through the gap between the auger and drive pulleys. Install new belt by reversing removal procedure and check auger belt adjustment as outlined in adjustment paragraph.

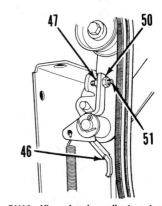

Fig. BN32—View showing adjustment of auger brake arm. Single drive belt for Models 624 and 824 is shown; other models have dual auger drive belts.

46. Brake arm
47. Jam nut
50. Jam nut
51. Adjusting screw

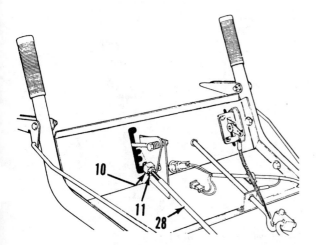

Fig. BN31—Bottom view of control panel showing speed control rod adjustment.

10. Control rod pivot
11. Jam nut
28. Speed control rod

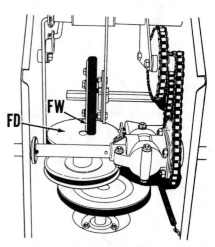

Fig. BN33—Bottom view of friction drive with bottom cover removed. Models 624 and 824 are shown; other models have dual auger drive belts. Clearance between friction wheel (FW) and friction disc (FD) should be 0.060-0.125 inch (1.5-3.2 mm).

FRICTION WHEEL

To renew rubber faced friction wheel, first tip snowthrower forward onto front of auger housing and be sure it is securely braced in this position. Remove lower cover from main frame and refer to Fig. BN33. Hold the hex shaft from turning with correct size open-end wrench, then remove the cap screws retaining wheel to adapter. The friction wheel can then be removed from over the end of the hex shaft. Install new wheel with cup side toward adapter.

DRIVE MECHANISM

Overhaul of the disc drive mechanism should be evident after inspection of unit and reference to Fig. BN33 and Fig. BN36. Friction wheel thrust bearing (50—Fig. BN36) should be installed on hub (51) using a locking compound. Refer to adjustment paragraph on reassembly.

AUGER GEARBOX

To remove auger gearbox, first remove auger belt and drive belt as outlined in preceding paragraph. Completely separate auger housing from snowthrower frame. Remove set screw (2—Fig. BN37) from auger drive pulley (1), then remove auger pulley and Woodruff key (12) from rear end of impeller shaft. Remove impeller shaft bearing retainer (4) and ball bearing (3). It may be necessary to heat bearing as locking compound is used on installation. Remove auger shaft bushings (5) from ends of auger shaft and bushing supports (6) from housing, then withdraw auger and blower fan assembly from housing. Remove shear bolts (19) and slide augers from auger shaft. Remove roll pin (13) from impeller and remove impeller from rear of auger gearbox input shaft. When reinstalling auger assembly, apply locking compound when placing rear ball bearing (3) on impeller shaft.

To disassemble gearbox, remove cover (1—Fig. BN38) and withdraw auger shaft and worm gear (6) as an assembly. If necessary to renew auger shaft (8) or gear, press shaft from gear and remove Woodruff key (9). Remove expansion

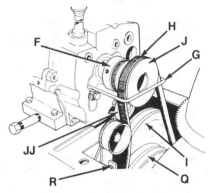

Fig. BN35—References for belt removal.

  F. Friction drive belt
  G. Belt guide
  H. Auger drive belt
  J. Engine pulley
 JJ. Drive belt idler
  Q. Auger drive pulley
  R. Auger brake arm

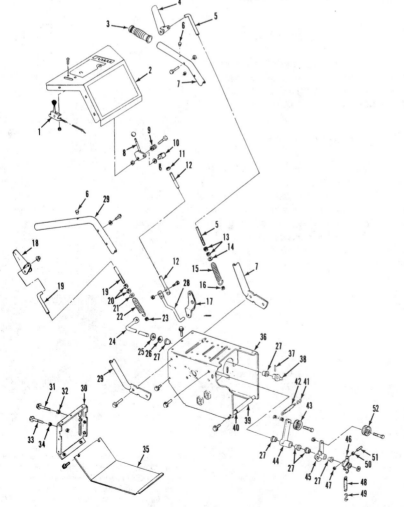

  1. Throttle assy.
  2. Console panel
  3. Handle grip
  4. Drive control lever
  5. Drive control rod
  6. Lever bumper
  7. Left handle
  8. Speed control lever
  9. Lever friction spring
 10. Speed control rod pivot
 11. Jam nut
 12. Speed control rod
 13. Jam nuts
 14. Lockwasher
 15. Drive control spring
 16. Adjusting nut
 17. Drive control arm
 18. Auger control lever
 19. Auger control rod
 20. Jam nuts
 21. Lockwasher
 22. Auger clutch control rod spring
 23. Adjusting nut
 24. Auger control arm
 25. Flat washer
 26. Spring washer
 27. Idler arm bushing
 28. Lower speed control rod
 29. Right handle
 30. Main frame back plate
 31. Adjusting bolt
 32. Locknut
 33. Adjusting bolt
 34. Locknut
 35. Main frame cover
 36. Main frame
 37. Drive pin
 38. Auger control arm
 39. Plastic tube
 40. Cross member
 41. Spring hook
 42. Drive idler spring
 43. Drive belt idler
 44. Drive idler arm
 45. Auger idler arm
 46. Auger brake arm
 47. Jam nut
 48. Auger idler spring
 49. Spring hook
 50. Jam nut
 51. Adjusting screw
 52. Auger belt idler

Fig. BN34—Exploded view showing handles, control linkage and main frame assembly.

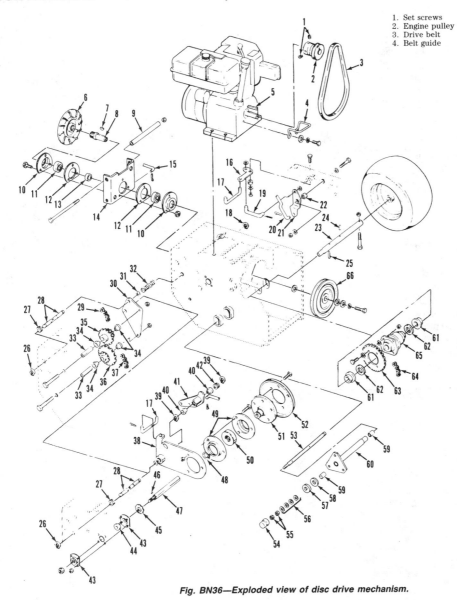

1. Set screws
2. Engine pulley
3. Drive belt
4. Belt guide
5. Square key
6. Disc drive plate
7. Woodruff key
8. Drive disc spindle
9. Tube
10. Bearing retainer
11. Ball bearing
12. Bearing retainer
13. Spacer
14. Drive disc pivot plate
15. Link
16. Speed control arm
17. Link
18. Snap ring
19. Link
20. Speed control rod
21. Speed control arm
22. Control arm pivot bushing
23. Axle shaft, L.H.
24. Drive pin
25. Drive pin
26. Snap ring
27. Nylon bushing
28. Support shaft
29. Roller chain
30. Support plate
31. Flat washer
32. Compression spring
33. Spacer
34. Sprocket bushings
35. Double sprocket
36. Double sprocket
37. Roller chain
38. Sliding plate
39. Snap ring
40. Control arm bushing
41. Drive control arm
42. Nylon bushing
43. Bearing retainer
44. Ball bearing
45. Hex shaft sprocket
46. Woodruff key
47. Hex shaft
48. Snap ring
49. Bearing retainers
50. Thrust bearing
51. Friction wheel hub
52. Friction wheel
53. Differential locking shaft
54. Dust cap
55. Jam nuts
56. Belleville washers
57. D-hole washer
58. Friction disc
59. Nylon bushing
60. Axle tube, R.H.
61. Flange bushing
62. Flat washer
63. Differential sprocket
64. Roller chain
65. Differential assy.
66. Drive pulley

Fig. BN36—Exploded view of disc drive mechanism.

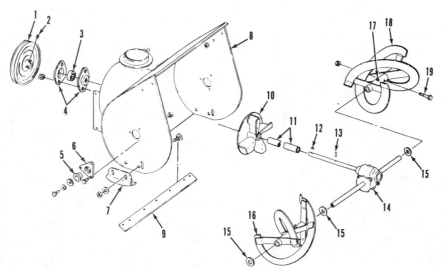

Fig. BN37—Exploded view of auger housing assembly. Refer to Fig. BN38 for gearbox assembly (14) exploded view. Models 824A, 1026 and 1032 have dual auger drive belts.

1. Auger drive pulley
2. Set screw
3. Ball bearing
4. Bearing retainers
5. Auger shaft bushing
6. Bushing support
7. Skid shoe
8. Auger housing
9. Scraper bar
10. Impeller
11. Sleeve
12. Woodruff key
13. Roll pin
14. Gearbox assy.
15. Flat washer
16. Auger, R.H.
17. Grease fitting
18. Auger, L.H.
19. Shear bolt

plug (22) and internal snap ring (21) from front of gearbox. The impeller shaft (15) with spacer (19), needle bearing (20), thrust race (18), thrust washer (17) and spacer (16) can then be removed out front opening of gearbox case. Removal of flange bushings (4), needle bearings (11 & 20) and seals (3 & 10) can now be accomplished. When installing new needle bearings, press only on lettered end of bearing cage. On reassembly, refill gearbox to level plug opening with SAE 90 gear lubricant.

DIFFERENTIAL

The limited slip differential (65—Fig. BN36) can be disassembled for inspection; if all parts are reusable, clean the parts thoroughly, apply grease on all parts and pack differential with 1-1/2 ounces (44.4 mL) of multipurpose grease. Refer to differential adjustment paragraph.

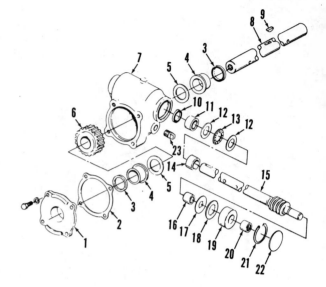

Fig. BN38—Exploded view of auger gearbox assembly.
1. Gearbox cover
2. Cover gasket
3. Oil seal
4. Flange bushing
5. Thrust bearing race
6. Worm gear
7. Gearbox case
8. Auger shaft
9. Woodruff key
10. Oil seal
11. Needle bearing
12. Thrust bearing race
13. Thrust bearing
14. Spacer
15. Impeller shaft & gear
16. Spacer
17. Thrust washer
18. Thrust race
19. Bearing spacer/retainer
20. Needle bearing
21. Internal snap ring
22. Expansion plug
23. Oil plug

CRAFTSMAN

SEARS, ROEBUCK & CO.
Sears Tower
Chicago, IL 60684

Model	Engine Make	Engine Model	Self-Propelled?	No. of Stages	Scoop Width
536.884220	Tecumseh	AH600	No	1	20 in. (508 mm)
536.884320	Tecumseh	AH600	No	1	20 in. (508 mm)

LUBRICATION

ENGINE

For the 2-stroke engine used on these models, recommended fuel:oil mixture ratio is 32:1. Use a high quality two-cycle oil or non-detergent SAE 20 or SAE 40 oil. The manufacturer cautions not to use BIA outboard motor oil or multiviscosity oils.

DRIVE MECHANISM

Frequently lubricate chute control rod where it passes through "U" bolt and the discharge chute with clinging type grease such as "Lubriplate."

ADJUSTMENTS

DRIVE BELT & PULLEYS

The auger drive belt does not require adjustment as the belt idler is spring loaded. Check belt pulleys for proper alignment by placing a straightedge across the outside edges of auger pulley and engine pulley. If out of alignment, loosen the two set screws on engine pulley and move pulley in or out until pulleys are flush with straightedge, then tighten set screws.

OVERHAUL

ENGINE

Engine make and model numbers are listed at beginning of section. Refer to Tecumseh 2-stroke engine section for engine overhaul information. Be sure to check drive pulley alignment after reinstalling engine.

DRIVE BELT

Remove belt cover (Fig. CR1), lift belt idler pulley (16) and remove old belt. Install new belt and check belt adjustment as previously outlined.

AUGER ASSEMBLY

To remove auger assembly, first refer to Fig. CR1. Remove belt cover (27), unscrew auger pulley (20) and remove left-hand auger bearing (22) from left side of auger housing. Slide the auger out of right side bearing assembly, tip right end of auger out and remove auger from housing. Reassemble in reverse order and check belt adjustment as outlined in previous paragraph.

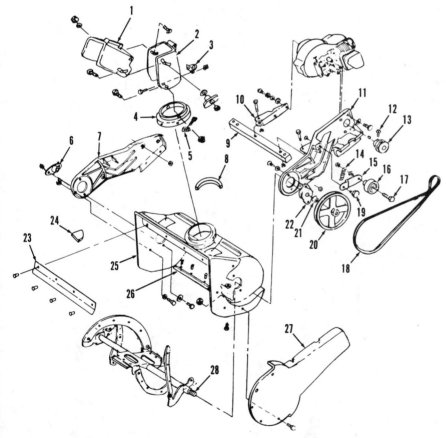

Fig. CR1—Exploded view of belt drive and auger mechanism. Idler arm spring (14) hooks to bracket on lower handle (not shown) and inner end of idler pulley bolt (17).

1. Chute deflector
2. Chute
3. Chute bracket
4. Chute ring
5. Chute guide
6. R.H. auger bearing
7. R.H. auger frame
8. Chute seal strip
9. Support tube
10. Engine bracket
11. L.H. auger frame
12. Set screw
13. Engine pulley
14. Idler arm spring
15. Idler arm
16. Idler pulley
17. Idler pulley bolt
18. Auger drive belt
19. Shoulder (pivot) bolt
20. Auger pulley
21. Spacer
22. L.H. auger bearing
23. Snow deflector
24. Auger seal
25. Auger housing
26. Scraper bar
27. Belt cover
28. Auger assy.

CRAFTSMAN

Model	Engine Make	Engine Model	Self-Propelled?	No. of Stages	Scoop Width
536.884800	Tecumseh	HS50	Yes	2	23 in. (584 mm)
536.886500	Tecumseh	HS50	Yes	2	23 in. (584 mm)
536.884900	Tecumseh	HM80	Yes	2	25 in. (635 mm)
536.886800	Tecumseh	HM80	Yes	2	25 in. (635 mm)
536.885900	Tecumseh	HM80	Yes	2	26 in. (660 mm)

LUBRICATION

ENGINE

Refer to Tecumseh engine section for engine lubrication information. Recommended fuel is unleaded or leaded regular gasoline.

DRIVE MECHANISM

Recommended lubrication period is after every 10 hours of operation and before storing unit at end of season. Oil the drive chains and sprockets, on track drive models, with SAE 10W-30 oil. The hex shaft assembly in the variable speed friction drive requires no lubrication except to wipe the hex shaft with SAE 10W-30 motor oil at end of season to prevent rusting. The hex shaft bearings are of the sealed type. Be careful not to get any grease or oil on the disc drive plate or friction wheel; clean these parts thoroughly if oil or grease comes in contact. Lubricate all moving part pivot points with SAE 10W-30 oil. On track drive models, lubricate weight transfer plate with clinging type grease such as "Lubriplate."

AUGER GEARBOX

The auger gearbox is factory lubricated and does not require lubricant changing. If for some reason oil should leak out or once each season, remove grease plug (17—Fig. CR15) and check for lubricant. If lubricant is not visible, use a wire as a dipstick and add lubricant as necessary. Recommended lubricants are Sunoco Prestige 740AEP, Shell Alvania EPR, Mobile Mobilplex EP23 and Texaco Marfak AP.

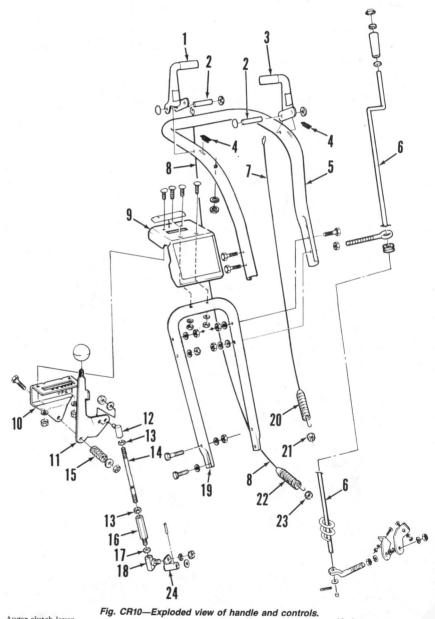

Fig. CR10—Exploded view of handle and controls.

1. Auger clutch lever
2. Lever pivot pin
3. Drive clutch lever
4. Lever bumper
5. Upper handle
6. Chute crank
7. Drive clutch cable
8. Auger clutch cable
9. Control panel
10. Speed control lever bracket
11. Speed control lever
12. Ball joint
13. Jam nut
14. Shift rod
15. Lever tension spring
16. Adapter
17. Lockwasher
18. Ball joint
19. Lower handle
20. Drive clutch spring
21. Adjusting nut
22. Auger clutch spring
23. Adjusting nut
24. Speed control lower lever

ADJUSTMENTS

AUGER HEIGHT

Lower the adjusting skids to raise auger assembly on rough surfaces. On paved surfaces, the skids can be raised to bring the auger assembly down. Keep in mind that the scraper bar below the auger should be 1/8 inch (3.2 mm) above sidewalk or other paved area to be cleaned. This clearance can be obtained by placing spacers under the scraper bar, then adjusting the skids so they contact paved surface.

To compensate for wear, the scraper bar position is adjustable, and the bar can be reversed or renewed. Be sure the bar is adjusted parallel with the working surface and for proper clearance, above skid shoes.

DRIVE BELTS

The traction drive belt is spring loaded and does not require adjustment; renew belt if it is slipping. The auger drive belt can be adjusted after removing belt cover, then proceed as follows: Loosen nut on idler pulley (18—Fig. CR11) and move idler toward belt about 1/8 inch (3 mm). Tighten nut, engage auger drive clutch lever and check belt adjustment;

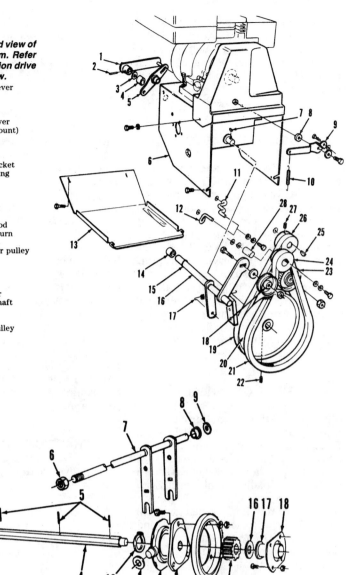

Fig. CR11—Exploded view of drive belt mechanism. Refer to Fig. CR12 for friction drive exploded view.

1. Auger clutch lever
2. Roll pin
3. Flat washer
4. Plastic bushing
5. Auger cable lever
6. Main (motor mount) frame
7. Hi-pro key
8. Spacer
9. Drive idler bracket
10. Drive idler spring
11. Belt guide
12. Belt guide
13. Bottom panel
14. Plastic bushing
15. Flat washer
16. Auger clutch rod
17. Auger idler return spring
18. Auger belt idler pulley
19. Drive belt
20. Drive pulley
21. Auger belt
22. Set screw
23. Drive belt idler
24. Engine crankshaft pulley
25. Woodruff key
26. Engine PTO pulley
27. Set screw
28. Spacer

Fig. CR12—Exploded view of disc drive mechanism.

1. Bearing plate
2. Bearing
3. Flat washer
4. Hex shaft
5. Roll pins
6. Locknut
7. Speed control yoke
8. Plastic bushing
9. Washer
10. Snap ring
11. Washer
12. Trunnion bearing
13. Friction wheel hub
14. Friction wheel
15. Pinion gear
16. Flat washer
17. Bearing
18. Bearing plate
19. Spacers
20. Traction pivot
21. Push nut
22. Bushing
23. Flat washer
24. Cotter pin
25. Traction shaft
26. Trunnion bearing
27. Traction spring bracket
28. Flat washer
29. Traction spring
30. Spring pin
31. Traction rod
32. Drive disc
33. Return spring
34. Hi-pro key
35. Washer
36. Drive disc shaft
37. Snap ring
38. Thrust bearing
39. Needle bearings
40. Washer
41. Axle bearing
42. Bolt
43. Driven gear
44. Axle
45. Grip rings
46. Klick pin

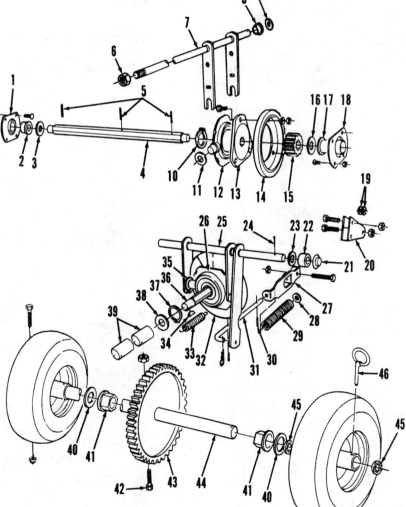

belt should deflect about 1/2 inch (12.7 mm) with moderate pressure. Reset idler pulley until proper belt tension is obtained with drive engaged. Belt guides (11 & 12) should be adjusted to clear belt 3/32 inch (2.4 mm) with auger drive clutch engaged.

CONTROL CABLES

Refer to Fig. CR10 and with drive clutch cable (7) "z" fitting disconnected from drive clutch lever (3), move lever forward to contact plastic bumper on handle; the center of the "z" fitting should be between the top and the center of the hole in the lever. If not, slide cable down through drive clutch spring (20), hold square part of cable end from turning and rotate adjusting nut (21) to obtain correct cable length. Then, pull cable up through spring and reconnect "z" fitting on cable to hole in lever. Repeat adjustment procedure for auger clutch cable (8).

FRICTION WHEEL

Remove bottom panel from snowthrower. Place speed control lever in first gear and check position of friction wheel on disc drive plate. The distance should be 3-3/8 inches (86 mm) as shown in Fig. CR13. If adjustment is necessary, loosen lower jam nut (13—Fig. CR10) on speed select rod. Remove ball joint (18) from speed control lower lever (24) and turn adapter (16) to obtain correct friction wheel position. Reinstall ball joint, tighten jam nut and reinstall bottom panel.

TRACKS

Refer to Fig. CR16 and loosen cap screw (13) on each side of track assembly. Rotate adjusting cam (14) equally on each side of assembly so deflected distance between top of track plate (12) and inside of track is not greater than 2 inches (50.8 mm). Uneven adjustment of adjusting cams will result in a twist

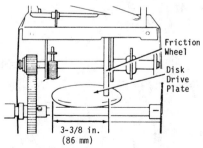

Fig. CR13—With speed control lever in first gear, friction wheel should be positioned as shown.

in the track. Repeat adjustment for opposite track assembly.

OVERHAUL

ENGINE

Engine make and model numbers are listed at beginning of section. Refer to Tecumseh engine section for engine overhaul.

AUGER DRIVE BELT

To remove auger drive belt, remove belt drive cover and loosen belt guides (11 & 12—Fig. CR11) and move them away from engine crankshaft pulley (24). Loosen auger belt idler pulley (18) and move it away from the belt. Engage auger drive clutch lever to move brake away from belt and slip the belt off auger pulley. Remove belt from engine drive pulley. Reverse removal procedure to install new belt and adjust belt and belt guides as outlined in a preceding paragraph.

TRACTION DRIVE BELT

To remove traction drive belt, remove belt cover and loosen the left-hand belt

guide (11—Fig. CR11) and move guide away from engine PTO pulley. Pull spring loaded traction idler pulley back and slip belt past idler pulley. Remove belt from engine pulley and up between the auger and traction drive pulleys. Reverse removal procedure to install new belt and adjust belt and belt guides as outlined in a preceding paragraph.

AUGER SHEAR BOLT

The augers are driven via the auger shaft through special shear bolts (16—Fig. CR14) that are designed to fracture to protect the machine if an object becomes lodged in the auger housing. Only original equipment replacement shear bolts should be used for safety reasons. When replacing shear bolt(s), be sure to install new shear bolt spacer (17). Using a hand grease gun, lubricate auger shaft grease fittings when new shear bolts are installed.

AUGER ASSEMBLY

To remove auger, remove auger drive belt, disconnect the chute crank from auger housing and separate the auger housing from main (motor mount) frame. Remove auger drive pulley (1—Fig. CR14) and impeller shaft bearing (2) from rear of housing. Unbolt auger shaft

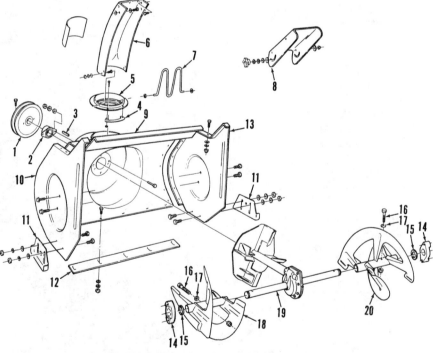

Fig. CR14—Exploded view of auger housing assembly.

1. Auger drive pulley	9. Center auger housing
2. Impeller shaft bearing	10. R.H. side plate
3. Square key	11. Skid shoe
4. Chute retainer	12. Scraper bar
5. Chute flange	13. L.H. side plate
6. Bottom chute extension	14. Auger shaft bushing
7. Hand guard	15. Spacer washer
8. Upper chute extension	16. Shear bolt
	17. Shear bolt spacer
	18. R.H. auger flight
	19. Gearbox assy.
	20. L.H. auger flight

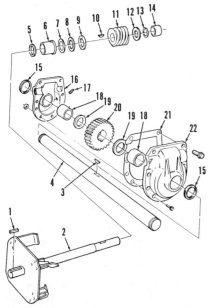

Fig. CR15—Exploded view of auger gearbox. Impeller is an integral part of impeller shaft (2).

1. Square key
2. Impeller & shaft assy.
3. Woodruff key
4. Auger shaft
5. Square cut "O" ring
6. Impeller shaft bushing
7. Thrust bearing race
8. Needle thrust bearing
9. Thrust bearing race
10. Woodruff key
11. Worm gear
12. Thrust washer
13. Snap ring
14. Impeller shaft bushing
15. Oil seals
16. L.H. housing half
17. Grease plug
18. Auger shaft bushing
19. Thrust washer
20. Gear
21. Gasket
22. R.H. housing half

bushings (14) at each side of housing and remove the assembly. Remove the auger drive shear bolts (16) and remove the auger flights from auger shaft.

With auger flights removed, refer to Fig. CR15 for exploded view of auger drive gearbox. To disassemble, remove the cap screws holding the gearbox housing halves (16 & 22) and separate the unit. Remove impeller shaft bushing (14), snap ring (13) and thrust washer (12) from impeller shaft. Press shaft from worm gear (11) and remove needle thrust bearing (8) and races (7 & 9), rear impeller shaft bushing (6) and "O" ring (5) from impeller shaft. Clean and inspect all parts and renew any excessively worn or damaged parts. To install auger shaft bushings (18), press bushings into case from inside and flush with inside of casting. Install new seals (15) in gearbox halves.

When reassembling, coat all parts with grease and pack gearbox with a maximum of 3-1/4 ounces (96 mL) of one of the following greases: Benalene #372, Shell Darina 1, Texaco Thermatex EP1, or Mobiltem 78.

FRICTION DRIVE

To renew friction wheel, proceed as follows: Stand snow thrower up on auger housing end and remove bottom panel (13—Fig. CR11). On track drive models, disconnect track connecting rod (2—Fig. CR16) at right track plate (1) and rotate right track assembly parallel to ground to gain clearance to hex shaft end bearing. Remove the three bolts securing friction wheel (14—Fig. CR12) to hub (13) and move speed control lever to first speed position. Loosen but do not remove the four nuts securing right hex shaft bearing plate (1). Move speed control lever to sixth speed position. Place tape over the four hex shaft bearing plate bolts inside main (motor mount) frame and remove the previously loosened retaining nuts. Remove bearing plate and slide hex shaft to right until friction wheel can be removed from end of hex shaft. Install replacement friction wheel retaining bolts loosely, then complete reassembly in reverse order of disassembly. Check friction wheel adjustment as outlined in a preceding paragraph.

TRACK ASSEMBLY

Refer to exploded view of track assembly in Fig. CR16. Disassembly and overhaul is evident after inspection of unit and reference to exploded view. Readjust track after servicing as outlined in a previous paragraph.

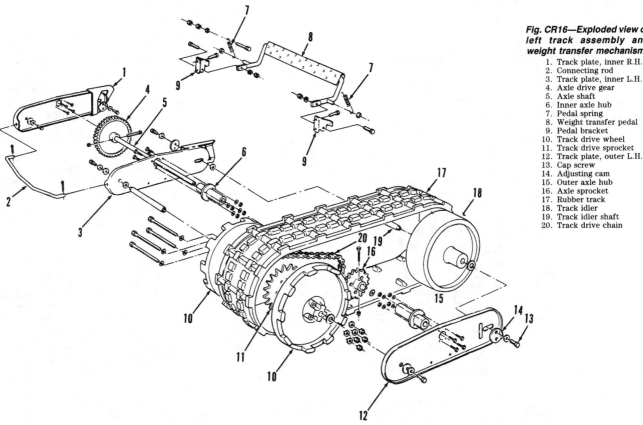

Fig. CR16—Exploded view of left track assembly and weight transfer mechanism.

1. Track plate, inner R.H.
2. Connecting rod
3. Track plate, inner L.H.
4. Axle drive gear
5. Axle shaft
6. Inner axle hub
7. Pedal spring
8. Weight transfer pedal
9. Pedal bracket
10. Track drive wheel
11. Track drive sprocket
12. Track plate, outer L.H.
13. Cap screw
14. Adjusting cam
15. Outer axle hub
16. Axle sprocket
17. Rubber track
18. Track idler
19. Track idler shaft
20. Track drive chain

CRAFTSMAN

Model	Engine Make	Engine Model	Self-Propelled?	No. of Stages	Scoop Width
536.885400	Tecumseh	HS40	Yes	2	20 in. (508 mm)

LUBRICATION

ENGINE

Refer to Tecumseh 4-stroke engine section for engine lubrication information. Recommended fuel is leaded or unleaded regular gasoline.

DRIVE MECHANISM

Recommended lubrication period is after every 10 hours of operation and before storing unit at end of season. Oil the drive chains and sprockets with SAE 10W-30 oil. The hex shaft assembly in the variable speed friction drive requires no lubrication except to wipe the hex shaft and gears with SAE 10W-30 motor oil at end of season to prevent rusting. The hex shaft bearings are of the sealed type. Be careful not to get any grease or oil on the disc drive plate or friction wheel; clean these parts thoroughly if oil or grease comes in contact with these parts.

Lubricate all moving part pivot points with SAE 10W-30 oil. Lubricate weight transfer plate and pivot points with clinging type grease such as "Lubriplate."

The auger flights do not have a grease fitting; before storage at end of season, remove the auger shear bolts and squirt oil into shear bolt holes. Turn the auger to distribute the oil throughout length of auger shaft.

AUGER GEARBOX

The auger gearbox is factory lubricated for life and unless grease leaks out, lubrication should not be necessary unless gearbox is being overhauled. Some models have grease plug (33—Fig. CR23) that can be removed. Use a wire as a dipstick to check for presence of grease.

ADJUSTMENTS

AUGER HEIGHT

Lower the adjusting skids to raise auger scraper bar on rough surfaces. On paved surfaces, the skids can be raised to bring the auger assembly down. Keep in mind that the scraper bar below the auger should be 1/8 inch (3.2 mm) above sidewalk or other paved area to be cleaned. This clearance can be obtained by placing spacers (the extra shear pins stored under snowthrower handle can be used for spacers) under the scraper bar, then adjusting the skids so they contact paved surface. To compensate for wear, the scraper bar position is adjustable, and the bar can be reversed or replaced. Be sure the bar is adjusted parallel with the working surface, and for proper clearance, above skid shoes.

DRIVE BELTS

The traction drive belt is spring loaded and does not require adjustment; check control cable adjustment, condition of idler spring and for free movement of idler arm if belt is slipping. Renew belt if all other drive mechanism parts check ok. The auger drive belt can be adjusted after removing belt cover, then proceed as follows: Loosen nut on auger belt idler (16—Fig. CR22) and move idler in slot of idler arm (14) toward belt about 1/8 inch (3 mm). Tighten nut, engage auger drive clutch lever and check belt adjustment; belt should

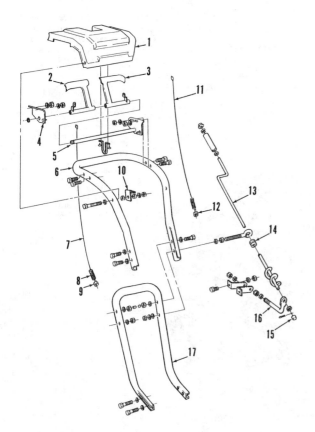

Fig. CR20—Exploded view of control and handle assembly.

1. Control panel
2. Auger clutch lever
3. Drive clutch lever
4. Lever rod bracket
5. Rod & bracket assy.
6. Upper handle
7. Auger cable
8. Auger clutch spring
9. Locknut
10. Panel bracket
11. Drive clutch cable & spring
12. Locknut
13. Chute crank assy.
14. Grommet
15. Plastic cap
16. Chute crank bracket
17. Lower handle

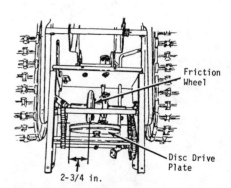

Fig. CR21—View showing measurement to check friction wheel adjustment.

deflect about 1/2 inch (12.7 mm) with moderate pressure. Reset idler pulley until proper belt tension is obtained with drive engaged. The belt guides at engine pulley should be adjusted to clear belt 3/32 inch (2.4 mm) with auger drive clutch engaged.

CONTROL CABLES

Refer to Fig. CR20 and with auger cable (7) "z" fitting disconnected from auger clutch lever (2), move lever forward to contact plastic bumper and pull upward on cable; the center of the "z" fitting should be between the top and the center of the hole in the lever. If not, slide cable down through auger clutch spring (8) and turn lock nut (9) to obtain proper adjustment. Then, pull cable up through spring and reconnect "z" fitting on cable in hole of lever. Repeat the procedure for drive clutch cable (11) adjustment.

FRICTION WHEEL

Remove bottom panel from snowthrower. Place shifter lever in first (Position 1) gear and check position of friction wheel on disc drive plate. Distance from left outer side of disc drive plate to right side of friction wheel should be 2-3/4 inches (70 mm) as shown in Fig. CR21. If necessary to adjust position of friction wheel, loosen bolts retaining upper speed shift control lever (29—Fig. CR22) to shift lever adapter (32). Then move friction wheel to proper position and tighten lever bolts.

TRACK DRIVE

To check track tension, pull up gently on center of track at midpoint between track rollers. Distance between track and top of track support frame should then be no more than 1-1/4 inches (32 mm). To adjust track, loosen jam nuts (3—Fig. CR24) on adjusting bolt (11) at rear end of track frame (4) and tighten or loosen adjusting bolt as necessary. Adjust opposite track in same manner. Be sure adjusting bolt jam nuts are securely tightened.

OVERHAUL

ENGINE

Engine make and model numbers are listed at beginning of section. Refer to Tecumseh engine section for engine overhaul.

AUGER DRIVE BELT

To remove auger drive belt, remove belt drive cover (2—Fig. CR22) and loosen the right-hand belt guide (12) and move it away from drive pulley. Loosen auger belt idler (16) and move it away from the belt. Engage auger drive clutch lever to move brake away from belt and slip the belt out from between auger pulley and brake. Remove the top two bolts securing auger housing to motor mount frame. Place weight transfer system pedal in transport position (latched in top notch). Slightly loosen bottom two bolts attaching auger housing to motor mount frame. Separate auger housing and motor mount frame by

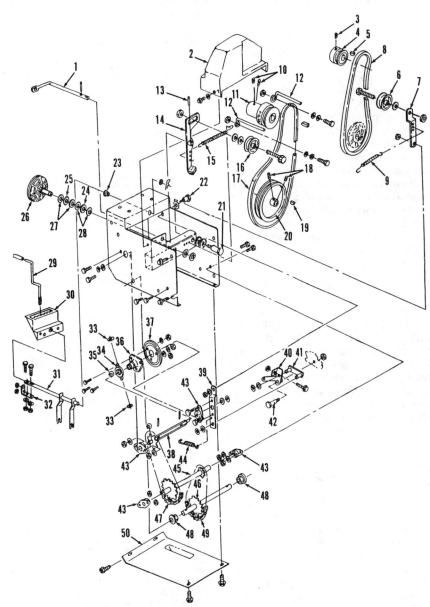

Fig. CR22—Exploded view of drive mechanism. The traction drive is from engine PTO pulley (4) and the auger is driven from engine crankshaft pulley (11). Drive disc (26) and shaft are separate items on some models.

1. Auger clutch lever
2. Belt cover
3. Set screw
4. Engine PTO pulley
5. Woodruff key
6. Drive idler pulley
7. Drive idler arm
8. Drive belt
9. Idler arm spring
10. Set screws
11. Engine crankshaft pulley
12. Belt guide
13. Roll pin
14. Auger brake & idler arm
15. Auger brake arm spring
16. Auger belt idler
17. Auger belt
18. Set screws
19. Woodruff key
20. Auger drive pulley
21. Shoulder bolt
22. Shoulder bolt
23. Plastic bushing
24. Thrust washer
25. Needle thrust bearing
26. Drive disc
27. Thrust washers
28. Needle thrust bearing
29. Shift lever
30. Shift bracket
31. Shift yoke
32. Shift lever/yoke adapter
33. Flat washers
34. Trunion bearing
35. Snap ring
36. Friction wheel hub
37. Friction wheel
38. Hex shaft
39. Drive clutch lever
40. Drive lever assy.
41. Drive lever link
42. Shoulder bolt
43. Self-centering bearing
44. Return spring
45. Jackshaft assy.
46. Wheel axle & sprocket
47. Roller chain
48. Axle bearing
49. Roller chain
50. Bottom cover panel

hinging on bottom two bolts. Remove belt from engine auger drive pulley. Install new belt by reversing removal procedure. Place weight transfer system in lowest notch in bracket to pivot auger housing and motor mount frame back into position. Adjust belt as outlined in preceding paragraph.

TRACTION DRIVE BELT

To remove traction drive belt, first remove upper belt cover, pull traction idler pulley back and slip belt past idler pulley, then remove belt from engine pulley (It may be necessary to loosen and remove engine auger drive pulley for clearance to remove belt). Place speed selector in sixth gear position and remove belt from between rubber drive disc and combination drive pulley and drive disc plate. Install new belt by reversing removal procedure. If engine auger drive pulley was removed, be sure auger drive pulleys are aligned and drive key is flush with end of engine shaft before tightening engine pulley set screw. Adjust belts as outlined in belt adjustment paragraph.

AUGER SHEAR BOLT

The augers are driven from the auger shaft through special bolts (37—Fig. CR23) that are designed to break to protect the machine if an object becomes lodged in the auger housing. Only original equipment replacement shear bolts should be used for safety reasons. Before installing new bolt, squirt oil between auger flight and auger shaft and turn auger flight to distribute oil throughout the full length of shaft.

AUGER ASSEMBLY

To remove auger, first remove auger drive belt as previously outlined. Remove auger drive pulley and impeller shaft bearing (6—Fig. CR23) from rear of housing, unbolt auger shaft bearings (10) at each side of housing and remove the auger and gearbox assembly. Remove the auger drive shear bolts (37) and slide the auger flights from auger shaft. Remove impeller bolts (34) and pull impeller from shaft. Remove the bolts holding the case halves (20 & 32) together and disassemble gearbox. Clean and inspect all parts and renew any worn or damaged parts. Coat the parts with clean grease and pack the gearbox with approximately 3 ounces (85 grams) grease on assembly. Some recommended greases are Benalene #372, Shell Darina 1, Texaco Thermatex EP1, and Mobiltem 78.

FRICTION DRIVE WHEEL

To replace friction wheel, proceed as follows: Stand snowthrower up on auger housing end and remove bottom cover panel (50—Fig. CR22). Unbolt bearings (43) from engine frame at each end of hex shaft (38). Slide the hex shaft with friction wheel, trunnion bearing and end bearings out of snowthrower. Be careful not to lose flat washers (33) from studs on trunnion bearing (34). Unbolt friction wheel (37) from hub and install new wheel (cup side away from

Fig. CR23—Exploded view of auger housing, gearbox assembly and augers.

1. Deflector chute
2. Hand guard (not all models)
3. Chute extension
4. Mounting flange
5. Flange retainer
6. Impeller shaft bearing
7. Auger housing
8. Left side plate
9. Right side plate
10. Auger shaft bearing
11. Skid shoe
12. Scraper bar
13. Oil seal
14. Bushing
15. Thrust washers
16. Needle bearing
17. Roll pin
18. Thrust collar
19. Impeller shaft
20. Right case half
21. Oil seal
22. Bushing
23. Thrust washer
24. Woodruff key
25. Worm gear
26. Bushing
27. Gasket
28. Auger shaft
29. Thrust washer
30. Bushing
31. Oil seal
32. Left case half
33. Grease plug
34. Impeller bolts
35. Impeller assy.
36. Locknuts
37. Auger shear bolts
38. Left auger assy.
39. Right auger assy.

Fig. CR24—Exploded view of track drive assembly. Track drive wheels (15) mount on axle shaft (46—Fig. CR22). Foot pedal (8) operates weight transfer system.

1. Mounting bracket
2. Pivot pin
3. Jam nuts
4. Track frame
5. Snap ("E") rings
6. Roll pin
7. Foot pedal spring
8. Foot pedal
9. Frame support bearing
10. Idler axle shaft
11. Track adjusting bolt
12. Spacer
13. Track idler wheel
14. Rubber track
15. Track drive wheel

hub) by reversing removal procedure. Check friction wheel adjustment as outlined in preceding paragraph before reinstalling bottom panel.

TRACK ASSEMBLY

Refer to exploded view of track assembly in Fig. CR24. Disassembly and overhaul is evident after inspection of unit and reference to exploded view. Readjust track as outlined in a previous paragraph.

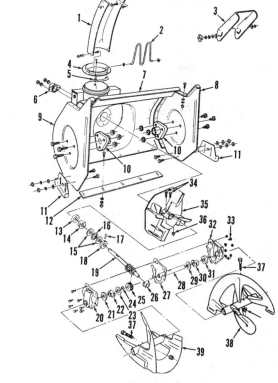

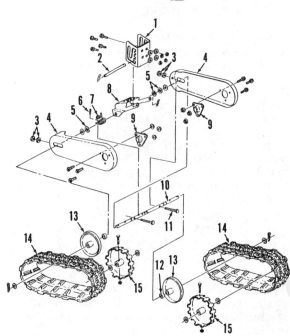

CRAFTSMAN

Model	Engine Make	Engine Model	Self-Propelled?	No. of Stages	Scoop Width
536.885000	Tecumseh	HM100	Yes	2	32 in. (813 mm)
536.887000	Tecumseh	HM100	Yes	2	32 in. (813 mm)

LUBRICATION

ENGINE

Refer to Tecumseh engine section for engine lubrication information. Recommended fuel is regular or low-lead gasoline.

DRIVE MECHANISM

Recommended lubrication period is after every 10 hours of operation and before storing unit at end of season. Oil the drive chains and sprockets with SAE 10W-30 oil. Using a hand grease gun, refer to Fig. CR32 and lubricate the grease fittings (4 & 23) on the spindle assembly, the shaft and gear assembly and the hex shaft once each season. Be careful not to get any grease or oil on the friction wheel drive plate or friction wheel; clean these parts thoroughly if oil or grease comes in contact. Lubricate all moving part pivot points with SAE 10W-30 oil. Using a hand operated grease gun, lubricate the auger shaft grease fittings (17—Fig. CR35) after every 10 hours of operation and each time an auger shear bolt (15) is replaced. On track drive, coat weight transfer plate parts with clinging grease such as "Lubriplate" every 10 hours of operation and before storage.

AUGER GEARBOX

Once each season, remove grease plug (18—Fig. CR36) and, if grease is not visible, use a wire as a dipstick to check for grease in gearbox. If gearbox is dry, fill to plug hole level with a recommended lubricant such as Sunoco Prestige 740 AAEP, Shell Alvania EPR, Mobile Mobilplex EP23 or Texaco Marfak AP.

ADJUSTMENTS

AUGER HEIGHT

Lower the adjusting skids to raise auger scraper bar on rough surfaces. On paved surfaces, the skids can be raised to bring the auger assembly down. Keep in mind that the scraper bar below the auger should be 1/8 inch (3 mm) above sidewalk or other paved area to be cleaned. This clearance can be obtained by placing spacers under the scraper bar, then adjusting the skids so they contact paved surface.

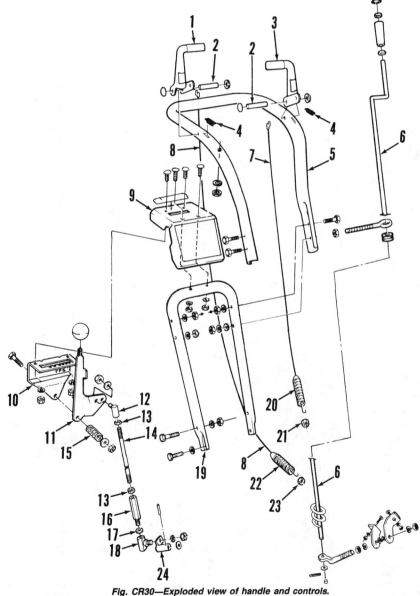

Fig. CR30—Exploded view of handle and controls.

1. Auger clutch lever
2. Lever pivot pin
3. Drive clutch lever
4. Bumper pad
5. Upper handle
6. Chute crank
7. Traction drive cable
8. Auger clutch cable
9. Control panel
10. Speed control bracket
11. Speed control lever
12. Ball joint
13. Jam nut
14. Speed control rod
15. Lever tension spring
16. Adapter
17. Lockwasher
18. Ball joint
19. Lower handle
20. Drive cable spring
21. Drive cable nut
22. Auger clutch spring
23. Locknut
24. Shifter bracket

To compensate for wear, the scraper bar position is adjustable, and the bar can be reversed for longer wear or replaced when both edges are worn. Be sure the bar is adjusted parallel with the working surface and for proper clearance above skid shoes.

DRIVE BELTS

The traction drive belt is spring loaded and does not require adjustment; replace belt if it is slipping. The auger drive belt can be adjusted after removing belt cover, then proceed as follows: Loosen nut on auger idler pulley (17—Fig. CR31) and move idler toward belt about 1/8 inch (3 mm). Tighten nut, engage auger drive clutch lever and check belt adjustment; belt should deflect about 1/2 inch (12.7 mm) with moderate pressure. Reset idler pulley until proper belt tension is obtained with drive engaged.

CONTROL CABLES

Refer to Fig. CR30 and with "z" fitting on upper end of auger clutch cable (8) disconnected from lever (1), move lever forward to contact plastic bumper (4); the center of the "z" fitting should be between the top and the center of the hole in the lever when pulling slack up out of cable. If not, push cable through auger clutch spring (22) and locknut (23) on lower end of cable to align "z" fitting with hole in lever. Then, pull cable up through springs and reconnect "z" fitting to lever. Repeat this procedure to adjust traction drive cable (8).

FRICTION WHEEL

Tip snowthrower forward onto front of auger housing. Remove rear panel and bottom panel from main frame. Position shifter lever in first (1) gear and check position of friction wheel on friction wheel drive plate. The left side of friction wheel should be 2-5/8 inches (66.7 mm) from outer edge of friction wheel drive plate as shown in Fig. CR33. If necessary to adjust position of friction wheel, refer to Fig. CR30 and loosen jam nut (13) on speed control rod at top of adapter (16). Remove ball joint (18) from shifter bracket (24) and turn adapter (16) to obtain correct friction wheel position with ball joint reinstalled in shifter bracket. Tighten jam nut and reinstall panels.

TRACKS

Refer to Fig. CR37 and loosen cap screws (13) on each side of track assembly. Rotate adjusting cam (14) equally at each side of assembly so deflected distance between top of side plate and inside of track is not greater than 2 inches (50.8 mm). Uneven adjustment of adjusting cams will result in a twist in the track. Repeat adjustment for opposite track assembly.

OVERHAUL

ENGINE

Engine make and model numbers are listed at beginning of section. Refer to Tecumseh engine section for engine overhaul.

FRICTION DRIVE

Overhaul is evident after inspection of unit and reference to the exploded view of friction drive transmission in Fig. CR32. Refer also to following paragraph on drive belt removal. During reassembly, be careful not to get any grease on the friction wheel drive plate or friction wheel and drive belts. On reassembly, check adjustment of drive belts and friction wheel as outlined in preceding paragraphs.

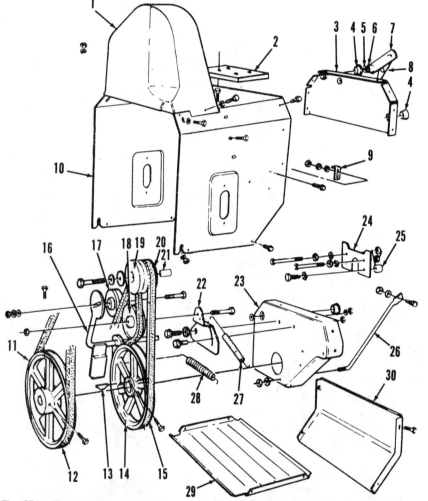

Fig. CR31—Exploded view of belt drive mechanism. Refer to Fig. CR32 for friction wheel drive components.

1. Belt cover
2. Engine riser plate
3. Rear main frame plate
4. Plastic bushings
5. Flat washer
6. Roll pin
7. Auger clutch lever
8. Clutch cable lever
9. Cable adjustment bracket
10. Main (motor mount) frame
11. Auger pulley
12. Auger belt
13. Drive pulley key
14. Traction drive pulley
15. Traction drive belt
16. Auger idler bracket & brake
17. Auger idler pulley
18. Drive idler pulley
19. Auger drive pulley
20. Traction drive pulley
21. Engine pulley spacer
22. Drive idler bracket
23. Drive spindle plate
24. Belt guide plate
25. Spacer
26. Plate support rod
27. Drive idler spring
28. Auger brake spring
29. Bottom cover panel
30. Rear cover panel

DRIVE BELTS

To remove auger belt, remove upper (belt) cover and remove auger drive engine pulley and belt from engine. Disconnect deflector chute crank mechanism at deflector. Place blocks under rear end of snowthrower frame and remove the two top bolts and loosen the bottom two bolts holding auger housing to engine mount frame. Separate auger housing from engine mount by pivoting on bottom bolts. Bend belt retainer tabs (Fig. CR34) away from auger pulley only as necessary to remove belt. Install new belt on auger pulley and while holding it tightly in pulley groove, bend retainer tabs back into position leaving 1/16 to 1/8 inch (1.5-3 mm) clearance between tabs and belt. Pull up on belt, engage auger clutch and swing engine mount frame back into position. Replace two top housing bolts and tighten the two lower bolts. Install auger drive belt on engine pulley and reinstall pulley on engine shaft. Adjust auger belt as outlined in a preceding paragraph.

To remove traction drive belt, follow steps outlined to remove auger belt except for removing belt from auger drive pulley. Pull spring loaded traction idler pulley back and slip belt past idler pulley and off of engine pulley. It may be necessary to remove auger brake assembly to remove belt from traction drive pulley. Install new belt and reassemble in reverse of disassembly procedure. Adjust belts as previously outlined.

AUGER SHEAR BOLT

The augers are driven by the auger shaft via shear bolts (15—Fig. CR35) that are designed to shear to protect the machine if an object becomes lodged in the auger housing. Only original equipment replacement shear bolts should be used for safety reasons. When replacing a shear bolt, be sure to install new shear bolt spacer (14) and, using a hand operated grease gun, lubricate auger shaft at grease fittings (17).

AUGER ASSEMBLY

To remove auger, first remove auger drive belt as previously outlined. Remove auger drive pulley (1—Fig. CR35) and impeller shaft bearing (2) from rear of housing. Unbolt auger shaft bearings at each side of housing and remove the assembly. Remove the auger drive shear bolts (15) and slide the auger flights from auger shaft.

With auger flights removed, refer to Fig. CR36 for exploded view of auger drive gearbox. Disassembly and overhaul procedure is evident after inspection of unit and reference to Fig. CR36. When reassembling, use a maximum of 3-1/4 ounces (92 grams) of grease. Recommended lubricants are Sunoco Prestige 740 AEP, Shell Alvania EPR, Mobile Mobilplex EP23 or Texaco Marfak AP.

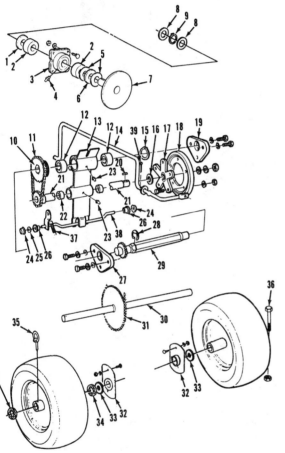

Fig. CR32—Exploded view of friction wheel drive mechanism. Friction wheel spindle housing (3) mounts in spindle plate (23—Fig. CR31).

1. Seal
2. Needle roller bearing
3. Spindle housing
4. Grease fitting
5. Thrust bearing races
6. Needle thrust bearing
7. Friction wheel drive plate & spindle
8. Thrust washer
9. Snap ring
10. Hex shaft drive sprocket
11. Roller chain
12. Hex shaft sliding bearing
13. Speed control bracket
14. Speed control yoke
15. Snap ring
16. Thrust washer
17. Friction wheel hub
18. Friction wheel
19. Hex shaft end bearing
20. Square key
21. Gear & shaft assy.
22. Bearing
23. Grease fitting
24. Push on end cap
25. Flat washer
26. Bushing
27. Hex shaft end bearing
28. Snap ring
29. Hex shaft
30. Axle shaft
31. Axle drive chain
32. Axle bearing bracket
33. Thrust washer
34. Grip type retainer
35. Klick pin
36. Bolt
37. Yoke drive
38. Yoke return spring
39. Cotter pin

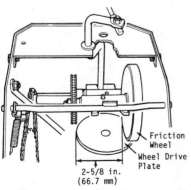

Fig. CR33—With shift (speed control) lever in first speed position, friction wheel position should be as shown.

Fig. CR34—Bend auger belt retainer tabs back to remove auger belt. On reassembly, bend tabs to clear belt as shown.

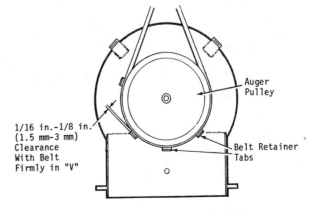

FRICTION DRIVE

To replace friction wheel, refer to Fig. CR32 and proceed as follows: Stand snowthrower up on auger housing end and remove rear panel and bottom panel. Hold hex shaft (29) from turning with a 11/16-inch open-end wrench and remove the nut and washers holding friction wheel (18) and hub assembly (17) on shaft. Remove the friction wheel and hub assembly. Unbolt wheel from hub and install new friction wheel (cup side away from hub). Reassemble by reversing disassembly procedure, being sure that square key (20) is in place in friction wheel hub and that thrust washer (16) is completely on the shoulder of the shaft, not trapped between the shoulder and threaded part of shaft.

TRACK ASSEMBLY

Refer to exploded view of track assembly in Fig. CR37. Disassembly and overhaul is obvious after inspection of unit and reference to exploded view. Readjust track as outlined in previous paragraph.

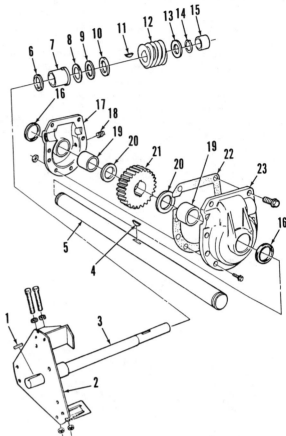

Fig. CR36—Exploded view of auger gearbox and impeller assembly.

1. Square key
2. Impeller fan
3. Impeller shaft
4. Woodruff key
5. Auger shaft
6. Square cut "O" ring
7. Impeller shaft bearing
8. Thrust bearing race
9. Needle thrust bearing
10. Thrust bearing race
11. Key
12. Worm gear
13. Thrust washer
14. Snap ring
15. Impeller shaft bearing
16. Auger shaft seal
17. L.H. housing half
18. Grease plug
19. Auger shaft bushing
20. Thrust washer
21. Auger shaft gear
22. Gasket
23. R.H. housing half

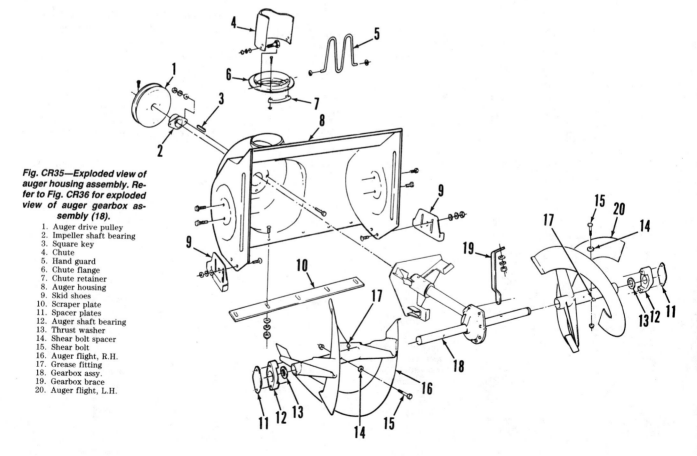

Fig. CR35—Exploded view of auger housing assembly. Refer to Fig. CR36 for exploded view of auger gearbox assembly (18).

1. Auger drive pulley
2. Impeller shaft bearing
3. Square key
4. Chute
5. Hand guard
6. Chute flange
7. Chute retainer
8. Auger housing
9. Skid shoes
10. Scraper plate
11. Spacer plates
12. Auger shaft bearing
13. Thrust washer
14. Shear bolt spacer
15. Shear bolt
16. Auger flight, R.H.
17. Grease fitting
18. Gearbox assy.
19. Gearbox brace
20. Auger flight, L.H.

Fig. CR37—Exploded view of left track assembly and weight transfer mechanism.

1. Track plate, inner R.H.
2. Connecting rod
3. Track plate, inner L.H.
4. Axle drive gear
5. Axle shaft
6. Inner axle hub
7. Pedal spring
8. Weight transfer pedal
9. Pedal bracket
10. Track drive wheel
11. Track drive sprocket
12. Track plate, outer L.H.
13. Cap screw
14. Adjusting cam
15. Outer axle hub
16. Axle sprocket
17. Rubber track
18. Track idler
19. Track idler shaft
20. Track drive chain

CRAFTSMAN

Model 247.886510	Engine Make Tecumseh	Engine Model HSSK50	Self- Propelled? Yes	No. of Stages 2	Scoop Width 23 in. (584 mm)

LUBRICATION

ENGINE

Refer to Tecumseh 4-stroke engine section for engine lubrication information. Recommended fuel is regular or unleaded regular gasoline.

DRIVE MECHANISM

Lubricate drive chains after every 10 hours of operation or at least once each season using a good quality chain oil. Auger shaft bushings and traction drive shaft bushings should be oiled after every 10 hours of operation with SAE 30 motor oil. Lubricate all control pivot and slide points for smooth operation. Once each season or when necessary to replace auger flight shear bolts, remove shear bolts and squirt oil in hole and at ends of auger flights, then turn auger flight to distribute oil. Remove wheels once each season and grease axle shafts.

AUGER GEARBOX

The auger gearbox is factory lubricated and should not require further lubrication unless disassembled for overhaul. Recommended lubricant is Shell Alvania grease EPR00.

MAINTENANCE

Inspect belts and pulleys and renew belts if excessively worn, frayed or burned. Renew pulleys if belt grooves are worn or damaged, or if bearings in idler pulleys are loose or rough. Check drive chains and sprockets and renew if excessively worn or damaged.

ADJUSTMENTS

AUGER HEIGHT

Auger height can be adjusted by loosening skid shoe mounting bolts and repositioning the skid shoes. Auger height should be at lowest setting (skid shoes raised) when operating over smooth, paved surfaces and toward the highest setting (skid shoes lowered) when operating over rough surfaces. Be sure skid shoes are adjusted equally.

SHIFT ROD

Disconnect shift control rod (6—Fig. CR40) from shift lever (5) and place both the transmission and shift lever in neutral position. Turn shift control rod in or out of ferrule (14) at bottom of rod so upper end can be inserted in hole in shift lever. Secure rod in lever with cotter pin.

DRIVE CLUTCH

Unhook spring (8—Fig. CR40), at lower end of clutch cable, from drive clutch arm (11). Turn adjusting nut (9) so hook of spring is aligned with center of hole in drive clutch arm (11) with drive control lever (1) released. Hook spring into drive clutch arm (spring should be loose) and engage clutch grip. The clutch should be engaged and move the snowthrower without belt slippage.

Fig. CR40 — Exploded view of handle assembly and controls.

1. Drive control lever
2. Auger control lever
3. Auger control cable
4. Drive control cable
5. Shift control lever
6. Shift control rod
7. Chute crank
8. Drive clutch spring
9. Adjusting nut
10. Shoulder bolt
11. Drive clutch arm
12. Drive clutch link rod
13. Drive return spring
14. Shift rod ferrule
15. Shift link arm
16. Auger cable ferrule
17. Auger clutch bracket
18. Drive pulley support
19. Frame cover
20. Auger return spring

AUGER CLUTCH

To check adjustment, remove belt cover and hold auger control lever against handle grip. The outer end of the slot in the end of auger brake linkage (25—Fig. CR41) should be against the spacer on the auger belt idler as shown in Fig. CR42. If not, disconnect auger control cable ferrule (16—Fig. CR40) from auger clutch bracket (17). Turn ferrule on cable end until proper adjustment is obtained with ferrule connected to auger clutch bracket.

OVERHAUL

ENGINE

Engine make and model numbers are listed at beginning of this section. Re-

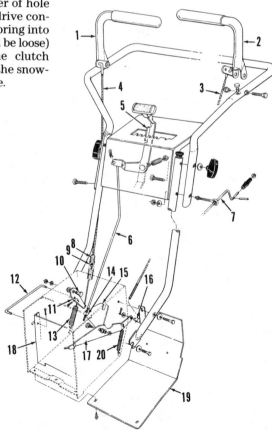

fer to Tecumseh 4-stroke engine section for engine overhaul information.

DRIVE BELTS

Remove the deflector chute crank at chute and remove upper belt cover. Remove the two shoulder bolt belt guides (4—Fig. CR41) at engine pulley. Remove the top bolts attaching auger housing to snowthrower frame assembly and loosen but do not remove the two bottom bolts. Lift up on auger belt to pull auger housing off of snowthrower frame and separate auger housing from the main unit.

To remove auger drive belt, remove the four shoulder bolt belt guides (11—Fig. CR45) at auger pulley. The auger belt can now be removed from the bottom pulley. To remove traction drive belt, disconnect drive idler spring (8—Fig. CR41) at engine plate and remove belt from engine pulley and bottom drive pulley. Install new belt(s) and reassemble by reversing disassembly procedure. Be sure pin on end of auger clutch bracket engages slot in auger brake bracket as shown in Fig. CR43 before installing auger housing mounting bolts. Adjust auger and drive clutches as outlined in preceding paragraphs.

FRICTION WHEEL

To renew friction disc, proceed as follows: Move shift lever to reverse position. Tip snowthrower up so it rests on front of auger housing and remove bottom cover, then remove friction wheel (38—Fig. CR44) from wheel adapter (40). Place new wheel with cup side away from adapter and install retaining bolts finger tight. Turn the wheel to be sure it does not wobble, then tighten retaining bolts. Replace bottom cover.

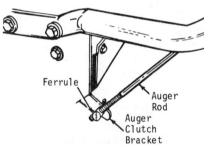

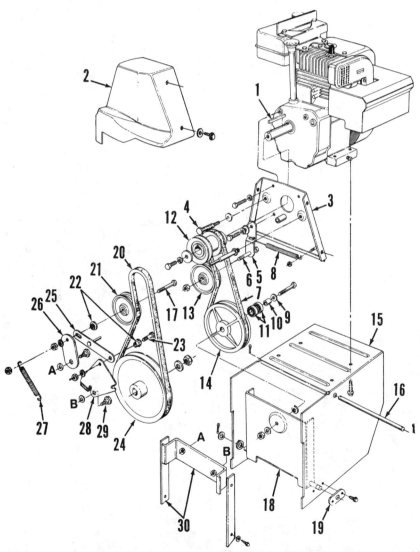

Fig. CR41—Exploded view of drive belt and auger belt drive components and engine/frame assembly. Back side of drive pulley (14) has drive surface for friction wheel; depressing drive clutch lever moves pulley against friction wheel (38—Fig. CR44).

1. Engine pulley key
2. Belt cover
3. Engine plate
4. Shoulder bolt belt guide
5. Drive clutch idler bracket
6. Idler bolt
7. Drive belt
8. Drive idler spring
9. Belleville washer
10. Spacer
11. Ball bearings
12. Engine pulley
13. Drive belt idler
14. Drive pulley
15. Main frame
16. Sliding bracket rod
17. Auger idler bolt
18. Drive pulley support
19. Support axle bracket
20. Auger belt
21. Auger belt idler
22. Shoulder spacers
23. Brake pivot bolt
24. Auger pulley
25. Brake linkage
26. Auger idler bracket
27. Auger idler spring
28. Brake bracket
29. Shoulder bolt
30. Blower housing parts

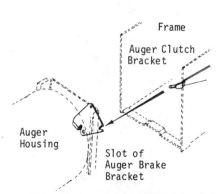

Fig. CR42—End of slot in brake linkage (25—Fig. CR41) should be against spacer on the auger belt idler when auger control lever is depressed against snowthrower handle. Refer to text for adjustment procedure.

Fig. CR43—Pin on end of auger clutch bracket must engage slot in auger brake bracket when reassembling snowthrower.

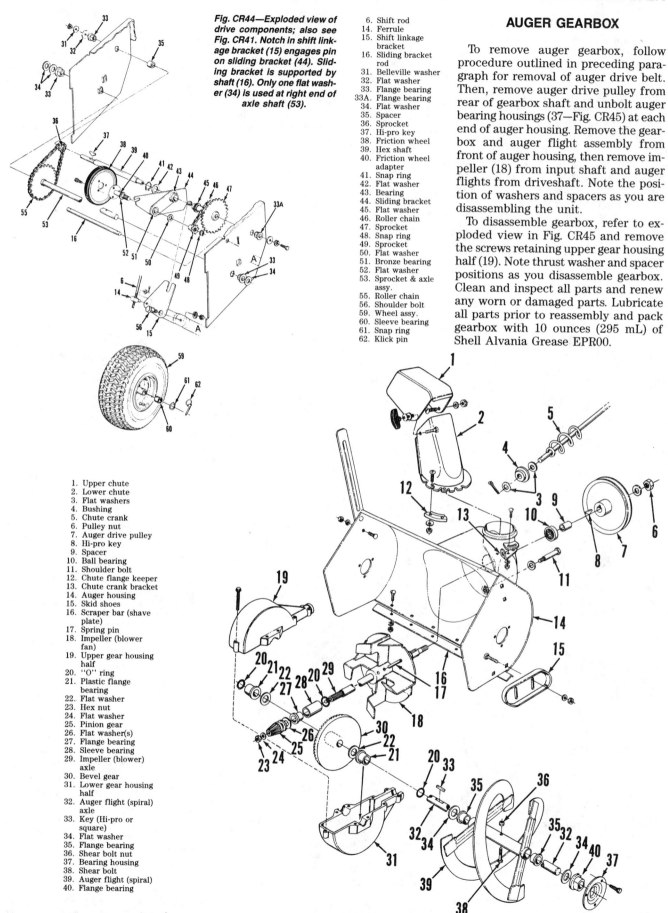

6. Shift rod
14. Ferrule
15. Shift linkage bracket
16. Sliding bracket rod
31. Belleville washer
32. Flat washer
33. Flange bearing
33A. Flange bearing
34. Flat washer
35. Spacer
36. Sprocket
37. Hi-pro key
38. Friction wheel
39. Hex shaft
40. Friction wheel adapter
41. Snap ring
42. Flat washer
43. Bearing
44. Sliding bracket
45. Flat washer
46. Roller chain
47. Sprocket
48. Snap ring
49. Sprocket
50. Flat washer
51. Bronze bearing
52. Flat washer
53. Sprocket & axle assy.
55. Roller chain
56. Shoulder bolt
59. Wheel assy.
60. Sleeve bearing
61. Snap ring
62. Klick pin

AUGER GEARBOX

To remove auger gearbox, follow procedure outlined in preceding paragraph for removal of auger drive belt. Then, remove auger drive pulley from rear of gearbox shaft and unbolt auger bearing housings (37—Fig. CR45) at each end of auger housing. Remove the gearbox and auger flight assembly from front of auger housing, then remove impeller (18) from input shaft and auger flights from driveshaft. Note the position of washers and spacers as you are disassembling the unit.

To disassemble gearbox, refer to exploded view in Fig. CR45 and remove the screws retaining upper gear housing half (19). Note thrust washer and spacer positions as you disassemble gearbox. Clean and inspect all parts and renew any worn or damaged parts. Lubricate all parts prior to reassembly and pack gearbox with 10 ounces (295 mL) of Shell Alvania Grease EPR00.

1. Upper chute
2. Lower chute
3. Flat washers
4. Bushing
5. Chute crank
6. Pulley nut
7. Auger drive pulley
8. Hi-pro key
9. Spacer
10. Ball bearing
11. Shoulder bolt
12. Chute flange keeper
13. Chute crank bracket
14. Auger housing
15. Skid shoes
16. Scraper bar (shave plate)
17. Spring pin
18. Impeller (blower fan)
19. Upper gear housing half
20. "O" ring
21. Plastic flange bearing
22. Flat washer
23. Hex nut
24. Flat washer
25. Pinion gear
26. Flat washer(s)
27. Flange bearing
28. Sleeve bearing
29. Impeller (blower) axle
30. Bevel gear
31. Lower gear housing half
32. Auger flight (spiral) axle
33. Key (Hi-pro or square)
34. Flat washer
35. Flange bearing
36. Shear bolt nut
37. Bearing housing
38. Shear bolt
39. Auger flight (spiral)
40. Flange bearing

Fig. CR45—Exploded view of auger housing, gearbox and auger flight assembly.

CRAFTSMAN

Model 247.886700	Engine Make Tecumseh	Engine Model HMSK80	Self- Propelled? Yes	No. of Stages 2	Scoop Width 26 in. (660 mm)

LUBRICATION

ENGINE

Refer to Tecumseh 4-stroke engine section for engine lubrication information. Recommended fuel is regular or unleaded regular gasoline.

DRIVE MECHANISM

Lubricate drive chains after every 10 hours of operation or at least once each season using a good-quality chain oil. Auger shaft bushings and traction drive shaft bushings should be oiled after every 10 hours of operation with SAE 30 motor oil. Lubricate all control pivot and slide points to ensure smooth operation. Once each season or when necessary to replace auger flight shear bolts, remove shear bolts and squirt oil in hole and at ends of auger flights, then turn auger flight to distribute oil. Keep weight transfer moving parts lubricated with a clinging grease such as "Lubriplate."

AUGER GEARBOX

The auger gearbox is factory lubricated and should not require further lubrication unless disassembled for overhaul. Recommended lubricant is Shell Alvania grease EPR00.

TRANSMISSION

The gear type transmission is factory lubricated and should not require further lubrication unless disassembled for overhaul. Refer to Peerless transmission section of this manual for information.

MAINTENANCE

Inspect belts and pulleys and renew belts if excessively worn, frayed or burned. Renew pulleys if belt grooves are worn or damaged or if bearings in idler pulleys are loose or rough. Check drive chains and sprockets and renew if excessively worn or damaged.

ADJUSTMENTS

AUGER HEIGHT

Auger height can be adjusted by loosening skid shoe mounting bolts and repositioning the skid shoes. Auger height should be at lowest setting (skid shoes raised) when operating over smooth, paved surfaces and toward the highest setting (skid shoes lowered) when operating over rough surfaces. Be sure skid shoes are adjusted equally.

SHIFT ROD

Snowthrower should roll easily with shift lever in neutral position, and the wheels should lock up with shift lever in reverse. If not, loosen jam nuts (11 & 13—Fig. CR50) (bottom nut is left hand thread) at each end of shift rod (12). Move transmission shift arm (15) to find neutral position and turn shift rod to bring gearshift lever (6) in line with neutral "N" position on console.

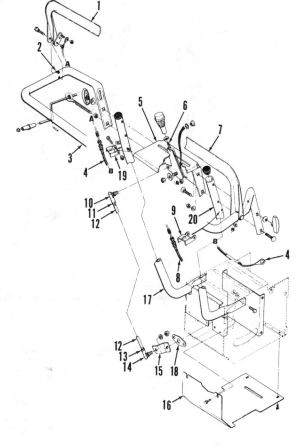

Fig. CR50—Exploded view of handle assembly and snow-thrower controls.

1. Drive clutch lever
2. Rubber bumper
3. Upper handle
4. Drive clutch cable
5. Control panel
6. Gearshift lever
7. Auger clutch lever
8. Auger clutch cable
9. Cable bracket
10. Ball joint assy.
11. Jam nut
12. Shift rod
13. Jam nut
14. Ball joint assy.
15. Shift arm
16. Bottom cover
17. Lower handle
18. Shift support
19. Cable bracket

CLUTCH CABLES

There should be a small amount of slack in the clutch control cables with clutch levers fully forward against bumpers on handles. To check adjustment, place a nickel on top of the rubber bumper at either handle and release the clutch lever. All slack in the cable should be removed without compressing the rubber bumper. If not, loosen and adjust jam nuts on cable housing at cable bracket (9 & 19—Fig. CR50) to obtain correct adjustment, then tighten jam nuts.

OVERHAUL

ENGINE

Refer to Tecumseh 4-stroke engine section for engine overhaul information.

DRIVE BELTS

Remove the deflector chute crank at chute and remove upper belt cover. Loosen the upper left-hand belt guide and remove the right-hand belt guide at

engine crankshaft pulley and roll belt off of pulley. Unhook the auger control cable from pin on auger brake link (7—Fig. CR52) after removing hairpin clip from brake link pin. Place weight transfer lever in "packed snow" position. Remove the top bolts attaching auger housing to snowthrower frame assembly and loosen but do not remove the two bottom bolts. Lift up on auger belt to pull auger housing off of snowthrower frame and separate auger housing from the main unit.

To remove auger drive belt, remove shoulder bolt belt guides (21—Fig. CR52) at auger pulley. Roll belt off of bottom (auger) pulley. Remove the second hairpin clip from brake link pin, push idler pulley to right and work belt out between link pin and idler.

To remove traction drive belt, disconnect auger cable (16—Fig. CR51) extension spring at idler arm (7) and unhook return spring (11) from arm. Loosen the two set screws (3 & 5) in engine PTO (camshaft) pulley (4) and remove pulley with belt. Remove transmission pulley (14) with belt.

Install new belt(s) and reassemble by reversing disassembly procedure. Adjust auger and drive clutches as outlined in preceding paragraphs.

TRANSMISSION

The snowthrower is equipped with a Peerless Model 700-031 transmission with six forward speeds and one reverse speed. With transmission removed, refer to Peerless transmission section in this manual for overhaul information.

AUGER GEARBOX

To remove auger gearbox, follow procedure outlined in a preceding paragraph for removal of auger drive belt. Then, remove auger drive pulley (16—Fig. CR52) from rear of gearbox shaft and unbolt auger bearing supports (48) at each end of auger housing. Remove the gearbox and auger flight assembly from front of housing, then remove impeller (37) from input shaft and auger flights from driveshaft. Note the position of washers and spacers as you are disassembling the unit.

To disassemble gearbox, refer to exploded view in Fig. CR52 and remove the screws retaining upper gear housing half (26). Note thrust washer and spacer positions as you disassemble gearbox. Clean and inspect all parts and renew any worn or damaged parts. Lubricate all parts prior to reassembly and pack gearbox with 10 ounces (295 mL) of Shell Alvania Grease EPR00.

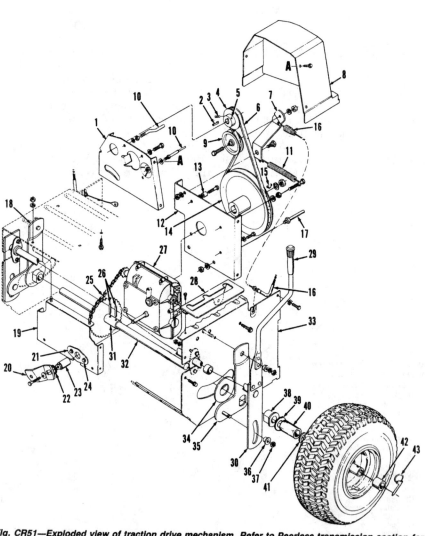

Fig. CR51—Exploded view of traction drive mechanism. Refer to Peerless transmission section for information on gear type transmission. Traction drive is powered from pulley (4) on engine auxiliary PTO shaft.

1. Engine plate	23. Shift arm extension	32. Wheel axle
2. Engine pulley key	24. Hex head screw	33. Main frame assy.
3. Set screw	25. Roller chain	34. Slide washer
4. Auxiliary PTO pulley	26. Spring pins	35. Side arm assy.
5. Set screw	27. Transmission assy.	36. Shoulder spacer
6. Drive belt	28. Weight transfer plate	37. Locknut
7. Idler arm	29. Handle grip	38. Flange bearing
8. Belt cover	30. Weight transfer handle	39. Flat washer
9. Flat idler	31. Sprocket & hub assy.	40. Spacer
10. Belt guard bolts	13. Spacer	41. Flat washer
11. Return spring	14. Transmission pulley	42. Sleeve bearing
12. Transmission support plate	15. Key	43. Klick pin
	16. Auger cable & spring	
	17. Belt guard bolt	
	18. Side arm assy.	
	19. Back plate	
	20. Shift arm	
	21. Shift support	
	22. Flat washer	

1. Upper chute
2. Lower chute
3. Chute flange keeper
4. Chute stop
5. Chute crank bracket
6. Idler bracket
7. Brake link
8. Auger brake spring
9. Shoulder spacer
10. Flat idler
11. Auger drive belt
12. Square key
13. Engine pulley
14. Hi-pro key
15. Shoulder spacer
16. Auger drive pulley
17. Shoulder bolt
18. Brake bracket
19. Spacer
20. Ball bearing
21. Shoulder bolt belt guide
22. Bushing
23. Chute crank rod
24. Auger (blower) housing
25. Skid shoe
26. Upper gear housing half
27. "O" ring
28. Plastic flange bushing
29. Flat washer
30. Locknut
31. Flat washer
32. Pinion gear
33. Flat washers
34. Flange bearing
35. Sleeve bearing
36. Impeller (blower) axle
37. Impeller (blower fan)
38. Spring pin
39. Lower gear housing half
40. Auger (spiral) axle
41. Flat washer
42. Square key
43. Spacer
44. Flange bearing
45. Auger flight (spiral assy.)
46. Shear bolt
47. Flange bearing
48. Bearing support
49. Scraper bar (shave plate)
50. Bevel gear

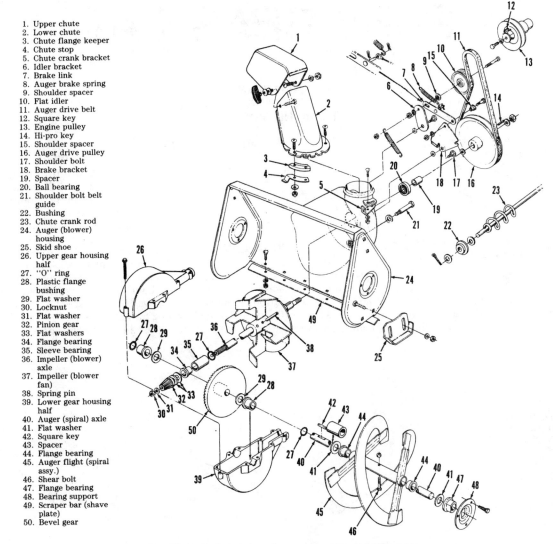

Fig. CR52—Exploded view of auger assembly and drive parts.

CRARY BEAR CAT

CRARY COMPANY
P.O. Box 1779
Fargo, ND 58107

Model	Engine Make	Engine Model	Self-Propelled?	No. of Stages	Scoop Width
624	Tecumseh	HSK60	Yes	2	24 in. (610 mm)
824	Tecumseh	HMSK80	Yes	2	24 in. (610 mm)
1028	Tecumseh	HMSK100	Yes	2	28 in. (711 mm)

LUBRICATION

ENGINE

Refer to Tecumseh engine section for engine lubrication requirements. Recommended fuel is unleaded gasoline.

DRIVE MECHANISM

The snowthrower should be greased at least once each season or after every 20 hours of operation. Remove the round access plugs (5—Fig. CB11) on top of main frame in front of engine and on the rear cover plate for access to lubrication fittings (zerks) on impeller jackshaft and ground drive jackshaft. Use a hand operated grease gun at these fittings, then replace the plugs. On Models 624 and 824, with belt guard removed, grease the slip clutch casting; be care-

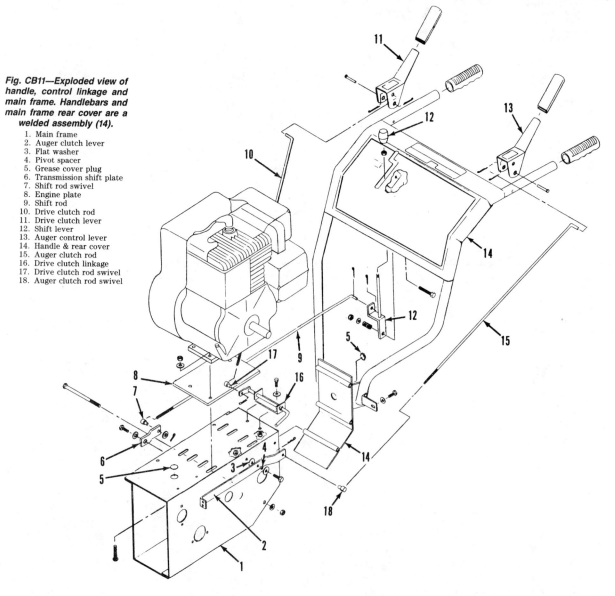

Fig. CB11—Exploded view of handle, control linkage and main frame. Handlebars and main frame rear cover are a welded assembly (14).

1. Main frame
2. Auger clutch lever
3. Flat washer
4. Pivot spacer
5. Grease cover plug
6. Transmission shift plate
7. Shift rod swivel
8. Engine plate
9. Shift rod
10. Drive clutch rod
11. Drive clutch lever
12. Shift lever
13. Auger control lever
14. Handle & rear cover
15. Auger clutch rod
16. Drive clutch linkage
17. Drive clutch rod swivel
18. Auger clutch rod swivel

ful not to get grease on slip clutch (torque clutch) discs. Grease axle and auger bearings at each side. Oil chains each season or more frequent if needed. Through access cover on bottom of transmission case, apply grease to ring and pinion gears after 100 hours or after 4 seasons of use. The Peerless transmission is lubricated at the factory and does not require service lubrication; refer to Peerless transmission section. Oil chains and apply grease behind the shift lever once each season. Lubricate chute disc, top of bucket extension and plastic chute bearings with lithium grease once each season.

MAINTENANCE

Check all bolts after every 40 hours of operation or at least once each season for proper tightness. Run machine a few minutes at medium speed and a few minutes at idle to prevent freeze-up of feeder, impeller or starter. Inspect belts, guides, pulleys, etc., for wear and tightness; replace belts if cracked, frayed or glazed.

ADJUSTMENTS

AUGER HEIGHT

Auger height can be adjusted by loosening skid shoe mounting bolts and repositioning skid shoes. Skid shoes should be located in the upper position (bucket down) when operating on smooth, hard surfaces and in the lower position (bucket up) when operating over rough surfaces such as unpaved driveways.

AUGER TORQUE CLUTCH

The snowthrower is equipped with a torque (slip) clutch to protect the auger from damage should it hit an obstruction. If unit stops, check for obstruction; stop engine, remove spark plug wire and check auger. To check automatic torque clutch, push down on feeder blade; a steady force of 175-200 pounds (79.4-90.7 kg) should not cause the clutch to slip, but a sudden application of this amount of force should cause slippage. To adjust clutch, refer to Fig. CB13 and turn tension adjusting nut (23 or 35) clockwise to increase clutch tension or counterclockwise to decrease tension.

WHEEL DRIVE

To check wheel drive, block wheels off of ground and with engine running, engage and disengage wheel drive clutch several times. If wheels do not turn when clutch is engaged, or if wheels do not stop turning when clutch is disengaged, refer to Fig. CB11 and adjust drive clutch rod (10) as follows: Remove spring clip pin and washer and pull swivel (17) from linkage (16) arm. Turn swivel clockwise (shorten rod) to increase clutch pressure or counterclockwise (lengthen rod) to allow clutch to disengage. Insert swivel in plate and install washer and clip pin. Remove block from under transmission and recheck clutch operation with wheels on ground; repeat adjustment operation if needed. If clutch rod adjustment does not provide enough range for proper clutch operation, check belt adjustment.

AUGER DRIVE CLUTCH

If auger drive belt slips, refer to Fig. CB11 and decrease clutch rod (15) length by turning swivel (18) clockwise; lengthen clutch rod if auger does not stop turning when clutch is disengaged. Check clutch operation while throwing snow and readjust clutch rod if necessary.

TRANSMISSION SHIFT ROD

Check to see that transmission is in neutral with shift lever in neutral position; the snowthrower should roll back and forth easily. If not, remove clip pin and washer from swivel (7—Fig. CB11) and remove swivel from shift plate (6). Move shift plate to transmission neutral position; the unit should roll back and forth easily. With transmission in neutral, place handlebar shift lever (12) at neutral position and turn swivel to lengthen or shorten rod until swivel will fit into shift plate hole. Then, reinstall washer and clip pin and recheck adjustment.

DRIVE CHAINS

All drive chains can be adjusted after removing the belt guard, then proceeding as follows:

The auger drive chain tension can be adjusted after loosening the four clutch bracket (15—Fig. CB13) retaining bolts on Models 624 and 824 or the right-hand and left-hand clutch brackets (26 & 28) retaining bolts on Model 1024. The mounting holes are elongated for adjust-

ment purposes; push bracket up to tighten chain and tighten retaining bolts to 25 ft.-lbs. (34.2 N·m). Check to see that the clutch shaft remains parallel and the sprockets are aligned.

Adjust impeller drive chain tension by loosening the nuts on jackshaft housing (17—Fig. CB14) and housing plate (1) and moving the jackshaft assembly as required to adjust chain tension. Make sure the exposed slot on each side of jackshaft housing is the same and tighten nuts to 25 ft.-lbs. (34.2 N·m).

To adjust auger drive chain, loosen nuts holding the two auger bearing stub shaft plates (4—Fig. CB12) on each side of feeder housing, rotate plates forward to tighten chain, then retighten the nuts.

To adjust transmission drive chain, loosen the nuts holding transmission jackshaft housing (6—Fig. CB14) and plate (4) mounting bolts, then slide jackshaft to rear to tighten chain. Make sure the slots at each side are the same and tighten nuts to 25 ft.-lbs. (34.2 N·m).

BELT ADJUSTMENT

When replacing belts or when necessary to adjust stretched or worn belts, remove the belt guard and proceed as follows: Disconnect swivel ends (17 & 18—Fig. CB11) on auger and drive clutch rods. Pull all belt slack up over engine pulley; if one belt has more slack length, loosen engine mounting bolts and slide engine so the belts have equal slack length. Tighten engine mounting bolts to 25 ft.-lbs. (34.2 N·m). Reinstall clutch rods with swivel ends adjusted so both belts are snug. Adjust wire belt guides to clear belts by 1/8 to 3/16 inch (3.18-4.76 mm) and reinstall the belt guard.

OVERHAUL

ENGINE

Engine make and model numbers are listed at the front of this section. Refer to Tecumseh engine section for engine overhaul procedure.

BELT DRIVES

Belts can be replaced after removing the belt guard. Refer to belt adjustment paragraph, loosen wire belt guides and belt adjustment, then remove the worn or damaged belts. Install new belts and readjust as previously outlined.

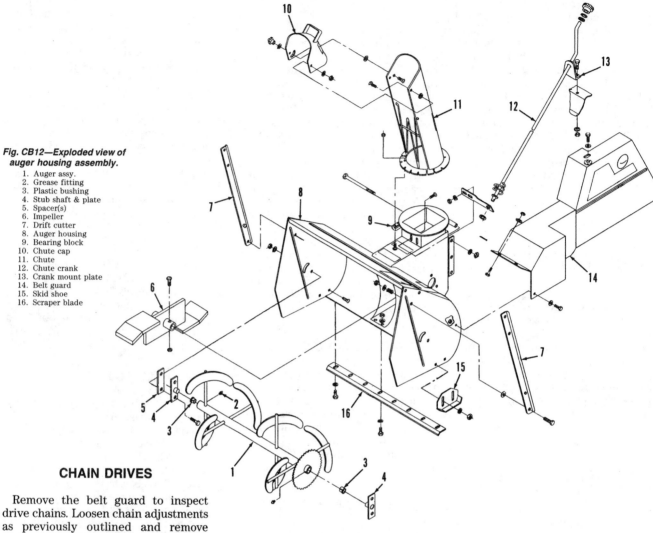

Fig. CB12—Exploded view of auger housing assembly.

1. Auger assy.
2. Grease fitting
3. Plastic bushing
4. Stub shaft & plate
5. Spacer(s)
6. Impeller
7. Drift cutter
8. Auger housing
9. Bearing block
10. Chute cap
11. Chute
12. Chute crank
13. Crank mount plate
14. Belt guard
15. Skid shoe
16. Scraper blade

CHAIN DRIVES

Remove the belt guard to inspect drive chains. Loosen chain adjustments as previously outlined and remove chains from sprockets. Disconnect chain master links to facilitate chain removal. To replace drive sprockets, refer to exploded views in Figs. CB13 and CB14, then disassemble as necessary.

Fig. CB13—Exploded view of auger drive clutch. Impeller shaft assembly (1 through 9) is same design for all models although parts vary in size by model application.

1. Bearing flange
2. Bearing w/lock collar
3. Set screw
4. Impeller shaft
5. Impeller drive chain
6. Set screw
7. Sprocket
8. Square key
9. Torque clutch chain
10. Auger drive chain
11. Roll pins
12. Clutch shaft
13. Oilite bushing
14. Grease fitting
15. Clutch bracket
16. "O" ring
17. Clutch plate
18. Sprocket
19. Clutch disc
20. Clutch plate
21. Clutch spring
22. Fender washers
23. Tension adjusting nut
24. Roll pins
25. Clutch shaft
26. R.H. clutch bracket
27. Bearing, flange & collar assy.
28. L.H. clutch bracket
29. Clutch plate
30. Bronze bushing
31. Clutch disc
32. Sprocket
33. Clutch plate
34. Clutch spring
35. Tension adjusting nut

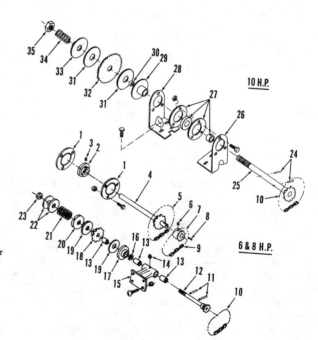

10 H.P.

6 & 8 H.P.

TRANSMISSION

A Peerless Model 700 transmission with five forward speeds and one reverse speed is used in the drive mechanism. Refer to Fig. CB14 for view showing mounting of transmission in snowthrower frame. With transmission removed from snowthrower, remove transmission shift plate (6—Fig. CB11) and refer to Peerless transmission section for overhaul information.

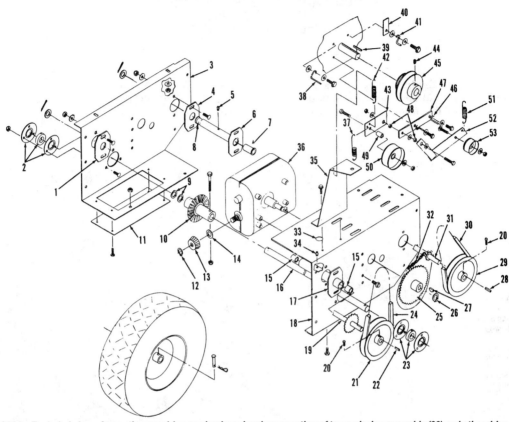

Fig. CB14—Exploded view of snowthrower drive mechanism showing mounting of transmission assembly (36) and other drive parts.

1. Housing plate	14. Washer spacer	26. Snap ring	38. Belt guide
2. Bearing & flange assy.	15. Bronze bushing	27. Woodruff key	39. Square key
3. Main frame R.H. side	16. Wheel axle shaft	28. Square key	40. Shield tab
4. Transmission jackshaft housing plate	17. Impeller jackshaft housing	29. Transmission jackshaft pulley	41. Belt guide
5. Grease fitting	18. Main frame L.H. side	30. Transmission jackshaft belt	42. Auger return spring
6. Transmission jackshaft housing	19. Impeller jackshaft	31. Transmission jackshaft	43. Idler extension plate
7. Bronze bushing	20. Set screw	32. Transmission drive chain	44. Set screw
8. Bronze bushing	21. Impeller jackshaft pulley	33. Hole for lubrication	45. Engine pulley
9. Thrust washers	22. Square key	34. Grease fitting	46. Belt guide
10. Bevel gear	23. Bearing & flange assy.	35. Belt guard bracket	47. Idler plate
11. Lower cover plate	24. Impeller drive belt	36. Transmission assy.	48. Pivot bolt spacers
12. Snap ring	25. Transmission drive sprocket	37. Auger drive spring	49. Spacer
13. Pinion gear			50. Idler pulley
			51. Drive spring
			52. Idler pulley arm
			53. Idler pulley

CUB CADET

CUB CADET POWER EQUIPMENT
P.O. Box 360930
Cleveland, Ohio 44136

Model	Engine Make	Engine Model	Self-Propelled?	No. of Stages	Scoop Width
321	Tecumseh	AH600	No	1	21 in. (533 mm)

Fig. CC1—Exploded view of drive mechanism. Belt tension spring (7) hooks in notch at bottom of idler bracket (11).

1. Fuel tank cap
2. Upper cover (cowling)
5. Auger clutch cable
6. Chute crank
7. Belt tension spring
8. Engine support bracket
9. Engine pulley half
10. Flat washer
11. Idler bracket & brake
12. Spacer
13. Flat washer
14. Belt cover
15. Drive belt
16. Drive pulley
17. Bearing retainer cup
18. Ball bearing
19. Auger (blower) housing
20. Scraper bar (shave plate)
21. Bearing & collar kit
22. Rubber auger flights (spirals)
23. Rubber paddles
24. Auger assy.
25. Idler pulley

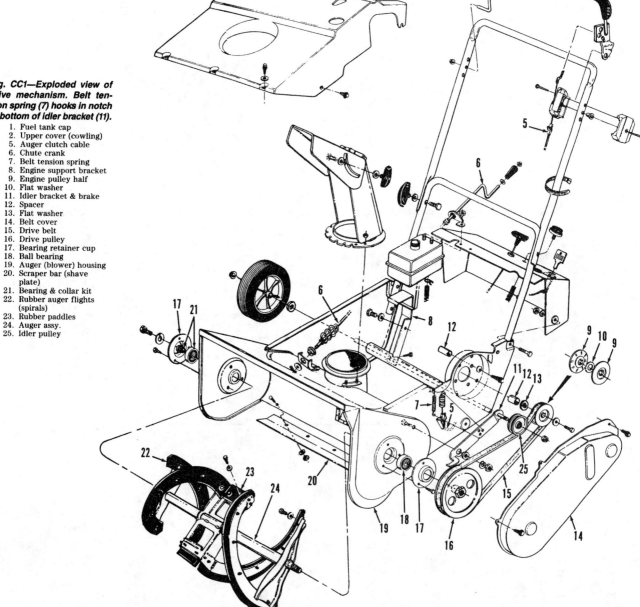

LUBRICATION

ENGINE

Refer to Tecumseh engine section for engine lubrication requirements.

The engine is a 2-stroke type which is lubricated by mixing oil with fuel. Recommended fuel:oil mixture ratio is 32:1 using a good-quality 2-cycle air-cooled engine oil.

DRIVE MECHANISM

Lubricate control lever pivot points, idler clutch and brake bracket pivot and control cable periodically to ensure smooth operation.

MAINTENANCE

Inspect the drive belt and pulleys; renew belt if frayed, cracked, excessively worn or otherwise damaged. Renew pulleys if excessively worn or damaged. Inspect the rubber auger flights and renew if excessively worn or otherwise damaged. Be sure auger clutch cable (5—Fig. CC1) operates freely. Lubricate or replace cable as necessary. Spring on end of auger clutch cable is not available as a separate item. Renew cable if spring is weak or broken. Check to see that idler clutch bracket operates smoothly and belt tension spring (7) is not stretched or broken.

ADJUSTMENTS

There are three holes in the belt idler bracket for belt tension adjustment. To adjust, first remove belt cover (14—Fig. CC1). Check to be sure that idler bracket (11) is moving smoothly on pivot and lubricate if necessary. If belt slips (auger seems to slow down or hesitate at constant engine speed), connect the spring on end of clutch cable (5) to next higher hole in idler bracket (11). If belt tension is not sufficient with spring in highest hole in bracket, it will be necessary to install a new drive belt. Too much belt tension will cause auger to turn with clutch bail released.

OVERHAUL

ENGINE

Engine make and model are listed at front of this section. Refer to Tecumseh 2-stroke engine section for engine overhaul information.

DRIVE MECHANISM

Refer to exploded view of drive mechanism in Fig. CC1. Overhaul procedure is evident after inspection of unit and reference to exploded view.

CUB CADET

Model	Engine Make	Engine Model	Self-Propelled?	No. of Stages	Scoop Width
420	Tecumseh	HSSK40	Yes	2	20 in. (508 mm)
524	Tecumseh	HSSK50	Yes	2	24 in. (610 mm)
826	Tecumseh	HMSK80	Yes	2	26 in. (660 mm)

LUBRICATION

ENGINE

Refer to Tecumseh 4-stroke engine section for engine lubrication information. Recommended fuel is regular or unleaded regular gasoline.

DRIVE MECHANISM

Lubricate drive chains after every 25 hours of operation or at least once each season using a good-quality chain oil. Auger shaft bushings and traction drive shaft bushings should be oiled after every 25 hours of operation with SAE 10W-30 motor oil. Lubricate all control pivot and slide points for smooth operation. Once each season or when necessary to replace auger flight shear bolts, remove shear bolts and squirt oil in hole and at ends of auger flights, then turn auger flight to distribute oil. On Models 420 and 524, remove wheels once each season and grease axle shafts. On track drive Model 826, keep weight transfer linkage points greased with a clinging lubricant such as Lubriplate and oil track drive chain with SAE 10W-30 motor oil after every 25 hours of operation.

AUGER GEARBOX

The auger gearbox is factory lubricated and should not require further lubrication unless disassembled for overhaul. Recommended lubricant is 10 ounces (295 mL) of Shell Alvania grease EPR00.

MAINTENANCE

Inspect belts and pulleys and renew belts if excessively worn, frayed or burned. Renew pulleys if belt grooves are worn or damaged, or if bearings in idler pulleys are loose or rough. Check drive chains and sprockets and renew if excessively worn or damaged.

ADJUSTMENTS

AUGER HEIGHT

Auger height can be adjusted by loosening skid shoe mounting bolts and repositioning the skid shoes. Auger height should be at lowest setting (skid shoes raised) when operating over smooth, paved surfaces and toward the highest setting (skid shoes lowered) when operating over rough surfaces. Be sure skid shoes are adjusted equally.

SHIFT ROD

On all models, disconnect shift control rod (6—Fig. CC10) from shift lever (5) and place both the transmission and shift lever in neutral position. Turn shift control rod in or out of ferrule (14) at bottom of rod so upper end can be inserted in hole in shift lever. Secure rod in lever with cotter pin.

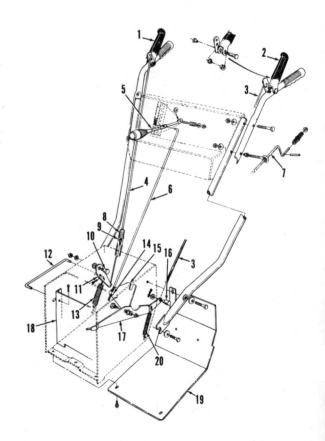

Fig. CC10 — Exploded view of handle assembly and controls.
1. Drive control lever
2. Auger control lever
3. Auger control rod
4. Drive control rod
5. Shift control lever
6. Shift control rod
7. Chute crank
8. Drive clutch spring
9. Adjusting nut
10. Shoulder bolt
11. Drive clutch bracket
12. Drive clutch rod
13. Drive spring
14. Ferrule
15. Shift bracket
16. Ferrule
17. Auger clutch bracket
18. Drive pulley support
19. Frame cover
20. Auger spring

DRIVE CLUTCH

With drive control lever (1—Fig. CC10) released, drive spring (8) at bottom of drive control rod (4) should be loose. If not, unhook spring (8) from drive clutch bracket (11). Turn adjusting nut (9), located inside spring, so hook of spring is aligned with center of hole in drive clutch bracket then rehook spring in bracket. With drive control lever held against handle grip, the drive clutch should engage and move the snowthrower without belt slippage.

AUGER CLUTCH

To check adjustment, remove belt cover and hold auger control lever against handle grip. The outer end of the slot in the end of auger brake linkage (25—Fig. CC11) should be against the spacer on the auger belt idler as shown in Fig. CC12. If not, disconnect auger control rod ferrule (Fig. CC12) from auger clutch bracket and turn ferrule until proper adjustment is obtained with ferrule reconnected to auger clutch bracket.

TRACK DRIVE

Drive track tension on Model 826 is properly adjusted when the track can be lifted approximately 1/2 inch (13 mm) at midpoint between track rollers. Adjustment is provided by adjuster cam (21—Fig. CC15) on each side of left and right track assembly. If necessary to adjust track tension, loosen the cap screws retaining adjuster cams (21) on each side of track. Turn the adjuster cams equally with a screwdriver placed between tab on adjuster cam and heavy washer, then tighten the cap screws.

OVERHAUL

ENGINE

Engine make and model numbers are listed at beginning of this section. Refer to Tecumseh 4-stroke engine section for engine overhaul information.

DRIVE BELTS

Remove the deflector chute crank at chute and remove upper belt cover. Remove the two shoulder bolt guides at engine crankshaft pulley. On track drive Model 826, place weight transfer lever in "Packed Snow" position. Remove the top bolts attaching auger housing to snowthrower frame assembly and loosen but do not remove the two bottom

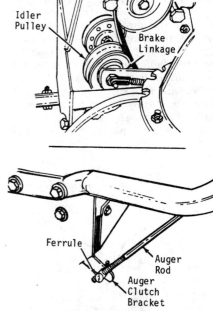

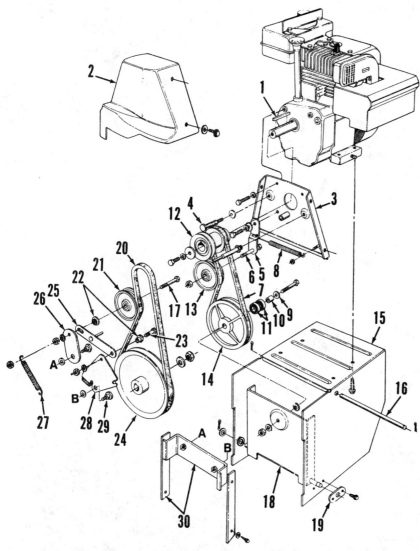

Fig. CC11—Exploded view of drive belt and auger belt drive components and engine/frame assembly. Back side of drive pulley (14) has drive surface for friction wheel; depressing drive control lever moves pulley against friction wheel (38—Fig. CC14).

1. Engine pulley key	8. Drive idler spring	16. Sliding bracket rod
2. Belt cover	9. Belleville washer	17. Auger idler bolt
3. Engine plate	10. Spacer	18. Drive pulley support
4. Shoulder bolt belt guide	11. Ball bearings	19. Support axle bracket
5. Drive clutch idler bracket	12. Engine pulley	20. Auger belt
6. Idler bolt	13. Drive belt idler	21. Auger belt idler
7. Drive belt	14. Drive pulley	22. Shoulder spacers
	15. Main frame	

23. Brake pivot bolt	27. Auger idler spring
24. Auger pulley	28. Brake bracket
25. Brake linkage	29. Shoulder bolt
26. Auger idler bracket	30. Blower housing parts

Fig. CC12—End of slot in brake linkage (25—Fig. CC11) should be against spacer on the auger belt idler when auger control lever is depressed against snowthrower handle. Refer to text for adjustment procedure.

bolts. Lift up on auger belt to pull auger housing off of snowthrower frame and separate auger housing from the main unit.

To remove auger drive belt, refer to Fig. CC13 and remove the four shoulder bolt belt guides at auger pulley. The auger belt can now be removed from the bottom pulley. To remove traction drive belt, disconnect drive idler spring (8—Fig. CC11) at engine plate and remove belt from engine pulley and bottom drive pulley. Install new belt(s) and reassemble by reversing disassembly procedure. Be sure pin on end of auger clutch bracket engages slot in auger brake bracket (see Fig. CC13) before installing auger housing mounting bolts. Adjust auger and drive clutches as outlined in preceding paragraphs.

FRICTION WHEEL

To renew friction wheel, proceed as follows: Move shift lever to reverse position. Tip snowthrower up so it rests on front of auger housing and remove bottom cover, then remove friction wheel (38—Fig. CC14) from wheel adapter (40). Place new wheel with cup side away from adapter and install retaining bolts finger tight. Turn the wheel to be sure it does not wobble, then securely tighten retaining bolts. Replace bottom cover.

TRACK DRIVE

To renew drive track or chain inside track assembly on Model 826, block up snowthrower frame so track assemblies are off the ground. Loosen the cap screws retaining track adjuster cams (21—Fig. CC15) at each side of assembly and turn adjusters so all track tension is removed. Roll the track off of track rollers and side supports. Remove roller chain (14) by disconnecting master link. Reassemble by reversing disassembly procedure, making sure to install

Fig. CC14—Exploded view of drive components; also see Fig. CC11. Notch in shift linkage bracket (15) engages pin on sliding bracket (44). Sliding bracket is supported by sliding bracket rod (16). On Model 420, only one flat washer (34) is used at right end of axle shaft (53).

6. Shift rod
14. Ferrule
15. Shift linkage bracket
16. Sliding bracket rod
31. Belleville washer
32. Flat washer
33. Flange bearing
33A. Flange bearing
34. Flat washer(s)
35. Spacer
36. Sprocket
37. Hi-pro key
38. Friction wheel
39. Hex shaft
40. Friction wheel adapter
41. Snap ring
42. Flat washer
43. Bearing
44. Sliding bracket
45. Flat washer
46. Roller chain
47. Sprocket
48. Snap ring
49. Sprocket
50. Flat washer
51. Bronze bearing
52. Flat washer
53. Sprocket & axle assy.
54. Sprocket (Model 524)
55. Roller chain
56. Shoulder bolt
57. Spacer (Model 524)
58. Flat washer (Model 524)
59. Wheel assy. (Model 524)
60. Sleeve bearing (Model 524)
61. Snap ring
62. Klick pin
63. Wheel assy. (Model 420)

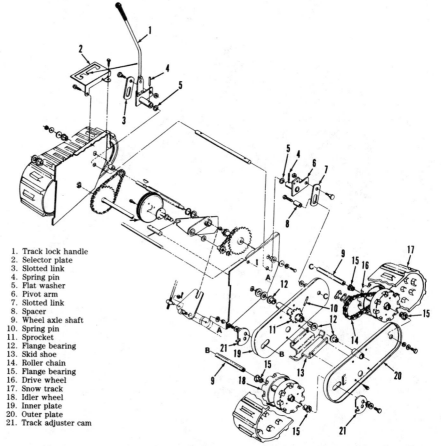

1. Track lock handle
2. Selector plate
3. Slotted link
4. Spring pin
5. Flat washer
6. Pivot arm
7. Slotted link
8. Spacer
9. Wheel axle shaft
10. Spring pin
11. Sprocket
12. Flange bearing
13. Skid shoe
14. Roller chain
15. Flange bearing
16. Drive wheel
17. Snow track
18. Idler wheel
19. Inner plate
20. Outer plate
21. Track adjuster cam

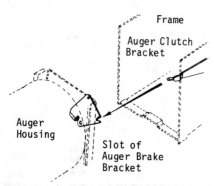

Fig. CC13—Pin on end of auger clutch bracket must engage slot in auger brake bracket when reassembling snowthrower.

Fig. CC15—Exploded view of track drive assembly for Model 826. Drive components are similar to wheel drive models shown in Fig. CC14. Track lock handle changes auger down force and has three positions: "Transport," "Normal Snow" and "Packed Snow."

track with cleats pointing toward front of snowthrower. Adjust track tension as outlined in adjustment paragraph.

AUGER GEARBOX

To remove auger gearbox, follow procedure outlined in preceding paragraph for removal of auger drive belt.

Then, remove auger drive pulley (7—Fig. CC16) from rear of gearbox shaft and unbolt auger bearing housings (37) at each end of auger housing. Remove the gearbox and auger flight assembly from front of auger housing, then remove impeller (18) from input shaft and auger flights from driveshaft. Note the position of washers and spacers as you are disassembling the unit.

To disassemble gearbox, refer to exploded view in Fig. CC16 and remove the screws retaining upper gear housing half (19). Note position of thrust washers and spacers as you disassemble gearbox. Clean and inspect all parts and renew any worn or damaged parts. Lubricate all parts prior to reassembly and pack gearbox with 10 ounces (295 mL) of Shell Alvania Grease EPR00.

1. Upper chute
2. Lower chute
3. Flat washers
4. Bushing
5. Chute crank
6. Pulley nut
7. Auger drive pulley
8. Hi-pro key
9. Spacer
10. Ball bearing
11. Shoulder bolt
12. Chute flange keeper
13. Chute crank bracket
14. Auger housing
15. Skid shoes
16. Scraper bar (shave plate)
17. Spring pin
18. Impeller (blower fan)
19. Upper gear housing half
20. "O" ring
21. Plastic flange bearing
22. Flat washer
23. Hex nut
24. Flat washer
25. Pinion gear
26. Flat washer(s)
27. Flange bearing
28. Sleeve bearing
29. Impeller (blower) axle
30. Bevel gear
31. Lower gear housing half
32. Auger flight (spiral) axle
33. Key (Hi-pro or square)
34. Flat washer
35. Flange bearing
36. Shear bolt nut
37. Bearing housing
38. Shear bolt
39. Auger flight (spiral)
40. Flange bearing

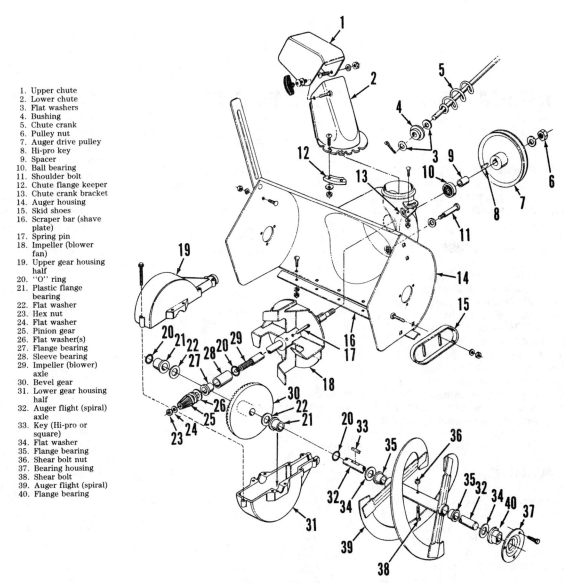

Fig. CC16—Exploded view of auger housing, gearbox and auger flight assembly.

CUB CADET

Model	Engine Make	Engine Model	Self-Propelled?	No. of Stages	Scoop Width
1028	Tecumseh	HMSK100	Yes	2	28 in. (711 mm)

LUBRICATION

ENGINE

Refer to Tecumseh 4-stroke engine section for engine lubrication information. Recommended fuel is regular or unleaded regular gasoline.

DRIVE MECHANISM

Lubricate drive chains and sprockets at least once each season or after every 25 hours of operation with a SAE 30 engine oil. Auger shaft bushings, traction drive shaft bushings and axle shaft bushings should be oiled with SAE 30 engine oil at least once each season or after every 25 hours of operation. Remove wheels and grease axle shaft with multipurpose automotive grease at least once each season. Lubricate control rod and lever pivot points to ensure smooth operation.

AUGER GEARBOX

The auger gearbox is lubricated with Shell Alvania grease EPROO. Remove filler plug (30—Fig. CC23) and check lubricant level using a wire as a dipstick. Read lubricant level on wire. Add lubricant through plug opening to bring gearbox level up to about 1/2 full. Capacity is 4 ounces (120 mL), do not overfill.

MAINTENANCE

Inspect drive belts and pulleys and renew belts if cracked, frayed, burned or otherwise damaged. Renew pulleys if belt grooves are excessively worn or damaged or if idler pulley bearings are loose or rough.

ADJUSTMENTS

AUGER HEIGHT

Auger height can be adjusted by loosening skid shoe mounting bolts and repositioning the skid shoes. Auger should be in lowest setting (skid shoes raised) when operating over smooth paved surfaces or raised (skid shoes lowered) when operating over rough surfaces.

CLUTCH RODS

With auger clutch lever (26—Fig. CC20) released against bumper (7), lower end of auger clutch spring (19) should fit loosely in the hole at rear of auger clutch bracket (16). If not, disconnect spring and slide spring upward on auger clutch rod (22) to expose adjusting nut (18) and turn nut up or down so bottom of hook in spring will be even with middle of hole in bracket when lifting clutch rod up to take slack out of rod

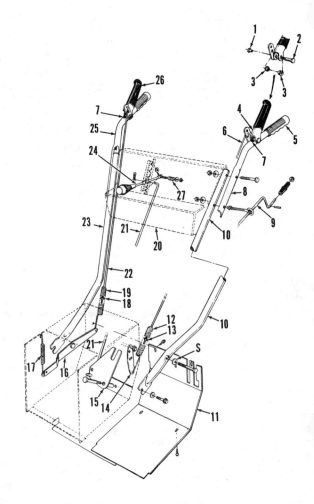

Fig. CC20—Exploded view of handle and controls.

1. Shoulder nut
2. Shoulder bolt
3. Plastic bushing
4. Drive clutch lever
5. Handle grip
6. Drive clutch rod
7. Bumper
8. Upper L.H. handle
9. Chute crank
10. Lower L.H. handle
11. Main frame cover
12. Drive clutch spring
13. Adjusting nut
14. Drive clutch bracket
15. Shift linkage bracket
16. Auger clutch bracket
17. Auger clutch spring
18. Adjusting nut
19. Auger clutch spring
20. Handle panel
21. Shift rod
22. Auger clutch rod
23. Lower R.H. handle
24. Shift lever
25. Upper R.H. handle
26. Auger clutch lever
27. Shift lever spring

and spring. There must not be any tension on clutch rod spring with the clutch lever in disengaged position. Repeat adjustment procedure for drive clutch rod (6).

SHIFT ROD

Move shift lever to neutral position on handle panel; sliding bracket assembly (13—Fig. CC22) should be aligned with slot (S—Fig. CC20) in main frame cover (11). Also, with the drive clutch lever (4) in engaged position, the snowthrower should roll back and forth with shift lever in neutral. If necessary to adjust shift rod, disconnect shift rod (21) from shift lever (24) at control console. Move shift lever to neutral position and move shift rod until edge of sliding bracket is visible through slot (S) in main frame cover. If cover is removed, sliding bracket should be aligned on center of neutral stop bolt (19—Fig. CC22). Thread shift rod in or out of ferrule (41) in shift bracket (40) at bottom end of rod until top end aligns with hole in shift lever. Reconnect shift rod and check adjustment.

Shift lever spring (27—Fig. CC20) should be tight enough to hold lever in handle panel detent notches, yet allow lever to move without binding on detents.

OVERHAUL

ENGINE

Engine make and model numbers are listed at beginning of this section. Refer to Tecumseh 4-stroke engine section for engine overhaul information.

DRIVE BELTS

To renew drive belts, it is first necessary to split the snowthrower between the auger housing and main frame assembly as follows: Disconnect the deflector chute crank at the chute end and remove drive belt cover. Remove shoulder bolt (17—Fig. CC21) and Belleville washer at left side of engine pulley. Remove shoulder bolt (4) and spacer (5) at right hand side of engine pulley, being careful not to lose the Belleville (spring cup) washer located between idler bracket and engine plate. Slip auger drive belt off engine pulley and loosely install shoulder bolt, roller and Belleville washer at right side of engine pulley. Remove the top cap screws retaining auger housing flanges (24) to

main frame (21). Two people are required to separate the auger housing from main frame. With someone holding handles, push in on auger idler pulley to lift brake pad (BR) out of pulley groove, then lift auger housing from main frame. At this point, either the auger belt or drive belt, or both, can be renewed.

To remove auger belt after separating front and rear units, remove the five shoulder bolts (5—Fig. CC23) and Belleville washers from auger housing and remove belt from pulley. An alternate method is to remove the cap screw holding auger drive pulley to impeller shaft and remove the pulley and belt; be careful not to lose the drive key and Belleville washer. Install new belt by reversing removal procedure. If auger pulley was removed, be sure key is in place on shaft and install Belleville washer with cup side toward pulley. Tighten pulley retaining cap screw securely.

To remove traction drive belt, unhook drive idler spring (16—Fig. CC21) from engine plate, then remove the belt from drive pulley. Remove shoulder bolt (4),

spacer (5) and Belleville washer at right side of engine pulley and remove the belt. Install new belt by reversing removal procedure.

Be sure that idler pulley and/or brake pad are not behind the drive pulley when assembling auger housing to main frame.

DRIVE MECHANISM

To overhaul drive mechanism, first remove drive belt as outlined in a preceding paragraph, then remove engine assembly and traction drive pulley (19—Fig. CC21). Further disassembly procedure is evident after inspection of unit and reference to Fig. CC22. Be sure not to overtighten the hex shaft self-aligning bearing (51—Fig. CC22) as hex shaft (50) needs to flex in the bearing for free movement of sliding bracket (13). If adjusting auger clutch rod (37) with main frame cover removed, forward point of sliding bracket (13) should be centered on head of neutral stop bolt (19) when shift lever on handle panel is in neutral position.

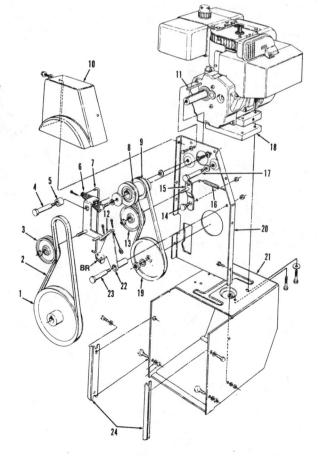

Fig. CC21—Exploded view of belt drive system. Auger pulley (1) is also shown in Fig. CC23 and traction drive pulley (19) mounts on drive plate spindle (35—Fig. CC22).

1. Auger pulley
2. Auger belt
3. Auger belt idler
4. Shoulder bolt
5. Spacer
6. Auger idler bracket
7. Auger clutch rod
8. Engine pulley
9. Traction drive belt
10. Belt cover
11. Square key
12. Auger brake linkage
13. Traction drive idler
14. Shoulder bolt
15. Drive idler bracket
16. Drive idler spring
17. Shoulder bolt
18. Engine spacer plate
19. Traction drive pulley
20. Engine bracket assy.
21. Main (motor mount) frame
22. Brake bracket assy.
23. Shoulder bolt
24. Auger housing flanges

FRICTION WHEEL

To remove rubber faced friction wheel, first tip snowthrower forward onto front of auger housing and remove the cover from main frame. Friction wheel (5—Fig. CC22) can then be unbolted from friction wheel adapter (4) and removed from the end of hex shaft. Install new friction wheel with cup side facing toward adapter and install retaining lockwashers and nuts finger tight. Turn the wheel to be sure it does not wobble (cocked on hub), then securely tighten nuts.

AUGER SHEAR BOLTS

The auger flights are driven through shear bolts (37—Fig. CC23) which will shear when a solid obstruction is encountered to prevent damage to snowthrower. If this occurs, be sure to replace the bolts with factory replacement parts; substitute bolts may not shear and cause extensive damage. Before installing new bolts, oil the auger flight bushings and be sure flights spin freely on the auger shaft.

AUGER GEARBOX

To remove auger gearbox, first remove auger belt as outlined in a preceding paragraph. If auger pulley (1—Fig. CC23) was not removed, remove it at this time. Remove outer bearing retainer (3), then remove set screws and unlock bearing (4) from gearbox input shaft. Unbolt auger shaft bearing flanges (39) at each end of auger housing, remove flange bearings (38) and thrust washers (34) and withdraw the gearbox with impeller (10) and auger flights (36). Remove shear bolts (37), auger flights, inner

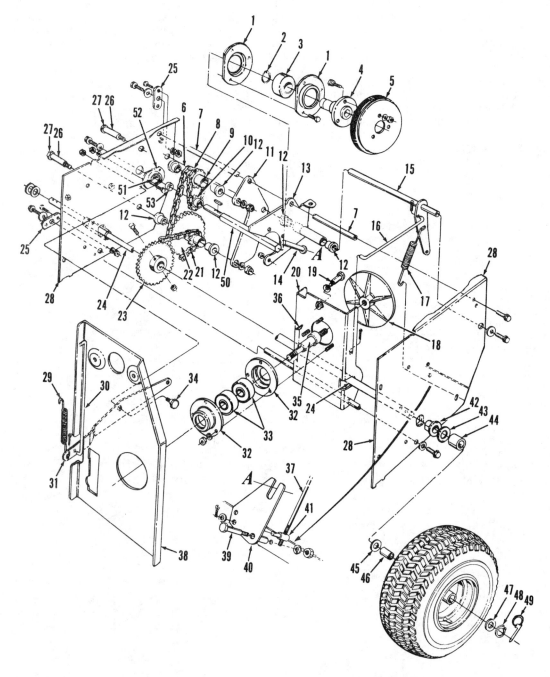

Fig. CC22—Exploded view of friction drive and chain speed reduction drive components.

1. Bearing housing
2. Snap ring
3. Ball bearing
4. Friction wheel adapter
5. Friction wheel
6. Roller chain
7. Sliding support axle
8. Roller chain
9. Double sprocket
10. Hex shaft sprocket
11. Chain support bracket
12. Flange bearing
13. Sliding bracket assy.
14. Pin (part of 13)
15. Drive clutch bracket
16. Drive clutch rod
17. Drive clutch spring
18. Aluminum drive plate
19. Neutral stop bolt
20. Drive pivot plate
21. Roller chain
22. Double sprocket
23. Axle sprocket
24. Axle shaft
25. Pivot shaft bracket
26. Sprocket hub tube
27. Sprocket bolt
28. Main frame assy.
29. Auger clutch spring
30. Auger clutch rod
31. Auger clutch lever
32. Bearing housing
33. Ball bearings
34. Shoulder bolt
35. Drive plate spindle
36. Hi-pro key
37. Auger clutch rod
38. Engine bracket assy.
39. Shoulder bolt
40. Shift bracket
41. Ferrule
42. Axle flange bearing
43. Spacer washer
44. Spacer
45. Wheel spacer washer
46. Wheel bushing
47. Wheel spacer washer
48. Snap ring
49. Klick pin
50. Hex shaft
51. Bearing
52. Bearing housing
53. Spacer

thrust washers and spacers (33) from auger shaft. Remove the carriage bolt and spring pins (9), then pull impeller from shaft. Remove the bolts holding the gearbox halves (19 and 31) together and disassemble gearbox. Clean and inspect all parts and renew any excessively worn or damaged parts. Worm gear (27) can be removed from hub (28).

Reassemble by reversing disassembly procedure and lubricate gearbox using 4 ounces (120 mL) of Shell Alvania Grease EPROO.

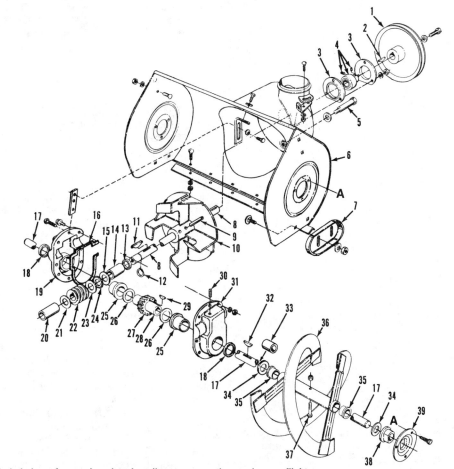

1. Auger pulley
2. Hi-pro key
3. Bearing retainer
4. Self-aligning bearing
5. Shoulder bolt
6. Auger housing
7. Skid shoe
8. Impeller shaft
9. Spring pin
10. Impeller
11. Hi-pro key
12. Snap ring
13. Oil seal
14. Sleeve bearing
15. Flat washer
16. Gasket
17. Auger shaft
18. Oil seal
19. Housing half
20. Sleeve bushing
21. Flat washer
22. Worm gear
23. Flat washer
24. Thrust bearing
25. Flange bearing
26. Flat washer
27. Worm gear
28. Gear hub
29. Woodruff key
30. Filler plug
31. Housing half
32. Woodruff key
33. Spacer
34. Thrust washer
35. Plastic bushing
36. Auger flight
37. Shear bolt
38. Flange bearing
39. Bearing flange

Fig. CC23—Exploded view of auger housing, impeller, auger gearbox and auger flights.

JOHN DEERE

DEERE & COMPANY
John Deere Road
Moline, Illinois 61265

Model	Engine Make	Engine Model	Self-Propelled?	No. of Stages	Scoop Width
320	Jacobsen	J501	No	1	20 in. (508 mm)

LUBRICATION

ENGINE

Refer to Jacobsen engine section. The engine is a two-stroke type lubricated by mixing oil with the fuel. John Deere recommends a 50:1 mixture ratio when using John Deere 2-cycle engine oil. When using other oils, refer to instructions on the oil package for mixture ratio: use a good quality oil designed for two-stroke engines.

DRIVE MECHANISM

Lubricate belt idler arm pivot with SAE 30 oil after every 25 hours of operation. If choke control cable is in metal conduit, lubricate with SAE 30 engine oil prior to season and after every 25 hours of operation.

MAINTENANCE

Inspect drive belt and pulleys. Renew belt if frayed, cracked, burned or otherwise damaged. Inspect impeller paddles and renew if excessively worn or damaged.

OVERHAUL

ENGINE

Refer to Jacobsen engine section for engine overhaul information.

DRIVE BELT

Remove the belt guard at left side of unit. Lift idler pulley and slip belt from engine and driven pulleys. Check condition of pulley grooves and idler pulley bearing; renew as necessary. If engine pulley is removed from crankshaft, position pulley so outside face is plus or minus 1/16 inch (1.6 mm) from end of crankshaft. Apply thread locking compound through set screw hole and tighten set screw to a torque of 80 inch-pounds (9 N·m).

IMPELLER ASSEMBLY

Impeller paddles (32—Fig. JD2) are a matched set. If old paddles are to be reinstalled, they should be marked end for end before removal. New paddles are connected together and must be marked prior to cutting apart so ends will not be swapped during installation. Tighten paddle screws in order, working from one end to the other to prevent waves from forming in the paddles. Tighten paddle screws to 40-50 in.-lbs. (4.7-5.6 N·m).

Impeller halves can be removed after removing impeller paddles. Remove belt guard (23) and impeller drive belt (22). Unscrew impeller pulley (21) (left-hand threads). If bearing removal is necessary, remove rivets and bearing retainer (30). When installing new bearing, use hardened bolts with self-locking nuts to replace rivets. Inspect components and renew as required.

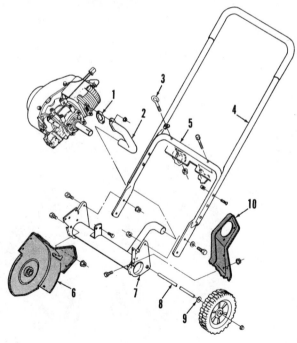

Fig. JD1—Exploded view of handle, muffler and wheel assembly.
1. Manifold gasket
2. Exhaust pipe
3. Eyebolt
4. Upper handle
5. Lower handle
6. R.H. impeller frame
7. Muffler assy.
8. Axle
9. Spacer
10. L.H. frame assy.

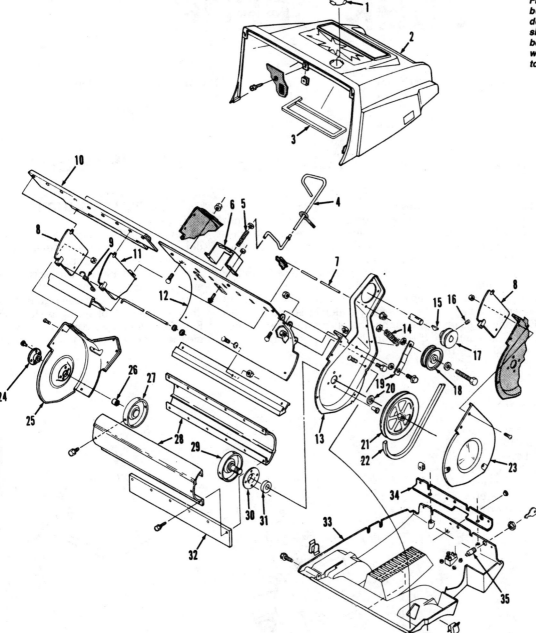

Fig. JD2—Exploded view of belt drive, impeller and deflector assembly. Right side impeller frame (25) and bearing assembly may differ with production types. Refer to Fig. JD1 for handle, muffler and wheel assembly.

1. Carburetor access cap
2. Upper cowling
3. Foam gasket
4. Vane control rod
5. Vane control spring
6. Control rod bracket
7. Vane rod
8. End vanes
9. Snow guide spring
10. Vane control panel
11. Inner vanes
12. Impeller housing
13. L.H. frame assy.
14. Belt idler spring
15. Morton type key
16. Set screw
17. Crankshaft pulley
18. Idler pulley
19. Idler arm
20. Spacer
21. Impeller pulley
22. Drive belt
23. Belt guard
24. R.H. impeller bearing
25. R.H. impeller frame
26. Impeller bushing
27. Impeller end plate
28. Impeller drum halves
29. L.H. impeller end plate assy.
30. Bearing retainer
31. Impeller bearing
32. Impeller paddle
33. Lower cowling
34. Control mounting bracket
35. Ignition switch

JOHN DEERE

Model	Engine Make	Engine Model	Self-Propelled?	No. of Stages	Scoop Width
322	Tecumseh	AH600	No	1	20 in. (508 mm)

LUBRICATION

ENGINE

Refer to Tecumseh 2-stroke engine section for engine lubrication information. John Deere recommends a 32:1 fuel:oil mix ratio when using John Deere 2-cycle engine oil. Recommended fuel is regular or unleaded gasoline. After every 25 hours of operation, apply a small amount of SAE 30 engine oil on choke knob (24—Fig. JD6) at control panel and work knob up and down to distribute oil.

DRIVE MECHANISM

Drive belt idler pivot arm should be lubricated at pivot stud (12—Fig. JD8) with SAE 30 engine oil after every 25 hours of operation. To gain access to pivot arm, remove belt cover from left side of snowthrower housing. Be careful not to get any oil on drive belt or pulleys.

MAINTENANCE

Inspect drive belt and renew if frayed, cracked, burned or excessively worn. Pulleys should be renewed if belt groove is excessively worn or pulley is otherwise damaged. Check auger clutch cable; lubricate or renew cable if binding. Check belt idler pulley for bearing being loose or rough and renew idler pulley assembly if defect is noted. Check the rubber impeller paddles and renew if excessively worn or damaged.

ADJUSTMENTS

IMPELLER CLUTCH

If impeller fails to turn or belt slips under load, check to be sure idler arm is working freely on pivot stud (12—Fig. JD8). Check idler arm return spring (13) and clutch cable spring (12—Fig. JD6). With control lever engaged against handle, check length of spring as shown in Fig. JD5. If spring length is not between measurements shown, release control

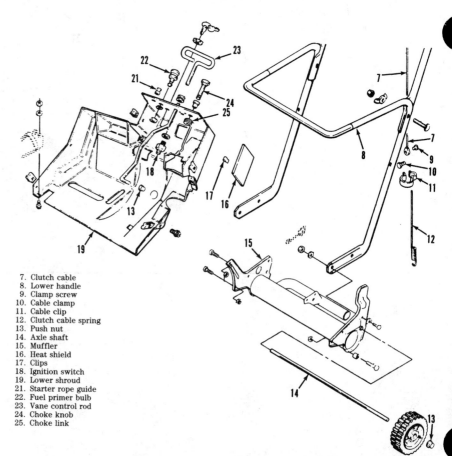

7. Clutch cable
8. Lower handle
9. Clamp screw
10. Cable clamp
11. Cable clip
12. Clutch cable spring
13. Push nut
14. Axle shaft
15. Muffler
16. Heat shield
17. Clips
18. Ignition switch
19. Lower shroud
21. Starter rope guide
22. Fuel primer bulb
23. Vane control rod
24. Choke knob
25. Choke link

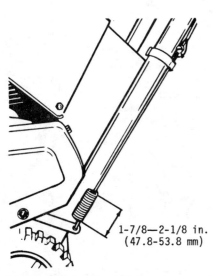

1-7/8—2-1/8 in. (47.8-53.8 mm)

Fig. JD5—With impeller clutch lever in engaged position, length of clutch cable spring should be as shown.

Fig. JD6—Exploded view of handle and controls. Heat shield (16) fastens to inside wall of lower shroud with clips (17). Choke knob (24) screws into coil end of choke link (25).

lever, loosen clamp screw (9—Fig. JD6) holding control cable (7) to spring (12) and lengthen or shorten spring as necessary.

BELT GUIDES

With impeller clutch lever in engaged position, gap between belt guides (9 & 14—Fig. JD8) and drive belt should be 1/16 inch (1.6 mm). If gap is not correct, loosen guide retaining nuts, reset gap and tighten nuts securely.

OVERHAUL

ENGINE

Refer to Tecumseh 2-stroke engine section for engine overhaul information.

DRIVE MECHANISM

Overhaul of basic unit should be evident after inspection of unit and reference to Fig. JD8. Impeller pulley (19) screws onto impeller shaft (23—Fig. JD7) with left-hand threads. If engine or impeller pulley has been removed, place straightedge on pulleys to check alignment. Move engine crankshaft pulley, if necessary, to bring alignment to within 0.030 inch (0.76 mm).

DRIVE BELT

To renew drive belt, remove cover from left side of snowthrower housing and, using pliers, pull fixed belt guide pin (21—Fig. JD8) from housing (8—Fig. JD7). Loosen belt guides (9 & 14—Fig. JD8). Remove belt from under idler pulley and from engine pulley, then lift belt from impeller pulley. Install new belt and tap fixed guide pin back into auger housing. Be sure that groove in pin is away from belt. Adjust engine pulley belt guides as outlined in adjustment paragraph.

IMPELLER PADDLES

Impeller paddles (22—Fig. JD7) can be renewed after removing the row of cap screws retaining each paddle in the impeller hub assembly, then pulling paddle from hub. Paddles must be replaced in pairs to avoid an unbalanced condition in impeller. Be sure that curve (cup) in paddle faces direction of impeller rotation.

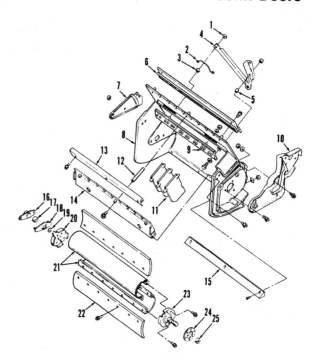

Fig. JD7—Exploded view of impeller and discharge vane assembly.

1. Self-thread nut
2. Detent spring
3. Bushing
4. Vane control lever
5. Bushing
6. Rear vane retainer
7. Right bracket
8. Impeller housing
9. Vane adjustment bar
10. Left bracket
11. Vanes
12. Spacer tube
13. Front vane retainer
14. Front vane mount
15. Scraper bar
16. Bearing retainer
17. Bearing
18. Bearing flange
19. Flat washer
20. Right impeller shaft
21. Impeller drum
22. Impeller paddle
23. Left impeller shaft
24. Bearing retainer
25. Bearing

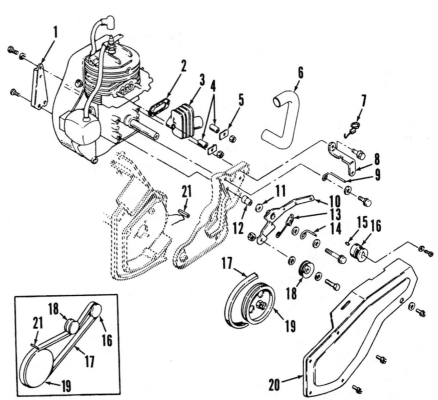

Fig. JD8—Exploded view showing engine and impeller drive mechanism. Spring (7) hooks in notch of belt cover bracket (8) and around exhaust pipe (6) to hold pipe in position.

1. Engine bracket
2. Manifold gasket
3. Exhaust manifold
4. Spacers
5. Nut retainer tab
6. Exhaust pipe
7. Pipe retainer spring
8. Belt cover bracket
9. Belt guide
10. Belt idler arm
11. Flat washer
12. Pivot stud
13. Idler arm return spring
14. Belt guide
15. Woodruff key
16. Engine pulley
17. Impeller belt
18. Idler pulley
19. Impeller pulley
20. Belt cover
21. Belt guide pin

JOHN DEERE

Model	Engine Make	Engine Model	Self-Propelled?	No. of Stages	Scoop Width
520	Tecumseh	H35	No	2	20 in. (508 mm)
522	Tecumseh	HSSK50	Yes	2	22 in. (559 mm)

LUBRICATION

ENGINE

Refer to Tecumseh 4-stroke engine section for engine lubrication information. Recommended fuel is unleaded or leaded regular grade gasoline.

DRIVE MECHANISM

Auger gearbox oil level should be maintained level with pipe plug (9—Fig. JD25) in gearbox with Grade GL4 gear lubricant. After every 10 hours of operation, lubricate all linkage pivot points and discharge chute pivot with SAE 30 motor oil.

On Model 522, lubricate traction drive chain after every 10 hours operation with a light coat of multipurpose grease. Remove bottom panel to gain access to lubrication points. The three-speed gear-type transmission is lubricated at the factory and does not require lubrication except during overhaul; refer to Peerless transmission section for service information.

Once each year, remove chute and wheels and grease chute flange and axles with a clinging type grease. Once each year or each time it is necessary to replace an auger shear pin, use a hand-operated grease gun and lubricate the augers at grease fittings.

MAINTENANCE

Inspect belts and pulleys and renew belts if frayed, cracked, burned or excessively worn. Check pulley grooves for excessive wear or damage and renew if defect is found. Check idler pulley bearings and renew idler pulley assembly if bearings are loose or rough. On Model 522, inspect traction drive chain and sprockets and renew if excessively worn or damaged.

ADJUSTMENTS

AUGER HEIGHT

Auger height on Model 520 is nonadjustable as there are no skid shoes. The rubber auger flights propel the snowthrower by contacting surface.

Auger height on Model 522 can be adjusted by loosening skid shoe mounting bolts and repositioning the skid shoes. Auger should be in lowest position (skid shoes raised on housing) when operating over smooth surfaces, and in highest position when operating over rough surfaces.

SHIFT CONTROL

To check shift control linkage on Model 522, place blocks under snowthrower so wheels are off the ground. With speed control lever (L—Fig. JD20) in neutral position, the wheels should be free to turn by hand. If not, loosen screws (A) retaining the speed control indicator bracket (B) to rear of snowthrower and move the bracket side-to-side until neutral position is found where wheels can be rotated. Tighten the bracket retaining screws and be sure that all gears will engage. The transmission must disengage when shift lever is in neutral position.

AUGER DRIVE BELT

If auger drive belt slips under load, the drive belt tension can be adjusted as follows: Loosen locknut on auger idler pulley (see Fig. JD21), push idler against belt while holding auger drive

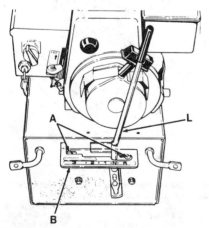

Fig. JD20—View showing adjustment points for speed control lever (L) on Model 522. Refer to text.

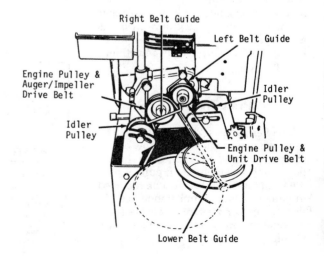

Fig. JD21—View of drive idler system with belt cover removed. Model 520 does not have traction drive components. Belt guides may differ in design because of production changes.

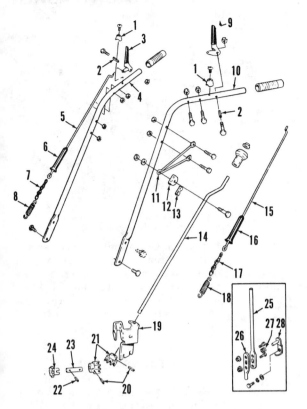

Fig. JD22—Exploded view of Model 522 handle and controls. Model 520 does not have right handle drive clutch mechanism or shift lever controls.

1. Lever bumper
2. Spacer
3. Drive control lever
4. Right handle
5. Drive control rod
6. Plastic sheath
7. Control rod chain
8. Drive control spring
9. Auger control lever
10. Left handle
11. Chute crank bracket
12. Chute crank clamp
13. Clamp liner
14. Chute crank
15. Control rod
16. Plastic sheath
17. Control rod chain
18. Auger control spring
19. Chute gear bracket
20. Roll pins
21. Chute bevel gears
22. Roll pin
23. Chute sprocket shaft
24. Chute sprocket
25. Upper speed control lever
26. Lower speed control lever
27. Spring
28. Transmission shift arm

ENGINE

Engine make and model numbers are listed at beginning of this section. Refer to Tecumseh 4-stroke engine section for engine overhaul information.

AUGER GEARBOX

Refer to Fig. JD25 for exploded view of auger gearbox. To remove auger gearbox, first remove auger drive belt. Block up under snowthrower frame, disconnect chute rotating linkage and unbolt auger housing from snowthrower frame. Remove set screw from auger drive pulley (25) and Woodruff key (23) from rear end of auger shaft. Remove snap rings at outer end of auger shaft and on early production units, unbolt auger bear-

control lever in engaged position and tighten locknut. With auger drive control lever engaged, auger brake pad (13—Fig. JD23) should clear auger drive belt. With auger drive control lever in disengaged position, there should be some slack in chain (17—Fig. JD22) connecting control rod (15) to spring (18). If necessary, slide sheath (16) up and connect chain in different link. The auger should not turn with control lever in disengaged position.

TRACTION DRIVE BELT

If traction drive belt slips on Model 522, loosen locknut on drive idler pulley (see Fig. JD21), push idler against belt while holding drive clutch lever in engaged position and tighten locknut. With drive control lever in disengaged position, there should be some slack in chain (7—Fig. JD22) connecting control rod (5) to spring (8). If necessary, slide sheath (6) up and connect chain in different link.

BELT GUIDES

With auger clutch lever and on Model 522 drive clutch lever held in engaged position, all belt guides should clear the drive belts from 1/16 to 1/8 inch (1.6-3.2 mm). Upper belt guides are bolted to engine and lower guides are mounted on rear of auger housing.

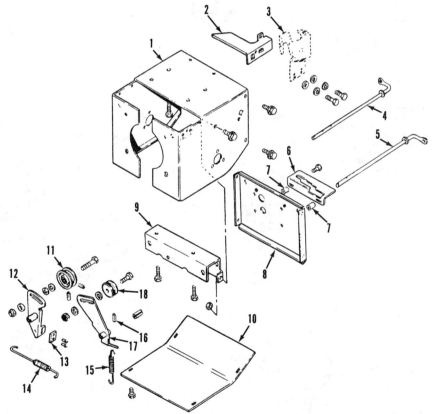

Fig. JD23—Exploded view of belt idler clutch system. Model 520 does not have shift guide bracket (6), drive belt idler (18), drive idler bracket (17) and drive control shaft (5). Idler brackets are attached to front ends of control shafts by spring pins (16).

1. Main frame
2. Belt cover
3. Chute gear bracket
4. Auger control shaft
5. Drive control shaft
6. Speed control guide bracket
7. Spacer
8. Main frame rear cover
9. Transmission support
10. Bottom cover
11. Auger belt idler
12. Auger idler bracket
13. Auger brake pad
14. Auger return spring
15. Drive return spring
16. Spring pin
17. Drive idler bracket
18. Drive belt idler

ings from ends of auger housing. Withdraw auger and impeller fan assembly from housing, remove shear pins and slide augers from auger shaft. Remove set screw from impeller fan (18) and pull fan from rear of auger gearbox input shaft.

Disassembly and reassembly of auger gearbox are evident after inspection of unit and reference to exploded view in

Fig. JD26—Exploded view of Model 520 auger flights. Snowthrower is propelled forward by contact of rubber flight edges with ground surface.

1. Shear pin
2. Grease fitting
3. Auger flight support
4. Spiral blade, R.H.
5. Rubber flights
6. Tubular rivet
7. Spiral blade, L.H.
8. Auger flight support
9. Flat washer
10. Snap ring

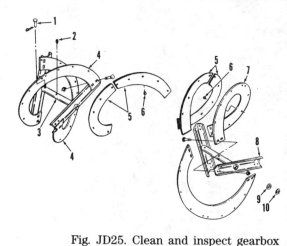

Fig. JD25. Clean and inspect gearbox parts and renew any excessively worn or damaged parts. During reassembly, refill gearbox with API grade GL-4 gear lubricant (worm gear oil). Lubricant capacity is approximately 3 ounces (90 mL).

Fig. JD24—Drawing showing Model 522 traction drive components. Refer to Peerless transmission section for information on gear type transmission (8).

1. Drive axle & sprocket
2. Roller chain
3. Spacer
4. Washer
5. Axle bearing
6. Transmission pulley
7. Set screw
8. Transmission

TRANSMISSION

Model 522 is equipped with a Peerless Series 700 transmission with three forward and one reverse speeds. With transmission removed from snowthrower, refer to Peerless transmission section for overhaul information.

1. Right gearbox half
2. Gasket
3. Auger shaft bushing
4. Auger shaft
5. Worm drive gear
6. Woodruff key
7. Auger shaft bushing
8. Left gearbox half
9. Pipe plug
10. Oil seal
11. Impeller shaft bushing
12. Thrust washer
13. Impeller worm shaft
14. Woodruff key
15. Thrust collar
16. Driv-lok pin
17. Oil seal
18. Impeller fan
19. Scraper bar
20. Auger housing
21. Ball bearing
22. Bearing retainer
23. Woodruff key
24. Set screws
25. Auger drive pulley
26. Housing end plate
27. Auger shaft bearing
28. Skid shoe
29. Impeller shaft bushing

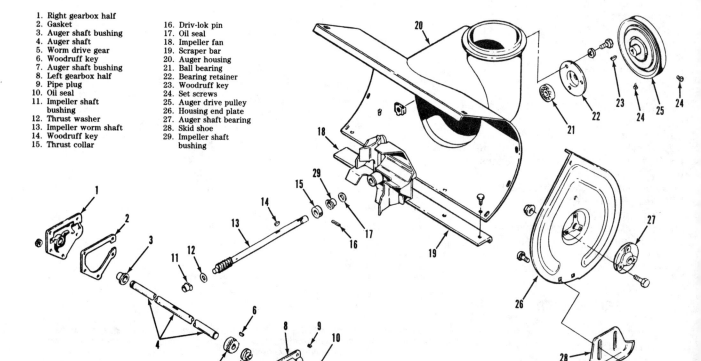

Fig. JD25—Exploded view of auger housing and gearbox. Model 520 auger housing is not fitted with skid shoes.

DRIVE BELTS

Remove top belt cover, loosen belt guides and slip the auger belt from engine crankshaft pulley or the traction drive belt (Model 522) off of engine camshaft (PTO) pulley. Tip snowthrower forward onto front of auger housing; support upper edge of housing with wood block and be sure unit is sitting solid. Remove lower cover and loosen belt guides. Pivot belt guides clear of drive belt(s). Lift auger drive belt off of pulley and remove it upward from the auger pulley. On Model 522, remove both belts upward between the auger and traction drive pulleys. It will be necessary to hold auger clutch lever down to free auger belt from under auger drive brake.

Install new belt(s) by reversing removal procedures and check adjustments as outlined in appropriate belt adjustment paragraph.

JOHN DEERE

Model	Engine Make	Engine Model	Self-Propelled?	No. of Stages	Scoop Width
524	Tecumseh	HSSK50	Yes	2	24 in. (610 mm)

LUBRICATION

ENGINE

Refer to Tecumseh 4-stroke engine section for engine lubrication information. Recommended fuel is unleaded or leaded regular grade gasoline.

DRIVE MECHANISM

Auger gearbox oil level should be maintained at a level even with pipe plug (5—Fig. JD34) opening in gearbox with 85W/140 Grade GL5 gear lubricant (worm gear oil). After every 10 hours of operation, lubricate all linkage pivot points and discharge chute pivot with SAE 30 motor oil. Lubricate traction drive chain after every 10 hours operation with a light coat of multipurpose grease. Remove bottom panel to gain access to lubrication points.

The auger flights are driven from the auger gearbox shaft by shear pins which may fracture if the auger encounters a solid obstruction. If this happens, turn auger so holes in auger tube and gearbox shaft are aligned and install a new factory replacement shear pin.

ADJUSTMENTS

AUGER HEIGHT

Auger height can be adjusted by loosening skid shoe mounting bolts and repositioning the skid shoes. Auger should be in lowest position (skid shoes raised on housing) when operating over smooth surfaces, and in highest position when operating over rough surfaces.

SHIFT CONTROL

Place the shift lever in 5th gear position and loosen the two bolts clamping upper speed control rod (28—Fig. JD30)

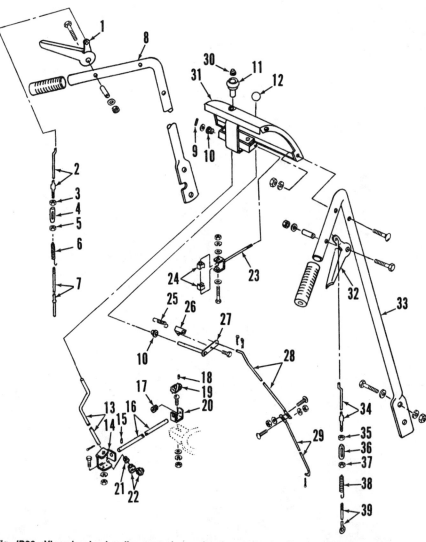

Fig. JD30—View showing handles, console panel and control parts. Eye in end of lower auger clutch rod (39) and auger clutch spring (38) loops over auger clutch rod and idler bracket (9—Fig. JD32). Eye in lower end of drive clutch spring hooks into hole of drive control bellcrank (32—Fig. JD33).

1. Drive clutch lever
2. Upper drive clutch rod
3. L.H. thread nut
4. Turnbuckle
5. R.H. thread nut
6. Drive clutch spring
7. Lower drive clutch rod
8. Left handle
9. Spring pin
10. Plastic bushing
11. Chute crank knob
12. Speed lever knob
13. Chute crank
14. Chute gear bracket
15. Spring pin
16. Chute gear shaft
17. Plastic bushing
18. Spring pin
19. Chute worm gear
20. Worm gear bracket
21. Plastic bushing
22. Bevel gears
23. Speed control lever
24. Pivot blocks
25. Lever detent spring
26. Spring link
27. Speed control bellcrank
28. Upper speed control rod
33. Right handle
34. Upper auger clutch rod
35. L.H. thread nut
36. Turnbuckle
37. R.H. thread nut
38. Auger clutch spring
39. Lower auger clutch rod
29. Lower speed control rod
30. Chute crank bushing
31. Console panel
32. Auger clutch lever

to lower speed control rod (29). The spring loaded friction wheel shift arm assembly (35—Fig. JD33) will move the lower speed control rod into proper position for 5th gear, then tighten the clamp bolts.

AUGER DRIVE BRAKE

With auger clutch lever released, brake pad (12—Fig. JD32) should contact auger belt and stop it from turning. With clutch lever in engaged position, the brake pad should clear the auger drive belt. Some adjustment can be obtained by loosening nut (10) and moving idler in slot of auger clutch rod and idler bracket (9). If brake clearance is correct, check auger drive clutch adjustment.

AUGER DRIVE CLUTCH

There should be 1/16 inch free play of auger clutch spring (38—Fig. JD30) on auger clutch rod (9—Fig. JD32). To adjust, loosen nuts (35 & 37—Fig. JD30) and rotate turnbuckle (36) until correct spring to shaft lever clearance is obtained. Tighten the turnbuckle locknuts and recheck adjustment.

TRACTION DRIVE CLUTCH

The bottom end of lower drive clutch rod (7—Fig. JD30) should be flush with bottom end of drive clutch spring (6) with the drive clutch lever engaged.

This pulls the friction wheel against friction disc to drive the snowthrower. To adjust, loosen the locknuts at control rod turnbuckle and rotate the turnbuckle. Tighten the turnbuckle locknuts and recheck adjustment. The traction drive belt idler is spring loaded and adjustment of belt is not necessary.

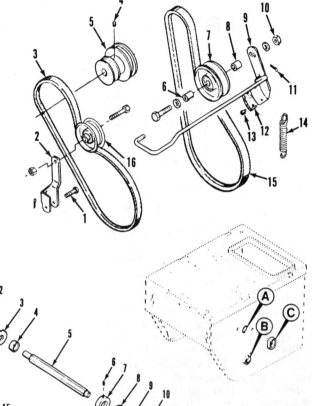

Fig. JD32—Exploded view of belt drive mechanism.

1. Pivot pin
2. Drive idler bracket
3. Traction drive belt
4. Set screw
5. Engine pulley
6. Spacer
7. Auger belt idler
8. Spacer
9. Auger clutch rod & idler bracket
10. Nut
11. Cotter pin
12. Brake pad
13. Rivet
14. Auger idler return spring
15. Auger drive belt
16. Drive belt idler

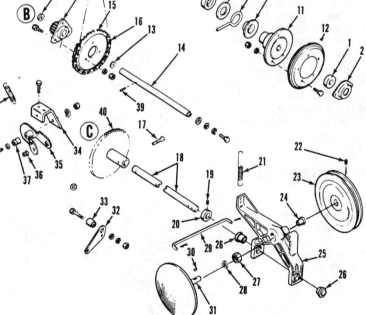

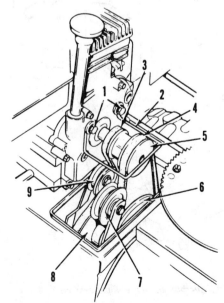

Fig. JD31—Drawing of drive belt mechanism with belt cover removed.

1. Belt guide screws
2. Belt guide
3. Traction drive belt
4. Auger drive belt
5. Engine pulley
6. Auger pulley
7. Auger belt idler
8. Traction drive pulley
9. Traction belt idler

Fig. JD33—Exploded view of friction disc drive.

1. Hex shaft bearing
2. Bearing retainer
3. Hex shaft sprocket
4. Spacer
5. Hex shaft
6. Set screw
7. Collar
8. Thrust washer
9. Slide pin
10. Thrust collar
11. Friction wheel hub
12. Friction wheel
13. Washer
14. Gear & hub
15. Jack shaft sprocket
16. Roller chain
17. Bolt
18. Axle
19. Set screw
20. Collar
21. Disc return spring
22. Set screw
23. Drive pulley
24. Bushing
25. Pivot arm assy.
26. Bushing
27. Bearing
28. Thrust washer
29. Pivot rod
30. Woodruff key
31. Friction drive disc
32. Drive control bellcrank
33. Pivot bushing
34. Shift arm bracket
35. Shift arm assy.
36. Bushing
37. Spacer bushing
38. Friction wheel return spring
39. Roll pin
40. Gear

FRICTION WHEEL

Other than traction drive clutch adjustment, there is no adjustment for friction wheel to disc clearance.

OVERHAUL

ENGINE

Engine make and model numbers are listed at beginning of this section. Refer to Tecumseh 4-stroke engine section for engine overhaul information.

AUGER GEARBOX

Refer to Fig. JD34 for exploded view of auger gearbox. To remove auger gearbox, first remove auger drive belt as outlined in a following paragraph. Block up under snowthrower frame, disconnect auger rotating linkage and unbolt auger housing from snowthrower frame. Remove set screw (31) from auger pulley (30) and woodruff key (17) from rear end of auger shaft. Unbolt auger bearings (26) and caps (24) from ends of auger housing and withdraw auger and blower fan assembly from housing. Remove the shear pins (13) and slide augers from auger shaft. Remove set screw (22) from impeller fan (21) and pull fan from rear of impeller shaft and worm gear (16).

Disassembly and reassembly of auger gearbox are evident after inspection of unit and reference to exploded view in Fig. JD34. Clean and inspect gearbox parts and renew any excessively worn or damaged parts. On assembly, refill gearbox with API grade GL-5 85W/140 gear lubricant (worm gear oil). Lubricant capacity is approximately 3 ounces (90 mL).

AUGER DRIVE BELT

Slip auger drive belt (4—Fig. JD31) from auger belt idler (7) and off of engine pulley (5). Slip belt from under brake pad, then off of auger pulley (6). Engage traction lever to move traction drive pulley back for clearance and remove belt from between auger pulley (6) and traction drive pulley (8). Install new belt by reversing removal procedure and check auger drive adjustment.

TRACTION DRIVE BELT

Remove auger drive belt as outlined in preceding paragraph. Pull traction belt idler (9—Fig. JD31) away from belt and move belt from engine pulley (5)

groove. Pull auger idler back to provide clearance to remove belt off lower pulley, slip belt off traction drive pulley (8) and then off engine pulley. Engage traction lever if necessary to gain clearance and remove drive belt upward between auger pulley and drive pulley. Install new belt by reversing removal procedure and check traction drive adjustment.

FRICTION WHEEL

To renew friction wheel (12—Fig. JD33), tilt snowthrower forward and rest top of auger housing on solid block. Remove the cover from bottom of main frame and remove right wheel from axle. Unbolt right hex shaft bearing retainer (2) and slide hex shaft (5) out to right side of main frame. Withdraw the friction wheel and hub assembly from bottom opening and unbolt wheel from hub. Install new friction wheel by reversing disassembly procedure.

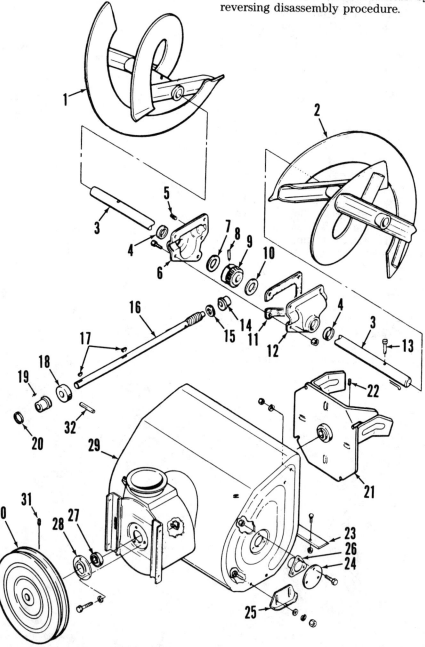

Fig. JD34—Exploded view of auger housing, gearbox and auger flight assembly.

1. L.H. auger flight	10. Thrust washer	17. Woodruff keys	25. Skid shoe
2. R.H. auger flight	11. Gasket	18. Thrust collar	26. Auger bearing
3. Auger shaft	12. R.H. gearcase half	19. Flange bushing	27. Ball bearing
4. Oil seal	13. Shear pin	20. Oil seal	28. Bearing retainer
5. Pipe plug	14. Flange bushing	21. Impeller fan	29. Auger housing
6. L.H. gearcase half	15. Thrust washer	22. Set screw	30. Auger pulley
7. Thrust washer	16. Impeller shaft &	23. Scraper bar	31. Set screw
8. Driv-lok pin	worm gear	24. Bearing cap	32. Pin
9. Worm wheel			

DEUTZ-ALLIS

DEUTZ-ALLIS
P.O. Box 933
Milwaukee, WI 53201

Model	Engine Make	Engine Model	Self-Propelled?	No. of Stages	Scoop Width
SNO-WHIZ	Tecumseh	AH520	No	1	20 in. (508 mm)
SNO-WHIZ II	Tecumseh	AH600	No	1	20 in. (508 mm)

LUBRICATION

ENGINE

Refer to Tecumseh 2-stroke engine section for engine lubrication information. Recommended fuel is regular or unleaded gasoline. After every 25 hours of operation, apply a small amount of SAE 30 engine oil on choke knob (24—Fig. DA2) at control panel and work knob up and down to distribute oil.

DRIVE MECHANISM

Drive belt idler pivot arm should be lubricated at pivot stud (12—Fig. DA4) with SAE 30 engine oil after every 25 hours of operation. To gain access to pivot arm, remove belt cover from left side of snowthrower housing. Be careful not to get any oil on drive belt or pulleys.

MAINTENANCE

Inspect drive belt and renew if frayed, cracked, burned or excessively worn. Pulleys should be renewed if belt groove is excessively worn or pulley is otherwise damaged. Check auger clutch cable; lubricate or renew cable if binding. Check belt idler pulley for bearing being loose or rough and renew idler pulley assembly if defect is noted. Check the rubber impeller paddles and renew if excessively worn or damaged.

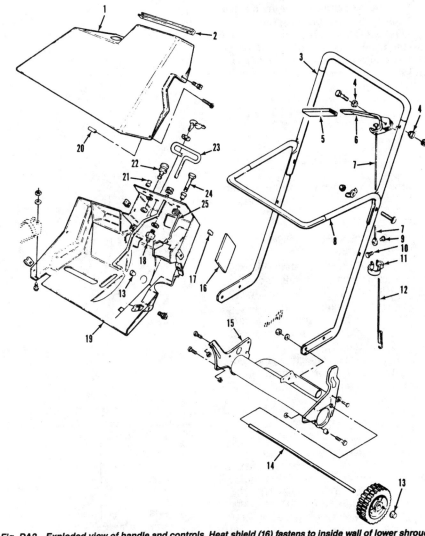

Fig. DA2—Exploded view of handle and controls. Heat shield (16) fastens to inside wall of lower shroud with clips (17). Choke knob (24) screws into coil end of choke link (25).

1. Upper shroud	8. Lower handle	14. Axle shaft
2. Trim channel	9. Clamp screw	15. Muffler
3. Upper handle	10. Cable clamp	16. Heat shield
4. Spacer	11. Cable clip	17. Clip
5. Vinyl lever cover	12. Clutch spring	18. Ignition switch
6. Clutch control lever	13. Push nut	19. Lower shroud
7. Clutch cable		

20. Spacer	
21. Starter rope guide	
22. Fuel primer bulb	
23. Vane control rod	
24. Choke knob	
25. Choke link	

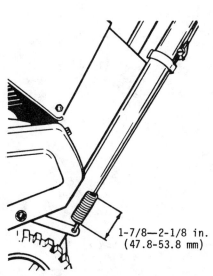

1-7/8—2-1/8 in.
(47.8-53.8 mm)

Fig. DA1—With impeller clutch lever in engaged position, length of clutch cable spring should be as shown.

ADJUSTMENTS

IMPELLER CLUTCH

If impeller fails to turn or belt slips under load, check to be sure idler arm is working freely on pivot stud (12—Fig. DA4). Check idler arm return spring (13) and clutch cable spring (12—Fig. DA2). With control lever engaged against handle, check length of spring as shown in Fig. DA1. If spring length is not between measurements shown, release control lever, loosen clamp screw (9—Fig. DA2) holding control cable (7) to spring (12) and lengthen or shorten spring as necessary.

BELT GUIDES

With impeller clutch lever in engaged position, gap between belt guides (9 & 14—Fig. DA4) and drive belt should be 1/16 inch (1.6 mm). If gap is not correct, loosen guide retaining nuts, reset gap and tighten nuts securely.

OVERHAUL

ENGINE

Refer to Tecumseh 2-stroke engine section for engine overhaul information.

DRIVE MECHANISM

Overhaul of basic unit should be evident after inspection of unit and reference to Fig. DA4. Impeller pulley (19) screws onto impeller shaft (23—Fig. DA3) with left-hand threads. If engine or impeller pulley has been removed, place straightedge on pulleys to check alignment. Move engine crankshaft pulley, if necessary, to bring alignment to within 0.030 inch (0.76 mm).

DRIVE BELT

To renew drive belt, remove cover from left side of snowthrower housing and, using pliers, pull fixed belt guide pin (21—Fig. DA4) from housing (8—Fig. DA3). Loosen belt guides (9 & 14—Fig. DA4). Remove belt from under idler pulley and from engine pulley, then lift belt from impeller pulley. Install new belt and tap fixed guide pin back into auger housing. Be sure that groove in pin is away from belt. Adjust engine pulley belt guides as outlined in adjustment paragraph.

IMPELLER PADDLES

Impeller paddles (22—Fig. DA3) can be renewed after removing the row of cap screws retaining each paddle in the impeller hub assembly, then pulling paddle from hub. Paddles must be replaced in pairs to avoid an unbalanced condition in impeller. Be sure that curve (cup) in paddle faces direction of impeller rotation.

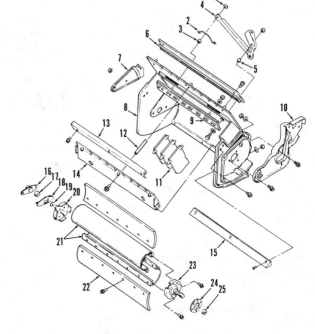

Fig. DA3—Exploded view of impeller and discharge vane assembly.

1. Self-thread nut
2. Detent spring
3. Bushing
4. Vane control lever
5. Bushing
6. Rear vane retainer
7. Right bracket
8. Impeller housing
9. Vane adjustment bar
10. Left bracket
11. Vanes
12. Spacer tube
13. Front vane retainer
14. Front vane mount
15. Scraper bar
16. Bearing retainer
17. Bearing
18. Bearing flange
19. Flat washer
20. Right impeller shaft
21. Impeller drum
22. Impeller paddle
23. Left impeller shaft
24. Bearing retainer
25. Bearing

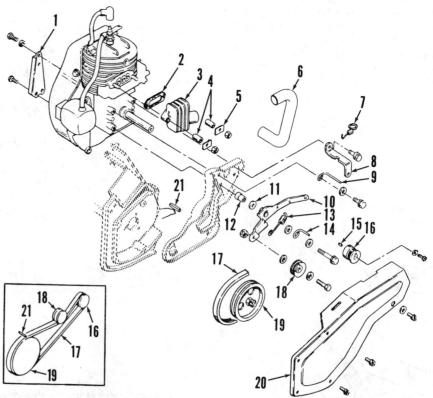

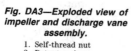

Fig. DA4—Exploded view showing engine and impeller drive mechanism. Spring (7) hooks in notch of belt cover bracket (8) and around exhaust pipe (6) to hold pipe in position.

1. Engine bracket
2. Manifold gasket
3. Exhaust manifold
4. Spacers
5. Nut retainer tab
6. Exhaust pipe
7. Pipe retainer spring
8. Belt cover bracket
9. Belt guide
10. Belt idler arm
11. Flat washer
12. Pivot stud
13. Idler arm return spring
14. Belt guide
15. Woodruff key
16. Engine pulley
17. Impeller belt
18. Idler pulley
19. Impeller pulley
20. Belt cover
21. Belt guide pin

DEUTZ-ALLIS

Model	Engine Make	Engine Model	Self-Propelled?	No. of Stages	Scoop Width
420	Tecumseh	HS35	No	2	20 in. (508 mm)
522	Tecumseh	HSSK50	Yes	2	22 in. (559 mm)
722	Tecumseh	HMSK70	Yes	2	22 in. (559 mm)

LUBRICATION

ENGINE

Refer to Tecumseh 4-stroke engine section for engine lubrication information. Recommended fuel is unleaded or leaded regular grade gasoline.

DRIVE MECHANISM

Auger gearbox oil level should be maintained level with plug opening (9—Fig. DA15) in gearbox with Grade GL4 gear lubricant (worm gear oil). After every 10 hours of operation, lubricate all linkage pivot points and discharge chute pivot with SAE 5W-30 motor oil.

Once each year, remove chute and wheels and grease chute flange and axles with a clinging type grease. Once each year or each time it is necessary to replace an auger shear pin, use a hand-operated grease gun and lubricate the augers at grease fittings.

On Models 522 and 722, lubricate traction drive roller chain after every 10 hours of operation with SAE 5W-30 motor oil. Remove bottom panel to gain access to lubrication points. The three-speed gear-type transmission is lubricated at the factory and does not require lubrication except during overhaul; refer to Peerless transmission section for service information.

MAINTENANCE

Inspect belts and pulleys and renew belts if frayed, cracked, burned or excessively worn. Check pulley grooves for excessive wear or damage and renew if defect is found. Check idler pulley bearings and renew idler pulley assembly if bearings are loose or rough. Inspect traction drive chain and sprockets on Models 522 and 722 and renew if excessively worn or damaged.

ADJUSTMENTS

AUGER HEIGHT

Auger height on Model 420 is nonadjustable as there are no skid shoes. The rubber auger flights propel the snowthrower by contacting ground surface.

Auger height on Models 522 and 722 can be adjusted by loosening skid shoe mounting bolts and repositioning the skid shoes. Auger should be in lowest position (skid shoes raised on housing) when operating over smooth surfaces, and in highest position when operating over rough surfaces.

SHIFT CONTROL

To check shift control linkage on Models 522 and 722, place blocks under snowthrower so wheels are off the ground. With speed control lever (L—Fig. DA10) in neutral position, the wheels should be free to turn by hand. If not, loosen screws (A) retaining speed control indicator bracket (B) to rear of snowthrower and move the bracket side-to-side until neutral position is found where wheels can be rotated. Tighten the bracket retaining screws and be sure that all gears will engage. If speed control lever strikes handles at either the reverse or third speed position, loosen screws (C) holding speed control lever (L), move lever as necessary and tighten the screws. Recheck adjustment; the transmission must dis-

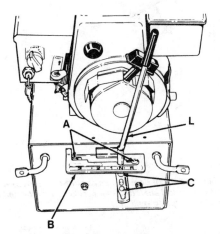

Fig. DA10—View showing adjustment points for speed control lever (L) on Models 522 and 722. Refer to text.

Fig. DA11—View of drive idler system with belt cover removed. Model 420 does not have traction drive components.

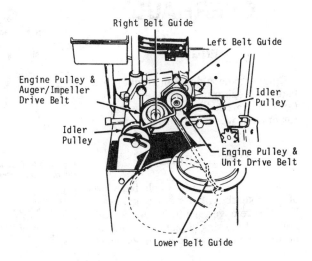

engage when speed control lever is in neutral position and reach all shift positions without contacting snowthrower handles.

AUGER DRIVE BELT

If auger drive belt slips under load, the drive belt tension can be adjusted as follows: Loosen locknut on auger idler pulley (see Fig. DA11), move idler 1/4 inch (6.4 mm) toward belt and tighten locknut. Note that auger brake pad (13—Fig. DA13) must clear belt with auger control lever in engaged position. With auger control lever in disengaged position, there should be some slack in chain (15—Fig. DA12) connecting control rod (13) to spring (16). If necessary, slide sheath (14) up and connect chain in different link. The auger should not turn with control lever in disengaged position.

TRACTION DRIVE BELT

If traction drive belt slips on Model 522 or Model 722, loosen locknut on drive idler pulley (see Fig. DA11), move idler 1/4 inch (6.4 mm) toward belt and tighten locknut. With drive control lever (2—Fig. DA12) in disengaged position, there should be some slack in chain (15) connecting drive control rod to spring. If necessary, slide sheath up and connect chain in different link.

BELT GUIDES

With auger and, on Model 522 or Model 722, drive clutch levers in engaged position, all belt guides should clear the drive belts from 1/16 to 1/8 inch (1.6-3.2 mm). Upper belt guides are bolted to engine and lower guides are mounted on rear of auger housing.

OVERHAUL

ENGINE

Engine make and model numbers are listed at beginning of this section. Refer to Tecumseh 4-stroke engine section for engine overhaul information.

AUGER GEARBOX

Refer to Fig. DA15 for exploded view of auger gearbox. To remove auger gearbox, first remove auger drive belt as outlined in a following paragraph. Block up under snowthrower frame, disconnect

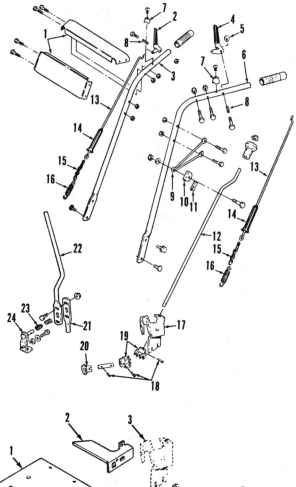

Fig. DA12—Exploded view of Model 722 handle and controls; Model 522 is similar. Model 420 does not have right-handle drive control mechanism or shift lever controls.

1. Handle panels
2. Drive control lever
3. Right handle
4. Auger control lever
5. Locknut
6. Left handle
7. Lever bumper
8. Spacer
9. Chute crank bracket
10. Chute crank guide
11. Crank guide liner
12. Chute control crank
13. Clutch control rod
14. Plastic sheath
15. Clutch control chain
16. Clutch control spring
17. Chute gear support
18. Roll pins
19. Chute crank gears
20. Chute crank sprocket
21. Lower speed control lever
22. Upper speed control lever
23. Spring
24. Transmission shift arm

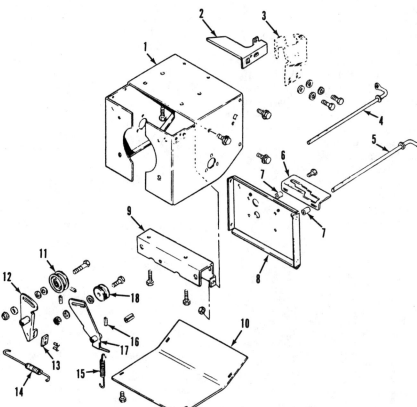

Fig. DA13—Exploded view of belt idler clutch system. Model 420 does not have shift guide bracket (6), drive belt idler (18), drive idler bracket (17) and drive control shaft (5). Idler brackets are attached to front ends of control shafts by spring pins (16).

1. Main frame
2. Belt cover
3. Chute gear bracket
4. Auger control shaft
5. Drive control shaft
6. Speed control guide bracket
7. Spacer
8. Main frame rear cover
9. Transmission support
10. Bottom cover
11. Auger belt idler
12. Auger idler bracket
13. Auger brake pad
14. Auger return spring
15. Drive return spring
16. Spring pin
17. Drive idler bracket
18. Drive belt idler

auger rotating linkage and unbolt auger housing from snowthrower frame. Remove set screws from auger drive pulley (25) and Woodruff key (23) from rear end of auger shaft. Remove snap rings (10—Fig. DA16) at outer ends of auger shaft and unbolt auger bearings from

ends of auger housing. Withdraw auger and impeller fan assembly from housing, remove shear pins (1) and slide augers from auger shaft. Remove set screw from impeller fan (18—Fig. DA15) and pull fan from rear of auger gearbox input shaft.

Disassembly and reassembly of auger gearbox are evident after inspection of unit and reference to exploded view in Fig. DA15. Clean and inspect gearbox parts and renew any excessively worn or damaged parts. During reassembly, refill gearbox with API grade GL-4 gear lubricant (worm gear oil). Lubricant capacity is approximately 3 ounces (90 mL).

TRANSMISSION

These models are equipped with a Peerless Series 700 transmission with three forward and one reverse speeds. With transmission removed from snowthrower, refer to Peerless transmission section for overhaul information.

DRIVE BELTS

Remove top belt cover, loosen belt guides and slip the auger belt from engine crankshaft pulley or the traction drive belt (Models 522 & 722) off of engine camshaft (PTO) pulley. Tip snowthrower forward onto front of auger

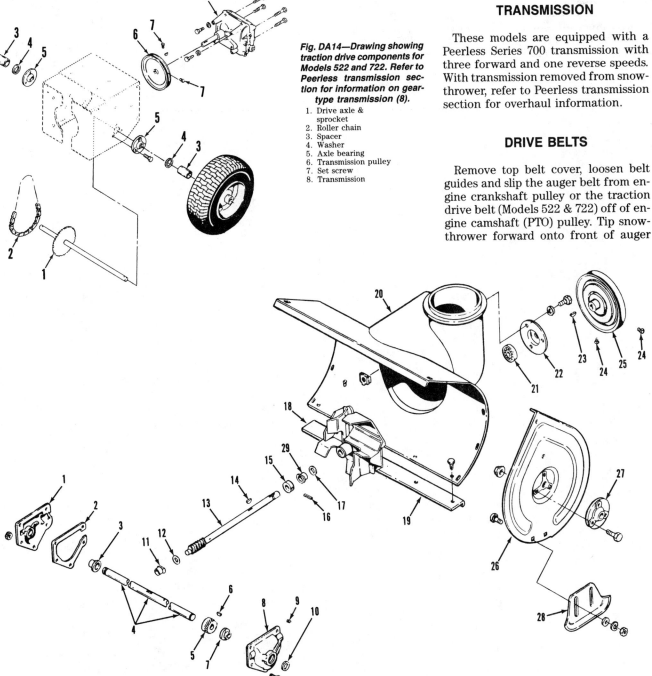

Fig. DA14—Drawing showing traction drive components for Models 522 and 722. Refer to Peerless transmission section for information on gear-type transmission (8).

1. Drive axle & sprocket
2. Roller chain
3. Spacer
4. Washer
5. Axle bearing
6. Transmission pulley
7. Set screw
8. Transmission

Fig. DA15—Exploded view of auger housing and gearbox. Model 420 auger housing is not fitted with skid shoes.

1. Right gearbox half	8. Left gearbox half	12. Thrust washer	17. Oil seal	22. Bearing retainer	27. Auger shaft bearing
2. Gasket	9. Pipe plug	13. Impeller worm shaft	18. Impeller fan	23. Woodruff key	28. Skid shoe
3. Auger shaft bushing	10. Oil seal	14. Woodruff key	19. Scraper bar	24. Set screws	29. Impeller shaft
4. Auger shaft	11. Impeller shaft bushing	15. Thrust collar	20. Auger housing	25. Auger drive pulley	bushing
5. Worm drive gear		16. Driv-lok pin	21. Ball bearing	26. Housing end plate	
6. Woodruff key					
7. Auger shaft bushing					

housing; support upper edge of housing with wood block and be sure unit is sitting solid. Remove lower cover and loosen belt guides. Pivot belt guides clear of drive belt(s). Lift auger drive belt off of pulley and remove it upward from the auger pulley. On Models 522 and 722, remove both belts upward between the auger and traction drive pulleys. It will be necessary to hold auger clutch lever down to free auger belt from under auger drive brake.

Install new belt(s) by reversing removal procedures and check adjustments as outlined in appropriate belt adjustment paragraph.

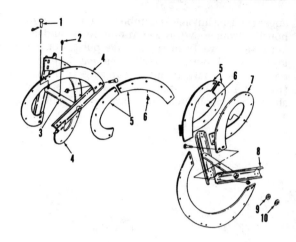

Fig. DA16—Exploded view of Model 420 auger flights. Snowthrower is propelled forward by contact of rubber flight edges with ground surface.

1. Shear pin
2. Grease fitting
3. Auger flight support
4. Spiral blades, R.H.
5. Rubber flights
6. Tubular rivet
7. Spiral blade, L.H.
8. Auger flight support
9. Flat washer
10. Snap ring

DEUTZ-ALLIS

Model	Engine Make	Engine Model	Self-Propelled?	No. of Stages	Scoop Width
524	Tecumseh	HSSK50	Yes	2	24 in. (610 mm)
724	Tecumseh	HMSK70	Yes	2	24 in. (610 mm)
828	Tecumseh	HMSK80	Yes	2	28 in. (711 mm)

LUBRICATION

ENGINE

Refer to Tecumseh 4-stroke engine section for engine lubrication information. Recommended fuel is unleaded or leaded regular grade gasoline.

DRIVE MECHANISM

Auger gearbox oil level should be maintained at a level even with plug (5—Fig. DA25) opening in gearbox with 85W/140 Grade GL5 gear lubricant (worm gear oil). After every 10 hours of operation, lubricate all linkage pivot points and discharge chute pivot with SAE 30 motor oil. Lubricate traction drive chain after every 10 hours of operation with a light coat of multipurpose grease. Remove bottom panel to gain access to lubrication points.

The auger flights are driven from the auger gearbox shaft by shear pins which may fracture if the auger encounters a solid obstruction. If this happens, turn auger so holes in auger tube and gearbox shaft are aligned and install a new factory replacement shear pin.

ADJUSTMENTS

AUGER HEIGHT

Auger height can be adjusted by loosening skid shoe mounting bolts and repositioning the skid shoes. Auger should be in lowest position (skid shoes raised on housing) when operating over smooth surfaces, and in highest position when operating over rough surfaces.

SHIFT CONTROL

Place the shift lever in 5th gear position and loosen the two nuts (F—Fig. DA21) clamping upper and lower speed control rods together. The spring loaded disc drive pivot assembly will move the lower rod (E) into proper position for 5th gear, then tighten nuts (F).

AUGER DRIVE BRAKE

With auger clutch lever released, brake pad (12—Fig. DA23) should contact auger belt and stop it from turning. With clutch lever in engaged position, the brake pad should clear the auger drive belt. Some adjustment can be obtained by loosening nut (10) and moving idler in slot of auger clutch rod and idler bracket (9). If brake clearance is correct, check auger drive clutch adjustment.

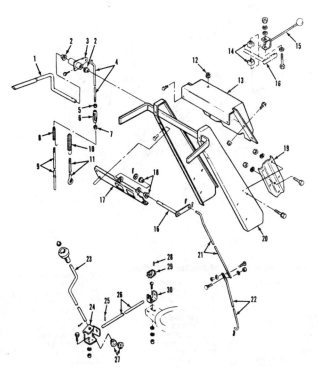

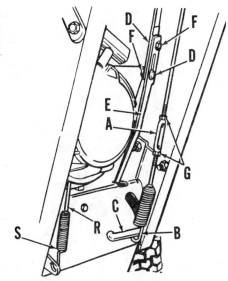

Fig. DA20—View showing handle and control parts. Auger clutch control parts not shown except for spring (10) and lower auger clutch rod (11); control lever is reverse of drive clutch lever (1) and upper control rod parts are same as for drive clutch (3, 4, 5, 6, & 7).

1. Drive clutch lever
2. Plastic bushing
3. Clutch bellcrank
4. Upper clutch rod
5. L.H. thread nut
6. Turnbuckle
7. R.H. thread nut
8. Drive clutch spring
9. Lower drive clutch rod
10. Auger clutch spring
11. Lower auger clutch rod
12. Chute crank plastic bushing
13. Handle console
14. Pivot blocks
15. Speed control lever
16. Speed control bellcrank
17. Speed control bracket
18. Plastic bushings
19. Handle bracket
20. Handle assy.
21. Upper speed control rod
22. Lower speed control rod
23. Chute crank
24. Bevel gear bracket
25. Roll pin
26. Chute control rod
27. Bevel gears
28. Roll pin
29. Chute worm gear
30. Worm gear bracket

Fig. DA21—Drawing showing control linkage adjustment points.

A. Turnbuckle
B. Auger clutch spring
C. Auger idler shaft
D. Carriage bolts
E. Lower speed control rod
F. Nuts
G. Turnbuckle nuts
R. Drive clutch lower rod
S. Drive clutch spring

AUGER DRIVE CLUTCH

There should be 1/16-inch free play of auger clutch spring (B—Fig. DA21) at auger idler shaft (C). To adjust, loosen turnbuckle nuts (G) and rotate turnbuckle (A) until correct spring to shaft lever clearance is obtained. Tighten the turnbuckle nuts and recheck adjustment.

TRACTION DRIVE CLUTCH

The bottom end of drive clutch lower rod (R—Fig. DA21) should be flush with bottom end of drive clutch spring (S) with the drive clutch lever engaged. This pulls the friction wheel against friction disc to drive the snowthrower. To adjust, loosen the locknuts at control rod turnbuckle and rotate the turnbuckle. Tighten the turnbuckle locknuts and recheck adjustment. The traction drive belt idler is spring loaded and adjustment of belt is not necessary.

FRICTION WHEEL

Other than traction drive clutch adjustment, there is no adjustment for friction wheel to disc clearance.

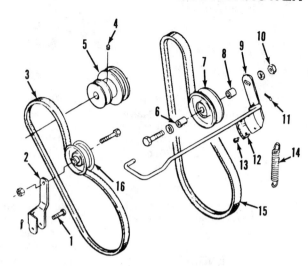

Fig. DA23—Exploded view of belt drive mechanism.

1. Pivot pin
2. Drive idler bracket
3. Traction drive belt
4. Set screw
5. Engine pulley
6. Spacer
7. Auger belt idler
8. Spacer
9. Auger clutch rod & idler bracket
10. Nut
11. Cotter pin
12. Brake pad
13. Rivet
14. Auger idler return spring
15. Auger drive belt
16. Drive belt idler

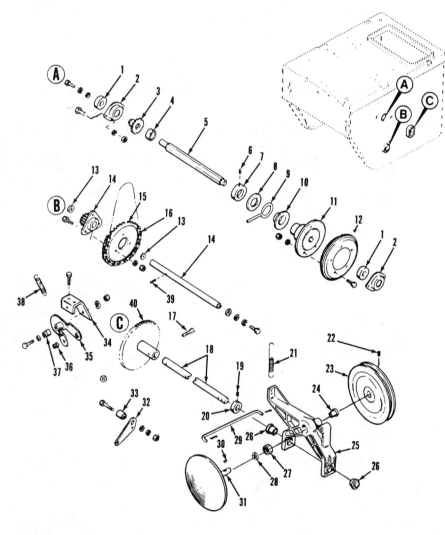

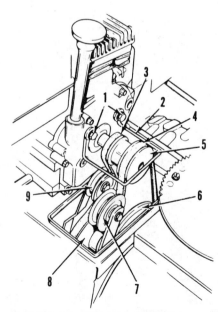

Fig. DA22—Drawing of drive belt mechanism with belt cover removed.

1. Belt guide screws
2. Belt guide
3. Traction drive belt
4. Auger drive belt
5. Engine pulley
6. Auger pulley
7. Auger belt idler
8. Traction drive pulley
9. Traction belt idler

Fig. DA24—Exploded view of friction disc drive.

1. Hex shaft bearing
2. Bearing retainer
3. Hex shaft sprocket
4. Spacer
5. Hex shaft
6. Set screw
7. Collar
8. Thrust washer
9. Slide pin
10. Thrust collar
11. Friction wheel hub
12. Friction wheel
13. Washers
14. Gear & hub
15. Jack shaft sprocket
16. Roller chain
17. Bolt
18. Axle
19. Set screw
20. Collar
21. Disc return spring
22. Set screw
23. Drive pulley
24. Bushing
25. Pivot arm assy.
26. Bushing
27. Bearing
28. Thrust washer
29. Pivot rod
30. Woodruff key
31. Friction drive disc
32. Drive control bellcrank
33. Pivot bushing
34. Shift arm bracket
35. Shift arm assy.
36. Bushing
37. Spacer bushing
38. Friction wheel return spring
39. Roll pin
40. Gear

OVERHAUL

ENGINE

Engine make and model numbers are listed at beginning of this section. Refer to Tecumseh 4-stroke engine section for engine overhaul information.

AUGER GEARBOX

Refer to Fig. DA25 for exploded view of auger gearbox. To remove auger gearbox, first remove auger drive belt as outlined in a following paragraph. Block up under snowthrower frame, disconnect auger rotating linkage and unbolt auger housing from snowthrower frame. Remove set screw (31) from auger pulley (30) and Woodruff key (17) from rear end of auger shaft. Unbolt auger bearings (26) and caps (24) from ends of auger housing and withdraw auger and blower fan assembly from housing. Remove the shear pins (13) and slide augers from auger shaft. Remove set screw (22) from impeller fan (21) and pull fan from rear of impeller shaft and worm gear (16).

Disassembly and reassembly of auger gearbox are evident after inspection of unit and reference to exploded view in Fig. DA25. Clean and inspect gearbox parts and renew any excessively worn or damaged parts. On assembly, refill gearbox with API grade GL-5 85W/140 gear lubricant. Lubricant capacity is approximately 3 ounces (90 mL).

AUGER DRIVE BELT

Slip auger drive belt (4—Fig. DA22) from auger belt idler (7) and off of engine pulley (5). Slip belt from under brake pad, then off of auger pulley (6). Engage traction lever on Model 522 and Model 722 to move traction drive pulley back for clearance and remove belt from between auger pulley (6) and traction drive pulley (8). Install new belt by reversing removal procedure and check auger drive adjustment.

TRACTION DRIVE BELT

Remove auger drive belt as outlined in preceding paragraph. Pull traction belt idler (9—Fig. DA22) away from belt and move belt from engine pulley (5) groove. Pull auger idler back to provide clearance to remove belt off lower pulley, slip belt off traction drive pulley (8) and then off engine pulley. Engage traction lever if necessary to gain clearance and remove drive belt upward between

auger pulley and drive pulley. Install new belt by reversing removal procedure and check traction drive adjustment.

FRICTION WHEEL

To renew friction wheel (12—Fig. DA24), tilt snowthrower forward and rest top of auger housing on solid block. Remove the cover from bottom of main frame and remove right wheel from axle. Unbolt right hex shaft bearing retainer (2) and slide hex shaft (5) out to right side of main frame. Withdraw the friction wheel and hub assembly from bottom opening and unbolt wheel from hub. Install new friction wheel by reversing disassembly procedure.

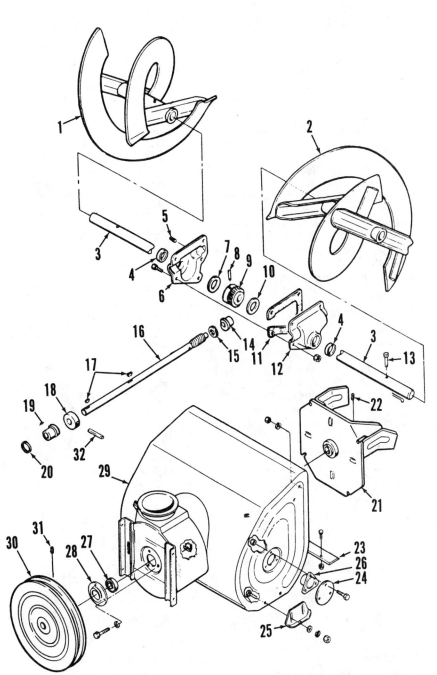

Fig. DA25—Exploded view of auger housing, gearbox and auger flight assembly.

1. L.H. auger flight	10. Thrust washer	17. Woodruff keys	25. Skid shoe
2. R.H. auger flight	11. Gasket	18. Thrust collar	26. Auger bearing
3. Auger shaft	12. R.H. gearcase half	19. Flange bushing	27. Ball bearing
4. Oil seal	13. Shear pin	20. Oil seal	28. Bearing retainer
5. Pipe plug	14. Flange bushing	21. Impeller fan	29. Auger housing
6. L.H. gearcase half	15. Thrust washer	22. Set screw	30. Auger pulley
7. Thrust washer	16. Impeller shaft &	23. Scraper bar	31. Set screw
8. Driv-lok pin	worm gear	24. Bearing cap	32. Pin
9. Worm wheel			

FORD

FORD TRACTOR OPERATIONS
2500 E. Maple Rd.
Troy, MI 48084

Model	Engine Make	Engine Model	Self-Propelled?	No. of Stages	Scoop Width
ST-220	Tecumseh	AH520	No	1	20 in. (508mm)
ST-320	B&S	62032	No	1	20 in. (508 mm)
ST-320B	Tecumseh	AH600	No	1	20 in. (508 mm)

LUBRICATION

ENGINE

Refer to Briggs & Stratton or Tecumseh engine section for engine lubrication information. Recommended fuel is unleaded or leaded regular gasoline mixed with a good-quality 2-cycle air-cooled engine oil. Fuel and oil should be mixed in a 32:1 ratio.

DRIVE MECHANISM

Lubricate auger shaft bearing (13—Fig. F1) at right side of auger housing after every 10 hours of operation using SAE 30 oil. Use sufficient oil so oil just runs from hole; felt plug (14) retains oil between lubrication intervals. Bearing at left end of auger is a sealed prelubricated type.

1. Auger flight
2. Impeller paddle
3. Auger plate half
4. Auger support shaft
5. Thrust washer (2-3)
6. Auger stop spring
7. Scraper bar
8. Deflector vane
9. Vane control link
10. Top cover
11. R.H. end plate
12. Rivet
13. Auger shaft bearing
14. Felt plug
15. Vane control handle
16. Control rod
17. Vane pivot assy.
18. Axle frame
19. Support channel
20. Wheel support clamp
21. Housing wrap sheet
22. Scraper baffle
23. Rivet
24. Bearing retainer
25. Ball bearing
26. Woodruff key
27. Wear strip
28. Pivot bushing
29. Wire belt guide
30. Idler arm
31. Auger brake spring
32. Belt retainer
33. Skid block
34. Auger pulley
35. Drive belt
36. Belt cover
37. Snap ring
38. Idler pulley
39. Engine pulley
40. Idler washers
41. Lower belt guide
42. Upper belt guide
43. L.H. end plate
44. Axle shaft
45. Wheel brace
46. Primer tube
47. Lower cover
48. Ignition switch
49. Choke wire
50. Primer bulb
51. Engine spacer
52. Woodruff key
53. Panel brace
54. Brace
55. Brace
56. Recoil brace
57. Lower handle
58. Cable clamp
59. Clutch cable & chain
60. Clutch adjuster clip
61. Clutch spring
62. Clutch lever
63. Upper handle

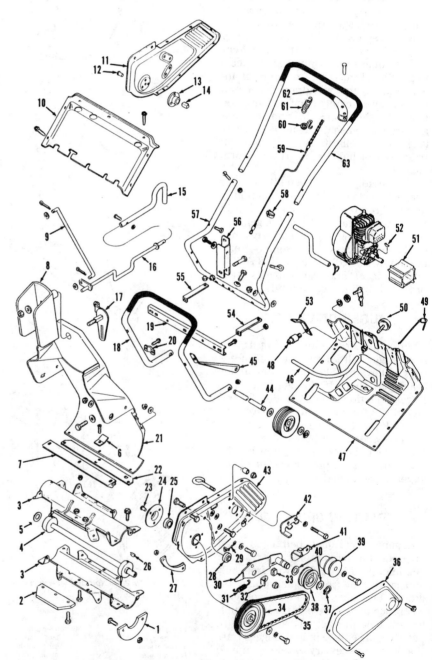

Fig. F1—Exploded view of Model ST-320 snowthrower. Model ST-220 and Model ST-320B are similar.

After each 10 hours of operation, remove belt cover and lubricate idler pulley (38) and shaft on idler arm (30) with SAE 30 oil. Turn pulley while applying lubricant to distribute oil evenly. Be careful not to allow oil to get on belt or pulleys. Idler arm pivot bushing (28) and skid block (33) should be lubricated with No. 2 wheel bearing grease.

MAINTENANCE

Inspect drive belt and renew if frayed, cracked, burned or otherwise damaged. Renew belt pulleys if belt groove is excessively worn or pulley is damaged. Inspect impeller for wear or damage. Impeller paddle (2—Fig. F1) and auger flights (1) can be renewed if necessary. Move deflector vane pivot assembly (17) from side-to-side. Deflector vane (8) locknut should be adjusted so that deflector moves without binding.

ADJUSTMENTS

DRIVE BELT

With auger clutch lever released, there should be no slack in the clutch cable. To increase belt tension, move open end of clutch adjuster clip (60—Fig. F1) down to next link in clutch cable and chain (59). If clip is in last link of chain, it may be necessary to move clip loop up to the next link in end of chain. To check adjustment, pull engine starter rope with ignition in "OFF" position. If auger turns, move adjustment clip upward in chain (closer to clutch lever). With clutch engaged, belt retainer (32) should clear belt by a maximum of 1/8 inch (3.2 mm) and retainer should be parallel to belt. Belt guides (29, 41 and 42) should be parallel to the belt and clear the belt when clutch is engaged.

OVERHAUL

ENGINE

Engine make and model information is listed at beginning of this section. Refer to appropriate Briggs & Stratton or Tecumseh engine section for engine overhaul information.

DRIVE MECHANISM

Drive belt (35—Fig. F1) can be removed and a new belt installed after removing belt cover (36) and lifting idler from belt. Refer to belt tension adjustment in previous ADJUSTMENT section.

If belt idler arm (30) is removed, lightly grease back side of arm where it contacts skid block (33) and pivot bushing (28) with No. 2 wheel bearing grease so idler arm will pivot freely after reassembly. Lubricate idler pulley shaft and bushing with No. 2 wheel bearing grease if pulley is removed.

AUGER ASSEMBLY

If excessive wear of auger flight and impeller paddle rubber fins is noted, the rubber fins should be replaced as a complete set. The fins can be replaced after removing the cap screws holding the fins between the two auger plate halves (3—Fig. F1).

If excessively worn or damaged, the auger assembly scraper bar (7) can be replaced after removing the three flat head socket screws that hold scraper bar to bottom of auger housing. Auger stop spring (6) mounted with center scraper bar bolt keeps auger from rotating backwards.

FORD

Model	Engine Make	Engine Model	Self-Propelled?	No. of Stages	Scoop Width
ST-420	B&S	100000	Yes	1	20 in. (508 mm)

LUBRICATION

ENGINE

Refer to Briggs & Stratton section for engine lubrication requirements. Recommended fuel is low-lead or regular gasoline.

DRIVE MECHANISM

Lubricate drive chain after each use with a suitable chain oil. Use a good quality multipurpose grease and lubricate auger shaft and jackshaft through grease fittings. Note that auger shaft grease fitting is accessible through hole in auger frame right side. Apply multipurpose grease to gear teeth and sliding surfaces of clutch dog. Lubricate pivot points of actuating linkage for smooth operation.

ADJUSTMENTS

HEIGHT

Auger height is adjusted by loosening skid shoe retaining screws and repositioning skid shoes. Low auger height may be used on smooth surfaces but height should be increased when operating on irregular surfaces.

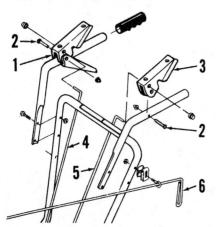

Fig. F10—Exploded view of control assembly.

1. Drive clutch control lever	4. Drive clutch control rod
2. Pivot rivet	5. Auger clutch control rod
3. Auger clutch control lever	6. Chute crank

NEUTRAL SPRING

Neutral spring (2–Fig. F14) should return clutch dog (11) to neutral when either forward or reverse control lever is released. Adjust spring by loosening nuts (N–Fig. F11) and turning screws on rear of main frame. Tighten nuts (N) and check adjustment.

DRIVE CLUTCH CONTROL ROD

With engine stopped and either forward or reverse control lever engaged by squeezing against handlebar, snowthrower should be locked in gear. If not, adjust control rod as follows: Refer to Fig. F12 and loosen the two screws (S) which clamp upper and lower drive clutch control rods together just enough so rods can move but remain aligned. Pull up reverse control lever and tighten the two rod screws finger tight. Depress forward control lever while rocking unit forward to ensure full clutch engagement. With forward control lever depressed against handlebar, tighten rod screws. Check adjustment.

AUGER CLUTCH CONTROL ROD

To adjust auger clutch control rod, loosen screw (S–Fig. F13) clamping upper and lower control rod together. Move auger clutch control lever to

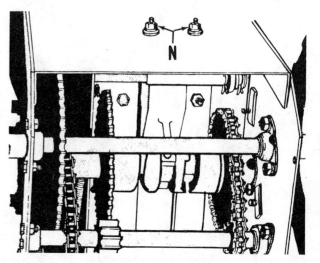

Fig. F11—Loosen nuts (N) and turn screws to adjust neutral spring as outlined in text.

disengaged position. Pull lower control rod until idler pulley contacts belt. Move clamp screw so it is midway between ends of control rods and tighten screw finger tight. Depress auger clutch control lever until lever almost contacts handlebar and tighten clamp screw. Check adjustment.

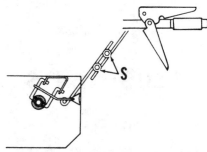

Fig. F12—Adjust drive clutch control rod using screws (S) as outlined in text.

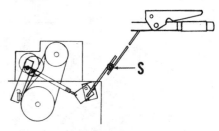

Fig. F13—Refer to text and adjust auger clutch control rod using screw (S).

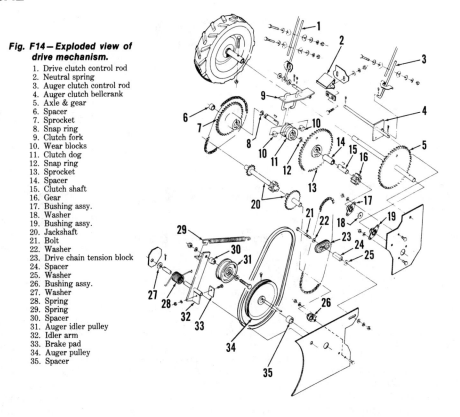

Fig. F14 — Exploded view of drive mechanism.

1. Drive clutch control rod
2. Neutral spring
3. Auger clutch control rod
4. Auger clutch bellcrank
5. Axle & gear
6. Spacer
7. Sprocket
8. Snap ring
9. Clutch fork
10. Wear blocks
11. Clutch dog
12. Snap ring
13. Sprocket
14. Spacer
15. Clutch shaft
16. Gear
17. Bushing assy.
18. Washer
19. Bushing assy.
20. Jackshaft
21. Bolt
22. Washer
23. Drive chain tension block
24. Spacer
25. Washer
26. Bushing assy.
27. Washer
28. Spring
29. Spring
30. Spacer
31. Auger idler pulley
32. Idler arm
33. Brake pad
34. Auger pulley
35. Spacer

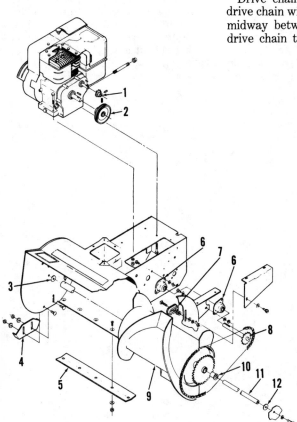

Fig. F15 — Exploded view of auger assembly.

1. Drive sprocket
2. Auger drive pulley
3. Washer
4. Skid shoe
5. Scraper blade
6. Bushing assy.
7. Auger chain tension block
8. Sprocket
9. Auger & sprocket assy.
10. Bushing
11. Auger shaft
12. Washer

DRIVE CHAIN TENSION

Drive chain tension is correct when drive chain will deflect 1/8 inch (3.2 mm) midway between sprockets. To adjust drive chain tension, loosen nut on left side of main frame retaining chain tension block. Move chain tension block as required for desired chain tension. Retighten nut.

AUGER CHAIN TENSION

Auger chain tension is correct when auger chain will deflect ¼ inch (6.4 mm) midway between sprockets. To adjust auger chain tension, loosen nut on left side of auger frame retaining chain tension block. Relocate chain tension block as required for desired chain tension. Retighten nut.

OVERHAUL

ENGINE

Engine make and model numbers are listed at front of section. Refer to engine section for overhaul procedures.

DRIVE MECHANISM

Overhaul procedure for drive mechanism is evident after inspection of unit and referral to Figs. F14 and F15. Inspect components for excessive wear or damage. Replace as needed. Refer to ADJUSTMENTS section after assembly.

FORD

Model	Engine Make	Engine Model	Self-Propelled?	No. of Stages	Scoop Width
ST-626	Tecumseh	H60	Yes	2	26 in. (660 mm)
ST-830	Tecumseh	HM80	Yes	2	30 in. (762 mm)

LUBRICATION

ENGINE

Refer to Tecumseh 4-stroke engine section for engine lubrication information. Recommended fuel is unleaded or leaded regular gasoline.

DRIVE MECHANISM

Lubrication should be done at start of each season and after every 40 hours of operation during the season. Lubricate all control linkage pivot points and the drive chain using SAE 30 engine oil. Lubricate discharge chute flange, chute sprocket and sprocket bearing, traction drive gears and friction wheel hex shaft with water-resistant grease. Remove auger shaft bushings and apply grease to inside of bushings and retainer cups; align slot of bushing with retainer tab when reinstalling. Using hand-operated grease gun, apply grease at auger lubrication fittings.

AUGER GEARBOX

Check auger gearbox oil level at start of season and after every 40 hours of operation. With snowthrower level, remove front cover lower retaining screw (S—Fig. F16); oil level should be at bottom of hole. Add lubricant (DX Fibrex grease number 00 or equivalent) as necessary through filler plug opening in top of gearbox.

MAINTENANCE

Inspect chain and sprockets and renew if excessively worn. Inspect belts and pulleys; renew belt if excessively worn, burned, frayed, cracked or otherwise damaged. Renew idler pulleys if bearings are loose or rough.

The augers are driven through special bolts designed to shear if auger encounters a solid obstruction. If bolt is sheared, turn auger to align bolt holes in auger and shaft and drive out remainder of old bolt. Apply grease at auger lubrication fittings and be sure auger turns freely on shaft, then install new bolt using only a factory replacement part.

ADJUSTMENTS

AUGER HEIGHT

Auger height is adjusted by loosening skid shoe retaining bolts and repositioning the skid shoes. Raise the skid shoes on auger housing to lower scraper bar for smooth surfaces; lower the skids to raise scraper bar when operating over rough surfaces such as gravel drives.

SPEED CONTROL ROD

To adjust speed control rod, detach upper speed control rod (6—Fig. F17) from speed control lever (4). Move speed control lever to reverse position on handle console and pull up on speed control rod. The upper end of rod should be aligned with hole in lever; if not, turn upper rod until end will fit into hole and reattach rod to lever.

AUGER BRAKE

If auger brake pad has been renewed or the brake arm bracket (17—Fig. F18) has been loosened, adjust brake as follows: With belt cover removed and auger control lever disengaged, loosen bracket (17) attaching bolts and push brake pad down securely in auger pulley belt groove and tighten the bolts.

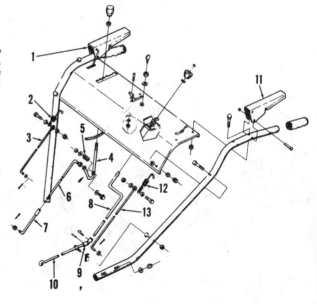

Fig. F17—Exploded view of handle and controls. Chute crank (8) has been relocated to bracket on left handle on later models.

1. Drive clutch lever
2. Drive control spring
3. Drive clutch control rod
4. Speed control lever
5. Lever friction spring
6. Upper speed control rod
7. Lower speed control rod
8. Chute crank
9. U-joint
10. Chute crank rod
11. Auger clutch lever
12. Auger control spring
13. Auger clutch control rod

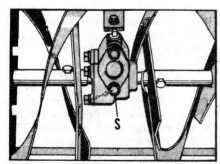

Fig. F16—Oil level in auger gearbox should be at bottom of lower cover screw (S) hole.

AUGER CLUTCH

First, if necessary, adjust auger brake as outlined in preceding paragraph, then proceed as follows: Loosen the bolt securing auger clutch control rod (13—Fig. F17) to spring (12) and place a 1/4-inch diameter round spacer between handle and front of auger clutch lever (11). Pull control rod upward in slot of spring until all slack is removed from auger clutch cable and tighten rod to spring bolt. Remove spacer from under auger clutch lever; with lever released, there should be a minimum of 1/8 inch (3.2 mm) slack in auger clutch cable. Spring (12) must extend when auger clutch lever is depressed against handle.

If proper adjustment cannot be obtained by above procedure, it will be necessary to loosen bellcrank pivot bracket (20—Fig. F18) bolts, reposition the bracket and tighten bolts. Then, repeat control adjustment procedure.

DRIVE CLUTCH CONTROL ROD

To adjust drive clutch control rod (3—Fig. F17), loosen bolt holding rod to spring (2). With drive clutch lever (1) released and pivoted forward against handle, pull rod up in slot of spring to remove all slack in rod, then tighten bolt. Do not stretch the spring or move clutch mechanism when removing slack from rod.

FRICTION WHEEL CLEARANCE

There should be from 3/32 to 5/32-inch (2.4-4.0 mm) clearance between friction wheel (21—Fig. F19) and friction drive

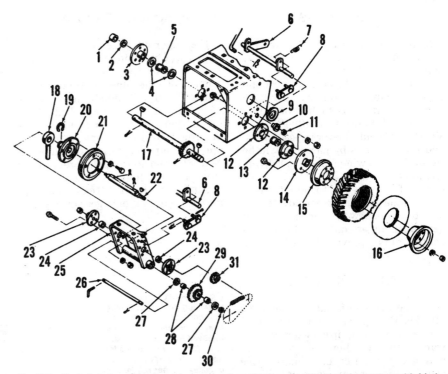

Fig. F19—Exploded view of friction drive mechanism. Friction wheel is driven by contact with friction drive plate (30—Fig. F20) attached to rear side of drive pulley (5—Fig. F20). Wheel (15 & 16), hub (3 or 14) and tire are a one-piece assembly on later models. Axle bushing (12 & 13) is a one-piece assembly with grease fitting on later models.

1. Hub cap
2. Washer
3. Wheel hub
4. Washers
5. Axle bushing
6. Drive clutch pivot
7. Drive clutch return spring
8. Clutch control link
9. Hex shaft access plug
10. Clutch pivot bushing
11. Tinnerman retainer
12. Axle bushing retainers
13. Axle bushing
14. Wheel hub
15. Inner tire rim
16. Outer tire rim
17. Axle shaft & gear
18. Speed control yoke
19. Snap ring
20. Friction wheel hub
21. Friction wheel
22. Hex shaft
23. Hex shaft bearing retainer
24. Hex shaft bearing
25. Friction wheel bracket
26. Bracket pivot shaft
27. Thrust washer
28. Roller bearings
29. Sprocket/gear assy.
30. Washer
31. Hex shaft sprocket

plate (30—Fig. F20) on drive pulley (5) when drive control lever is released. Normal wear is compensated for by

drive clutch control rod adjustment. However, further adjustment may be required. To check friction wheel clearance, tip snowthrower forward on front of auger housing and remove rear cover from main frame (motor mount). If clearance is not within specifications, detach the links connecting drive clutch pivot (6—Fig. F19) to friction wheel bracket (25) and reconnect to different set of holes. There are three sets of link attaching holes in bracket (25); move links to rear to decrease clearance or forward to increase gap between friction wheel and friction drive plate.

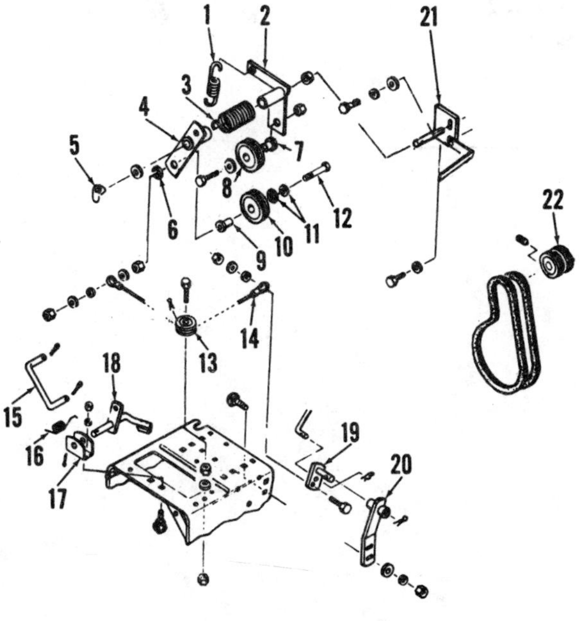

OVERHAUL

ENGINE

Engine make and model numbers are listed at beginning of this section. Refer to Tecumseh 4-stroke engine section of this manual for engine overhaul information.

Fig. F18—Exploded view of drive belt system.

1. Drive idler spring
2. Drive idler arm
3. Auger brake spring
4. Auger idler arm
5. Idler arm spring (early type)
6. Lockwasher
7. Idler spacer
8. Drive belt idler
9. Idler spacer
10. Auger belt idler
11. Spacer washers
12. Idler bolt
13. Auger cable pulley
14. Auger clutch cable
15. Brake link
16. Brake rod spring
17. Brake arm bracket
18. Auger brake arm
19. Auger bellcrank (early type)
20. Bellcrank bracket
21. Idler spindle plate
22. Engine pulley

DRIVE BELTS

To remove drive belts, remove belt cover and push idler pulleys away from belts and slide belts off of engine pulley. Tip the snowthrower forward onto auger housing and remove rear cover from main frame. Detach lower end of drive clutch control rod (3—Fig. F17). Move speed control lever to position "4" and remove belts by passing them between friction wheel (21—Fig. F19) and friction drive plate (30—Fig. F20) while turning friction wheel to move the belts through the gap. Reverse removal procedure to install new belts and recheck auger brake, auger control rod and drive control rod as previously outlined.

FRICTION WHEEL

To renew rubber faced friction wheel (21—Fig. F19), place speed control lever in reverse position, tip snowthrower forward onto auger housing and remove rear main frame cover. Remove drive chain by disconnecting master link. Detach lower end of drive clutch control rod (3—Fig. F17) from drive clutch pivot (6—Fig. F19). Remove left wheel from drive axle for access and pry out hex shaft access plug (9) in main frame at left end of hex shaft. Remove the two cotter pins from hex shaft. Remove nuts from both ends of hex shaft. Unbolt left bearing housing (23) and remove sprocket (31) and bearing housing. Slide hex shaft to left through hole in frame until friction wheel (21) and hub (20) can be withdrawn and unbolt friction wheel from hub. Install new wheel with cup side facing hub.

Reassemble by reversing disassembly procedure. When installing wheel on axle, first install key and be sure "V" of tire tread points forward at top of tire.

AUGER GEARBOX

The auger gearbox can be disassembled without removing auger housing from snowthrower main frame. Unbolt and remove the auger shaft bushing retainer (7—Fig. F20) and bushing (6) from each end of auger shaft. Unbolt the brace from top of gearbox and the gearbox front cover (18). Remove the cover and turn impeller fan (17) to spin worm gear (21) from housing, then remove key from impeller shaft (24). The gear housing and auger shaft (12) can then be removed from front of auger housing. Fur-ther disassembly is evident from inspection of unit and reference to Fig. F20. Clean and inspect all parts and renew any excessively worn or damaged parts.

Reassemble using new seals and by reversing disassembly procedure. When gearbox and auger shaft assembly are installed back in housing, turn impeller fan so slot for key is up. Hold key in place with needlenose pliers and tap key into place with drift inserted through filler plug opening in top of gearbox. Fill gearbox with lubricant as specified in auger gearbox lubrication paragraph.

Fig. F20—Exploded view of auger and gearbox assembly. Traction drive pulley and bearing assembly (5) can be removed from rear of impeller shaft (24) after removing friction drive plate (30) from rear of pulley and hex nut from impeller shaft.

1. Bearing retainer
2. Bearing
3. Auger drive pulley
4. Thrust washer
5. Traction drive pulley
6. Auger shaft bushing
7. Bushing retainer
8. Oil seal
9. Gearbox side cover
10. Bearing
11. Gasket
12. Auger shaft
13. Thrust washers
14. Worm wheel
15. Bearing
16. Oil seal
17. Impeller fan
18. Gearbox front cover
19. Gasket
20. Thrust washer
21. Worm gear
22. Thrust bearing races
23. Thrust bearing
24. Impeller shaft
25. Front impeller bearing
26. Auger shaft bearing
27. Oil seal
28. Auger flight assy.
29. Roll pins
30. Friction drive plate

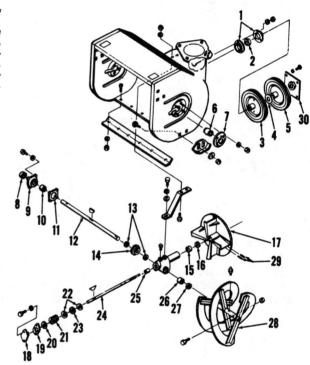

FORD

Model	Engine Make	Engine Model	Self-Propelled	No. of Stages	Scoop Width
ST-524	B&S	130000	Yes	2	24 in. (610 mm)
ST-526	B&S	130000	Yes	2	26 in. (660 mm)
ST-826	B&S	190000	Yes	2	26 in. (660 mm)

LUBRICATION

ENGINE

Refer to Briggs & Stratton engine section for engine lubrication requirements. Recommended fuel is regular or low lead gasoline.

DRIVE MECHANISM

After every ten hours of operation, lubricate all pivot points and axle bearings. Lubricate idler pulley bushings if so equipped. Do not allow oil on drive belts. After each season, lubricate auger shaft bearings with a good quality automotive wheel bearing grease.

MAINTENANCE

The augers are secured to the auger shaft by shear pins which will fracture when an auger contacts an obstruction. Use only original equipment type shear pins.

Inspect drive belts and pulley belt grooves. Renew belt if frayed, cracked, burnt or otherwise damaged.

ADJUSTMENTS

HEIGHT

To adjust auger height, loosen runner mounting bolts and reposition runners. Height should be increased when operating snowthrower over an irregular surface.

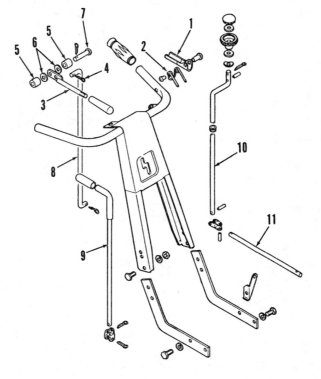

Fig. F-31—Exploded view of handle and controls used on Model ST-524.

1. Clutch handle
2. Spring
3. Lever
4. Cotter key
5. Spacer
6. Washer
7. Pin
8. Speed control rod
9. Auger clutch control rod
10. Chute crank
11. Chute rod

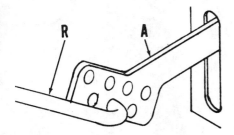

Fig. F30—Shift rod (R) is shown in center hole of shift arm (A). Refer to text for adjustment.

Fig. F32—Exploded view of handle and controls used on Models ST-526 and ST-826.

1. Clutch handle
2. Spring
3. Clutch handle
4. Auger clutch control rod
5. Cotter pin
6. Spacer
7. Shift lever
8. Upper shift rod
9. Lower shift rod
10. Spring
11. Chute crank
12. Clutch rod
13. U-joint
14. Roll pin
15. Support
16. Chute rod
17. Pivot bushing
18. U-joint

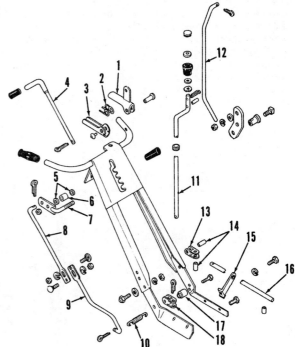

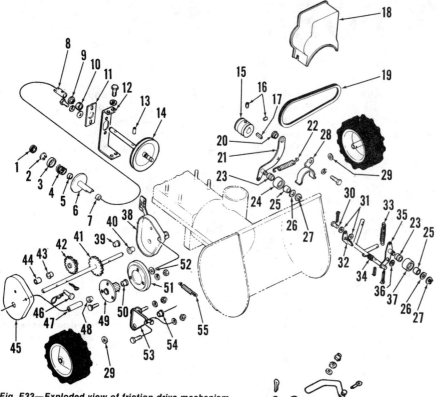

SHIFT LINKAGE

Model ST-524

To adjust shift linkage, disconnect shift rod (R–Fig. F30) from shift arm (A) and move shift control lever to neutral "N" position. Attach shift rod (R) in center hole of shift arm (A) as shown in Fig. F30. Move shift lever to forward and reverse positions while noting distance for each from neutral position. If required, relocate shift rod (R) in shift arm (A) so movement of shift lever from neutral to forward and from neutral to reverse is the same.

FRICTION DRIVE

Models ST-526 and ST-826

With engine off, snowthrower wheels should roll easily with speed control in neutral position.

Fig. F33—Exploded view of friction drive mechanism.

1. Bearing
2. Snap ring
3. Spacer
4. Spring
5. Bushing
6. Drive disc
7. Bushing
8. Shaft
9. Snap ring
10. Bearing
11. Bearing plate
12. Support
13. Roll pin
14. Pulley & shaft
15. Engine pulley
16. Set screws
17. Key
18. Belt guard
19. Belt
20. Pivot bushing
21. Idler arm
22. Spring
23. Inner race
24. Friction drive idler pulley
25. Bearing
26. Washer
27. Snap ring
28. Belt guide
29. Washer
30. Pin
31. Washers
32. Brake arm
33. Spring
34. Spring
35. Idler arm & rod
36. Washer
37. Auger drive idler pulley
38. Gearcase
39. Bushing
40. Bushing
41. Gear & shaft
42. Gear
43. Bushing
44. Spacer
45. Cover
46. Pin
47. Spacer
48. Bushing
49. Hub
50. Bushing
51. Friction wheel
52. Washer
53. Spindle
54. Bushing
55. Spring

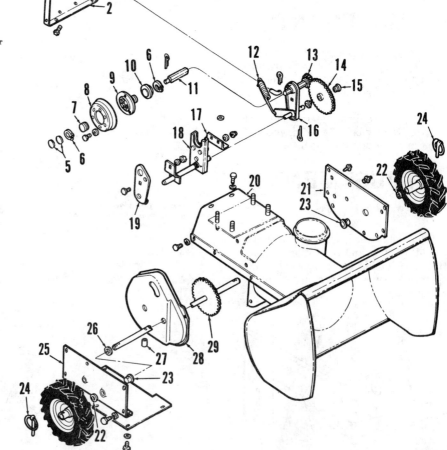

1. Bellcrank
2. Backing plate
3. Link
4. Bushing
5. Washer
6. Snap ring
7. Bearing
8. Roller
9. Hub
10. Bearing
11. Shaft
12. Spring
13. Gear
14. Gear
15. Bushing
16. Arm
17. Slide
18. Arm
19. Arm
20. Collector housing
21. L.H. frame
22. Washer
23. Bearing
24. "Klik" pin
25. R.H. frame
26. Washer
27. Pin
28. Cover

Fig. F34—Exploded view of traction drive mechanism used on Models ST-526 and ST-826.

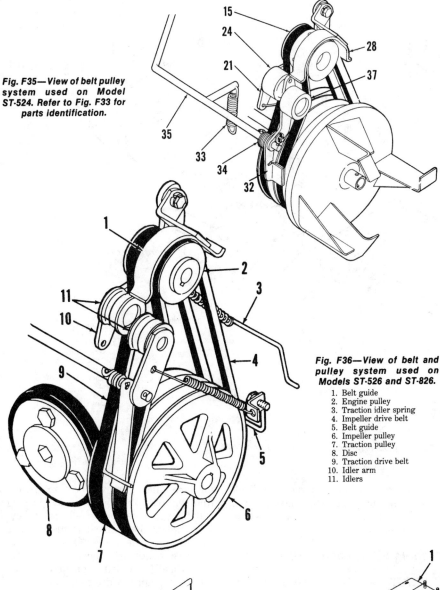

Fig. F35—View of belt pulley system used on Model ST-524. Refer to Fig. F33 for parts identification.

Fig. F36—View of belt and pulley system used on Models ST-526 and ST-826.

1. Belt guide
2. Engine pulley
3. Traction idler spring
4. Impeller drive belt
5. Belt guide
6. Impeller pulley
7. Traction pulley
8. Disc
9. Traction drive belt
10. Idler arm
11. Idlers

Refer to Fig. F32. Place shift lever in reverse position and loosen bolts at slip coupler of rods (8 and 9). Pry rods apart as far as they will go, then retighten bolts.

OVERHAUL

ENGINE

Engine make and model numbers are listed at the beginning of this section. Refer to Briggs & Stratton engine section for overhaul procedures.

DRIVE BELTS

Use the following procedures to renew excessively worn or damaged drive belts.

Model ST-524

AUGER DRIVE BELT. Remove belt guard and belt guide, then lift auger drive belt off engine pulley. Push belt down so belt is forced out of auger belt pulley groove. Remove auger shaft bushings (1–Fig. F37), move auger clutch control to engaged position and remove complete auger and impeller assembly.

Renew belt and reassemble by reversing disassembly procedure.

FRICTION DRIVE BELT. To remove friction drive belt, first follow procedure previously outlined and

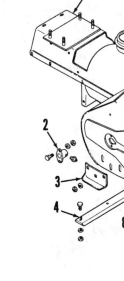

Fig. F37—Exploded view of auger components used on Model ST-524.

1. Bushing
2. Runner
3. Snap ring
4. Bushing
5. Scraper
6. Impeller
7. Roll pin
8. Shear pin
9. Washer
10. Auger R.H.
11. Auger gearbox
12. Auger L.H.

Fig. F38—Exploded view of auger components used on Models ST-526 and ST-826.

1. Collector housing
2. Bearing
3. Skid
4. Scraper
5. Impeller
6. Gearbox
7. Washers
8. R.H. auger
9. L.H. auger
10. Shear pin

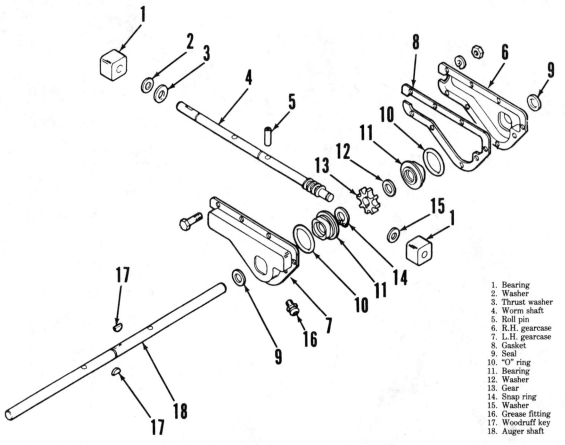

Fig. F39—Exploded view of gearcase used on ST-524.

1. Bearing
2. Washer
3. Thrust washer
4. Worm shaft
5. Roll pin
6. R.H. gearcase
7. L.H. gearcase
8. Gasket
9. Seal
10. "O" ring
11. Bearing
12. Washer
13. Gear
14. Snap ring
15. Washer
16. Grease fitting
17. Woodruff key
18. Auger shaft

remove complete auger and impeller assembly. Remove friction drive belt idler spring (22–Fig. F33) and lift drive belt off engine pulley. Drain gas tank, tilt snowthrower forward and remove bottom panel. Remove the four screws on sides of rear panel and loosen but do not remove the two screws at top of rear panel near engine. Push or tap on fric-

tion drive shaft (14) to move it back enough so drive belt can pass around side of pulley for removal. Reverse removal procedure to reinstall.

Models ST-526 and ST-826

FRICTION AND AUGER DRIVE BELTS. Refer to Fig. F36. Detach belt guard and remove friction drive belt idler spring (3) and belt guide (1). Disengage auger and friction drive belts from engine pulley (2). Tip snowthrower

forward and remove bottom panel. Remove friction drive belt by passing belt between friction wheel and pulley. To remove auger drive belt, move auger clutch control lever to "OFF" position and pass auger drive belt between friction wheel (8) and friction drive pulley (7). Reverse procedure to install belts.

AUGER GEARBOX

Model ST-524

Refer to Fig. F39 for an exploded view. Remove impeller and augers from shafts. Remove the eight cap screws holding case halves (6 and 7) out over auger shaft (18). Slide auger shaft out of right-hand case half (6) along with bearings (11), washer (12), worm wheel (13) and snap ring (14). Inspect all components for damage and excessive wear and renew as needed.

Reassembly is reverse of disassembly procedure. Renew "O" rings (10) and seals (9) during reassembly. Pump ½ pint (0.3 L) of wheel bearing grease back into gearbox through grease fitting (16).

Models ST-526 and ST-826

Refer to Fig. F40 for an exploded view. Remove auger and impeller

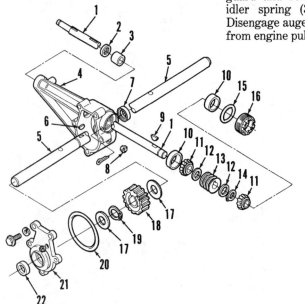

Fig. F40—Exploded view of gearcase used on Models ST-526 and ST-826.

1. Input shaft
2. Seal
3. Bushing
4. Case
5. Auger shaft
6. Woodruff key
7. Seal
8. Plug
9. Hypro key
10. Bearing cup
11. Bearing cone
12. Spacer
13. Worm gear
14. Snap ring
15. "O" ring
16. Adjuster plug
17. Thrust washer
18. Gear
19. Snap ring
20. "O" ring
21. Cover
22. Seal

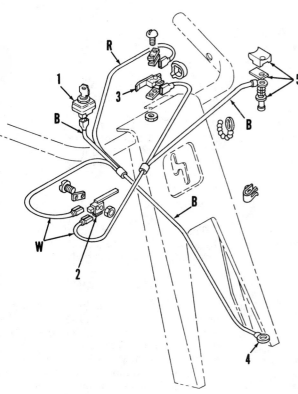

Fig. F41—View of safety interlock wiring. Wires are color coded black (B), white (W) and red (R).

1. Ignition switch
2. Auger interlock switch
3. Speed control interlock wiring
4. Engine terminal
5. Ground block

assembly from snowthrower. Remove impeller and augers from shafts and drain oil from gearbox. Remove gearbox side cover (21). Withdraw shaft (5) with gear (18). Unscrew plug (16) and remove gear/shaft assembly through front of gearbox. Inspect and renew worn or damaged components.

Reverse disassembly procedure to reassemble gearbox. Tighten plug (16) just enough to remove shaft (1) end play. Be sure shaft turns freely. Fill gearbox with SAE 30 oil to level of plug hole.

ELECTRICAL INTERLOCK SYSTEM

The electrical interlock system grounds out the engine ignition system if operator's safety lever on handlebar is released while auger clutch is engaged or drive lever is in gear. Interlock switches are actuated by auger and drive controls and are not adjustable. Be sure ground wire is properly attached to handlebar. Refer to Fig. F41 for wiring diagram.

FORD

Model	Engine Make	Engine Model	Self-Propelled?	No. of Stages	Scoop Width
ST-524A	B&S	130000	Yes	2	24 in. (610 mm)
ST-524B	Tecumseh	HS50	Yes	2	24 in. (610 mm)
ST-824B	B&S	190000	Yes	2	24 in. (610 mm)
ST-826A	B&S	190000	Yes	2	26 in. (660 mm)
ST-826B	Tecumseh	HM80	Yes	2	26 in. (660 mm)

LUBRICATION

ENGINE

Refer to appropriate Briggs & Stratton or Tecumseh 4-stroke engine section for engine lubrication information. Recommended fuel is leaded or unleaded regular gasoline.

DRIVE MECHANISM

Lubricate all pivot points and drive chains with a suitable lubricant after every 10 hours of operation. Use a grease gun to lubricate fittings for axle bushings, auger shaft bushings and countershaft bushings after every 10 hours of operation. Wheel hubs should be greased after each season of use.

AUGER GEARBOX

On 24-inch auger width models, the auger gearbox is fitted with a grease fitting and can be lubricated with a hand-operated grease gun using NLGI Grade 2 Lithium Base EP grease. For other models, check oil level in gearbox periodically. Oil level should be at bottom of auger gearbox pipe plug opening with unit setting level. Recommended lubricant for this type gearbox is SAE 30 motor oil.

TRANSMISSION

The gear-type transmission is lubricated at the factory and should not require further lubrication except at overhaul; refer to Peerless transmission section for service information.

MAINTENANCE

The auger flights are driven via the auger shaft by shear pins which will fracture when an auger contacts a solid obstruction. If necessary to renew shear pins, use only factory replacement pins.

Inspect drive belts and pulley belt grooves. Renew belts if frayed, cracked, burned or excessively worn. Renew pulleys if bent, excessively worn or otherwise damaged.

Inspect traction drive chain and sprockets. Renew chain if excessively worn, stretched or otherwise damaged. Renew sprockets if teeth are worn or otherwise damaged.

ADJUSTMENTS

AUGER HEIGHT

Auger height can be adjusted by loosening skid shoe mounting bolts and repositioning the skid shoes. Auger should be in lowest position (skid shoes raised) when operating over smooth surface and in highest position (skid shoes lowered) when operating over rough surfaces. Be sure skid shoes are equally adjusted on each side of auger housing.

CONTROL CABLES

The auger and traction drive control cables should have all slack removed when control lever is all the way against the rubber stop on the handle. If not, remove the spring clevis pin from the lower end of either control cable and turn the adjustment yoke as necessary to achieve proper adjustment. Reattach cable with spring clevis pin.

DRIVE BELTS

If auger drive belt slips, readjust position of idler pulley to increase belt tension. Remove belt guard and note position of idler pulley on auger idler arm. Loosen hex nut holding idler pulley to arm and move pulley toward belt about 1/4 inch (6.4 mm), tighten nut and recheck belt tension. Belt should release fully when control lever is released. It may be necessary to readjust control cable as described in preceding paragraph.

The traction drive belt can be adjusted following procedure outlined for auger

belt by moving idler pulley in slot of drive idler arm.

The wire belt guides should clear belts by about 1/8 inch (3.2 mm) when belt control levers are engaged.

DRIVE CHAIN

When pushing on chain at midpoint between transmission and axle sprockets, the chain should deflect about 1/8 inch (3.2 mm). To adjust chain, tip the snowthrower forward onto front of auger housing and loosen the three bolts retaining the axle bearing retainers on each side of main frame. Slide the axle shaft bearing assemblies in the slotted mounting holes of main frame until proper adjustment is reached. Be sure both bearings are the same distance from back of main frame and tighten the bearing retaining bolts.

OVERHAUL

ENGINE

Engine make and model numbers are listed at beginning of this section. Refer to Briggs & Stratton or Tecumseh 4-stroke engine section in this manual for engine overhaul information.

TRANSMISSION

All models are equipped with a gear-type transmission. Refer to Peerless section for transmission overhaul information.

AUGER DRIVE BELT

With belt drive cover removed, remove belt from engine pulley and slip belt out from between auger idler pulley and retainer. Push the idler arm in to move brake shoe away from auger pulley and remove belt from between auger pulley and lower belt guide. Install new belt by reversing removal procedure and check belt and cable adjustments as outlined in a previous paragraph.

TRACTION DRIVE BELT

With belt drive cover removed, remove auger drive belt from engine pulley. Move traction drive belt from engine pulley to a position between pulley and engine. Slip belt out from between traction drive idler pulley and retainer. Remove belt from transmission pulley. Install new belt by reversing removal procedure and replace auger belt on engine pulley. Check belt and cable adjustments as outlined in appropriate previous paragraph.

GILSON

LAWN-BOY PRODUCT GROUP
P.O. Box 152
Plymouth, WI 53073

Model	Engine Make	Engine Model	Self-Propelled?	No. of Stages	Scoop Width
ST-320	B&S	62032	No	1	20 in. (508 mm)
55130	Tecumseh	AH520	No	1	20 in. (508 mm)
55327	Tecumseh	AH520	No	1	20 in. (508 mm)
55328	Tecumseh	AH520	No	1	20 in. (508 mm)
55379	Tecumseh	AH600	No	1	20 in. (508 mm)
55380	Tecumseh	AH600	No	1	20 in. (508 mm)

Note: Models 55379 and 55380 are alike except 55380 is electric start.

LUBRICATION

ENGINE

Refer to appropriate Briggs & Stratton or Tecumseh engine section for engine lubrication information. Recommended fuel is regular or low-lead gasoline mixed with a good-quality 2-cycle air-cooled engine oil. Fuel and oil should be mixed in a 32:1 ratio.

DRIVE MECHANISM

Lubricate auger shaft bearing (13—Fig. G1) at right side of auger housing after every 10 hours of operation using SAE 30 oil. Use sufficient oil so oil just runs from hole; felt plug (14) retains oil between lubrication intervals. Bearing at left end of auger is a sealed prelubricated type.

After every 10 hours of operation, remove belt cover and lubricate idler pulley (38) and shaft on idler arm (30) with SAE 30 oil. Turn pulley while applying lubricant to distribute oil evenly. Be careful not to allow oil to get on belt or pulleys. Idler arm pivot bushing (28) and skid block (33) should be lubricated with No. 2 wheel bearing grease.

MAINTENANCE

Inspect drive belt and renew if frayed, cracked, burned or otherwise damaged. Renew belt pulleys if belt groove is excessively worn or pulley is damaged. In-spect impeller for wear or damage. Impeller paddle (2—Fig. G1) and auger flights (1) can be renewed if necessary. Move deflector vane pivot assembly (17) from side-to-side. Deflector vane (8) locknut should be adjusted so that deflector moves without binding.

ADJUSTMENTS

DRIVE BELT

With auger clutch lever released, there should be no slack in the clutch cable. To increase belt tension, move open end of clutch adjuster clip (60—Fig. G1) down to next link in clutch cable and chain (59). If clip is in last link of chain, it may be necessary to move clip loop up to the next link in end of chain. To check adjustment, pull engine starter rope with ignition in "OFF" position. If auger turns, move adjustment clip upward in chain (closer to clutch lever). With clutch engaged, belt retainer (32) should clear belt by a maximum of 1/8 inch (3.2 mm) and retainer should be parallel to belt. Belt guides (29, 41 and 42) should be parallel to the belt and clear the belt when clutch is engaged.

OVERHAUL

ENGINE

Engine make and model information are listed at beginning of this section.

Refer to appropriate Briggs & Stratton or Tecumseh engine section for engine overhaul information.

DRIVE MECHANISM

Drive belt (35—Fig. G1) can be removed and a new belt installed after removing belt cover (36) and lifting idler from belt. Refer to belt tension adjustment in previous ADJUSTMENT section.

If belt idler arm (30) is removed, lightly grease back side of arm where it contacts skid block (33) and pivot bushing (28) with No. 2 wheel bearing grease so idler arm will pivot freely after reassembly. Lubricate idler pulley shaft and bushing with No. 2 wheel bearing grease if pulley is removed.

AUGER ASSEMBLY

If excessive wear of auger flight and impeller paddle rubber fins is noted, the rubber fins should be replaced as a complete set. The fins can be replaced after removing the cap screws holding the fins between the two auger plate halves (3—Fig. G1).

If excessively worn or damaged, the auger assembly scraper bar (7) can be replaced after removing the three flat head socket screws that hold scraper bar to bottom of auger housing. Auger stop spring (6) mounted with center scraper bar bolt keeps auger from rotating backwards.

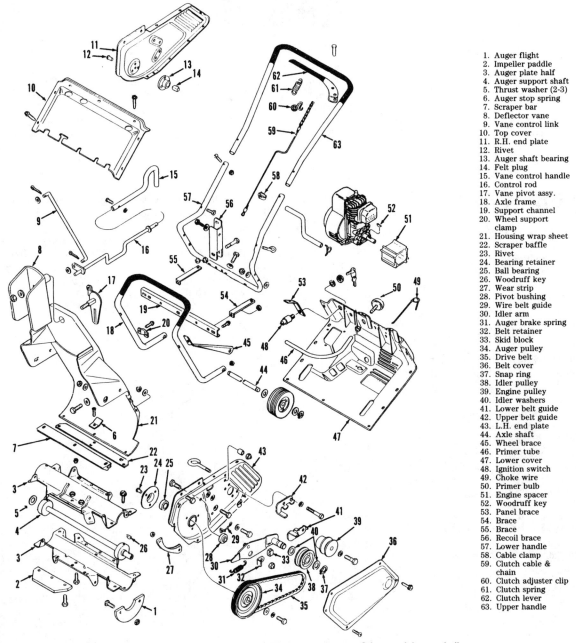

1. Auger flight
2. Impeller paddle
3. Auger plate half
4. Auger support shaft
5. Thrust washer (2-3)
6. Auger stop spring
7. Scraper bar
8. Deflector vane
9. Vane control link
10. Top cover
11. R.H. end plate
12. Rivet
13. Auger shaft bearing
14. Felt plug
15. Vane control handle
16. Control rod
17. Vane pivot assy.
18. Axle frame
19. Support channel
20. Wheel support clamp
21. Housing wrap sheet
22. Scraper baffle
23. Rivet
24. Bearing retainer
25. Ball bearing
26. Woodruff key
27. Wear strip
28. Pivot bushing
29. Wire belt guide
30. Idler arm
31. Auger brake spring
32. Belt retainer
33. Skid block
34. Auger pulley
35. Drive belt
36. Belt cover
37. Snap ring
38. Idler pulley
39. Engine pulley
40. Idler washers
41. Lower belt guide
42. Upper belt guide
43. L.H. end plate
44. Axle shaft
45. Wheel brace
46. Primer tube
47. Lower cover
48. Ignition switch
49. Choke wire
50. Primer bulb
51. Engine spacer
52. Woodruff key
53. Panel brace
54. Brace
55. Brace
56. Recoil brace
57. Lower handle
58. Cable clamp
59. Clutch cable & chain
60. Clutch adjuster clip
61. Clutch spring
62. Clutch lever
63. Upper handle

Fig. G1—Exploded view of Model ST-320 snowthrower. Other models are similar.

GILSON

Model	Engine Make	Engine Model	Self-Propelled	No. of Stages	Scoop Width
55036	B&S	92000	Yes	1	20 in. (508 mm)
55091	B&S	92000	Yes	1	20 in. (508 mm)
55131	B&S	92000	Yes	1	20 in. (508 mm)

LUBRICATION

ENGINE

Refer to Briggs and Stratton engine section for engine lubrication requirements. Recommended fuel is regular or low lead gasoline.

DRIVE MECHANISM

After every ten hours of operation, left auger shaft bushing should be lubricated through grease fitting in bushing retainer.

As required, lubricate drive clutch on wheel axle with engine oil so clutch will operate freely. Lubricate auger drive belt idler pulley and shaft after every ten hours of operation. Do not allow oil on belt. Lubricate wheel bushings after every ten hours of operation.

MAINTENANCE

Inspect drive belts and renew belt if frayed, cracked, burnt or otherwise damaged. Inspect belt pulley groove and renew pulley if damaged or excessively worn.

The discharge chute is rotated by a rope attached to the chute. Check condition of rope and renew if necessary. If chute is removed, note tabs on chute and mounting flange which must interlock during assembly. Chute should sit squarely on mounting flange.

ADJUSTMENT

DRIVE BELTS

Auger drive belt tension does not re-

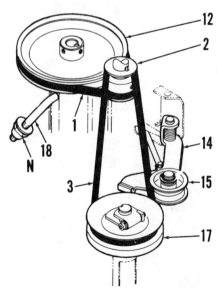

Fig. G10—Drawing of belt and pulley system. Turn nut (N) to adjust wheel drive belt tension. Refer to Fig. G11 for parts identification.

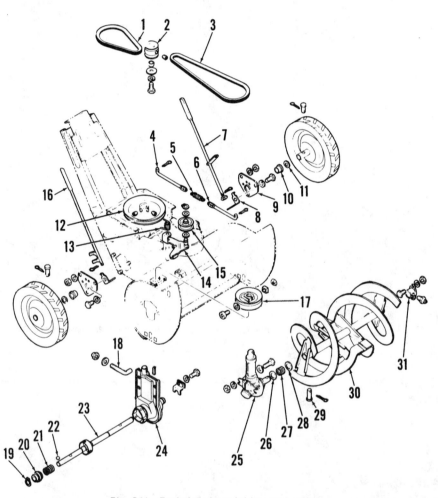

Fig. G11—Exploded view of drive mechanism.

1. Wheel drive belt
2. Engine pulley
3. Auger drive belt
4. Auger clutch rod
5. Spring
6. Auger clutch rod
7. Auger control rod
8. Bracket
9. Bushing retainer
10. Bushing
11. Washer
12. Wheel drive pulley
13. Spring
14. Idler arm
15. Idler
16. Wheel drive control rod
17. Auger drive pulley
18. Adjusting link
19. Snap ring
20. Clutch thimble
21. Spring
22. Clutch balls (3)
23. Axle
24. Gearbox
25. Auger gearbox
26. Washer
27. Spring
28. Washer
29. Shear pin
30. Auger
31. Bushing

quire adjustment as idler arm is spring-loaded. Wheel drive belt tension is adjusted by turning nut (N—Fig. G10) so belt is tight but not stretched.

OVERHAUL

ENGINE

Engine make and model numbers are listed at beginning of section. Refer to Briggs and Stratton engine section for engine overhaul.

DRIVE MECHANISM

All models are equipped with a ball type clutch mounted on axle (23—Fig. G11). The clutch control rod forces clutch thimble (20) in against spring (21) so clutch balls (22) engage clutch ring on gearbox (24) shaft.

Refer to Fig. G11 for exploded views of wheel and auger drive components. Overhaul is apparent after inspection of unit and referral to Fig. G11. Axle and auger gearboxes are available only as unit assemblies.

GILSON

Model	Engine Make	Engine Model	Self-Propelled?	No. of Stages	Scoop Width
55008	B&S	10000	Yes	2	18 in. (457 mm)
55037	B&S	130000	Yes	2	24 in. (610 mm)
55066	B&S	100000	Yes	2	18 in. (457 mm)
55092	B&S	130000	Yes	2	24 in. (610 mm)
55132	B&S	130000	Yes	2	24 in. (610 mm)
55342	Tecumseh	HS40	Yes	2	22 in. (559 mm)
55343	B&S	130202	Yes	2	22 in. (559 mm)

LUBRICATION

ENGINE

Refer to Briggs & Stratton or Tecumseh engine section for engine lubrication requirements. Recommended fuel is regular or low lead gasoline.

DRIVE MECHANISM

After every ten hours of operation lubricate all pivot points and axle bearings and on early models lubricate idler pulley bushings. Later models are equipped with sealed bearings in idler pulleys. Do not allow oil on drive belts. After each season lubricate auger shaft bearings with a good quality automotive wheel bearing grease.

MAINTENANCE

The augers are secured to auger shaft by shear pins which will fracture when an auger contacts an obstruction. Renew shear pins as required with factory authorized shear pins.

Inspect drive belts and pulley belt grooves. Renew belt if frayed, cracked, burnt or otherwise damaged.

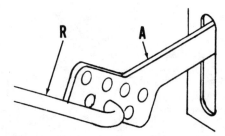

Fig. G15—Shift rod (R) is shown in center hole of shift arm (A). Refer to text for adjustment.

ADJUSTMENT

HEIGHT

To adjust auger height, loosen runner mounting bolts and reposition runners. Height should be increased when operating snowthrower over an irregular surface.

SHIFT LINKAGE

To adjust shift linkage, disconnect shift rod (R—Fig. G15) from shift arm (A) and move shift control lever to neutral "N" position. Attach shift rod (R) in center hole of shift arm (A) as shown in Fig. G15. Move shift lever to forward and reverse position while noting distance for each from neutral position. If required, relocate shift rod (R) in shift arm (A) so movement of shift

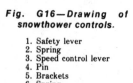

Fig. G16—Drawing of snowthower controls.

1. Safety lever
2. Spring
3. Speed control lever
4. Pin
5. Brackets
6. Spring
7. Pin
8. Speed control rod
9. Auger clutch control rod
10. Chute crank
11. Chute rod

lever from neutral to forward and from neutral to reverse is equidistant.

OVERHAUL

ENGINE

Engine make and model numbers are listed at beginning of section. Refer to Briggs and Stratton engine section for engine overhaul.

AUGER GEARBOX

The auger gearbox on all except 55342 and 55343 is available only as a unit assembly and must be serviced as a unit. Refer to Fig. G19B for exploded view of gearbox used on Models 55342 and 55343.

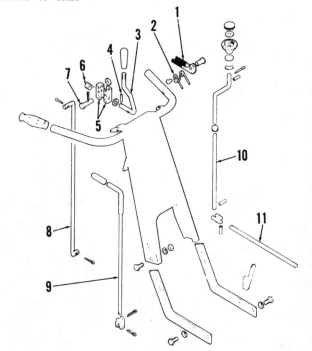

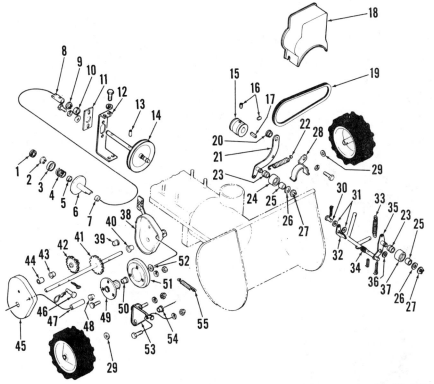

Fig. G17—Exploded view of friction drive mechanism. Later models are equipped with non--renewable sealed bearings in idler pulleys (24 & 37).

1. Bearing	16. Set screws	29. Washer	42. Gear
2. Snap ring	17. Key	30. Pin	43. Bushing
3. Spacer	18. Belt guard	31. Washers	44. Spacer
4. Spring	19. Belt	32. Brake arm	45. Cover
5. Bushing	20. Pivot bushing	33. Spring	46. Pin
6. Drive disc	21. Idler arm	34. Spring	47. Spacer
7. Bushing	22. Spring	35. Idler arm & rod	48. Bushing
8. Shaft	23. Inner race	36. Washer	49. Hub
9. Snap ring	24. Friction drive idler	37. Auger drive idler	50. Bushing
10. Bearing	pulley	pulley	51. Friction wheel
11. Bearing plate	25. Bearing	38. Gearcase	52. Washer
12. Support	26. Washer	39. Bushing	53. Spindle
13. Roll pin	27. Snap ring	40. Bushing	54. Bushing
14. Pulley & shaft	28. Belt guide	41. Gear & shaft	55. Spring
15. Engine pulley			

DRIVE BELTS

Use the following procedure to renew excessively worn or damaged drive belts.

AUGER DRIVE BELT. Remove belt guard and belt guide then lift auger drive belt off engine pulley. Push belt down so belt is forced out of auger belt pulley groove. Remove auger shaft bushings (1—Fig. G19), move auger clutch control to engaged position and remove complete auger and impeller assembly. Remove auger drive belt and install a new belt by reversing removal procedure.

FRICTION DRIVE BELT. To remove friction drive belt, first follow procedure previously outlined and remove complete auger and impeller assembly. Remove friction drive belt idler spring (22—Fig. G17) and lift drive belt off engine pulley. Drain gas

tank, tilt snowthrower forward and remove bottom panel. Remove four screws on sides of rear panel and

Fig. G18—Drawing of belt and pulley system. Refer to Fig. G17 for parts identification.

loosen but do not remove two screws at top of rear panel near engine. Push or tap on friction drive shaft (14) to

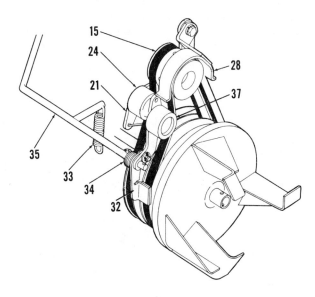

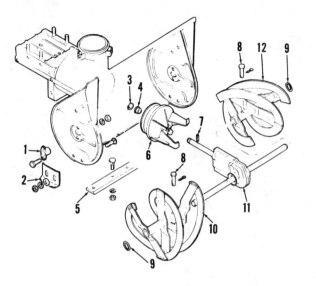

Fig. G19—Exploded view
of auger components.

1. Bushing
2. Runner
3. Snap ring
4. Bushing
5. Scraper
6. Impeller
7. Roll pin
8. Shear pin
9. Washer
10. Auger R.H.
11. Auger gearbox
12. Auger L.H.

move it back sufficiently so drive belt
can pass around side of pulley for
removal. Reverse removal procedure to
install belt.

ELECTRICAL INTERLOCK SYSTEM

These models are equipped with an
electrical interlock system to prevent
engine operation unless auger clutch is
disengaged and drive lever is in neutral. Interlock switches are actuated by
auger and drive controls and are not
adjustable. Be sure ground wire is properly attached to handlebar. Note wiring schematic in Fig. G19A.

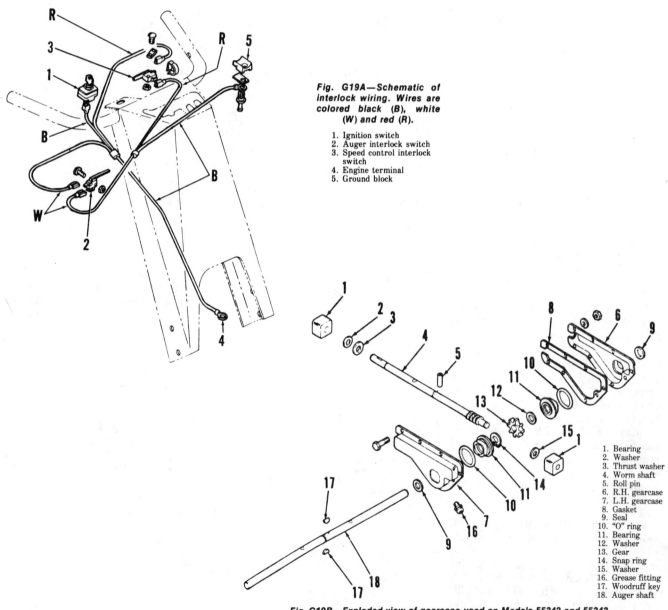

Fig. G19A—Schematic of
interlock wiring. Wires are
colored black (B), white
(W) and red (R).

1. Ignition switch
2. Auger interlock switch
3. Speed control interlock
   switch
4. Engine terminal
5. Ground block

1. Bearing
2. Washer
3. Thrust washer
4. Worm shaft
5. Roll pin
6. R.H. gearcase
7. L.H. gearcase
8. Gasket
9. Seal
10. "O" ring
11. Bearing
12. Washer
13. Gear
14. Snap ring
15. Washer
16. Grease fitting
17. Woodruff key
18. Auger shaft

Fig. G19B—Exploded view of gearcase used on Models 55342 and 55343.

GILSON

Model	Engine Make	Engine Model	Self-Propelled?	No. of Stages	Scoop Width
55010	B&S	130000	Yes	2	26 in. (660 mm)
55011	B&S	130000	Yes	2	26 in. (660 mm)
55016	B&S	190000	Yes	2	26 in. (660 mm)
55067	B&S	130000	Yes	2	26 in. (660 mm)
55068	B&S	130000	Yes	2	26 in. (660 mm)
55069	B&S	190000	Yes	2	26 in. (660 mm)
55093	B&S	130000	Yes	2	26 in. (660 mm)
55095	Tecumseh	HM100	Yes	2	24 in. (610 mm)
55133	B&S	130000	Yes	2	26 in. (660 mm)
55134	B&S	190000	Yes	2	26 in. (660 mm)
55135	Tecumseh	HM100	Yes	2	26 in. (660 mm)

LUBRICATION

ENGINE

Refer to Briggs and Stratton or Tecumseh engine section for engine lubrication requirements. Recommended fuel is regular or low lead gasoline.

DRIVE MECHANISM

After every ten hours of operation lubricate all pivot points and axle bear-

Fig. G20—View showing location of grease fitting (G) for drive gears.

ings and on early models lubricate idler pulley bushings. Later models are equipped with sealed bearings in idler pulleys. Do not allow oil on drive belts. Lubricate auger shaft through grease fittings in augers and drive gears through grease fittings shown in Figs. G20 and G21. Wheel hubs should be greased after each season.

AUGER GEARBOX

Oil level in auger gearbox should be checked periodically. Oil level should be even with oil plug hole (H—Fig. G21). Recommended oil is SAE 30.

MAINTENANCE

The augers are secured to auger shaft by shear pins which will fracture when an auger contacts an obstruction. Renew shear pins as required with factory authorized shear pins.

Inspect drive belts and pulley belt grooves. Renew belt if frayed, cracked, burnt or otherwise damaged.

On models equipped with inflatable

tires, recommended tire pressure is 20 psi (137.9 kPa).

ADJUSTMENTS

HEIGHT

To adjust auger height, loosen runner mounting bolts and reposition runners. Height should be increased when operating snowthrower over an irregular surface.

FRICTION DRIVE

With engine off, snowthrower wheels should roll easily with speed control in neutral position but it should not be possible to roll snowthrower backwards with speed control lever in position "3". Detach and turn control rod clevis (7—Fig. G22) as necessary to obtain aforementioned action.

Loosen bolts securing upper and lower shift rods (3 and 4). Push lower shift rod (4) in as far as possible and position speed control lever in reverse slot. Pry shift rods (3 and 4) apart as

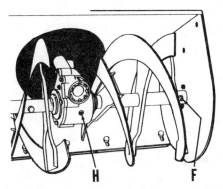

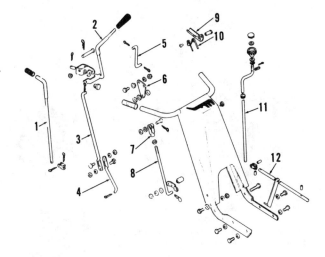

Fig. G22—Drawing of control assembly

1. Auger clutch control
2. Speed control lever
3. Upper shift rod
4. Lower shift rod
5. Link
6. Bellcrank
7. Clevis
8. Speed control rod
9. Safety lever
10. Spring
11. Chute crank
12. Chute rod

Fig. G21—Fill auger gearbox with oil until level with plug hole (H). Auger grease fittings (F) are located at end of each auger.

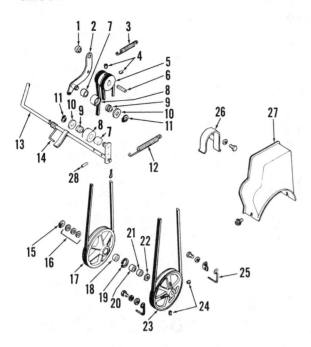

Detach belt guard then remove friction drive belt idler spring (3—Fig. G23) and belt guide (26). Disengage auger and friction drive belts from engine pulley then tip snowthrower forward and remove bottom panel. Remove friction drive belt by passing belt between friction wheel and pulley. To remove auger drive belt, move auger clutch control lever to "OFF" position and pass auger drive belt between friction wheel and friction drive pulley. Reverse procedure to install belts.

AUGER GEARBOX

To overhaul auger gearbox, remove auger and impeller assembly from snowthrower. Remove impeller and augers from shafts and drain oil from gearbox. Remove gearbox side cover (4—Fig. G27 or G28) while noting number of gaskets (5—Fig. G27) on early model gearbox for assembly. Withdraw shaft (13) with gear (11) then unscrew plug (22) and remove gear and

far as possible and retighten bolts being sure edges of shift rods are parallel.

OVERHAUL

ENGINE

Engine make and model numbers are listed at beginning of this section. Refer to Briggs and Stratton or Tecumseh engine section for engine overhaul.

DRIVE BELTS

Excessively worn or damaged drive belts should be renewed as follows:

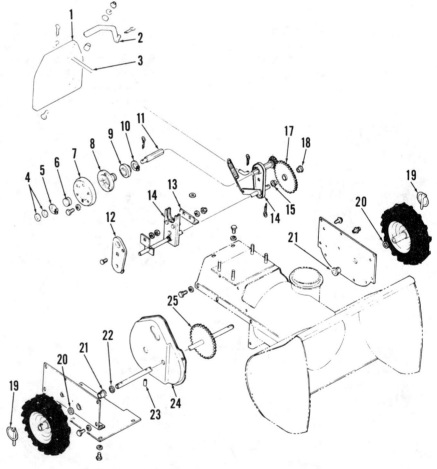

Fig. G25—Exploded view of wheel drive mechanism.

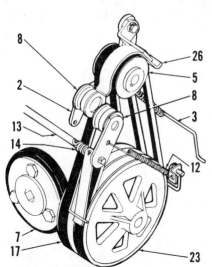

Fig. G24—Drawing of pulley and belt system. Refer to Fig. G23 for parts identification.

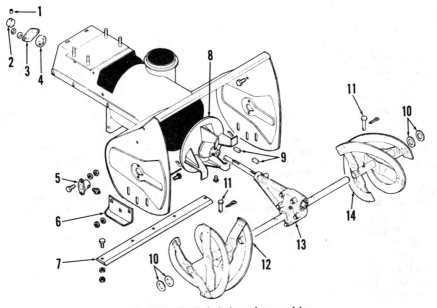

1. Set screw
2. Collar
3. Bearing retainer
4. Bearing
5. Bushing
6. Runner
7. Scraper
8. Impeller
9. Keys
10. Washer
11. Shear pin
12. Auger R.H.
13. Auger gearbox
14. Auger L.H.

Fig. G26—Exploded view of auger drive.

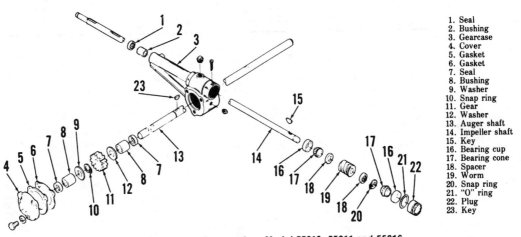

1. Seal
2. Bushing
3. Gearcase
4. Cover
5. Gasket
6. Gasket
7. Seal
8. Bushing
9. Washer
10. Snap ring
11. Gear
12. Washer
13. Auger shaft
14. Impeller shaft
15. Key
16. Bearing cup
17. Bearing cone
18. Spacer
19. Worm
20. Snap ring
21. "O" ring
22. Plug
23. Key

Fig. G27—Exploded view of auger gearbox used on Model 55010, 55011 and 55016.

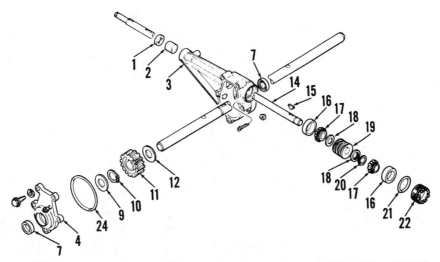

Fig. G28—Exploded view of auger gearbox used on all models except 55010, 55011 and 55016.
Refer to Fig. G27 for parts identification except for: 24. "O" ring.

shaft assembly through front of gearbox. Inspect and renew worn or damaged components.

Reverse disassembly procedure to reassemble gearbox. Tighten plug (22) just enough to remove shaft (14) end play. Be sure shaft (14) turns freely. Fill gearbox with SAE 30 oil to level of plug hole (H—Fig. G21).

ELECTRICAL INTERLOCK SYSTEM

These models are equipped with an electrical interlock system which grounds out the engine ignition system if operator's safety lever on handlebar is released while auger clutch is engaged or drive lever is in gear. Interlock switches are actuated by auger and drive controls and are not adjustable. Be sure ground wire is properly attached to handlebar.

Note wiring schematic in Fig. G29.

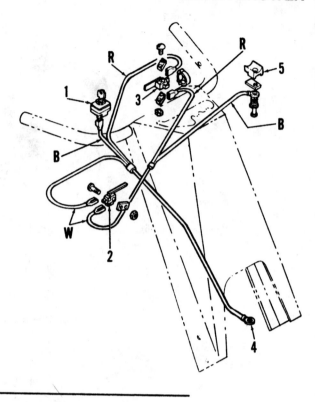

Fig. G29—Schematic of interlock wiring. Wires are colored black (B), white (W) and red (R).

1. Ignition switch
2. Auger interlock switch
3. Speed control interlock switch
4. Engine terminal
5. Ground block

GILSON

Model	Engine Make	Engine Model	Self-Propelled?	No. of Stages	Scoop Width
55012	B&S	190000	Yes	2	28 in. (711 mm)
55063	B&S	190000	Yes	2	32 in. (813 mm)
55064	B&S	190000	Yes	2	32 in. (813 mm)
55096	B&S	190000	Yes	2	28 in. (711 mm)
55097	Tecumseh	HM100	Yes	2	32 in. (813 mm)
55136	B&S	190000	Yes	2	28 in. (711 mm)
55137	Tecumseh	HM100	Yes	2	32 in. (813 mm)
55340A	Tecumseh	HMSK100	Yes	2	32 in. (813 mm)
55353	Tecumseh	HMSK100	Yes	2	32 in. (813 mm)

LUBRICATION

ENGINE

Refer to Briggs & Stratton section for engine lubrication requirements. Recommended lubrication requirements. Recommended fuel is regular or low-lead gasoline.

TRANSMISSION

The gear-type transmission is factory lubricated and should not require lubrication unless disassembled for overhaul.

Refer to Peerless transmission section for service information.

DRIVE MECHANISM

Lubricate all pivot points and drive chains with a suitable lubricant after every ten hours of operation. After every ten hours of operation force grease into grease fittings for axle bushings, auger shaft bushings and countershaft bushings. On early models lubricate idler pulley shafts and bushings but do not allow oil on belts; later models are equipped with sealed bearings. Wheel hubs should be greased after each season.

AUGER GEARBOX

Oil level in auger gearbox should be checked periodically. Oil level should be even with oil plug hole (H—Fig. G31). Recommended oil is SAE 30.

MAINTENANCE

The augers are secured to auger shaft by shear pins which will fracture when an auger contacts an obstruction. Renew shear pins as required with factory authorized shear pins.

Inspect drive belts and pulley belt grooves. Renew belt if frayed, cracked, burnt or otherwise damaged.

Inspect drive chains and sprockets. Renew drive chain if excessively stretched, worn or stiff. Sprockets should be renewed if teeth are worn or otherwise damaged.

Recommended tire pressure is 20 psi (137.9 kPa).

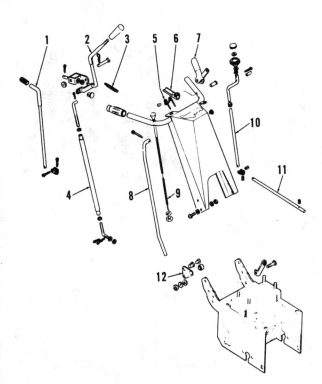

Fig. G30—Drawing of snowthrower controls. Remote throttle cable (9) is not used on all models.

1. Auger clutch rod
2. Shift lever
3. Spring
4. Shift rod
5. Spring
6. Safety lever
7. Clutch lever
8. Clutch rod
9. Throttle cable
10. Chute crank
11. Chute rod
12. Bellcrank

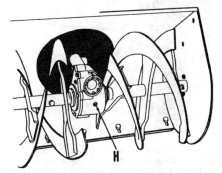

Fig. G31—View showing location of auger gearbox oil level hole (H).

ADJUSTMENTS

AUGER HEIGHT

To adjust auger height, loosen runner mounting bolts and reposition runners. Height should be increased when operating snowthrower over an irregular surface.

DRIVE CLUTCH

To adjust drive clutch, loosen set screw (S—Fig. G32) securing clutch control rod (R) in pivot block (B). Pull clutch rod (C) out until resistance is felt and making sure clutch control lever is in uppermost position, retighten set screw (S).

SHIFT LINKAGE

Before checking or adjusting shift linkage, be sure lock pins are inserted through holes in wheel hubs and axle. Shift linkage should be adjusted if unit does not roll easily with shift control lever in neutral or if unit is not difficult to push with shift control lever in third or reverse.

To adjust shift linkage, detach shift control rod end (E—Fig. G33) from shift arm (A) and push shift arm (A) up as far as possible. Move snowthrower forward and recheck position of shift arm to be sure it has gone into gear. Position shift lever in third gear then rotate rod end (E) until it will fit easily into hole in shift arm (A). Check adjustment as outlined in preceding paragraph.

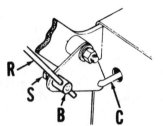

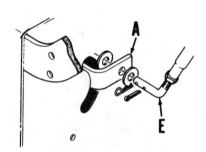

Fig. G32—Refer to text for drive clutch adjustment.

Fig. G33— With shift lever in third gear and shift arm (A) up as far as possible, rod end (E) should fit easily into shift arm hole. See text.

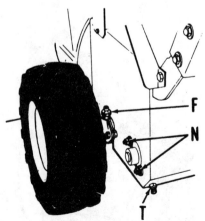

Fig. G34—View showing location of countershaft retaining nuts (N) and chain adjuster nut (T) located on both sides of main frame. Note location of grease fitting (F) on both ends of axle.

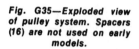

Fig. G35—Exploded view of pulley system. Spacers (16) are not used on early models.

1. Clutch rod
2. Clutch spring
3. Pivot bushing
4. Bellcrank
5. Engine pulley
6. Belt guide
7. Belt guard
8. Idler arm rod
9. Spring
10. Bushings
11. Clutch idler arm
12. Spring
13. Brake finger
14. Auger idler arm
15. Washer
16. Spacer
17. Clutch idler pulley
18. Auger idler pulley
19. Spacer
20. Spring
21. Drive pulley
22. Snap ring
23. Auger pulley

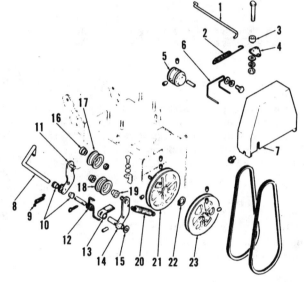

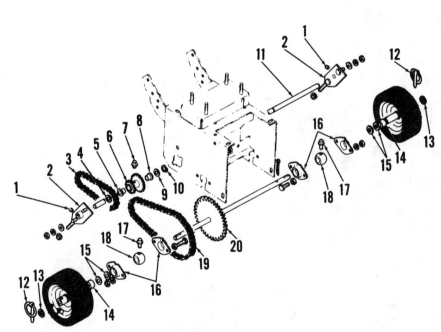

Fig. G36—Exploded view of wheel drive.

1. Set screw	6. Sprocket cluster	11. Countershaft	16. Bushing retainer
2. Countershaft plate	7. Grease fitting	12. Lock pin	17. Grease fitting
3. Intermediate chain	8. Bushing	13. Retainer	18. Bushing
4. Washer	9. Washer	14. Spacer	19. Drive chain
5. Bushing	10. Snap ring	15. Washers	20. Axle

CHAIN

All models are equipped with two drive chains which transmit power from transmission to countershaft to axle. Chains should deflect slightly at midpoint when properly tensioned. The countershaft is mounted in moveable plates which are used to adjust chain tension. Loosen nuts (N—Fig. G34) securing plates to main frame then turn adjusting nuts (T) at each end of countershaft to obtain desired chain tension. Check to be sure sprockets are aligned after chain tension adjustment and retighten nuts (N).

OVERHAUL

ENGINE

Engine make and model numbers are listed at beginning of this section. Refer to Briggs and Stratton or Tecumseh engine section for engine overhaul.

TRANSMISSION

All models are equipped with a Peerless series 700 transmission. Refer to Peerless transmission section for transmission overhaul information.

DRIVE BELTS

The drive belts may be extracted after draining fuel tank, removing belt guard and belt guide then tipping snowthrower forward. Refer to Fig. G37.

AUGER GEARBOX

To overhaul auger gearbox, remove auger and impeller assembly from snowthrower. Remove impeller and augers from shafts and drain oil from gearbox. Remove gearbox side cover (4—Figs. G39 or G40) while noting number of gaskets (5—Fig. G39) on Model 55012 gearbox for assembly. Withdraw shaft (13) with gear (11) then unscrew plug (22) and remove gear and shaft assembly through front of gearbox. Inspect and renew worn or damaged components.

Reverse disassembly procedure to reassemble gearbox. Tighten plug (22) just enough to remove shaft (14) end play. Be sure shaft (14) turns freely. Fill gearbox with SAE 30 oil to level of plug hole (H—Fig. G31).

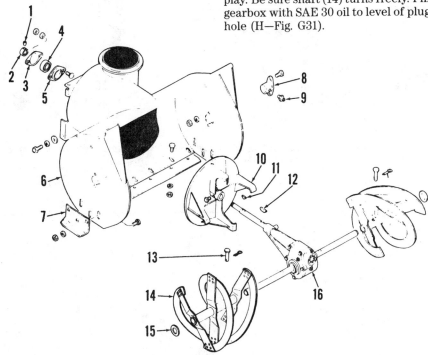

Fig. G38—Exploded view of auger.

1. Set screw	5. Bearing retainer	9. Grease fitting	13. Shear pin
2. Collar	6. Auger frame	10. Impeller	14. Auger
3. Bearing retainer	7. Runner	11. Key	15. Washer
4. Bearing	8. Bushing & housing	12. Key	16. Auger gearbox

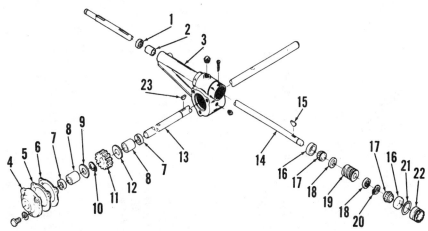

Fig. G37—View of pulley and belt system. Refer to Fig. G35 for parts identification.

Fig. G39—Exploded view of auger gearbox used on Model 55012.

1. Seal	7. Seal	13. Auger shaft	
2. Bushing	8. Bushing	14. Impeller shaft	19. Worm
3. Gearcase	9. Washer	15. Key	20. Snap ring
4. Cover	10. Snap ring	16. Bearing cup	21. "O" ring
5. Gasket	11. Gear	17. Bearing cone	22. Plug
6. Gasket	12. Washer	18. Spacer	23. Key

ELECTRICAL INTERLOCK SYSTEM

All models are equipped with an interlock switch which is actuated by the drive clutch. The switch is closed when auger clutch is engaged to ground out ignition system through safety control lever on handlebar. Refer to Fig. G41.

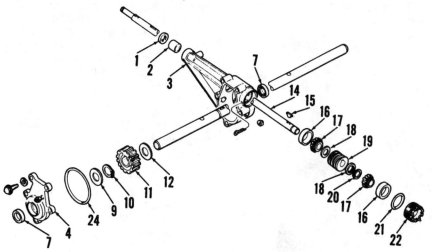

Fig. G40—Exploded view of auger gearbox used on all models except 55012. Refer to Fig. G39 for parts identification except for: 24. "O" ring.

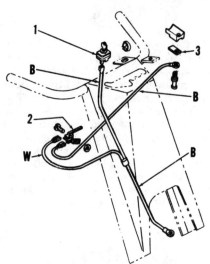

Fig. G41—Schematic of interlock wiring. Wires are colored black (B) and white (W).

1. Ignition switch
2. Clutch interlock switch
3. Ground plate

GILSON

Model	Engine Make	Engine Model	Self-Propelled?	No. of Stages	Scoop Width
55323A	B&S	130000	Yes	2	24 in. (610 mm)
55324A	B&S	190000	Yes	2	26 in. (660 mm)
55333B	Tecumseh	HS50	Yes	2	24 in. (610 mm)
55334	Tecumseh	HMSK80	Yes	2	26 in. (660 mm)
55350	B&S	130000	Yes	2	24 in. (610 mm)
55351	B&S	130000	Yes	2	24 in. (610 mm)
55352	B&S	190000	Yes	2	26 in. (660 mm)

LUBRICATION

ENGINE

Refer to B&S or Tecumseh engine section for engine lubrication requirements. Recommended fuel is regular or low-lead gasoline.

DRIVE MECHANISM

After every ten hours of operation, lubricate all pivot points and axle bearings. Do not allow oil on drive belts or friction discs. After each season, lubricate auger shaft bearings with a good quality automotive wheel bearing grease.

MAINTENANCE

The augers are secured to the auger shaft with shear pins which will fracture when the auger contacts an obstruction. Use only original equipment type shear pins.

Inspect drive belts and belt pulley grooves frequently. Renew belt if frayed, cracked, burnt or otherwise damaged.

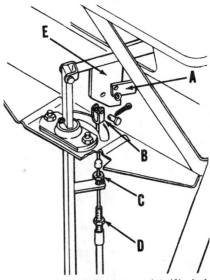

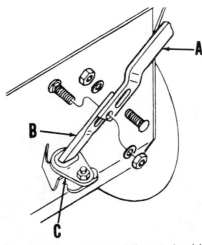

Fig. G51—View of upper shift control rod (A), shifter rod end (B) and bellcrank (C).

Fig. G52—View showing locator plate (A), clevis (B), upper adjusting nut (C), lower adjusting nut (D) and speed selector plate (E).

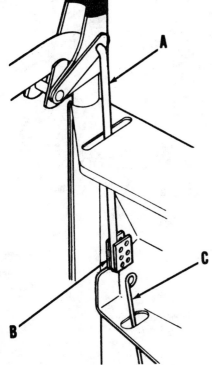

Fig. G50—View of clutch rod (A), adjusting plates (B) and clutch spring (C).

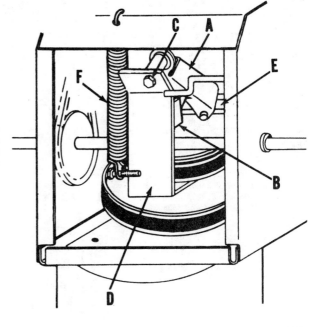

Fig. G53—View from underside showing fork assembly (A), stop plate (B), stop screw (C), disc carrier (D), hex shaft (E) and traction drive spring (F).

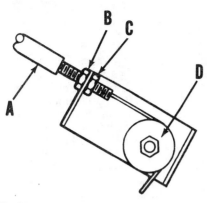

Fig. G54—View showing cable lower end (A), upper adjustment nut (B), lower adjustment nut (C) and roller (D).

Fig. G56—View of handle and control assembly used on Models 55333B and 55334.

1. Handle assy.
2. Shift lever assy.
3. Locator assy.
4. Long handle
5. Short handle
6. Shift bar assy.
7. Clutch cable
8. Clutch lever
9. Chute control crank
10. Chute shaft
11. U-joint
12. Roll pin
13. Control rod & locator plate
14. Rivet
15. Spacer
16. Spring

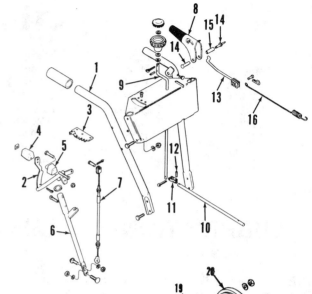

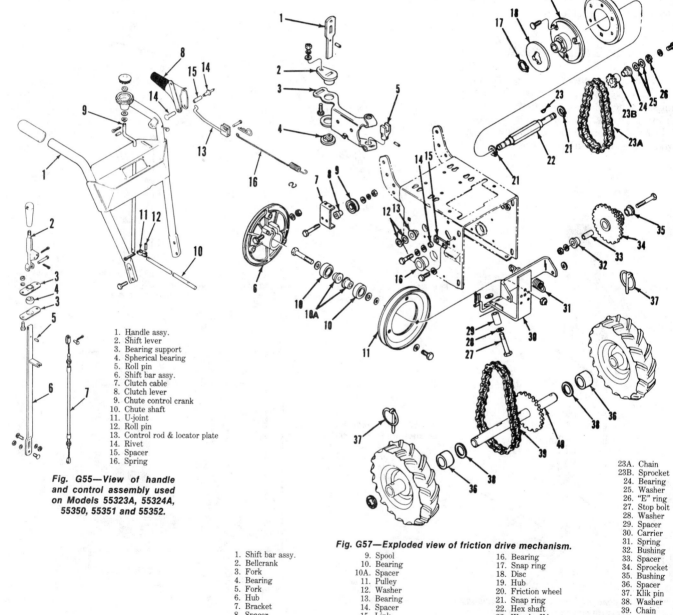

Fig. G55—View of handle and control assembly used on Models 55323A, 55324A, 55350, 55351 and 55352.

1. Handle assy.
2. Shift lever
3. Bearing support
4. Spherical bearing
5. Roll pin
6. Shift bar assy.
7. Clutch cable
8. Clutch lever
9. Chute control crank
10. Chute shaft
11. U-joint
12. Roll pin
13. Control rod & locator plate
14. Rivet
15. Spacer
16. Spring

Fig. G57—Exploded view of friction drive mechanism.

1. Shift bar assy.
2. Bellcrank
3. Fork
4. Bearing
5. Fork
6. Hub
7. Bracket
8. Spacer
9. Spool
10. Bearing
10A. Spacer
11. Pulley
12. Washer
13. Bearing
14. Spacer
15. Link
16. Bearing
17. Snap ring
18. Disc
19. Hub
20. Friction wheel
21. Snap ring
22. Hex shaft
23. Woodruff key
23A. Chain
23B. Sprocket
24. Bearing
25. Washer
26. "E" ring
27. Stop bolt
28. Washer
29. Spacer
30. Carrier
31. Spring
32. Bushing
33. Spacer
34. Sprocket
35. Bushing
36. Spacer
37. Klik pin
38. Washer
39. Chain
40. Axle

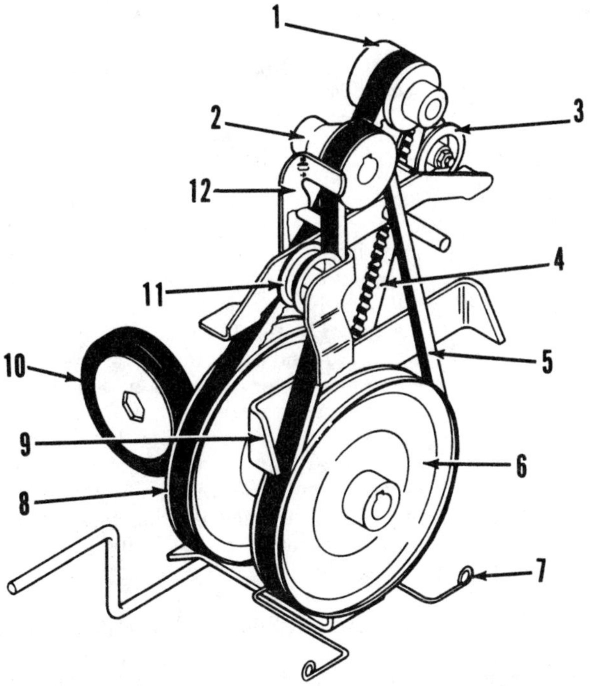

Fig. G58—View of belt pulley system.

1. Engine camshaft pulley
2. Engine crankshaft pulley
3. Traction idler pulley
4. Traction drive belt
5. Impeller drive belt
6. Impeller drive pulley
7. Belt finger
8. Traction drive pulley
9. Disc carrier
10. Friction wheel
11. Impeller idler pulley
12. Belt guide

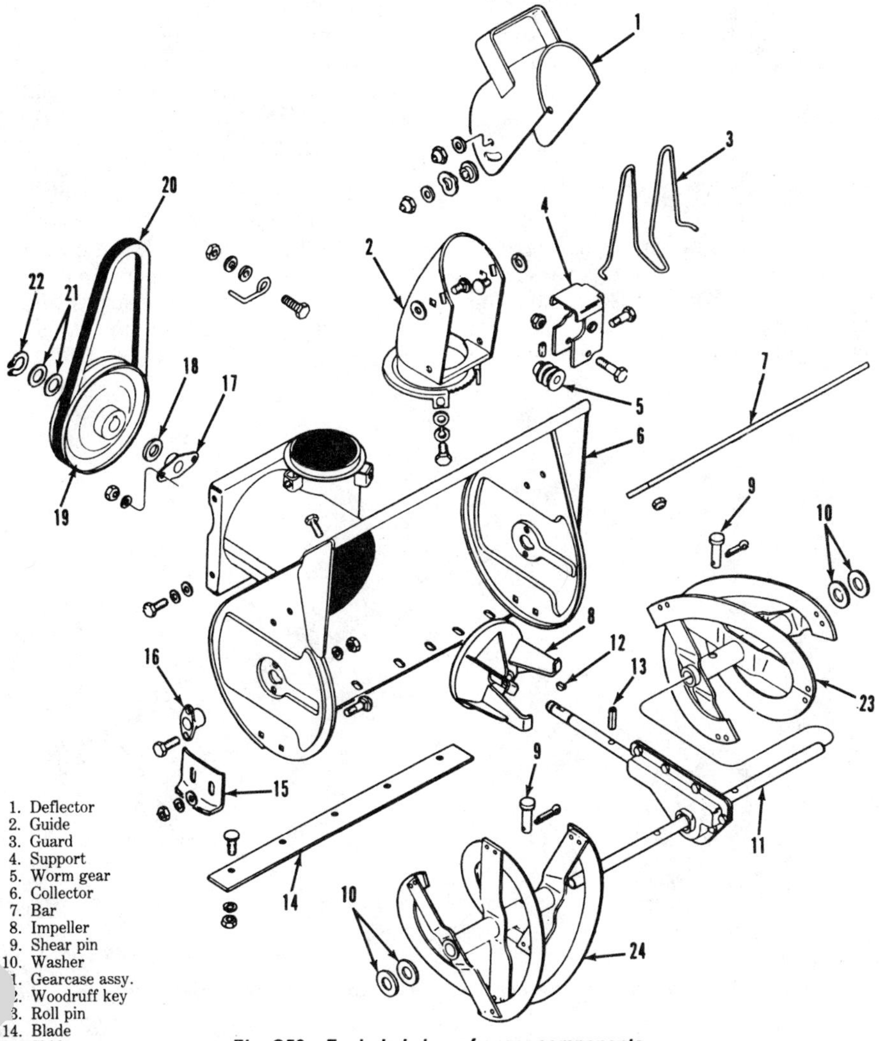

ADJUSTMENTS

HEIGHT

To adjust auger height, loosen runner mounting bolts and reposition runners. Increase height when operating snowthrower over an irregular surface.

CLUTCH CABLE

To adjust clutch cable, remove clevis pin and spring clip from adjusting plate (B–Fig. G50). With clutch lever in the straight up position, place end of spring (C) in a position that will just tighten the cable but not turn the hex shaft. Reinstall clevis pin and spring clip.

SHIFT LINKAGE

If adjustment is required, place speed control in outer reverse position and tip snowthrower on end. Remove inspection cover from bottom and check position of traction wheel. Traction wheel should be bottomed out against stop. If not, disconnect shift control rod (A–Fig. G51) from shifter rod end (B). Disconnect clevis (B–Fig. G52) from speed selector plate (E). Relocate speed selector plate on locator plate (A) as needed to lock shift lever into proper reverse position. Loosen handle assembly hardware and retighten with handle pushed as far forward as possible. Loosen bellcrank (C–Fig. G51) nut. Move fork assembly (A–Fig. G53) to the right until end of traction hub is resting against retaining ring. Reconnect shift control rod and shifter rod end. Retighten bellcrank nut. Reconnect clevis (B–Fig. G52) and locator plate (A). Fit of clevis in locator plate should be snug. If adjustment is necessary, refer to Fig. G54. Remove roller bracket assembly from outside of frame and turn lower adjustment nut (C) toward end of cable (A). Tighten upper adjustment nut (B) down against bracket until cable is tightened. Tighten lower adjustment nut against bracket and reinstall bracket on frame. Reinstall inspection cover.

OVERHAUL

ENGINE

Engine make and model number are listed at the beginning of this section. Refer to Briggs & Stratton or Tecumseh engine section for engine overhaul.

1. Deflector
2. Guide
3. Guard
4. Support
5. Worm gear
6. Collector
7. Bar
8. Impeller
9. Shear pin
10. Washer
11. Gearcase assy.
12. Woodruff key
13. Roll pin
14. Blade
15. Skid
16. Bearing
17. Bearing
18. Washer
19. Pulley
20. Belt
21. Washer
22. Snap ring
23. L.H. auger
24. R.H. auger

Fig. G59—Exploded view of auger components.

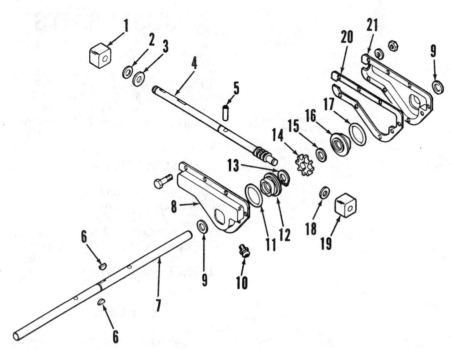

Fig. G60—Exploded view of gearcase assembly used on all models except 55324A, 55334 and 55352.

1. Bearing block	6. Woodruff key	12. Bearing	17. "O" ring
2. Washer	7. Auger shaft	13. Snap ring	18. Washer
3. Thrust washer	8. R.H. case	14. Worm wheel	19. Bearing block
4. Worm shaft	9. Seal	15. Washer	20. Gasket
5. Roll pin	10. Grease fitting	16. Bearing	21. L.H. case
	11. "O" ring		

AUGER GEARBOX

Models 55323B, 55333B, 55350 and 55351

Refer to Fig. G60 for an exploded view. Remove impeller and augers from shafts. Remove the eight cap screws holding case halves (8 and 21) together and slide left-hand side of case (21) out over auger shaft (7). Slide auger shaft out of right-hand case half (8) along with bearings (12 and 16), washer (15), worm wheel (14) and snap ring (13). Inspect all components for damage and/or wear and renew as needed.

Reassembly is reverse of disassembly procedure. Renew "O" rings (11) and seals (9) during reassembly. Pump ½ pint (0.3 L) of wheel bearing grease into gearbox through grease fitting (10).

Models 55324A, 55334 and 55352

Refer to Fig. G61 for an exploded view. Remove auger and impeller assembly from snowthrower collector. Remove impeller and augers from shafts and drain oil from gearbox. Remove gearbox side cover (22). Remove shaft (5) and gear (18), then unscrew plug (14) and remove gear and shaft assembly through front of gearbox. Inspect and renew worn or damaged components.

Reassemble by reversing disassembly procedure. Tighten plug (14) just enough to remove shaft (1) end play and lock with cotter key (16). Fill gearbox with SAE 30 engine oil to level of plug (15) hole.

TRACTION DRIVE BELT

Refer to Fig. G58 for view of belt and pulley assembly. Remove belt guard and lower inspection cover. Remove belt (4) from engine pulley (1) and slide belt out between idler pulley (3) and retainer. Place speed control in reverse position and disconnect traction drive spring (F–Fig. G53). Remove traction pulley bracket cap screws from outside of frame on both sides. Push traction pulley toward impeller pulley and slide belt out between pulley and friction wheel.

Reinstall new belt by reversing removal procedure.

IMPELLER DRIVE BELT

Refer to Fig. G58 for view of belt and pulley assembly. Remove belt guard and lower inspection cover. Loosen belt guide (12) and swing away from crankshaft pulley (2). Remove belt (5) from crankshaft pulley and slide out between idler pulley (11) and retainer. Unbolt and remove belt finger (7). Slip belt off impeller pulley (6) and slide out between pulley and disc carrier (9).

Reinstall by reversing removal procedure. Specified clearance between belt guard and engine crankshaft pulley, with traction drive control lever engaged, is 1/16 to 1/8 inch (1.6-3.2 mm). This same clearance is also specified between belt finger and impeller drive pulley.

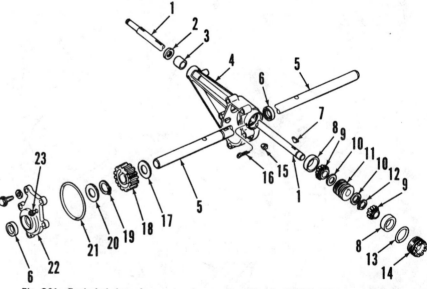

Fig. G61—Exploded view of auger gearbox used on Models 55324A, 55334 and 55352.

1. Input shaft	7. Woodruff key	12. Retaining ring	18. Gear
2. Seal	8. Bearing cup	13. "O" ring	19. Snap ring
3. Bushing	9. Bearing cone	14. Plug	20. Washer
4. Case	10. Spacer	15. Plug	21. "O" ring
5. Auger shaft	11. Worm gear	16. Key	22. Cover
6. Seal		17. Washer	23. Vent plug

GILSON

Model	Engine Make	Engine Model	Self-Propelled?	No. of Stages	Scoop Width
55358	Tecumseh	HSK60	Yes	2	24 in. (610 mm)
55369	B&S	190000	Yes	2	24 in. (610 mm)
55370	B&S	190000	Yes	2	26 in. (660 mm)
55371	Tecumseh	HMSK100	Yes	2	32 in. (813 mm)

LUBRICATION

ENGINE

Refer to appropriate Briggs & Stratton or Tecumseh 4-stroke engine section for engine lubrication information. Recommended fuel is regular or low-lead gasoline.

DRIVE MECHANISM

Lubricate all pivot points and drive chains with a suitable lubricant after every 10 hours of operation. Use a grease gun to lubricate fittings for axle bushings, auger shaft bushings and countershaft bushings after every 10 hours of operation. Wheel hubs should be greased after each season of use.

AUGER GEARBOX

On Models 55358 and 55369, the auger gearbox is fitted with a grease fitting and can be lubricated with a hand-operated grease gun using NLGI Grade 2 Lithium Base EP grease. For other models, check oil level in gearbox periodically. Oil level should be at bottom of auger gearbox pipe plug opening with unit setting level. Recommended lubricant for this type gearbox is SAE 30 motor oil.

TRANSMISSION

The gear-type transmission is lubricated at the factory and should not require further lubrication except at overhaul; refer to Peerless transmission section for service information.

MAINTENANCE

The auger flights are driven via the auger shaft by shear pins which will fracture when an auger contacts a solid obstruction. If necessary to renew shear pins, use only factory replacement pins.

Inspect drive belts and pulley belt grooves. Renew belts if frayed, cracked, burned or excessively worn. Renew pulleys if bent, excessively worn or otherwise damaged.

Inspect traction drive chain and sprockets. Renew chain if excessively worn, stretched or otherwise damaged. Renew sprockets if teeth are worn or otherwise damaged.

ADJUSTMENTS

AUGER HEIGHT

Auger height can be adjusted by loosening skid shoe mounting bolts and repositioning the skid shoes. Auger should be in lowest position (skid shoes raised) when operating over smooth surface and in highest position (skid shoes lowered) when operating over rough surfaces. Be sure skid shoes are equally adjusted on each side of auger housing.

CONTROL CABLES

The auger and traction drive control cables should have all slack removed when control lever is all the way against the rubber stop on the handle. If not, remove the spring clevis pin from the lower end of either control cable and turn the adjustment yoke as necessary to achieve proper adjustment. Reattach cable with spring clevis pin.

DRIVE BELTS

If auger drive belt slips, readjust position of idler pulley to increase belt tension. Remove belt guard and note position of idler pulley on auger idler arm. Loosen hex nut holding idler pulley to arm and move pulley toward belt about 1/4 inch (6.4 mm), tighten nut and recheck belt tension. Belt should release fully when control lever is released. It may be necessary to readjust control cable as described in preceding paragraph.

The traction drive belt can be adjusted following procedure outlined for auger belt by moving idler pulley in slot of drive idler arm.

The wire belt guides should clear belts by about 1/8 inch (3.2 mm) when belt control levers are engaged.

DRIVE CHAIN

When pushing on chain at midpoint between transmission and axle sprockets, the chain should deflect about 1/8 inch (3.2 mm). To adjust chain, tip the snowthrower forward onto front of auger housing and loosen the three bolts retaining the axle bearing retainers on each side of main frame. Slide the axle shaft bearing assemblies in the slotted mounting holes of main frame until proper adjustment is reached. Be sure both bearings are the same distance from back of main frame and tighten the bearing retaining bolts. On Model 55371, check drive chain idler sprocket (2).

OVERHAUL

ENGINE

Engine make and model numbers are listed at beginning of this section. Refer to Briggs & Stratton or Tecumseh 4-stroke engine section in this manual for engine overhaul information.

TRANSMISSION

All models are equipped with a gear-type transmission. Refer to Peerless section for transmission overhaul information.

AUGER DRIVE BELT

With belt drive cover removed, remove belt from engine pulley and slip belt out from between auger idler pulley and retainer. Push the idler arm in to move brake shoe away from auger pulley and remove belt from between auger pulley and lower belt guide. Install new belt by reversing removal procedure and check belt and cable adjustments as outlined in a previous paragraph.

TRACTION DRIVE BELT

With belt drive cover removed, remove auger drive belt from engine pulley. Move traction drive belt from engine pulley to a position between pulley and engine. Slip belt out from between traction drive idler pulley and retainer. Remove belt from transmission pulley. Install new belt by reversing removal procedure and replace auger belt on engine pulley. Check belt and cable adjustments as outlined in appropriate previous paragraph.

HOMELITE/JACOBSEN

HOMELITE DIVISION OF TEXTRON, INC.
P.O. Box 7047
14401 Carowinds Boulevard
Charlotte, NC 28217

Model	Engine Make	Engine Model	Self-Propelled?	No. of Stages	Scoop Width
SNO 320	Own	J501	No	1	20 in. (508 mm)
320, 320E	Own	*	No	1	20 in. (508 mm)

*Specifications and overhaul information not available at time of publication.

LUBRICATION

ENGINE

All models are equipped with 2-stroke cycle engines. Refer to Jacobsen engine section for Model SNO 320 engine. Information on engine for Models 320 and 320E was not available at time of publication. If Homelite or Jacobsen 2-cycle oil is used, follow directions on package. For other 2-cycle oil, use a fuel:oil mixture ratio of 16:1.

DRIVE MECHANISM

Remove upper cowling and lubricate belt idler arm pivot with SAE 30 oil after every 25 hours of operation. On models with engine choke cable enclosed in conduit, lubricate cable wire after every 25 hours of operation.

MAINTENANCE

Inspect drive belt and pulleys. Renew belt if frayed, cracked, burned or otherwise damaged. Inspect impeller blades and renew if excessively worn or damaged. Check filter in fuel line between tank and carburetor and renew if fuel will not flow freely through filter.

ADJUSTMENTS

CONTROL CABLE

Except on Model SNO 320, if drive belt slips or impeller continues to turn after control lever is released, control cable (24—Fig. HJ1) should be adjusted. If belt slips, loosen jam nuts (13) and turn adjuster (12) counterclockwise to shorten cable until no belt slippage occurs with control lever engaged. If impeller continues to turn after control lever is released, turn adjusting nut clockwise to lengthen cable until impeller does not rotate. After making adjustment, tighten jam nuts securely.

OVERHAUL

ENGINE

Refer to Jacobsen engine section for Model SNO 320 engine overhaul. Information on engine for later Models 320 and 320E was not available at time of publication (Use Champion CJ8Y spark plug or equivalent for late model engines).

IMPELLER ASSEMBLY

Impeller blades (1—Fig. HJ2) are a matched set. If old blades are to be reinstalled, they should be marked end for end before removal. New blades are connected together and must be marked prior to cutting apart so ends will not be swapped during installation. Tighten blade screws in order, working from one end to the other to prevent waves from forming in the blades, to 55 in.-lbs. (6.2 N·m). Impeller halves (3) can be removed after removing impeller blades. To renew left hand impeller bearing,

Fig. HJ1—Exploded view of handle, muffler and wheel assembly. Later production Models 320 and 320E units have clutch spring (14) at bottom end of cable (24). Model SNO 320 does not have impeller clutch.

1. Manifold gasket
2. Exhaust pipe
3. Eyebolt
4. Upper handle
5. Lower handle
6. R.H. side plate
7. Muffler assy.
8. Axle
9. Spacer
10. L.H. frame assy.
11. Cable clamp
12. Cable adjuster
13. Jam nuts
14. Clutch spring
15. Control lever
24. Control cable

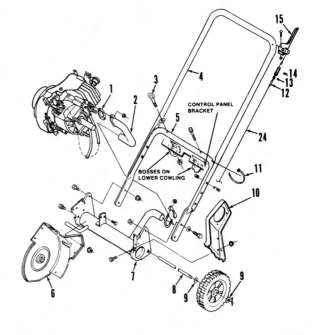

first remove impeller blades and halves, then remove belt cover (31) and impeller drive belt (30). Unscrew impeller pulley (29) (left-hand threads) and remove impeller plate and shaft (2) with bearing.

If bearing has not been removed previously, remove rivets from bearing retainer and left-hand side plate (20). When installing new bearing, use hardened bolts with self-locking nuts to re-

place rivets. Inspect components and renew as required.

Right-hand impeller bushing is renewable in right-hand side plate (5) on some models.

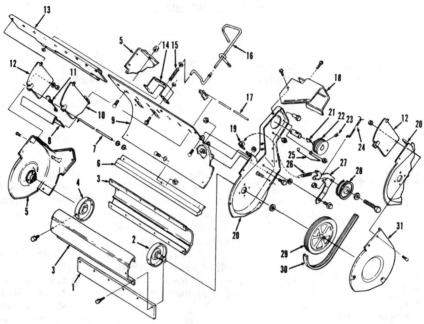

Fig. HJ2—Exploded view of impeller, snow deflector vanes, housing and belt drive on Models 320 and 320E. Belt drive components for Model SNO 320 are similar except for impeller clutch components. Impeller plate and shaft (2) are serviced with bearing assembly.

1. Impeller blades	9. Impeller housing	24. Clutch cable
2. Impeller plate & shaft	10. Deflector vanes (7)	25. Belt guide
3. Impeller half	11. R.H. deflector spring	26. Return spring
4. Impeller plate & pin	12. End vanes (2)	27. Idler clutch arm
5. R.H. side plate	13. Vane panel	28. Idler pulley
6. Scraper blade	14. Control bracket	29. Impeller pulley
7. Vane retaining rod	15. Tension spring	30. Drive belt
8. Vane guide	16. Vane control handle	31. Belt cover
	17. Vane control rod	
	18. Belt cover	
	19. Spring clip	
	20. L.H. side plate	
	21. Woodruff key	
	22. Engine pulley	
	23. Setscrew	

HOMELITE/JACOBSEN

Model	Engine Make	Engine Model	Self-Propelled?	No. of Stages	Scoop Width
420	Jacobsen	*	No	1	20 in. (508 mm)

*Engine service information not available at time of publication. Refer to engine information in this section for all available information.

LUBRICATION

ENGINE

If Homelite 2-cycle engine oil is used, refer to instructions on container for recommended fuel:oil mixture ratio. If other 2-cycle engine oils are used, Homelite recommends a 16:1 fuel:oil mixture ratio regardless of instructions on container. Unleaded regular gasoline is recommended fuel.

DRIVE MECHANISM

Remove upper cowling and apply a few drops of SAE 30 engine oil between belt idler arm and frame. Do not allow oil to get on belt or pulleys.

ADJUSTMENT

CLUTCH CONTROL CABLE

If impeller drive belt slips under load, loosen jam nut (N—Fig. HJ5) at top of clutch cable (11) and turn adjuster (A) counterclockwise until no belt slippage occurs. If impeller turns with clutch control bail (10) released, clutch cable is too tight; turn adjuster clockwise to loosen clutch cable.

OVERHAUL

ENGINE

Service information on engine was not available at time of publication. Recommended spark plug is Champion CJ-8Y. Recommended spark plug gap is 0.025 inch (0.64 mm).

DRIVE MECHANISM

Overhaul of drive mechanism is evident after inspection of unit and reference to Fig. HJ6. Impeller blades (3) are serviceable and must be installed as a pair to avoid an out of balance condition. Be sure spacers (5) are installed at all clamp bolt positions.

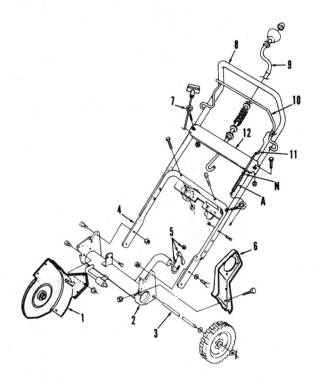

Fig. HJ5—Exploded view of handle assembly, muffler and clutch controls.

A. Adjuster
N. Jam nut
1. R.H. end plate
2. Muffler assy.
3. Axle
4. Lower handle
5. Muffler clamp
6. L.H. end plate
7. Starter rope
8. Upper handle
9. Chute crank
10. Clutch control bail
11. Clutch cable
12. Chute crank bracket

Illustrations Courtesy of Homelite Div. of Textron, Inc.

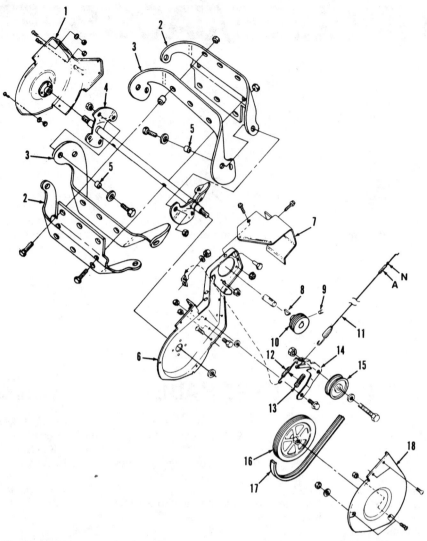

Fig. HJ6—Exploded view of impeller and drive mechanism. Impeller clamp plates (2) on some models may not be full width as shown.

A. Adjuster
N. Jam nut
1. R.H. side plate
2. Clamp plate
3. Impeller blade
4. Impeller shaft

5. Spacer
6. L.H. side plate
7. Belt cover
8. Woodruff key
9. Set screw

10. Engine pulley
11. Clutch cable & spring
12. Auger idler return spring

13. Spring cover
14. Idler arm
15. Idler pulley
16. Impeller pulley
17. Drive belt
18. Belt cover

Illustrations Courtesy of Homelite Div. of Textron, Inc.

HOMELITE/JACOBSEN

Model	Engine Make	Engine Model	Self-Propelled?	No. of Stages	Scoop Width
52006	B&S	100000	Yes	2	20 in. (508 mm)
52011	B&S	130000	Yes	2	20 in. (508 mm)
52012	B&S	130000	Yes	2	20 in. (508 mm)

LUBRICATION

ENGINE

Refer to Briggs & Stratton engine section for lubrication information. Recommended fuel is leaded or unleaded regular gasoline.

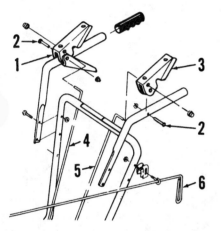

Fig. HJ10—Drawing of control assembly.

1. Drive clutch control lever
2. Pivot rivet
3. Auger clutch control lever
4. Drive clutch control rod
5. Auger clutch control rod
6. Chute crank

DRIVE MECHANISM

Lubricate drive chain after each use with a suitable chain oil. Use a good quality multipurpose grease and lubricate auger shaft and jackshaft through grease fittings. Note that auger shaft grease fitting is accessible through hole in auger frame right side. Apply multipurpose grease to gear teeth and sliding surfaces of clutch dog. Lubricate pivot points of actuating linkage to ensure smooth operation.

ADJUSTMENTS

AUGER HEIGHT

Auger height is adjusted by loosening skid shoe retaining screws and repositioning skid shoes. Low auger height may be used on smooth surfaces but height should be increased when operating on irregular surfaces.

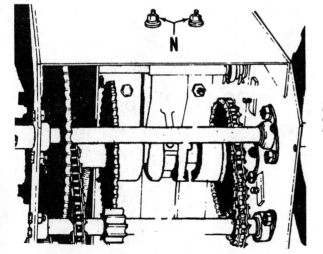

Fig. HJ11—Loosen nuts (N) and turn screws to adjust neutral spring as outlined in text.

NEUTRAL SPRING

Neutral spring (2—Fig. HJ14) should return clutch dog (11) to neutral when either forward or reverse control lever is released. Adjust spring by loosening nuts (N—Fig. HJ11) and turning screws on rear of main frame. Tighten nuts (N) and check adjustment.

DRIVE CLUTCH CONTROL ROD

With engine stopped and either forward or reverse control lever engaged by squeezing lever against handlebar, snowthrower should be locked in gear. If not, adjust control rod as follows: Refer to Fig. HJ12 and loosen the two screws (S) which clamp upper and lower drive clutch control rods together. Loosen screws just enough so rods can move but remain aligned. Pull up reverse control lever and tighten the two rod screws finger tight. Depress forward control lever while rocking unit forward to ensure full clutch engagement. With forward control lever depressed against handlebar, tighten rod screws and check adjustment.

AUGER CLUTCH CONTROL ROD

To adjust auger clutch control rod, loosen screw (S—Fig. HJ13) clamping upper and lower control rods together.

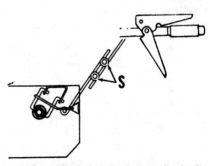

Fig. HJ12—Adjust drive clutch control rod using screws (S) as outlined in text.

Move auger clutch control lever to disengaged position. Pull lower control rod until idler pulley contacts belt. Move clamp screw so it is midway between ends of control rods and tighten screws finger tight. Depress auger clutch control lever until lever almost contacts handlebar. Then tighten clamp screw and check adjustment.

DRIVE CHAIN TENSION

Drive chain tension is correct when drive chain will deflect 1/8 inch (3.2 mm) midway between sprockets. To adjust drive chain tension, loosen nut retaining chain tension block located on left side of main frame and move chain tension block as required to obtain desired chain tension. Then tighten nut.

AUGER CHAIN TENSION

Auger chain tension is correct when auger chain will deflect 1/4 inch (6.4 mm) midway between sprockets. To adjust auger chain tension, loosen nut retaining chain tension block located on left side of auger frame and move chain tension block as required to obtain desired chain tension. Then tighten the nut.

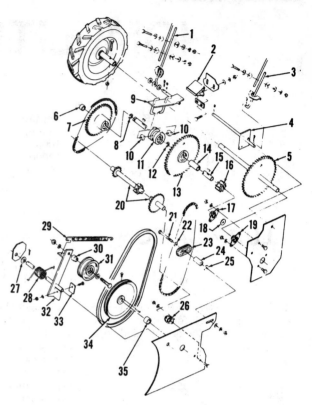

Fig. HJ14—Exploded view of drive mechanism.

1. Drive clutch control rod
2. Neutral spring
3. Auger clutch control rod
4. Auger clutch bellcrank
5. Axle & gear
6. Spacer
7. Sprocket
8. Snap ring
9. Clutch fork
10. Wear blocks
11. Clutch dog
12. Snap ring
13. Sprocket
14. Spacer
15. Clutch shaft
16. Gear
17. Bushing assy.
18. Washer
19. Bushing assy.
20. Jackshaft
21. Bolt
22. Washer
23. Drive chain tension block
24. Spacer
25. Washer
26. Bushing assy.
27. Washer
28. Spring
29. Spring
30. Spacer
31. Auger idler pulley
32. Idler arm
33. Brake pad
34. Auger pulley
35. Spacer

OVERHAUL

ENGINE

Engine make and model numbers are listed at front of this section. Refer to Briggs & Stratton engine section for engine overhaul information.

DRIVE MECHANISM

Overhaul of drive mechanism is evident after inspection of unit and reference to Figs. HJ14 and HJ15. Inspect components and renew any which are excessively worn or any other damage is noted. Refer to ADJUSTMENTS section after assembly.

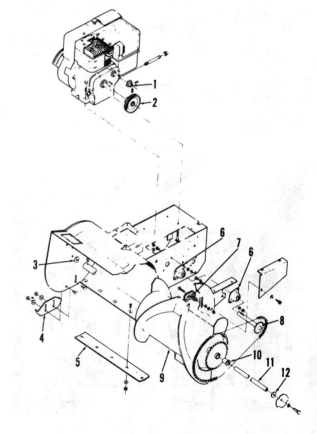

Fig. HJ15—Exploded view of auger assembly.

1. Drive sprocket
2. Auger drive pulley
3. Washer
4. Skid shoe
5. Scraper blade
6. Bushing assy.
7. Auger chain tension block
8. Sprocket
9. Auger & sprocket assy.
10. Bushing
11. Auger shaft
12. Washer

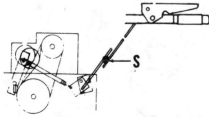

Fig. HJ13—Refer to text and adjust auger clutch control rod using screw (S).

HOMELITE/JACOBSEN

Model	Engine Make	Engine Model	Self-Propelled?	No. of Stages	Scoop Width
Jacobsen Models:					
52628	Tecumseh	H60	Yes	2	26 in. (660 mm)
52629	Tecumseh	H60	Yes	2	26 in. (660 mm)
52632	Tecumseh	HM80	Yes	2	30 in. (762 mm)
52633	Tecumseh	HM80	Yes	2	30 in. (762 mm)
52635	Tecumseh	HM80	Yes	2	26 in. (660 mm)
52636	Tecumseh	HM80	Yes	2	26 in. (660 mm)
52638	Tecumseh	H60	Yes	2	26 in. (660 mm)
52640	Tecumseh	HM80	Yes	2	26 in. (660 mm)
52642	Tecumseh	HM80	Yes	2	30 in. (762 mm)
Homelite/Jacobsen Models:					
524	Tecumseh	H50	Yes	2	24 in. (610 mm)
624	Tecumseh	H60	Yes	2	24 in. (610 mm)
626	Tecumseh	H60	Yes	2	26 in. (660 mm)
824	Tecumseh	HM80	Yes	2	24 in. (610 mm)
826	Tecumseh	HM80	Yes	2	26 in. (660 mm)
830	Tecumseh	HM80	Yes	2	30 in. (762 mm)

LUBRICATION

ENGINE

Refer to Tecumseh 4-stroke engine section for engine lubrication information. Recommended fuel is unleaded or leaded regular gasoline.

DRIVE MECHANISM

Lubrication should be done at start of each season and after every 40 hours of operation during the season. Lubricate all control linkage pivot points and the drive chain using SAE 30 engine oil. Lubricate discharge chute flange, chute sprocket and sprocket bearing, traction drive gears and friction wheel hex shaft with water resistant grease. Remove auger shaft bushings and apply grease to inside of bushings and retainer cups; align slot of bushing with retainer tab when reinstalling. Using hand-operated grease gun, apply grease at auger lubrication fittings.

AUGER GEARBOX

Check auger gearbox oil level at start of each season and after every 40 hours of operation. With snowthrower level, remove front cover lower retaining screw (S—Fig. HJ20); oil level should be at bottom of hole. Add lubricant (Homelite part no. JA-50065-0 or equivalent) as necessary through filler plug opening in top of gearbox.

MAINTENANCE

Inspect chain and sprockets and renew if excessively worn. Inspect belts and pulleys; renew belt if excessively worn, burned, frayed, cracked or otherwise damaged. Renew idler pulleys if bearings are loose or rough.

The augers are driven via special bolts designed to shear if auger encounters a solid obstruction. If bolt is sheared, turn auger to align bolt holes in auger and shaft and drive out remainder of old bolt. Apply grease at auger lubrication fittings and be sure auger turns freely on shaft, then install a new shear bolt using only a factory replacement part.

ADJUSTMENTS

AUGER HEIGHT

Auger height is adjusted by loosening skid shoe retaining bolts and repositioning the skid shoes. Raise the skid shoes on auger housing to lower scraper bar for smooth surfaces; lower the skids to raise scraper bar when operating over rough surfaces such as gravel drives.

SPEED CONTROL ROD

To adjust speed control rod, detach upper speed control rod (6—Fig. HJ21) from speed control lever (4). Move speed control lever to reverse position on handle console and pull up on speed control rod. The upper end of rod should be aligned with hole in lever; if not, turn upper rod until end will fit into hole and reattach rod to lever.

AUGER BRAKE

If auger brake pad has been renewed or the brake arm bracket (17—Fig. HJ22) has been loosened, adjust brake as follows: With belt cover removed and auger control lever disengaged, loosen bracket (17) attaching bolts and push brake pad down securely in auger pulley belt groove and tighten the bolts.

AUGER CLUTCH

First, if necessary, adjust auger brake as outlined in preceding paragraph, then proceed as follows: Loosen the bolt

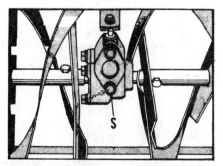

Fig. HJ20—Oil level in auger gearbox should be at bottom of lower cover screw (S) hole.

securing auger clutch control rod (13—Fig. HJ21) to spring (12) and place a 1/4 inch diameter round spacer between handle and front of auger clutch lever (11). Pull control rod upward in slot of spring until all slack is removed from auger clutch cable and tighten rod to spring bolt. Remove spacer from under auger clutch lever. With lever released, there should be a minimum of 1/8 inch (3.2 mm) slack in auger clutch cable. The spring (12) must extend when auger clutch lever is depressed against handle.

If proper adjustment cannot be obtained by above procedure, it will be necessary to loosen bellcrank pivot bracket (20—Fig. HJ22) bolts, reposition the bracket and tighten bolts. Then, repeat control adjustment procedure.

DRIVE CLUTCH CONTROL ROD

To adjust drive clutch control rod (3—Fig. HJ21), loosen bolt holding rod to spring (2). With drive control lever (1) released and pivoted forward against handle, pull rod up in slot of spring to remove all slack in rod, then tighten bolt. Do not stretch the spring or move the friction wheel bracket (25—Fig. HJ23) when removing slack from rod.

FRICTION WHEEL CLEARANCE

There should be from 3/32 to 5/32-inch (2.4-4 mm) clearance between friction wheel (21—Fig. HJ23) and face of drive pulley (5—Fig. HJ24) when drive control lever is released. Normal wear is normally compensated for by drive clutch control rod adjustment. However, further adjustment may be required. To check friction wheel clearance, tip snowthrower forward on front of auger housing and remove rear cover from main frame (motor mount). If clearance is not within specifications, detach the links connecting drive clutch pivot (6—Fig. HJ23) to friction wheel bracket (25) and reconnect to different set of holes. There are three sets of link attaching holes in bracket (25); move links to rear to decrease clearance or forward to increase gap between friction wheel and pulley drive face.

OVERHAUL

ENGINE

Engine make and model numbers are listed at beginning of this section. Refer to Tecumseh 4-stroke engine section of this manual for engine overhaul information.

DRIVE BELTS

To remove drive belts, remove belt cover and push idler pulleys away from belts and slide belts off of engine pulley. Tip the snowthrower forward onto auger housing and remove rear cover from main frame. Detach lower end of drive clutch control rod (3—Fig. HJ21). Move speed control lever to position "4" and remove belts by passing them between friction wheel (21—Fig. HJ23) and drive pulley (5—Fig. HJ24) while turning friction wheel to move the belts through the gap. Reverse removal procedure to install new belts and recheck auger brake, auger control rod and drive control rod as previously outlined.

FRICTION WHEEL

To renew rubber faced friction wheel (21—Fig. HJ23), place speed control lever in reverse position, tip snowthrower forward onto auger housing and remove rear main frame cover. Remove drive chain by disconnecting master link. Detach lower end of drive clutch control rod (3—Fig. HJ21) from drive clutch pivot (6—Fig. HJ23). Remove left wheel from drive axle for access and pry out hex shaft access plug (9) in main frame at left end of hex shaft. Remove the two cotter pins from hex shaft. Remove nuts from both ends of hex shaft. Unbolt left bearing retainer (23) and remove sprocket (31) and bearing retain-

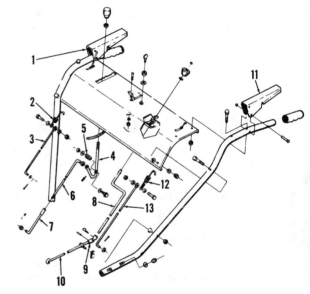

Fig. HJ21—Exploded view of handle and controls. Chute crank (8) has been relocated to bracket on left handle on later models.

1. Drive control lever
2. Drive control spring
3. Drive clutch control rod
4. Speed control lever
5. Lever friction spring
6. Upper speed control rod
7. Lower speed control rod
8. Chute crank
9. U-joint
10. Chute crank rod
11. Auger clutch lever
12. Auger control spring
13. Auger clutch control rod

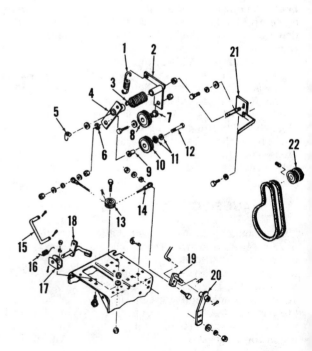

Fig. HJ22—Exploded view of drive belt system.

1. Drive idler spring
2. Drive idler arm
3. Auger brake spring
4. Auger idler arm
5. Idler arm spring
6. Lockwasher
7. Idler spacer
8. Drive belt idler
9. Idler spacer
10. Auger belt idler
11. Spacer washers
12. Idler bolt
13. Auger cable pulley
14. Auger clutch cable
15. Brake link
16. Brake rod spring
17. Brake arm bracket
18. Auger brake arm
19. Auger bellcrank
20. Bellcrank pivot bracket
21. Idler spindle plate
22. Engine pulley

Illustrations Courtesy of Homelite Div. of Textron, Inc.

er. Slide hex shaft to left through hole in frame until friction wheel (21) and hub (20) can be withdrawn and unbolt friction wheel from hub. Install new wheel with cup side facing hub.

Reassemble by reversing disassembly procedure. When installing wheel on axle, first install key and be sure "V" of tire tread points forward at top of tire.

AUGER GEARBOX

The auger gearbox can be disassembled without removing auger housing from snowthrower main frame. Unbolt and remove the auger shaft bushing retainer (7—Fig. HJ24) and bushing (6) from each end of auger shaft. Unbolt the brace from top of gearbox and gearbox front cover (18). Remove the cover and turn impeller fan (17) to spin worm gear (21) from housing, then remove key from impeller shaft (24). The gear housing and auger shaft (12) can then be removed from front of auger housing. Further disassembly is evident after inspection of unit and reference to Fig. HJ24. Clean and inspect all parts and renew any excessively worn or damaged parts.

Reassemble using new seals and by reversing disassembly procedure. When

gearbox and auger shaft assembly are installed back in housing, turn impeller fan so slot for key is up. Hold key in place with needlenose pliers and tap key

into place with drift inserted through filler plug opening in top of gearbox. Fill gearbox with lubricant as specified in gearbox lubrication paragraph.

Fig. HJ24—Exploded view of auger and gearbox assembly. Traction drive pulley and bearing assembly (5) can be removed from rear of impeller shaft (24) after removing friction drive plate (30) from rear of pulley and hex nut from impeller shaft.

1. Bearing retainer
2. Bearing
3. Auger drive pulley
4. Thrust washer
5. Traction drive pulley
6. Auger shaft bushing
7. Bushing retainer
8. Oil seal
9. Gearbox side cover
10. Bearing
11. Gasket
12. Auger shaft
13. Thrust washers
14. Worm wheel
15. Bearing
16. Oil seal
17. Impeller fan
18. Gearbox front cover
19. Gasket
20. Thrust washer
21. Worm gear
22. Thrust bearing races
23. Thrust bearing
24. Impeller shaft
25. Front impeller bushing
26. Auger shaft busing
27. Oil seal
28. Auger flight assy.
29. Roll pins
30. Friction drive plate

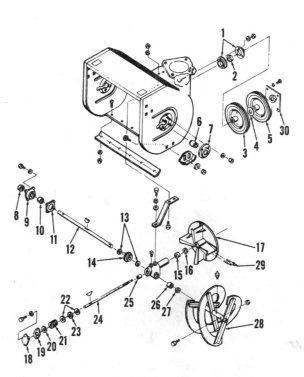

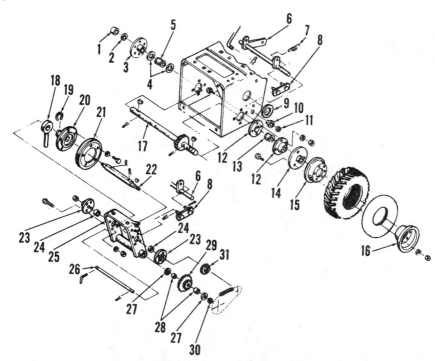

Fig. HJ23—Exploded view of friction drive mechanism. Friction wheel is driven by contact with friction drive plate (30—Fig. HJ24) attached to rear side of drive pulley (5—Fig. HJ24). Wheel (15 & 16), hub (3 or 14) and tire are a one-piece assembly on later models. Axle bushing (12 & 13) is a one-piece assembly with grease fitting on later models.

1. Hub cap
   Washer
   Wheel hub
   Washers
5. Axle bushing
6. Drive clutch pivot
7. Drive clutch return spring
8. Clutch control link
9. Hex shaft access plug
10. Clutch pivot bushing
11. Tinnerman retainer
12. Axle bushing retainers
13. Axle bushing
14. Wheel hub
15. Inner tire rim
16. Outer tire rim
17. Axle shaft & gear
18. Speed control yoke
19. Snap ring
20. Friction wheel hub
21. Friction wheel
22. Hex shaft
23. Hex shaft bearing retainer
24. Hex shaft bearing
25. Friction wheel bracket
26. Bracket pivot shaft
27. Thrust washer
28. Roller bearings
29. Sprocket/gear assy.
30. Washer
31. Hex shaft sprocket

HOMELITE/JACOBSEN

Model	Engine Make	Engine Model	Self-Propelled?	No. of Stages	Scoop Width
523	Tecumseh	HSK50	Yes	2	24 in. (610 mm)

LUBRICATION

ENGINE

Refer to Tecumseh engine section for engine lubrication information. Recommended fuel is regular or low-lead gasoline.

DRIVE MECHANISM

Recommended lubrication period is after every 10 hours of operation and before storing unit at end of season. Oil the drive chains and sprockets with SAE 10W-30 oil. The hex shaft assembly in the variable speed friction drive requires no lubrication except to wipe the hex shaft and gears with SAE 10W-30 motor oil at end of season to prevent rusting. The hex shaft bearings are of the sealed type. Be careful not to get any grease or oil on the disc drive plate or friction wheel; clean these parts thoroughly if oil or grease comes in contact.

Lubricate all moving part pivot points with SAE 10W-30 oil. Using a hand-operated grease gun, lubricate the auger shaft zerk fittings after every 10 hours of operation.

AUGER GEARBOX

The auger gearbox is factory lubricated and unless oil should leak out, lubrication should not be necessary unless gearbox is being overhauled. Once each season, remove plug (17—Fig. HJ35) and, if grease is not visible, use a wire as a dipstick to check for grease in gearbox. Recommended lubricants are

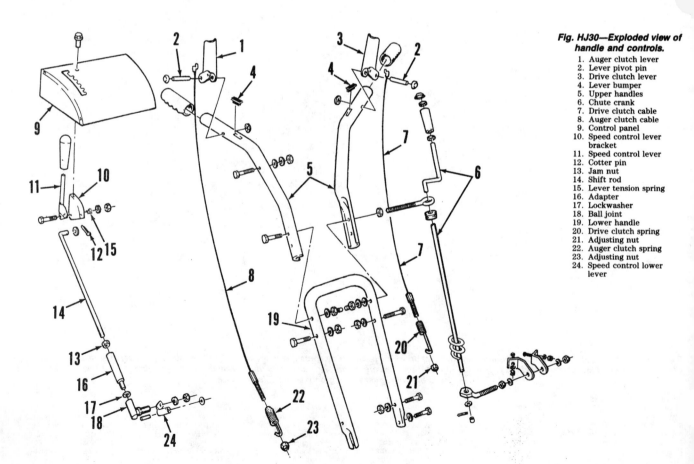

Fig. HJ30—Exploded view of handle and controls.

1. Auger clutch lever
2. Lever pivot pin
3. Drive clutch lever
4. Lever bumper
5. Upper handles
6. Chute crank
7. Drive clutch cable
8. Auger clutch cable
9. Control panel
10. Speed control lever bracket
11. Speed control lever
12. Cotter pin
13. Jam nut
14. Shift rod
15. Lever tension spring
16. Adapter
17. Lockwasher
18. Ball joint
19. Lower handle
20. Drive clutch spring
21. Adjusting nut
22. Auger clutch spring
23. Adjusting nut
24. Speed control lower lever

Benalene #372, Shell Darina 1, Texaco Thermatex EP1 and Mobiltem 78.

ADJUSTMENTS

AUGER HEIGHT

Lower the adjusting skids to raise auger scraper bar on rough surfaces. On paved surfaces, the skids can be raised to bring the auger assembly down. Keep in mind that the scraper bar below the auger should be 1/8 inch (3.2 mm) above sidewalk or other paved area to be cleaned. This clearance can be obtained by placing spacers under the scraper bar, then adjusting the skids so they contact paved surface.

To compensate for wear, the scraper bar position is adjustable, and the bar can be reversed for longer wear or renewed when both edges are worn. Be sure the bar is adjusted parallel with the working surface and for proper clearance, above skid shoes.

DRIVE BELTS

The traction drive belt is spring loaded and does not require adjustment; renew belt if it is slipping. If auger drive belt slips, adjust after removing belt cover as follows: Loosen nut on idler pulley (18—Fig. HJ31) and move idler in idler arm slot toward belt about 1/8 inch (3 mm). Tighten nut, engage auger drive clutch lever and check belt adjustment; belt should deflect about 1/2 inch (12.7 mm) with moderate pressure. Reset idler pulley until proper belt tension is obtained with drive engaged. Belt guides (11 & 12) should be adjusted to clear belt 3/32 inch (2.4 mm) with auger drive clutch engaged.

CONTROL CABLES

Refer to Fig. HJ30 and with drive clutch cable (7) "z" fitting disconnected from drive clutch lever (3), move lever forward to contact plastic bumper (4); the center of the "z" fitting should be between the top and the center of the hole in the lever. If not, slide cable down through drive clutch spring (20), hold square part of cable end from turning and rotate adjusting nut (21) to obtain correct cable length. Then, pull cable up through spring and reconnect "z" fitting on cable to hole in lever. Repeat adjustment procedure for auger clutch cable (8).

FRICTION WHEEL

Remove bottom panel from snowthrower. Place speed control lever in first gear and check position of friction wheel on disc drive plate. The right side of friction wheel should be 3-3/8 inches (86 mm) from outer edge of disc drive plate as shown in Fig. HJ33. If adjustment is necessary, loosen jam nut (13—Fig. HJ30) on speed select rod. Remove ball joint (18) from speed control lower lever (24) and turn adapter (16) to obtain correct friction wheel position with ball joint inserted in speed control lower lever. Reinstall ball joint, tighten jam nut and reinstall bottom panel.

OVERHAUL

ENGINE

Engine make and model number is listed at beginning of section. Refer to Tecumseh engine section for engine overhaul.

AUGER DRIVE BELT

To remove auger drive belt, remove belt drive cover and loosen belt guides (11 & 12—Fig. HJ31) and move them away from engine crankshaft pulley (24). Loosen auger belt idler pulley (18) and move it away from the belt. Engage auger drive clutch lever to move brake away from belt and slip the belt off auger pulley. Remove belt from engine drive pulley. Install new belt by reversing removal procedure. Adjust belt and belt guides as outlined in a previous paragraph.

TRACTION DRIVE BELT

To remove traction drive belt, remove belt cover and loosen the left-hand belt guide (11—Fig. HJ31) and move guide away from engine PTO pulley. Pull spring loaded traction idler pulley back and slip belt past idler pulley. Remove belt from engine pulley and up between the auger and traction drive pulleys. Install new belt by reversing removal procedure. Adjust belt and belt guide as outlined in a preceding paragraph.

AUGER SHEAR BOLT

The augers are driven via the auger shaft through special shear bolts (16—Fig. HJ34) that are designed to shear to protect the machine if an object becomes lodged in the auger housing. Only original equipment replacement shear bolts should be used for safety reasons. When replacing shear bolt(s), be sure to install new shear bolt spacer (17) and, using a hand-operated grease gun, lubricate auger shaft grease fittings.

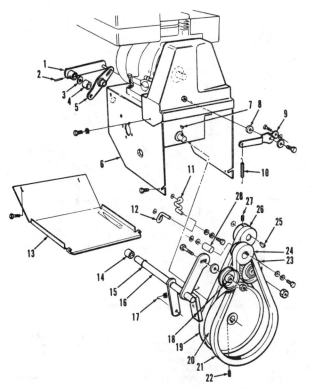

Fig. HJ31—Exploded view of drive belt mechanism. Refer to Fig. HJ32 for friction drive exploded view.

1. Auger clutch lever
2. Roll pin
3. Flat washer
4. Plastic bushing
5. Auger cable lever
6. Main (motor mount) frame
7. Hi-pro key
8. Spacer
9. Drive idler bracket
10. Drive idler spring
11. Belt guide
12. Belt guide
13. Bottom panel
14. Plastic bushing
15. Flat washer
16. Auger clutch rod
17. Auger idler return spring
18. Auger belt pulley
19. Drive belt
20. Drive pulley
21. Auger belt
22. Set screw
23. Drive belt idler
24. Engine crankshaft pulley
25. Woodruff key
26. Engine PTO pulley
27. Set screw
28. Spacer

Fig. HJ32—Exploded view of disc drive mechanism.

1. Bearing plate
2. Bearing
3. Flat washer
4. Hex shaft
5. Roll pins
6. Locknut
7. Speed control yoke
8. Plastic bushing
9. Washer
10. Snap ring
11. Washer
12. Trunnion bearing
13. Friction wheel hub
14. Friction wheel
15. Pinion gear
16. Flat washer
17. Bearing
18. Bearing plate
19. Spacers
20. Traction pivot
21. Push nut
22. Bushing
23. Flat washer
24. Cotter pin
25. Traction shaft
26. Trunnion bearing
27. Traction spring bracket
28. Flat washer
29. Traction spring
30. Spring pin
31. Traction rod
32. Drive disc
33. Return spring
34. Hi-pro key
35. Washer
36. Drive disc shaft
37. Snap ring
38. Thrust bearing
39. Needle bearings

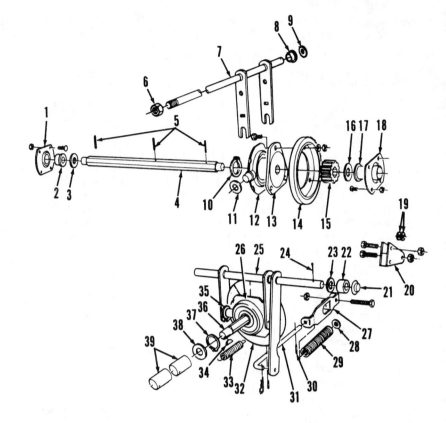

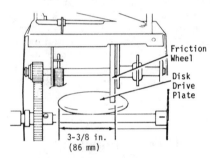

Fig. HJ33—With speed control lever in first gear, friction wheel should be positioned as shown.

AUGER ASSEMBLY

To remove auger, remove auger drive belt, disconnect the chute crank from auger housing and separate auger housing from main (motor mount) frame. Remove auger drive pulley (1—Fig. HJ34) and impeller shaft bearing (2) from rear of housing. Unbolt auger shaft bushings (14) at each side of housing and remove the assembly. Remove the auger drive shear bolts (16) and remove the auger flights from auger shaft.

With auger flights removed, refer to Fig. HJ35 for exploded view of auger drive gearbox. To disassemble, remove the cap screws holding the gearbox housing halves (16 & 22) and separate the unit. Remove impeller shaft bushing (14), snap ring (13) and thrust washer (12) from impeller shaft. Press shaft from worm gear (11) and remove needle

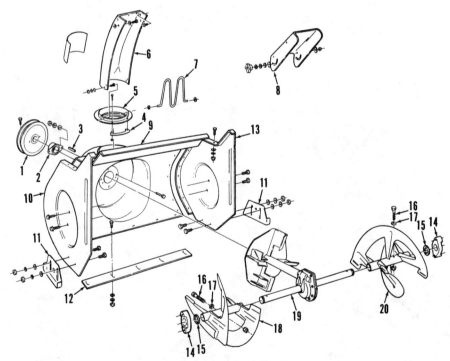

Fig. HJ34—Exploded view of auger housing assembly.

1. Auger drive pulley
2. Impeller shaft bearing
3. Square key
4. Chute retainer
5. Chute flange
6. Bottom chute extension
7. Hand guard
8. Upper chute extension
9. Center auger housing
10. R.H. side plate
11. Skid shoe
12. Scraper bar
13. L.H. side plate
14. Auger shaft bushing
15. Spacer washer
16. Shear bolt
17. Shear bolt spacer
18. R.H. auger flight
19. Gearbox assy.
20. L.H. auger flight

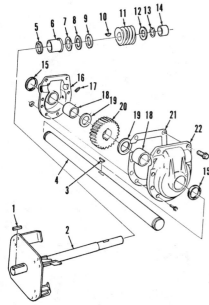

Fig. HJ35—Exploded view of auger gearbox. Impeller is an integral part of impeller shaft (2).

1. Square key	11. Worm gear
2. Impeller & shaft assy.	12. Thrust washer
3. Woodruff key	13. Snap ring
4. Auger shaft	14. Impeller shaft bushing
5. Square cut "O" ring	15. Oil seals
6. Impeller shaft bushing	16. L.H. housing half
7. Thrust bearing race	17. Grease plug
8. Needle thrust bearing	18. Auger shaft bushing
9. Thrust bearing race	19. Thrust washer
10. Woodruff key	20. Gear
	21. Gasket
	22. R.H. housing half

thrust bearing (8) and races (7 & 9), rear impeller shaft bushing (6) and "O" ring (5) from impeller shaft. Clean and inspect all parts and renew any excessively worn or damaged parts. To install auger shaft bushings (18), press bushings into case from inside and flush with inside of casting. Install new seals (15) in gearbox halves.

When reassembling, coat all parts with grease and pack gearbox with a total of 3-1/4 ounces (96 mL) of one of the following greases: Benalene #372, Shell Darina 1, Texaco Thermatex EP1, or Mobiltem 78.

FRICTION DRIVE

To renew friction wheel, proceed as follows: Stand snowthrower up on auger housing end and remove bottom panel (13—Fig. HJ31). Disconnect track connecting rod (2—Fig. HJ36) at right track plate (1) and rotate right track assembly parallel to ground to gain clearance to hex shaft end bearing. Remove the three bolts securing friction wheel (14—Fig. HJ32) to hub (13) and move speed control lever to first speed position. Loosen but do not remove nuts securing right hex shaft bearing plate (1). Move speed control lever to sixth speed position. Place tape over the four hex shaft bearing plate bolts inside motor mount and remove the previously loosened retaining nuts. Remove bearing plate and slide hex shaft to right until friction wheel can be removed. Install replacement friction wheel retaining bolts loosely, then complete reassembly in reverse order of disassembly. Check friction wheel adjustment as outlined in a previous paragraph.

Fig. HJ36—Exploded view of left track assembly and weight transfer mechanism.

1. Track plate, inner R.H.
2. Connecting rod
3. Track plate, inner L.H.
4. Axle drive gear
5. Axle shaft
6. Inner axle hub
7. Pedal spring
8. Weight transfer pedal
9. Pedal bracket
10. Track drive wheel
11. Track drive sprocket
12. Track plate, outer L.H.
13. Cap screw
14. Adjusting cam
15. Outer axle hub
16. Axle sprocket
17. Rubber track
18. Track idler
19. Track idler shaft
20. Track drive chain

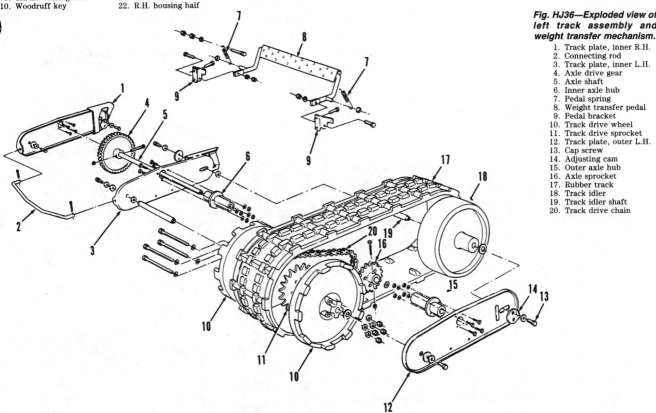

HONDA

AMERICAN HONDA MOTOR CO., INC.
100 W. Alondra Blvd.
Gardena, CA 90247

Model	Engine Model	Self-Propelled?	No. of Stages	Scoop Width
HS35	G150	No	1	20 in. (508 mm)

LUBRICATION

ENGINE

Refer to Honda engine section for lubrication requirements. Use regular or unleaded gasoline.

Before placing snowthrower in storage at end of season, remove spark plug and pour three tablespoons of engine oil in cylinder. Pull starter rope slowly to distribute oil, then reinstall spark plug and pull starter rope until compression is felt. This will close valves and protect cylinder from corrosion.

WHEELS AND CONTROLS

Once each season, lubricate wheels and clutch linkage pivot points with engine oil. Do not allow oil on auger drive belt. Lightly coat chute crank worm gear with grease.

AUGER BEARINGS

Auger bearings are sealed and no lubrication is required.

MAINTENANCE

Inspect drive belt and renew if frayed, cracked, burnt or otherwise damaged. Renew belt pulleys if grooves are excessively worn or damaged.

Inspect auger flighting (1—Fig. H1) for damage or wear which may reduce efficiency of unit. Service for limit of flighting width is 3.4 inch (87 mm) as shown in Fig. H1. Measurement (A) is taken between side of auger shaft and outside edge of flighting. If replacement is needed, install with symbols facing forward as shown.

Service limit (B) for housing guard beneath auger is 0.12-0.24 inch (3-6 mm). Renew as needed.

ADJUSTMENT

CLUTCH CABLE/BELT

If auger belt slips or if auger continues to turn after handle is released, then it will be necessary to adjust clutch cable. Loosen locknut (A—Fig. H2) and turn adjusting nut (B) as required to obtain proper adjustment. When clutch lever is held down, specified distance between top of belt at bottom of idler pulley and bottom of belt is 2.10 inch (53.3 mm).

OVERHAUL

ENGINE

Engine model number is listed at beginning of section. Refer to Honda engine section for overhaul procedures.

DRIVE MECHANISM

Inspect auger and idler bearings for smooth operation. Renew as needed.

Fig. H1—Measure thickness of auger flighting (1) as shown. Minimum service limit thickness (A) is 3.4 inch (87 mm). Minimum thickness (B) for lower guard (2) is 0.12-0.24 inch (3-6 mm).

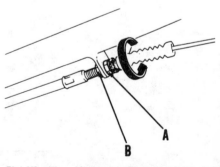

Fig. H2—View of clutch cable locknut (A) and adjusting nut (B).

HONDA

Model	Engine Make	Engine Model	Self-Propelled?	No. of Stages	Scoop Width
HS50T	Honda	G200	Yes	2	550 mm (21.7 in.)
HS50W	Honda	G200	Yes	2	600 mm (23.6 in.)
HS55T	Honda	GX140	Yes	2	550 mm (21.7 in.)
HS55W	Honda	GX140	Yes	2	600 mm (23.6 in.)
HS55K1T	Honda	GX140	Yes	2	550 mm (21.7 in.)
HS55K1W	Honda	GX140	Yes	2	600 mm (23.6 in.)
HS70T	Honda	G300	Yes	2	605 mm (23.8 in.)
HS70W	Honda	G300	Yes	2	605 mm (23.8 in.)
HS80T	Honda	GX240	Yes	2	605 mm (23.8 in.)
HS80W	Honda	GX240	Yes	2	605 mm (23.8 in.)

NOTE: The letter ''T'' or ''W'' on end of model number indicates whether unit is track drive (T) or wheel drive (W).

LUBRICATION

ENGINE

Refer to Honda engine service section for engine lubrication information. Recommended fuel is unleaded regular gasoline. Leaded regular gasoline may be used as a substitute.

DRIVE MECHANISM

Once each season, apply a light coat of oil on the friction wheel shaft (1—Fig. H11) so disc will slide easily on shaft when changing gears. Apply a light coating of oil on the discharge chute pivot, chute deflector pivot and to au-ger and traction drive pivot points. Apply grease to the chute worm gear and chute teeth (Fig. H13).

On all models except HS50, a sub-transmission is located between the drive disc and drive pulley. Recommended lubricant is SAE 10W-30 motor oil. Refill capacity is 0.14 L (0.15 qt.). On all models, a gear reduction gearcase is fitted to the left side of the main frame assembly. This transmission is packed with grease and does not require further lubrication unless unit is disassembled.

AUGER TRANSMISSION

Specified lubricant for the auger transmission (gearbox) is SAE 90 weight gear lubricant. Refill capacity is 0.2 L (0.21 qt.).

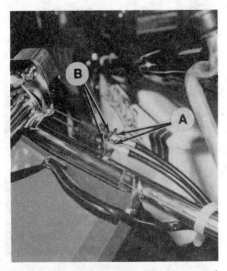

Fig. H10—View of auger and traction drive control cables on Model HS50. Control cable adjustment is similar on all models. Loosen jam nuts (A and B) and adjust cable housing position in bracket by turning both nuts up or down.

Fig. H11—View of disc drive components. Spring (7) holds drive disc (4) against friction wheel (3) except when shift lever is in neutral.

1. Friction wheel shaft
2. Shift arm
3. Friction wheel
4. Drive disc
5. Spring bolt
6. Nut
7. Spring

MAINTENANCE

Inspect drive belts and belt pulley grooves. Renew belt(s) if frayed, cracked, burned or otherwise damaged. Inspect friction wheel (3—Fig. H11) rubber ring for wear, cracking or other damage and renew if any such defects are noted.

The augers are driven via special bolts (S—Fig. HS12) through impeller fan and shaft and through auger and auger shaft. The bolts may shear if auger or impeller fan contacts a solid obstruction. If replacing a sheared bolt, use only original equipment type bolts. The shear bolt has a shoulder on most models, and care should be taken that it is installed in correct direction.

ADJUSTMENTS

AUGER HEIGHT

On track-type units, place height adjustment pedal located at rear of main frame in intermediate (mid) position. Auger clearance can be adjusted by loosening skid shoe mounting bolts and repositioning the shoes equally at each side of auger housing. For general use, clearance between auger housing guard and ground should be 6-8 mm (1/4 to 1/3 inch). On uneven or rough surfaces, adjust clearance to 10-30 mm (0.4-1.2 inches) and on flat, smooth surfaces, clearance can be reduced to 0-5 mm (0-13/64 inch). On track drive models, the height pedal can be used to vary auger clearance from the skid shoe setting.

BELT STOPPER AND GUIDES

All models are equipped with belt stopper/guide (3—Fig. H14) which keeps the drive belts from turning when clutch levers are released. Clearance between stopper/guide and belts should be adjusted with both clutch levers engaged. Recommended belt-to-stopper/guide clearance is as follows:

Models	Stopper Clearance
HS50, HS55	4-6 mm (0.16-0.24 in.)
HS70	1-3 mm (0.04-0.12 in.)
HS55K1, HS80	0.5-2.5 mm (0.02-0.1 in.)

Models HS55K1 and HS80 are also equipped with a belt guide at the top of main frame just below the belt stopper/guide and two auger belt supports bolted to auger housing at either lower side of auger pulley. Adjust the belt guide with auger clutch engaged to clear belt by 1-3 mm (0.04-0.12 in.). The belt supports can be adjusted only when auger housing assembly is removed. Hold the belt tightly in the pulley using hand pressure and adjust right belt support-to-belt clearance to 1-3 mm (0.04-0.12 in.) and left belt support-to-belt clearance to 3-5 mm (0.12-0.2 in.).

AUGER CLUTCH LEVER

First, check auger brake adjustment except on Model HS50, then proceed as follows: Engage auger clutch lever, hold or tie drive clutch lever down and release auger clutch lever. The distance between tip of auger clutch lever and handlebar grip should then be 15-20 mm

Fig. H12—View showing shear bolt (S) locations in impeller and augers.

Fig. H13—Lubricate auger chute worm gear and chute teeth (L).

Fig. H14—View of Model HS50 belt and pulley system. Other models are similar.

1. Traction drive belt
2. Auger/impeller belt
3. Belt stopper/guide
4. Engine drive pulley
5. Traction drive pulley
6. Auger/impeller pulley

(5/8 to 3/4 in.) on Models HS50, HS55 and HS70, and 10-15 mm (13/31 to 19/32 in.) on Models HS55K1 and HS80. If not, loosen the nuts holding lever to lever pivot arm, move lever to proper distance and tighten the nuts. Release drive clutch lever so auger clutch will return to disengaged position and check auger clutch adjustment.

AUGER BRAKE

All models except Model HS50 are equipped with an auger brake at lower end of the auger idler bracket. To check brake adjustment, engage auger with engine running at maximum speed, then release auger clutch lever. The auger should stop turning within 5 seconds. If not, loosen the adjustment on auger clutch cable until auger will stop turning within the 5 second limit, then recheck auger clutch lever adjustment.

AUGER CLUTCH

Model HS50

If auger belt slips, check auger clutch lever adjustment as previously outlined, then loosen jam nuts (A and B—Fig. H10) on the auger clutch cable and adjust the nuts to move cable housing up in bracket. Check auger clutch for belt slippage and to be sure belt stops turning when auger clutch lever is released. Readjust jam nuts as necessary for optimum auger clutch performance.

All Models Except HS50

First, check auger clutch lever and auger brake adjustments as previously outlined. With engine running, slowly depress auger clutch lever. The auger should start turning when rear tip of clutch lever is at the following specified distance from handlebar grip:

Model	Lever-to-Handlebar Distance
HS55	114-118 mm (4.49-4.65 in.)
HS55K1	135-137 mm (5.3-5.4 in.)
H70T	97.5-100.5 mm (3.86-3.98 in.)
H70W	107-110 mm (4.21-4.33 in.)
HS80T	140-142 mm (5.51-5.59 in.)
HS80W	142-144 mm (5.59-5.67 in.)

If auger control lever-to-handlebar distance is not as specified, remove the plastic belt cover in front of engine, loosen auger idler pulley bolt and move idler in slot of idler arm toward belt to increase auger clutch lever distance or away from belt to decrease distance. The auger belt should not slip when lever to handlebar distance is within specified limits.

DRIVE CLUTCH

On all models, if drive belt slips under load, refer to Fig. H10 and adjust clutch as follows: Loosen jam nuts (A and B) and adjust position of cable housing in bracket. Move cable up to increase belt tension, or down if drive belt continues to turn with drive clutch lever released. On Models HS55K1 and HS80, the drive clutch should start to engage when lever is depressed so distance between tip of lever and handlebar grip is 140-143 mm (5.5-5.6 in.).

SUB-TRANSMISSION CABLE

Except on Model HS50T, track-drive models are equipped with a 2-speed sub-transmission between the drive pulley and drive disc. The transmission is shifted by operation of the traction drive shift lever through a cable attached to the sub-transmission lever. With shift lever in third gear position, free play at lower end of cable should be 0-1.0 mm (0-0.04 in.) on Models HS55T and HS70T, and 1.0-2.0 mm (0.04-0.08 in.) on Models HS55K1T and HS80T. To adjust cable free play, select cable from sub-transmission and loosen jam nuts at cable bracket just below shift lever. Reposition the cable in cable bracket by turning the jam nuts until desired cable end free play is obtained, then tighten the nuts and recheck cable free play.

SHIFT ROD

Early Model HS50

On early Model HS50 without stop collar on friction wheel shaft (1—Fig. H11), the center of the friction wheel (3) should be located 28.5-31.5 mm (1.12-1.24 in.) from center of the drive disc (4) when shift lever is in low (1st) speed position. If not, detach the shift rod from shift lever and screw rod in or out of rod lower section to change friction wheel position. Lengthening the rod increases the distance of friction wheel from center of drive disc. Reconnect rod to shift lever and recheck adjustment.

All Later Models

All models except early Model HS50 are equipped with a stop collar on the friction wheel shaft (1—Fig. H11). Disconnect shift rod from shift lever and place shift lever in neutral position to move friction wheel off of drive disc. Shove shift rod down until friction wheel contacts stop collar. Move shift le-

ver to 3rd speed position. If shift rod will not enter hole in lever, loosen locknut and adjust length of shift rod so it can be reconnected to shift lever.

DRIVE DISC

With shift lever in neutral position, check clearance (C—Fig. H17) between friction wheel and drive disc. On Model HS50, this distance should be 1.0-1.3 mm (0.039-0.051 in.). On all other models, the friction wheel clearance should be 0.9-1.11 mm (0.035-0.043 in.). Also, on all models except Model HS50, there should be 0-1.0 mm (0-0.4 in.) free play of shift cable when shift lever is in first speed position. If necessary to adjust, locate the upper shift cable end, refer to Fig. H10, and adjust position of cable in bracket to obtain desired disc clearance in neutral and/or cable free play in first gear. Tighten the cable jam nuts and recheck adjustment. With shift lever in gear, spring (7—Fig. H11) should hold the drive disc (4) against friction disc (3) with a force of 24-28 kg (52.9-61.7 pounds). To adjust spring, loosen and adjust the nut (6) on spring bolt (5) located on each side of bracket. Do not turn the bolt as this will twist the spring.

TRACK DRIVE

Track adjustment can be checked by pushing down on center of track (A—Fig. H15). Track should deflect 19-25 mm (3/4-1 in.) when a force of 15 kg (33 lb.) is applied. If adjustment is required, loosen rear track wheel bolts (1) and adjusting bolt locknuts (A—Fig. H16). Left side track wheel bolt has left-hand threads. Tighten the adjusting nuts under the locknuts to increase track tension as required, then tighten wheel bolts and adjusting bolt locknuts.

OVERHAUL

ENGINE

Honda engine model numbers are listed at beginning of this section. Refer to Honda engine section of this manual for engine overhaul information.

DRIVE BELTS

Model HS50

On Model HS50, belts can be removed without first removing auger housing from main frame. Remove the plastic belt guard located in front of engine.

Pull auger idler pulley back and roll belt off of engine pulley, then work belt up between auger and traction drive pulley. With auger belt removed, roll drive belt off of engine drive pulley and let top end lay between engine drive pulleys. Remove belt from traction drive pulley and work the belt up between the auger and drive pulleys. Reverse removal procedures to reinstall belt(s) and check all lever and belt guard adjustments as outlined in ADJUSTMENTS section.

All Other Models

On all models except Model HS50, remove the auger housing assembly as outlined in following paragraph. Remove auger pulley cap screw and washer and remove pulley and auger belt from auger housing. Remove traction drive belt from engine pulley and drive pulley. Install new belts by reversing removal procedure. Check all adjustments.

AUGER HOUSING

To remove auger housing assembly from main frame, proceed as follows: Remove the auger worm gear bracket and move auger crank back out of way. Remove belt guard from main frame at front of engine. Remove belt stopper/guide (two stoppers, one at each side of engine pulley on Models HS55K1 and HS80) and roll auger drive belt off of engine pulley. Unbolt and remove auger housing from main frame. Be sure that auger brake pad (not on Model HS50) does not get between auger drive belt and pulley when installing auger housing assembly.

AUGER GEARBOX

With auger housing removed as previously outlined, proceed as follows: Remove the auger pulley cap screw and washer, auger pulley and pulley key from rear of impeller shaft, then unbolt and remove bearing retainers and impeller shaft bearing. Remove the cap screws and washers at each end of auger shaft, then unbolt auger shaft bearing retainers from side plates and remove the auger and impeller assembly from front of housing. Remove the auger shaft bearings and the auger shear bolts, then slide augers from each end of shaft. Remove the impeller fan shear bolt and pull impeller from shaft. Re-

move oil plug and drain oil from gearbox. Unbolt side cover and carefully remove cover from case. Slide the auger shaft and worm wheel out side of case. Note position of spacer at cover side of worm wheel. Unbolt and remove the impeller shaft bearing retainer and slide impeller shaft out rear of case. Further disassembly is obvious from inspection of unit.

Clean and inspect all components for excessive wear or damage and renew as necessary. Use new shaft oil seals and cover gasket when reassembling unit. Refill gearbox with 0.2 L (0.21 U.S. qt.) of SAE 90 gear oil. Reinstall assembly by reversing removal procedure.

DRIVE TRACKS

The tracks can be removed after loosening rear track wheel bolts (1—Fig. H15) and track adjusting bolt locknuts (A—Fig. H16) and pushing the wheels forward to loosen track. Left side track wheel bolt has left-hand threads. The remainder of the track drive can be disassembled after removing the rubber

track. If the track guide is removed from both sides, note that the guide must be installed with the long end of guide toward the rear. Adjust track as outlined in ADJUSTMENTS section.

Fig. H17—Check clearance (C) between friction wheel and drive disc with shift lever in neutral.

Fig. H15—Loosen rear track wheel bolt (1) and refer to Fig. H16 for adjustment of track tension. Track should deflect 19-25 mm (0.75-1.0 in.) (B) when 15 kg (33 lb.) of pressure is applied at (A).

Fig. H16—With rear track wheel bolts loosened, tracks may be adjusted with adjustment bolts (A).

INTERNATIONAL HARVESTER

Model	Engine Make	Engine Model	Self-Propelled?	No. of Stages	Scoop Width
10	Tecumseh	AH520	No	1	20 in. (508 mm)

LUBRICATION

ENGINE

The engine is a two-stroke type which is lubricated by mixing oil with the fuel at a recommended fuel:oil ratio of 32:1 using International Harvester or other good quality two-stroke oil. The fuel must be regular or low-lead gasoline.

DRIVE MECHANISM

Drive mechanism components are designed not to require lubrication for life of unit. Components should be inspected periodically, however, to insure against premature failure.

MAINTENANCE

Periodically inspect impeller drive belt. Renew belt if frayed, cracked, excessively worn or otherwise damaged. Drive belt may be removed after detaching belt cover and pushing idler pulley away from belt.

OVERHAUL

ENGINE

Engine make and model are listed in front of this section. Refer to Tecumseh engine section for overhaul procedures.

DRIVE MECHANISM

Overhaul of drive mechanism is apparent after inspection of unit and referral to exploded view in Fig. IH1.

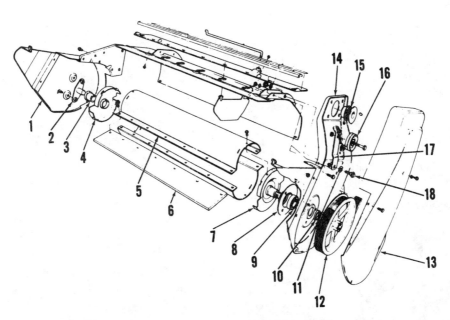

Fig. IH1—Exploded view of impeller assembly.

1. Side plate	6. Impeller blade	10. Spacer	14. Side plate
2. Impeller spindle	7. Impeller drive cap	11. Washer	15. Engine pulley
3. Bushing	8. Bearing retainer	12. Pulley	16. Idler pulley
4. End cap	9. Bearing	13. Cover	17. Idler arm
5. Impeller half			18. Shoulder screw

INTERNATIONAL HARVESTER

Model	Engine Make	Engine Model	Self-Propelled?	No. of Stages	Scoop Width
265	B&S	130000	Yes	2	26 in. (660 mm)
268	B&S	190000	Yes	2	26 in. (660 mm)
328	B&S	190000	Yes	2	32 in. (813 mm)

LUBRICATION

ENGINE

Refer to Briggs and Stratton engine for engine lubrication requirements.

Recommended fuel is regular or low-lead gasoline.

DRIVE MECHANISM

Periodically check and lubricate control linkage as required. The auger shaft is lubricated by forcing a good quality multi-purpose lithium grease such as IH 251H EP grease through grease fittings in augers. The auger shaft must be lubricated if a shear pin has failed. Rotate auger on shaft prior to installing shear pin so grease will be distributed. Lubricate shafts and gears of chute gearbox as required to prevent binding and hard turning.

AUGER GEARBOX

The auger gearbox should be filled with lubricant so oil is level with level plug hole (H—Fig. IH10). Recommended lubricant is IH 135H EP gear oil or an equivalent SAE 90 multi-purpose gear oil.

MAINTENANCE

Inspect drive belts and pulley belt grooves. Renew belt if frayed, cracked, burnt or otherwise damaged.

The augers are secured to auger shaft by shear pins which will fracture when auger contacts an obstruction. Renew shear pins as required with factory authorized shear pins.

Recommended tire pressure is 8 psi (55.2 kPa).

Fig. IH11—Exploded view of control components.

1. Auger clutch lever
2. Shift lever
3. Spring
4. Shift arm
5. Auger clutch rod
6. Collar
7. Eye bolt
8. Auger clutch arm
9. Spring
10. Shift control rod
11. Shift control rod
12. Throttle control
13. Drive clutch lever
14. Drive clutch rod
15. Bellcrank
16. Link
17. Eye bolt
18. Spring
19. Collar

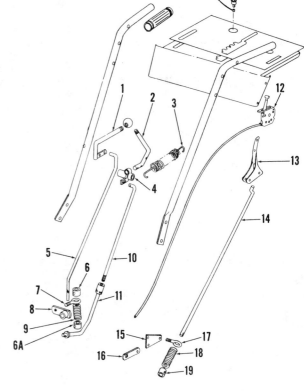

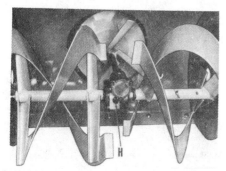

Fig. IH10—View of auger. Note location of oil level plug hole (H) on auger gearbox.

Fig. IH12—View of belt idler assembly.

1. Drive belt idler arm
2. Spring
3. Auger belt idler arm
4. Engine pulley
5. Drive belt idler pulley
6. Auger belt idler pulley

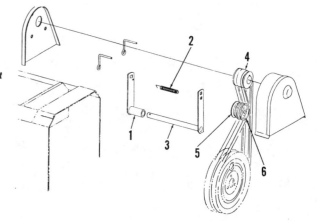

ADJUSTMENTS

HEIGHT

Auger height is adjusted by loosening runner mounting bolts and repositioning runners. Scraper bar should be 1/8 inch (3.2 mm) above a smooth surface with height increased if snowthrower is operating over an irregular surface.

FRICTION DRIVE

With speed control lever in neutral slot of control panel and clutch engaged, unit should roll forward and rearward easily. If not, detach upper end of control rod (10—Fig. IH11) from control arm (4). Remove bottom panel and move control rod up or down to center friction wheel (17—Fig. IH13) on friction disc (11). Rotate control rod (10—Fig. IH11) so rod end will mate with slot in control arm (4) and reattach control rod.

AUGER CLUTCH

To adjust auger clutch control rod, loosen set screw in upper collar (6—Fig. IH11) and move auger control lever to disengaged position. Push auger clutch control arm (8) down as far as possible, make sure upper collar (6) is against eye bolt (7) and tighten set screw in collar (6). Loosen set screw in lower collar (6A) then lightly force collar and spring (9) up against eye bolt (7) and retighten lower collar set screw.

OVERHAUL

ENGINE

Engine make and model numbers are listed at beginning of this section. Refer to Briggs and Stratton engine section for engine overhaul.

FRICTION DRIVE

Friction drive components are accessible after removal of engine and auger frame. Disassembly is apparent after inspection of unit and referral to Figs. IH12 and IH13. Note that friction disc (11) has left hand threads and bearings (5—Fig. IH13) are a press fit on shaft (9) and in bearing housing (10). Inspect all components for damage and excessive wear. Be sure hex shaft (18) is straight and hub (16) will slide freely on shaft. Renew chain if worn and stretched excessively. Bearings are sealed type and should be renewed if rotation is not smooth.

AUGER GEARBOX

To overhaul auger gearbox, first remove auger assembly from frame, then remove impeller and augers. Unscrew side cover (14—Fig. IH16) and withdraw auger shaft and gear assembly. Note shims (12) which should be retained for assembly. Unscrew front cover (1) and remove impeller shaft and worm gear assembly. Note shims (7) which should be retained for assembly. Disassemble components as required. Bearings (4) and worm gear (5) must be pressed off shaft.

Bearing or bushing adjustment is ac-

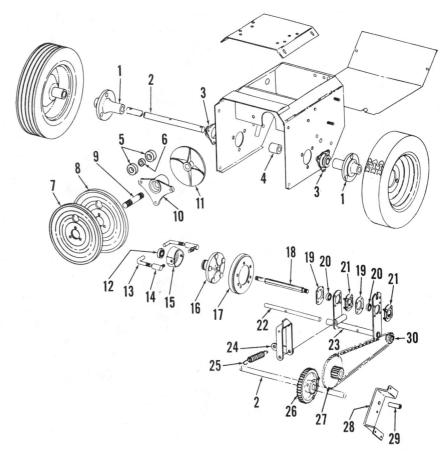

Fig. IH13—Exploded view of drive assembly.

1. Hub	9. Shaft	17. Friction wheel	25. Spring
2. Axle	10. Bearing housing	18. Hex shaft	26. Gear
3. Bearing assy.	11. Friction disc	19. Bearing retainer	27. Sprocket & gear
4. Spacer	12. Bearing	20. Bearing	28. Sprocket & gear
5. Bearings	13. Link	21. Bearing retainer	bracket
6. Spacer	14. Ball joint	22. Pivot shaft	29. Axle
7. Auger pulley	15. Shift collar	23. Carrier	30. Sprocket
8. Drive pulley	16. Friction wheel hub	24. Shift bracket	

Fig. IH14—View of friction wheel assembly.

155

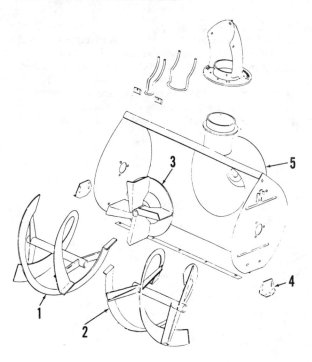

*Fig. IH15—View of auger
assembly.*

1. Auger R.H.
2. Auger L.H.
3. Impeller
4. Runner
5. Auger frame

complished as follows: Install impeller shaft with worm gear (5) and bearings (4) installed. Install a 0.010 inch shim (7) and front cover (1) then tighten front cover screws to 5 ft.-lbs. (6.8 N·m). Check impeller shaft for binding and add shims (7) if necessary so shaft rotates smoothly. When installing auger shaft assembly, install a 0.020 inch shim (12) and tighten side cover screws to 5 ft.-lbs. (6.8 N·m). If auger shaft does not rotate smoothly, increase thickness of shims.

Reverse disassembly procedure to reassemble auger gearbox. Fill gearbox with IH 135H EP gear oil or an equivalent SAE 90 multi-purpose gear oil to level of plug hole (H—Fig. IH10).

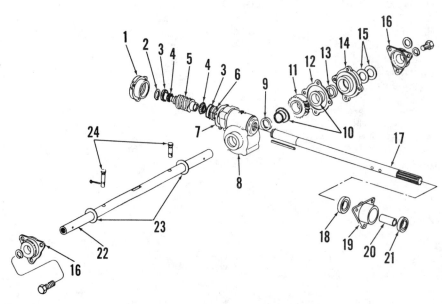

Fig. IH16—Exploded view of auger gearbox.

1. Front cover	7. Shim	13. Seal	19. Bearing housing
2. Snap ring	8. Gearbox	14. Side cover	20. Spacer
3. Bearing cup	9. Seal	15. Thrust washer	21. Bearing
4. Bearing cone	10. Bushing	16. Bearing assy.	22. Auger shaft
5. Worm gear	11. Gear	17. Impeller shaft	23. Thrust washers
6. Seal	12. Shim	18. Bearing	24. Shear pins

LAWN-BOY

LAWN-BOY PRODUCT GROUP
P.O. Box 152
Plymouth, WI 53073

Model	Engine Make	Engine Model	Self-Propelled?	No. of Stages	Scoop Width
320, 320E	Tecumseh	AH600	No	1	20 in. (508 mm)

LUBRICATION

ENGINE

Refer to Tecumseh engine section for engine lubrication requirements. Recommended fuel is unleaded or leaded regular gasoline mixed with a good-quality 2-cycle air-cooled engine oil. Fuel and oil should be mixed in a 32:1 ratio.

DRIVE MECHANISM

Lubricate auger shaft bearing (13—Fig. LB1) at right side of auger housing after every 10 hours of operation using SAE 30 oil. Use sufficient oil so oil just runs from hole; felt plug (14) retains oil between lubrication intervals. Bearing at left end of auger is a sealed prelubricated type.

After every 10 hours of operation, remove belt cover and lubricate idler pulley (38) and shaft on idler arm (30) with SAE 30 oil. Turn pulley while applying lubricant to distribute oil evenly. Be careful not to allow oil to get on belt or pulleys. Idler arm pivot bushing (28) and skid block (33) should be lubricated with No. 2 wheel bearing grease.

MAINTENANCE

Inspect drive belt and renew if frayed, cracked, burned or otherwise damaged. Renew belt pulleys if belt groove is excessively worn or pulley is damaged. Inspect impeller for wear or damage. Impeller paddle (2—Fig. LB1) and auger flights (1) can be renewed if necessary. Move deflector vane pivot assembly (17) from side-to-side. Deflector vane (8) locknut should be adjusted so that deflector moves without binding.

ADJUSTMENTS

DRIVE BELT

With auger clutch lever released, there should be no slack in the clutch cable. To increase belt tension, move open end of clutch adjuster clip (60—Fig. LB1) down to next link in clutch cable and chain (59). If clip is in last link of chain, it may be necessary to move clip loop up to the next link in end of chain. To check adjustment, pull engine starter rope with ignition in "OFF" position. If auger turns, move adjustment clip upward in chain (closer to clutch le-

ver). With clutch engaged, belt retainer (32) should clear belt by a maximum of 1/8 inch (3.2 mm) and retainer should be parallel to belt. Belt guides (29, 41 and 42) should be parallel to the belt and clear the belt when clutch is engaged.

OVERHAUL

ENGINE

Engine make and model information is listed at beginning of this section. Refer to Tecumseh 2-stroke engine section in this manual for engine overhaul information.

DRIVE MECHANISM

Drive belt (35—Fig. LB1) can be removed and a new belt installed after removing belt cover (36) and lifting idler from belt. Refer to belt tension adjustment in previous ADJUSTMENT section.

If belt idler arm (30) is removed, lightly grease back side of arm where it contacts skid block (33) and pivot bushing (28) with No. 2 wheel bearing grease so idler arm will pivot freely after reassembly. Lubricate idler pulley shaft and bushing with No. 2 wheel bearing grease if pulley is removed.

AUGER ASSEMBLY

If excessive wear of auger flight and impeller paddle rubber fins is noted, the rubber fins should be replaced as a complete set. The fins can be replaced after removing the cap screws holding the fins between the two auger plate halves (3—Fig. LB1).

If excessively worn or damaged, the auger assembly scraper bar (7) can be replaced after removing the three flat head socket screws that hold scraper bar to bottom of auger housing. Auger stop spring (6) mounted with center scraper bar bolt keeps auger from rotating backwards.

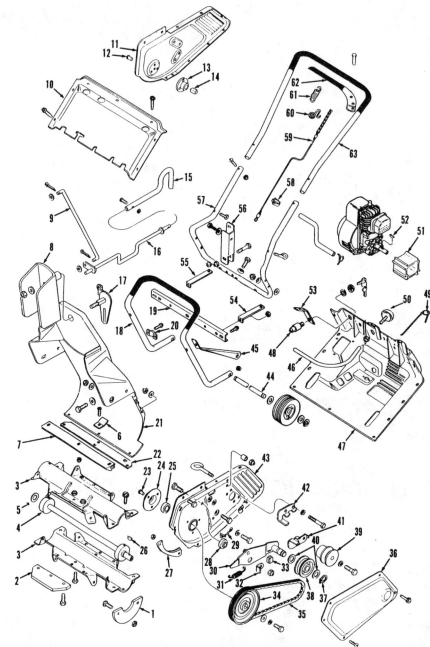

1. Auger flight
2. Impeller paddle
3. Auger plate half
4. Auger support shaft
5. Thrust washer (2-3)
6. Auger stop spring
7. Scraper bar
8. Deflector vane
9. Vane control link
10. Top cover
11. R.H. end plate
12. Rivet
13. Auger shaft bearing
14. Felt plug
15. Vane control handle
16. Control rod
17. Vane pivot assy.
18. Axle frame
19. Support channel
20. Wheel support clamp
21. Housing wrap sheet
22. Scraper baffle
23. Rivet
24. Bearing retainer
25. Ball bearing
26. Woodruff key
27. Wear strip
28. Pivot bushing
29. Wire belt guide
30. Idler arm
31. Auger brake spring
32. Belt retainer
33. Skid block
34. Auger pulley
35. Drive belt
36. Belt cover
37. Snap ring
38. Idler pulley
39. Engine pulley
40. Idler washers
41. Lower belt guide
42. Upper belt guide
43. L.H. end plate
44. Axle shaft
45. Wheel brace
46. Primer tube
47. Lower cover
48. Ignition switch
49. Choke wire
50. Primer bulb
51. Engine spacer
52. Woodruff key
53. Panel brace
54. Brace
55. Brace
56. Recoil brace
57. Lower handle
58. Cable clamp
59. Clutch cable & chain
60. Clutch adjuster clip
61. Clutch spring
62. Clutch lever
63. Upper handle

Fig. LB1—Exploded view of Model 320 snowthrower. Model 320E is equipped with electric start.

LAWN-BOY

Model	Engine Make	Engine Model	Self-Propelled?	No. of Stages	Scoop Width
ST524	Tecumseh	HS50	Yes	2	20 in. (508 mm)
ST824	Tecumseh	HM80	Yes	2	24 in. (610 mm)
ST826	Tecumseh	HM80	Yes	2	26 in. (660 mm)
ST1032	Tecumseh	HM100	Yes	2	32 in. (813 mm)

LUBRICATION

ENGINE

Refer to appropriate Tecumseh 4-stroke engine section for engine lubrication information. Recommended fuel is leaded or unleaded regular gasoline.

DRIVE MECHANISM

Lubricate all pivot points and drive chains with a suitable lubricant after every 10 hours of operation. Use a grease gun to lubricate fittings for axle bushings, auger shaft bushings and countershaft bushings after every 10 hours of operation. Wheel hubs should be greased after each season of use.

AUGER GEARBOX

On Models ST524 and ST824, the auger gearbox is fitted with a grease fitting and can be lubricated with a hand-operated grease gun using NLGI Grade 2 Lithium Base EP grease. For other models, check oil level in gearbox periodically. Oil level should be at bottom of auger gearbox pipe plug opening with unit setting level. Recommended lubricant for this type gearbox is SAE 30 motor oil.

TRANSMISSION

The gear-type transmission is lubricated at the factory and should not require further lubrication except at overhaul; refer to Peerless transmission section for service information.

MAINTENANCE

The auger flights are driven via the auger shaft by shear pins which will fracture when an auger contacts a solid obstruction. If necessary to renew shear pins, use only factory replacement pins.

Inspect drive belts and pulley belt grooves. Renew belts if frayed, cracked, burned or excessively worn. Renew pulleys if bent, excessively worn or otherwise damaged.

Inspect traction drive chain and sprockets. Renew chain if excessively worn, stretched or otherwise damaged. Renew sprockets if teeth are worn or otherwise damaged.

ADJUSTMENTS

AUGER HEIGHT

Auger height can be adjusted by loosening skid shoe mounting bolts and repositioning the skid shoes. Auger should be in lowest position (skid shoes raised) when operating over smooth surface and in highest position (skid shoes lowered) when operating over rough surfaces. Be sure skid shoes are equally adjusted on each side of auger housing.

CONTROL CABLES

The auger and traction drive control cables should have all slack removed when control lever is all the way against the rubber stop on the handle. If not, remove the spring clevis pin from the lower end of either control cable and turn the adjustment yoke as necessary to achieve proper adjustment. Reattach cable with spring clevis pin.

DRIVE BELTS

If auger drive belt slips, readjust position of idler pulley to increase belt tension. Remove belt guard and note position of idler pulley on auger idler arm. Loosen hex nut holding idler pulley to arm and move pulley toward belt about 1/4 inch (6.4 mm), tighten nut and recheck belt tension. Belt should release fully when control lever is released. It may be necessary to readjust control cable as described in preceding paragraph.

The traction drive belt can be adjusted following procedure outlined for auger belt by moving idler pulley in slot of drive idler arm.

The wire belt guides should clear belts by about 1/8 inch (3.2 mm) when belt control levers are engaged.

DRIVE CHAIN

When pushing on chain at midpoint between transmission and axle sprockets, the chain should deflect about 1/8 inch (3.2 mm). To adjust chain, tip the snowthrower forward onto front of auger housing and loosen the three bolts retaining the axle bearing retainers on each side of main frame. Slide the axle shaft bearing assemblies in the slotted mounting holes of main frame until proper adjustment is reached. Be sure both bearings are the same distance from back of main frame and tighten the bearing retaining bolts. On Model ST1032, check drive chain idler sprocket.

OVERHAUL

ENGINE

Engine make and model numbers are listed at beginning of this section. Re-fer to Tecumseh 4-stroke engine section for engine overhaul information.

TRANSMISSION

All models are equipped with a gear-type transmission. Refer to Peerless section for transmission overhaul information.

AUGER DRIVE BELT

With belt drive cover removed, remove belt from engine pulley and slip belt out from between auger idler pulley and retainer. Push the idler arm in to move brake shoe away from auger pulley and remove belt from between auger pulley and lower belt guide. In-stall new belt by reversing removal procedure and check belt and cable adjustments as outlined in a previous paragraph.

TRACTION DRIVE BELT

With belt drive cover removed, remove auger drive belt from engine pulley. Move traction drive belt from engine pulley to a position between pulley and engine. Slip belt out from between traction drive idler pulley and retainer. Remove belt from transmission pulley. Install new belt by reversing removal procedure and replace auger belt on engine pulley. Check belt and cable adjustments as outlined in appropriate previous paragraph.

MTD

MTD PRODUCTS INC.
P.O. Box 36900
Cleveland, Ohio 44136

Model	Engine Make	Engine Model	Self-Propelled?	No. of Stages	Scoop Width
100	Tecumseh	AH520	No	1	20 in. (508 mm)

LUBRICATION

ENGINE

Refer to Tecumseh engine section for engine lubrication requirements.

The engine is a two-stroke type engine which is lubricated by mixing oil with fuel. MTD recommends a fuel:oil ratio of 32:1 using a good quality oil designed for two-stoke engines mixed with regular gasoline.

DRIVE MECHANISM

Lubricate shafts and bushings as required to prevent binding and galling. Lubricate control lever pivot point periodically for smooth operation.

On early models, periodically lubricate clutch bushing (5—Fig. M4) with oil. Do not allow oil on clutch shoes or drum as clutch slippage will result.

MAINTENANCE

On early models inspect drive chain and sprockets. On later models inspect drive belt and pulleys and renew belt if frayed, cracked, excessively worn or otherwise damaged.

On later models the auger is secured to drive pulley shaft by a shear pin which will fracture when auger contacts an obstruction. Renew shear pin with a suitable replacement. Do not use ordinary fasteners which may damage drive components.

ADJUSTMENTS

CHAIN

On early models a chain is used to drive auger. Remove chain guard and top panel, then loosen engine mount bolts and locating bolt on right side of engine. Adjust engine location so chain will deflect ½ inch (12.7 mm) with thumb pressure. Retighten engine mounting bolts.

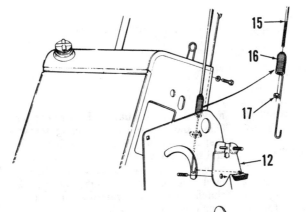

Fig. M1—View of auger control clutch rod and spring on later models. Refer to text for adjustment.

12. Idler arm & brake
15. Auger clutch control rod
16. Spring
17. Nut

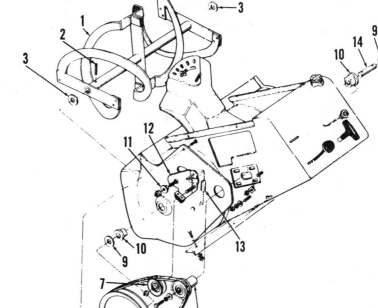

Fig. M2—Exploded view of auger drive components on later models.

1. Auger
2. Pin
3. Washer
4. Guide wheel
5. Auger pulley
6. Belt
7. Idler pulley
8. Engine pulley
9. Washer
10. Bushing
11. Spacer
12. Idler arm & brake
13. Spring
14. Shaft

161

CLUTCH ROD

On later models the clutch rod (15—Fig. M1) should be adjusted so there is 1/8 inch (3.2 mm) end play between rod and clutch spring (16) with auger clutch lever against handlebar in disengaged position. To adjust end play, detach clutch rod (15) and turn rod in nut (17) located in spring (16).

OVERHAUL

ENGINE

Engine make and model are listed in front of section. Refer to Tecumseh section for engine overhaul.

DRIVE MECHANISM

Overhaul of drive mechanism is evident after inspection of unit and referral to Fig. M2 or M3.

CENTRIFUGAL CLUTCH

Early models are equipped with a centrifugal clutch as shown in Fig. M4. The clutch may be removed after re-moving chain guard, disconnecting chain and unscrewing retaining screw. Inspect components for excessive wear and damage. Clutch shoes (1) are available only as a set. Bushing (5) should be lubricated with a light oil. Check clutch drum for warpage or out-of-round due to overheating.

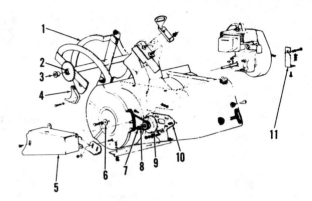

Fig. M3—Exploded view of auger drive components on early models. Refer to Fig. M4 for clutch components.

1. Auger
2. Auger shaft
3. Bushing
4. Chain cleaner
5. Chain guard
6. Washer
7. Spring washer
8. Washer
9. Clutch
10. Engine plate
11. Engine bracket

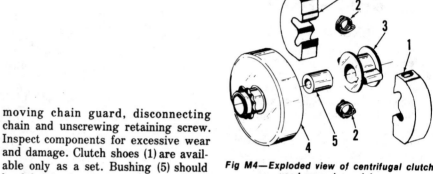

Fig M4—Exploded view of centrifugal clutch used on early models.

1. Clutch shoes
2. Springs
3. Hub
4. Clutch drum
5. Bushing

MTD

Model	Engine Make	Engine Model	Self-Propelled?	No. of Stages	Scoop Width
150	Tecumseh	AH600	No	1	21 in. (533 mm)
151	Tecumseh	AH600	No	1	21 in. (533 mm)
160	Tecumseh	AH750	No	1	21 in. (533 mm)
180	Tecumseh	AH600	No	1	21 in. (533 mm)
181	Tecumseh	AH600	No	1	21 in. (533 mm)
190	Tecumseh	AH750	No	1	21 in. (533 mm)

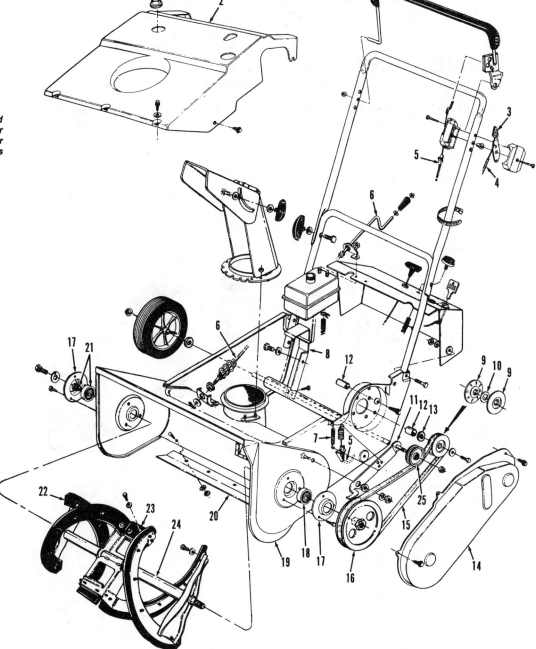

Fig. M6—Typical exploded view of drive mechanism for all models. Impeller for Models 150, 151 and 160 is shown in Fig. M7.

1. Fuel tank cap
2. Upper cover (cowling)
3. Throttle lever (Models 160 & 190)
4. Throttle cable (Models 160 & 190)
5. Auger clutch cable
6. Chute crank
7. Belt tension spring
8. Engine support bracket
9. Engine pulley half
10. Flat washer
11. Idler bracket & brake
12. Spacer
13. Flat washer
14. Belt cover
15. Drive belt
16. Drive pulley
17. Bearing retainer cup
18. Ball bearing
19. Auger (blower) housing
20. Scraper bar (shave plate)
21. Bearing & collar kit
22. Rubber auger flights (spirals)
23. Rubber paddles
24. Auger assy.
25. Idler pulley

LUBRICATION

ENGINE

The engines used on these models are 2-stroke type which are lubricated by mixing oil with fuel. MTD recommends a fuel:oil ratio of 32:1 when using a good-quality oil designed for 2-stroke engines. Mix with leaded or unleaded regular gasoline.

DRIVE MECHANISM

Lubricate control lever pivot points, idler clutch and brake bracket pivot and control cable periodically to ensure smooth operation.

MAINTENANCE

Inspect the drive belt and pulleys; renew belt if frayed, cracked, excessively worn or otherwise damaged. Renew pulleys if excessively worn or damaged. Inspect the rubber auger flights and renew if excessively worn or otherwise damaged. Be sure auger clutch cable (5—Fig. M6) operates freely. Lubricate or replace cable as necessary. Spring on end of auger clutch cable is not available as a separate item. Renew cable if spring is weak or broken. Check to see that idler clutch bracket operates smoothly and bracket return spring (7) is not stretched or broken.

ADJUSTMENTS

There are three holes in the belt idler bracket for belt tension adjustment. To adjust, first remove belt cover (14—Fig. M6). Check to be sure that idler bracket (11) is moving smoothly on pivot and lubricate if necessary. If belt slips (auger seems to slow down or hesitate at constant engine speed), connect the spring on end of clutch cable (5) to next higher hole in idler bracket (11). Renew clutch cable if spring on end of cable is

weak or broken. If belt tension is not sufficient with spring in highest hole in bracket, it will be necessary to install a new drive belt. Too much belt tension will cause auger to turn with clutch bail released.

OVERHAUL

ENGINE

Engine make and model are listed at front of this section. Refer to Tecumseh 2-stroke engine section for engine overhaul information on Model AH600. For overhaul information on Tecumseh Model AH750, refer to Intertec publication *Small Air-Cooled Engines Service Manual.*

DRIVE MECHANISM

Refer to exploded view of drive mechanism in Fig. M6. Overhaul procedure is evident after inspection of unit and reference to exploded view.

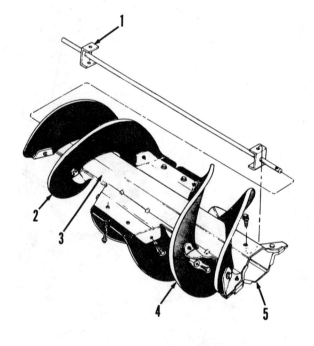

Fig. M7—Impeller assembly for Models 150, 151 and 160. Other components are similar to those shown in Fig. M6.

1. Impeller shaft
2. R.H. rubber flight
3. Impeller half
4. L.H. rubber flight
5. Impeller half

MTD

Model	Engine Make	Engine Model	Self-Propelled?	No. of Stages	Scoop Width
225	Tecumseh	HSSK40	Yes	1	20 in. (508 mm)
230	Tecumseh	HS50	Yes	1	20 in. (508 mm)
430	Tecumseh	HS50	Yes	1	21 in. (533 mm)

LUBRICATION

ENGINE

Refer to appropriate Tecumseh engine section for engine lubrication information. Recommended fuel is unleaded or leaded regular gasoline.

DRIVE MECHANISM

Lubricate drive chains with good-quality chain oil at least once a season. Auger shaft and axle bushings should be oiled and axle greased at least once a season or after every 25 hours of operation. Lubricate control rod and lever pivot points for smooth operation.

MAINTENANCE

Inspect chains and sprockets and renew if excessively worn. Inspect belts and renew if cracked, excessively worn or otherwise damaged. Check pulleys and renew if excessively worn or otherwise damaged. Inspect control cables and renew if frayed or binding.

Recommended tire pressure is 7-10 psi (46-69 kPa).

ADJUSTMENTS

AUGER HEIGHT

Auger height can be adjusted by loosening skid shoe mounting bolts and repositioning the skid shoes. Auger should be in lowest position (skid shoes raised) when operating over smooth surface and in highest position (skid shoes lowered) when operating over rough, irregular surface.

SHIFT LINKAGE

Detach shift control rod (4—Fig. M10) from shift control lever (3). With both shift control lever and shift control rod in neutral position (roll unit back and forth to locate control rod neutral point), thread shift rod in or out of ferrule at bottom end of rod until end of rod aligns with hole in shift lever. Reattach control rod to shift control lever.

DRIVE CLUTCH

Hold drive clutch control lever (1—Fig. M10) so grip is against snowthrower handle grip. Slack should then be taken out of control cable wire (2) so that control wire is straight. If necessary, adjust clutch cable with adjusting nuts (16) at cable bracket.

AUGER CLUTCH

To check for correct adjustment, squeeze auger clutch lever against snowthrower handle. While holding lever in this position, check clearance between coils of compression spring (10—Fig. M10). Clearance between coils should be 0.015 inch (0.38 mm). Adjust locknut (12) as required to obtain correct adjustment.

AUGER DRIVE CHAIN

Auger drive chain tension on Models 225 and 230 can be adjusted after loosening cap screws in each end of auger shaft, then turning adjusting cam (Fig. M11) to take slack out of chain.

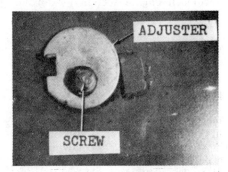

Fig. M11—View showing auger chain adjuster cam on Models 225 and 230.

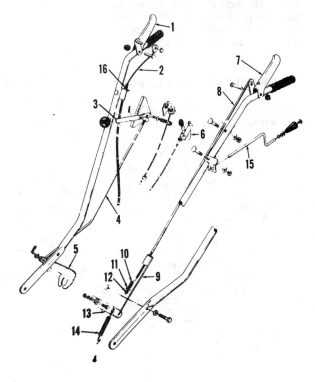

Fig. M10—Exploded view of early control assembly. Later models are similar except for changes in chute crank (15) and Model 225 does not have throttle assembly (6).

1. Drive clutch control lever
2. Clutch cable
3. Shift control lever
4. Shift control rod
5. Shift arm
6. Throttle assy.
7. Auger clutch control lever
8. Auger clutch rod
9. Auger clutch rod
10. Spring
11. Washer
12. Locknut
13. Lever
14. Spring
15. Discharge chute crank
16. Nut

Tighten the cap screws and recheck chain tension.

On Model 430, refer to Fig. M13; loosen nut (37) and reposition chain idler to remove slack from chain.

REAR AXLE CHAIN

To adjust rear axle chain, tip snowthrower forward so it rests on auger housing. Loosen the two nuts holding wheel axle bushing (4—Fig. M12) at right side of machine. Lift up on right wheel to take slack out of chain and retighten the two nuts.

OVERHAUL

ENGINE

Engine make and model numbers are listed at front of this section. Refer to Tecumseh engine section for engine overhaul information.

DRIVE BELTS

Remove the belt cover, then remove top belt keeper plate. Tip snowthrower forward so it rests on auger housing. Remove bottom cover retaining screws and lift cover out from slots in frame. Remove the large shoulder bolt which acts a belt keeper. Remove the belt guard plate. The auger drive belt can then be removed by slipping it off of engine pulley, lifting belt from bottom pulley, then sliding the belt from bottom of chaincase. The traction drive belt can then be removed in similar manner. Reinstall belts by reversing removal procedure.

TRACTION DRIVE CHAIN

Overhaul procedure is evident after inspection of unit and reference to Fig.

M12. However, during reassembly, it will be necessary to check for proper alignment of sprockets (8) and (12). Place a straightedge across faces of the sprockets; if not in line, position of the wheel axle sprocket can be changed by altering shims (1) between the wheel hub and bushing (2).

Fig. M12—Exploded view of traction drive assembly.

1. Shim
2. Bushing
3. Pin
4. Bushing
5. Shift arm
6. Pivot bolt
7. Snap ring
8. Sprocket
9. Snap ring
10. Washer
11. Bushing
12. Sprocket
13. Bracket
14. Shift shaft
15. Shift fork
16. Reverse gear
17. Shift hub
18. Forward sprocket
19. Washers
20. Gear
21. Pin
22. Spacer
23. Sprocket
24. Bearing
25. Spacer
26. Spacer
27. Bearing retainer
28. Bracket
29. Washers
30. Shaft
31. Sprocket
32. Pin
33. Shaft
34. Sprocket & pulley assy.

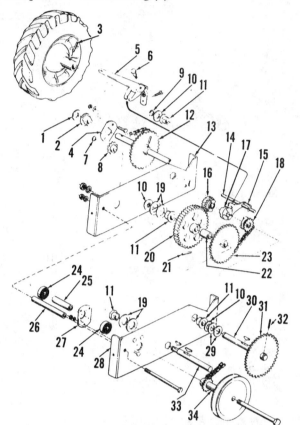

Fig. M13—Exploded view of Model 430 drive mechanism and auger housing assembly. On Models 225 and 230, auger sprocket (26) is an integral part of auger assembly (36) and drive chain is located inside left end plate of auger housing. Refer to Fig. M11 for auger chain adjustment on Models 225 and 230.

1. Drive idler arm
2. Drive clutch cable
3. Engine plate
4. Spring
5. Engine pulley
6. Drive idler pulley
7. Auger idler pulley
8. Drive pulley
9. Auger pulley
10. Auger idler arm
11. Auger clutch arm
12. Bearing retainer
13. Bearing
14. Snap ring
15. Shaft
16. Bolt
17. Sprocket
18. Bearing
19. Chain cover
20. Bearing retainer
21. Washer
22. Bushing
23. Bushing retainer
24. Snap ring
25. Washer
26. Sprocket
27. Pin
28. Skid shoe
29. Washer
30. Spacer
31. Idler
32. Belleville washer
33. Washer
34. Spring pin
35. Plate
36. Auger assy.
37. Nut

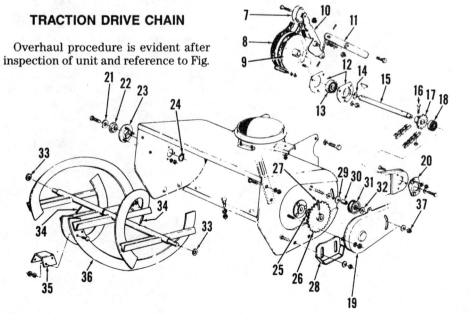

MTD

Model	Engine Make	Engine Model	Self-Propelled	No. of Stages	Scoop Width
600	Tecumseh	HS50	Yes	2	24 in. (610 mm)
650	Tecumseh	H70	Yes	2	24 in. (610 mm)
750	Tecumseh	HS70	Yes	2	26 in. (660 mm)
800*	Tecumseh	HM80	Yes	2	26 in. (660 mm)
850*	Tecumseh	HM80	Yes	2	26 in. (660 mm)
950	Tecumseh	HM80	Yes	2	33 in. (838 mm)
950*	Tecumseh	HM100	Yes	2	33 in. (838 mm)

*Models prior to 1986.

LUBRICATION

ENGINE

Refer to Tecumseh engine section for engine lubrication requirements.

Recommended fuel is regular or low-lead gasoline.

DRIVE MECHANISM

Lubricate drive chains at least once each season with good quality chain oil Auger shaft and axle bushings should be oiled and axle should be greased at least once each season or after 25 hours of operation. Lubricate control rod and lever pivot points for smooth operation.

AUGER GEARBOX

The auger gearbox is lubricated with either SAE 90 oil or Plastilube. On models equipped with a plug (20—Fig. M24), fill gearbox with SAE 90 until oil level reaches plug opening. If gearbox is not equipped with plug (20), then gearbox is lubricated with Plastilube during assembly and does not normally require new lubricant.

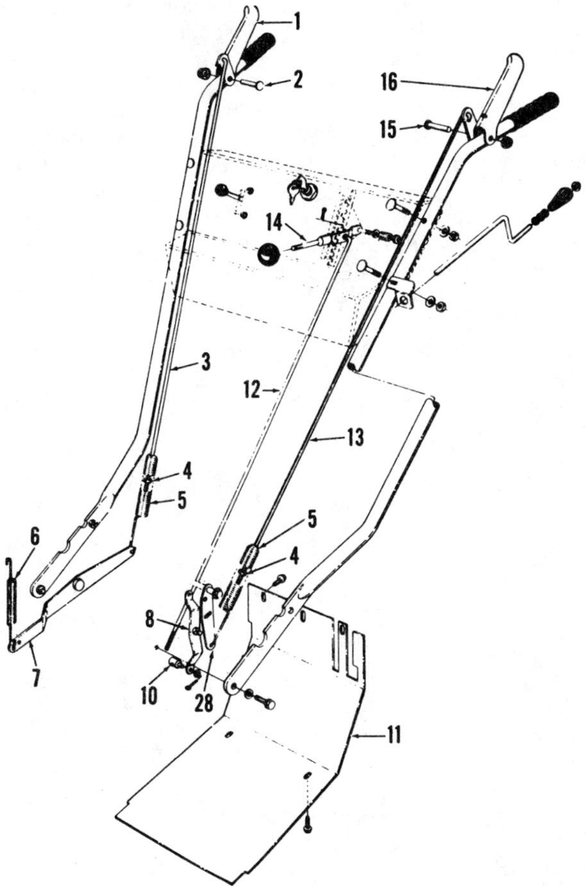

MAINTENANCE

Inspect chain and sprockets and re new if excessively worn. Inspect belts and pulleys and renew belts if frayed, cracked, burnt or otherwise damaged. Check throttle cable for binding and stiff operation.

Recommended tire pressure is 7-10 psi (48-69 kPa).

ADJUSTMENTS

HEIGHT

Auger height may be adjusted by loosening skid shoe mounting bolts and repositioning skid shoe. Skid shoes should be located in lowest setting when operating over smooth surfaces and in highest setting when operating over rough, irregular surfaces.

SHIFT CONTROL ROD

Detach shift control rod (12—Fig. M20) from shift control lever (14). On later models with locating slot in lower frame cover as shown in Fig. M21, move shift rod so shift plate (27—Fig. M23) is centered in slot. On early models without locating slot, remove lower frame cover and move shift control rod so friction wheel is centered

Fig. M20—Drawing of control assembly.

1. Auger clutch control lever
2. Pin
3. Auger clutch control rod
4. Nut
5. Spring
6. Spring
7. Lever
8. Lever
10. Pivot nut
11. Lower panel
12. Shift rod
13. Drive clutch control rod
14. Shift control lever
15. Pivot pin
16. Drive clutch control lever
28. Clutch control arm

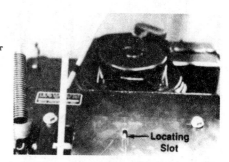

Fig. M21—View showing location of locating slot used for shift control adjustment on later models.

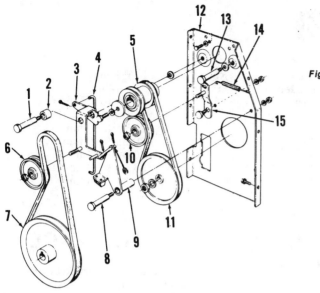

Fig. M22—Exploded view of pulley system.

1. Pivot bolt
2. Spacer
3. Auger idler bracket
4. Auger clutch rod
5. Engine pulley
6. Auger idler pulley
7. Auger pulley
8. Pivot bolt
9. Auger pulley brake
10. Drive idler pulley
11. Drive pulley
12. Engine plate
13. Belt guide bolt
14. Spring
15. Drive idler arm

on neutral stop bracket (N—Fig. M23). On all models, position shift control lever in neutral detent then rotate shift control rod in ferrule at bottom of rod so upper rod end will fit easily into shift control lever. Attach rod then check adjustment by noting if unit will roll forward and backward with shift control lever in neutral.

CLUTCH RODS

Adjustment of auger and drive clutch control rods is accomplished by turning nut at spring end of rod. There should be a small amount of end play between rod and spring (5—Fig. M20) with clutch disengaged so idler will relieve pressure against drive belt. With clutch engaged, drive belt should tighten around pulleys without slippage.

OVERHAUL

ENGINE

Engine make and model numbers are listed at front of this section. Refer to Tecumseh engine section for engine overhaul procedure.

AUGER GEARBOX

Auger gearbox may be serviced after removing complete auger assembly from frame then detaching impeller and augers. Overhaul of gearbox is evident after referral to Fig. M24 and inspection of unit. Note thickness of washers during disassembly and reassembly.

All gearboxes may be filled with 4 ounces of Plastilube grease while models equipped with plug (20) may be lubricated with oil as outlined in LUBRICATION section. Models without plug (20) must be lubricated with grease.

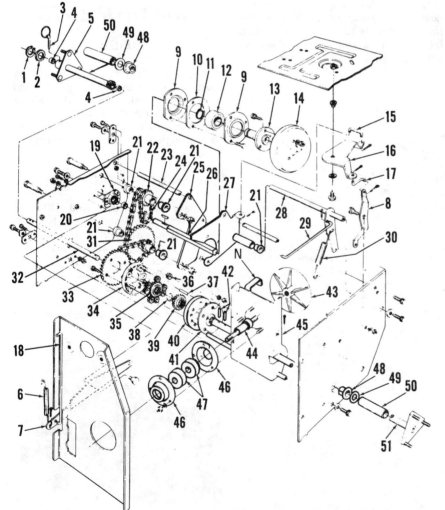

1. Snap ring	27. Shift plate
2. Washer	28. Shaft
3. Lock pin	29. Link
4. Bushing	30. Spring
5. Right axle	31. Sprocket
6. Spring	32. Differential lock
7. Lever	shaft
8. Lever	33. Sprocket
9. Bearing retainer	34. Gear housing
10. Spacer	35. Sun gear
11. Snap ring	36. Planetary gear
12. Bearing	37. Shaft
13. Friction hub	38. Spacer
14. Friction wheel	39. Gear
15. Link	40. Gear housing
16. Bellcrank	41. Plate
17. Link	42. Pins
18. Auger clutch rod	43. Friction disc
19. Bushing plate	44. Shaft
20. Bushing	45. Drive support
21. Bushing	46. Bearing housing
22. Sprocket	47. Bearings
23. Sprocket	48. Bushing
24. Sprocket	49. Washer
25. Support	50. Spacer
26. Hex shaft	51. Left axle

Fig. M23—Exploded view of drive mechanism used on Models 850, 950 and 960. Other models are similar but differential components (34 thru 41) and spacers (50) are not used.

DRIVE MECHANISM

Refer to Figs. M22 and M23 for exploded views of drive components. A differential is used on Models 850, 950 and 960. On all models a lock pin may be used to lock wheel and axle together so both wheels are driven. The lock pin is located in right wheel of Models 850, 950 and 960 and in left wheel of all other models. Do not overtighten locknuts securing bearing retaining plates (9). Bearing (12) is self-centering and must be able to move in retaining plates or bearing will bind on friction hub (13). Install friction wheel (14) with cupped side towards hub (13). Be sure auger idler bracket (3—Fig. M22) has sufficient travel to operate properly. If not, grind approximately 1/16 inch (1.6 mm) from end of spacer (2) and recheck movement.

DRIVE BELTS

To remove drive belts, disconnect chute linkage and remove belt cover. Unscrew shoulder bolts adjacent to engine pulley while noting roller mounted on right side bolt. Lift up on auger idler pulley (6—Fig. M22) so brake disengages auger drive pulley then lift up auger frame slightly and separate auger and main frames. Unscrew shoulder bolts which serve as belt guides and remove drive belts. Reverse removal· procedure to install belts.

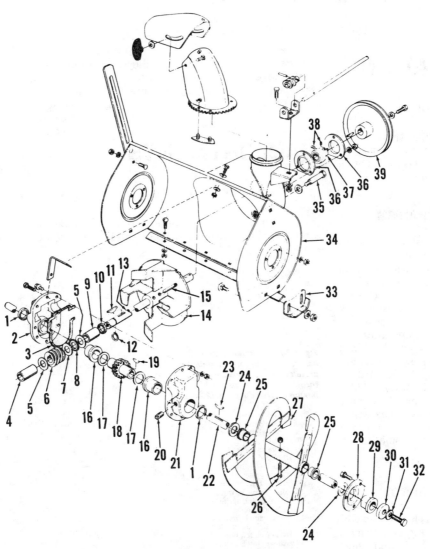

Fig. M24—Exploded view of auger assembly.

1. Seal	9. Bushing	17. Washer	25. Bushing
2. Gearbox half R.H.	10. Seal	18. Gear	26. Bolt
3. Gasket	11. Key	19. Key	27. Auger
4. Bushing	12. Snap ring	20. Plug	28. Bushing retainer
5. Washer	13. Drive shaft	21. Gearbox half L.H.	29. Bushing
6. Worm gear	14. Impeller	22. Auger shaft	30. Washer
7. Washer	15. Pin	23. Key	31. Belleville washer
8. Thrust bearing	16. Bushing	24. Washer	32. Screw

MTD

Model	Engine Make	Engine Model	Self-Propelled?	No. of Stages	Scoop Width
800*	Tecumseh	HMSK80	Yes	2	26 in. (660 mm)
840	Tecumseh	HMSK80	Yes	2	26 in. (660 mm)
841	Tecumseh	HMSK80	Yes	2	26 in. (660 mm)
850*	Tecumseh	HMSK100	Yes	2	26 in. (660 mm)
960*	Tecumseh	HMSK100	Yes	2	33 in. (838 mm)

*1986 and later production models.

LUBRICATION

ENGINE

Refer to Tecumseh 4-stroke engine section for engine lubrication information. Recommended fuel is leaded or unleaded regular gasoline.

DRIVE MECHANISM

Lubricate drive chains and sprockets at least once each season or after every 25 hours of operation with a SAE 30 engine oil. Auger shaft bushings, traction drive shaft bushings and axle shaft bushings should be oiled with SAE 30 engine oil at least once each season or after every 25 hours of operation. Remove wheels and grease axle shaft with multipurpose automotive grease at least once each season. On Model 960, remove snap ring (48—Fig. M33) and klick pin (49), slide wheel and differential tube out and apply grease on axle shaft (24). Lubricate control rod and lever pivot points to ensure smooth operation.

AUGER GEARBOX

The auger gearbox is lubricated with Shell Alvania grease EPROO. Remove filler plug (30—Fig. M34) and check lubricant level using a wire as a dipstick. Read lubricant level on wire. Add lubricant through plug opening to bring gearbox level up to about 1/2 full. Capacity is 4 ounces (120 mL), do not overfill.

MAINTENANCE

Inspect drive belts and pulleys and renew belts if cracked, frayed, burned or otherwise damaged. Renew pulleys if belt grooves are excessively worn or damaged or if idler pulley bearings are loose or rough.

ADJUSTMENTS

AUGER HEIGHT

Auger height can be adjusted by loosening skid shoe mounting bolts and repositioning the skid shoes. Auger should be in lowest setting (skid shoes raised) when operating over smooth paved surfaces or raised (skid shoes lowered) when operating over rough surfaces.

CLUTCH RODS

With auger clutch lever (26—Fig. M30) released against bumper (7), lower end of auger clutch spring (19) should fit loosely in the hole at rear of auger

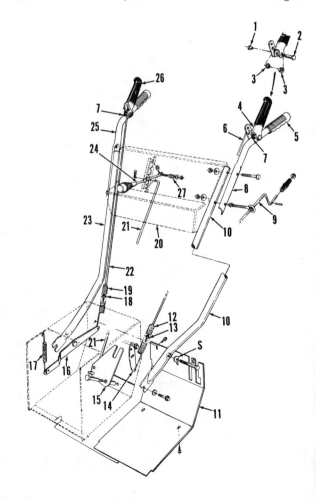

Fig. M30—Exploded view of handle and controls. Models produced in 1986 also had a throttle control assembly mounted in handle panel (20).

1. Shoulder nut
2. Shoulder bolt
3. Plastic bushing
4. Drive clutch lever
5. Handle grip
6. Drive clutch rod
7. Bumper
8. Upper L.H. handle
9. Chute crank
10. Lower L.H. handle
11. Main frame cover
12. Drive clutch spring
13. Adjusting nut
14. Drive clutch bracket
15. Shift linkage bracket
16. Auger clutch bracket
17. Auger clutch spring
18. Adjusting nut
19. Auger clutch spring
20. Handle panel
21. Shift rod
22. Auger clutch rod
23. Lower R.H. handle
24. Shift lever
25. Upper R.H. handle
26. Auger clutch lever
27. Shift lever spring

clutch bracket (16). If not, disconnect spring and slide spring upward on auger clutch rod (22) to expose adjusting nut (18) and turn nut up or down so bottom of hook in spring will be even with middle of hole in bracket when lifting clutch rod up to take slack out of rod and spring. There must not be any tension on clutch rod spring with the clutch lever in disengaged position. Repeat adjustment procedure for drive clutch rod (6).

SHIFT ROD

Move shift lever to neutral position on handle panel; sliding bracket assembly (13—Fig. M32 or M33) should be aligned with slot (S—Fig. M30) in main frame cover (11). Also, with the drive clutch lever (4) in engaged position, the snowthrower should roll back and forth with shift lever in neutral. If necessary to adjust shift rod, disconnect shift rod (21) from shift lever (24) at control console. Move shift lever to neutral position and move shift rod until sliding bracket is aligned with slot in main frame cover. If cover is removed, sliding bracket should be aligned on center of neutral stop bolt (19—Fig. M32 or M33). Thread shift rod in or out of ferrule (41) in shift bracket (40) at bottom end of rod until top end aligns with hole in shift lever. Reconnect shift rod and check adjustment.

Shift lever spring (27—Fig. M30) should be tight enough to hold lever in handle panel detent notches, yet allow lever to move without binding on detents.

OVERHAUL

ENGINE

Engine make and model numbers are listed at beginning of this section. Refer to Tecumseh 4-stroke engine section for engine overhaul information.

DRIVE BELTS

To renew drive belts, it is first necessary to split the snowthrower between the auger housing and main frame assembly as follows: Disconnect the deflector chute crank at the chute end and remove drive belt cover. Remove shoulder bolt (17—Fig. M31) and Belleville washer at left side of engine pulley. Remove shoulder bolt (4) and spacer (5) at right hand side of engine pulley, being careful not to lose the

Belleville (spring cup) washer located between idler bracket and engine plate. Slip auger drive belt off engine pulley and loosely install shoulder bolt, roller and Belleville washer at right side of engine pulley. Remove the top cap screws retaining auger housing flanges (24) to main frame (21). Two people are required to separate the auger housing from main frame. With someone holding handles, push in on auger idler pulley to lift brake pad (BR) out of pulley groove, then lift auger housing from main frame. At this point, either the auger belt or drive belt, or both, can be renewed.

To remove auger belt after separating front and rear units, remove the five shoulder bolts (5—Fig. M34) and Belleville washers from auger housing and remove belt from pulley. An alternate method is to remove the cap screw holding auger drive pulley to impeller shaft and remove the pulley and belt; be careful not to lose the drive key and Belleville washer. Install new belt by reversing removal procedure. If auger pulley was removed, be sure key is in place on shaft and install Belleville washer with cup side toward pulley. Tighten pulley retaining cap screw securely.

To remove traction drive belt, unhook drive idler spring (16—Fig. M31) from engine plate, then remove the belt from drive pulley. Remove shoulder bolt (4), spacer (5) and Belleville washer at right side of engine pulley and remove the belt. Install new belt by reversing removal procedure.

Be sure that idler pulley and/or brake pad are not behind the drive pulley when assembling auger housing to main frame.

DRIVE MECHANISM

To overhaul drive mechanism, first remove drive belt as outlined in a preceding paragraph, then remove engine assembly and traction drive pulley (19—Fig. M31). Further disassembly procedure is evident after inspection of unit and reference to view in Fig. M33 for Model 960 or Fig. M32 for all other models.

On Model 960, when renewing bushing (56—Fig. M33) in differential tube (55), the bushing must be drilled for klick pin (49) after installing in tube. Installing the pin through the hole in differential tube locks the two drive wheels together. With the pin inserted

Fig. M31—Exploded view of belt drive system. Auger pulley (1) is also shown in Fig. M34 and traction drive pulley (19) mounts on drive plate spindle (35—Fig. M32 or Fig. M33).

1. Auger pulley
2. Auger belt
3. Auger belt idler
4. Shoulder bolt
5. Spacer
6. Auger idler bracket
7. Auger clutch rod
8. Engine pulley
9. Traction drive belt
10. Belt cover
11. Square key
12. Auger brake linkage
13. Traction drive idler
14. Shoulder bolt
15. Drive idler bracket
16. Drive idler spring
17. Shoulder bolt
18. Engine spacer plate
19. Traction drive pulley
20. Engine bracket assy.
21. Main (motor mount) frame
22. Brake bracket assy.
23. Shoulder bolt
24. Auger housing flanges

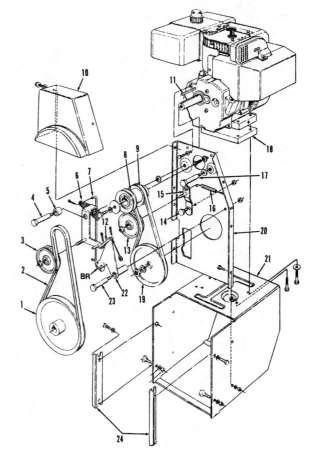

in the hole in outer end of axle, the differential is free to operate. If differential unit is disassembled for inspection or overhaul, clean all parts thoroughly and renew any excessively worn or damaged parts. Lubricate differential at reassembly with EP lithium base grease.

On all models, be sure not to over tighten the hex shaft self-aligning bearing (51—Fig. M32 or M33) as hex shaft (50) needs to flex in the bearing for free movement of sliding bracket (13). If adjusting auger clutch rod (37) with main frame cover removed, forward point of sliding bracket (13) should be centered

on head of neutral stop bolt (19) when shift lever on handle panel is in neutral position.

FRICTION WHEEL

To remove rubber faced friction wheel, first tip snowthrower forward onto front of auger housing and remove the cover from main frame. Friction wheel (5—Fig. M32 or M33) can then be unbolted from friction wheel adapter (4) and removed from the end of hex shaft. Install new friction wheel with cup side facing toward adapter and in-

stall retaining lockwashers and nuts finger tight. Turn the wheel to be sure it does not wobble (cocked on hub), then securely tighten nuts.

AUGER SHEAR BOLTS

The auger flights are driven through shear bolts (37—Fig. M34) which will shear when a solid obstruction is encountered to prevent damage to snowthrower. If this occurs, be sure to replace the bolts with factory replacement parts; substitute bolts may not shear and cause extensive damage. Before in-

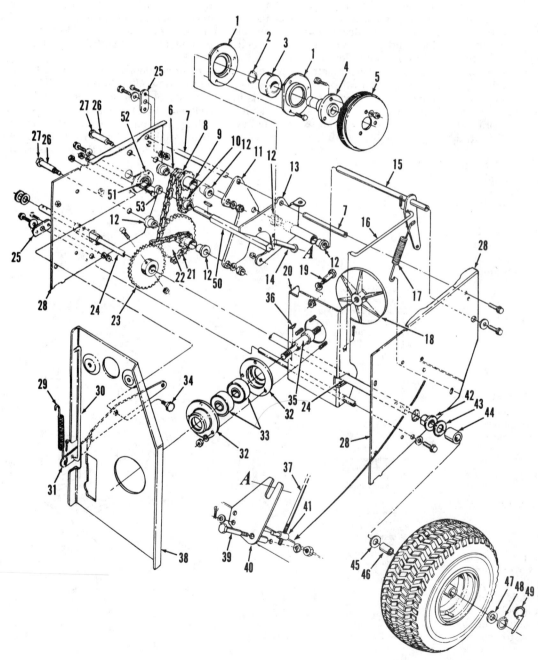

Fig. M32—Exploded view of friction drive and chain speed reduction drive components for all models except Model 960.

1. Bearing housing
2. Snap ring
3. Ball bearing
4. Friction wheel adapter
5. Friction wheel
6. Roller chain
7. Sliding support axle
8. Roller chain
9. Double sprocket
10. Hex shaft sprocket
11. Chain support bracket
12. Flange bearing
13. Sliding bracket assy.
14. Pin (part of 13)
15. Drive clutch bracket
16. Drive clutch rod
17. Drive clutch spring
18. Aluminum drive plate
19. Neutral stop bolt
20. Drive pivot plate
21. Roller chain
22. Double sprocket
23. Axle sprocket
24. Axle shaft
25. Pivot shaft bracket
26. Sprocket hub tube
27. Sprocket bolt
28. Main frame assy.
29. Auger clutch spring
30. Auger clutch rod
31. Auger clutch lever
32. Bearing housing
33. Ball bearings
34. Shoulder bolt
35. Drive plate spindle
36. Hi-pro key
37. Auger clutch rod
38. Engine bracket assy.
39. Shoulder bolt
40. Shift bracket
41. Ferrule
42. Axle flange bearing
43. Spacer washer
44. Spacer
45. Wheel spacer washer
46. Wheel bushing
47. Wheel spacer washer
48. Snap ring
49. Klick pin
50. Hex shaft
51. Bearing
52. Bearing housing
53. Spacer

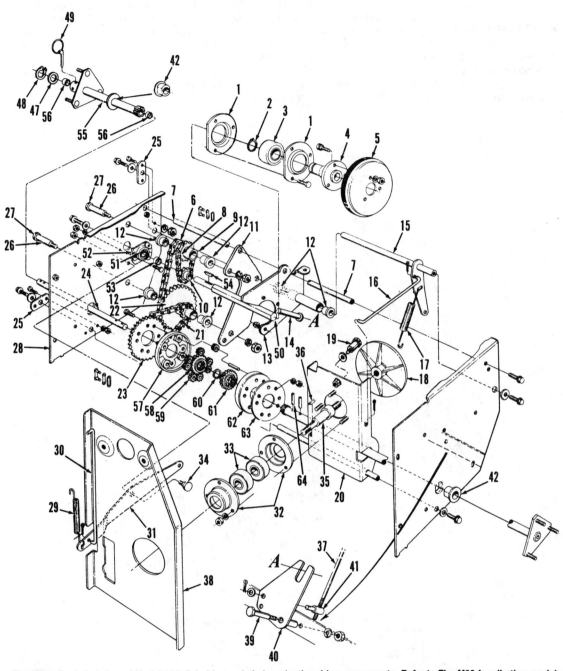

Fig. M33—Exploded view of Model 960 disk drive and chain reduction drive components. Refer to Fig. M32 for all other models.

1. Bearing housing
2. Snap ring
3. Ball bearing
4. Friction wheel adapter
5. Friction wheel
6. Roller chain
7. Sliding support axle
8. Roller chain
9. Double sprocket
10. Hex shaft sprocket
11. Chain support bracket
12. Flange bearing
13. Sliding bracket assy.
14. Pin (part of 13)
15. Drive clutch bracket
16. Drive clutch rod

17. Drive clutch spring
18. Aluminum drive plate
19. Neutral stop bolt
20. Drive pivot plate
21. Roller chain
22. Double sprocket
23. Axle sprocket
24. Axle shaft
25. Pivot shaft bracket
26. Sprocket hub tube
27. Sprocket bolt
28. Main frame assy.
29. Auger clutch spring

30. Auger clutch rod
31. Auger clutch lever
32. Bearing housings
33. Ball bearings
34. Shoulder bolt
35. Drive plate spindle
36. Hi-pro key
37. Auger clutch rod
38. Engine bracket assy.
39. Shoulder bolt
40. Shift bracket
41. Ferrule
42. Axle flange bearing

47. Wheel spacer washer
48. Snap ring
49. Klick pin
50. Hex shaft
51. Bearing
52. Bearing housing
53. Spacer
54. Hi-pro key
55. Differential tube
56. Bushing
57. Differential housing half
58. Gear
59. Gear
60. Flat washer
61. Gear
62. Differential housing half
63. Support plate
64. Spring pin

stalling new bolts, oil the auger flight bushings and be sure flights spin freely on the auger shaft.

AUGER GEARBOX

To remove auger gearbox, first remove auger belt as outlined in a preceding paragraph. If auger pulley (1—Fig. M34) was not removed, remove it at this time.

Remove outer bearing retainer (3), then remove set screws and unlock bearing (4) from gearbox input shaft. Unbolt auger shaft bearing flanges (39) at each end of auger housing, remove flange bearings (38) and thrust washers (34) and withdraw the gearbox with impeller (10) and auger flights (36). Remove shear bolts (37), auger flights, inner thrust washers and spacers (33) from auger shaft. Remove the carriage bolt and

spring pins (9), then pull impeller from shaft. Remove the bolts holding the gearbox halves (19 & 31) together and disassemble gearbox. Clean and inspect all parts and renew any excessively worn or damaged parts. Worm gear (27) can be removed from hub (28).

Reassemble by reversing disassembly procedure and lubricate gearbox using 4 ounces (120 mL) of Shell Alvania Grease EPROO.

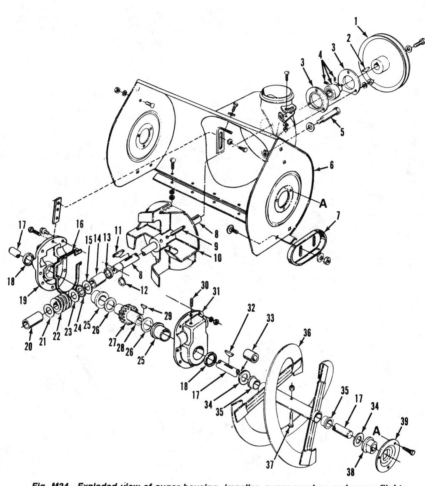

Fig. M34—Exploded view of auger housing, impeller, auger gearbox and auger flights.

1. Auger pulley	11. Hi-pro key	21. Flat washer
2. Hi-pro key	12. Snap ring	22. Worm gear
3. Bearing retainer	13. Oil seal	23. Flat washer
4. Self-aligning bearing	14. Sleeve bearing	24. Thrust bearing
5. Shoulder bolt	15. Flat washer	25. Flange bearing
6. Auger housing	16. Gasket	26. Flat washer
7. Skid shoe	17. Auger shaft	27. Worm gear
8. Impeller shaft	18. Oil seal	28. Gear hub
9. Spring pin	19. Housing half	29. Woodruff key
10. Impeller	20. Sleeve bushing	30. Filler plug

31. Housing half	
32. Woodruff key	
33. Spacer (except 960)	
34. Thrust washer	
35. Plastic bushing	
36. Auger flight	
37. Shear bolt	
38. Flange bearing	
39. Bearing flange	

MTD

Model	Engine Make	Engine Model	Self-Propelled?	No. of Stages	Scoop Width
440	Tecumseh	HSK40	Yes	2	20 in. (508 mm)
450	Tecumseh	HSSK50	Yes	2	20 in. (508 mm)
550, 552	Tecumseh	HSSK50	Yes	2	24 in. (610 mm)
586, 588	Tecumseh	HMSK80	Yes	2	26 in. (660 mm)

LUBRICATION

ENGINE

Refer to Tecumseh 4-stroke engine section for engine lubrication information. Recommended fuel is regular or unleaded regular gasoline.

DRIVE MECHANISM

Lubricate drive chains after every 25 hours of operation or at least once each season using a good-quality chain oil. Auger shaft bushings and traction drive shaft bushings should be oiled after every 25 hours of operation with SAE 30 motor oil. Lubricate all control pivot and slide points for smooth operation. Once each season or when necessary to replace auger flight shear bolts, remove shear bolts and squirt oil in hole and at ends of auger flights, then turn auger flight to distribute oil. On Models 550 and 586, remove wheels once each season and grease axle shafts. On track drive Models 552 and 588, keep weight transfer linkage points greased with a clinging lubricant such as Lubriplate and oil track drive chain with SAE 10W-30 motor oil after every 25 hours of operation.

AUGER GEARBOX

The auger gearbox is factory lubricated and should not require further lubrication unless disassembled for overhaul. Recommended lubricant is 10 ounces (295 mL) of Shell Alvania grease EPR00.

MAINTENANCE

Inspect belts and pulleys and renew belts if excessively worn, frayed or burned. Renew pulleys if belt grooves are worn or damaged, or if bearings in idler pulleys are loose or rough. Check drive chains and sprockets and renew if excessively worn or damaged.

ADJUSTMENTS

AUGER HEIGHT

Auger height can be adjusted by loosening skid shoe mounting bolts and repositioning the skid shoes. Auger height should be at lowest setting (skid shoes raised) when operating over

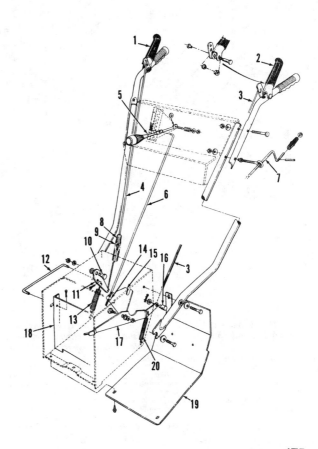

Fig. M40—Exploded view of handle assembly and controls.

1. Drive control lever
2. Auger control lever
3. Auger control rod
4. Drive control rod
5. Shift control lever
6. Shift control rod
7. Chute crank
8. Drive clutch spring
9. Adjusting nut
10. Shoulder bolt
11. Drive clutch bracket
12. Drive clutch rod
13. Drive spring
14. Ferrule
15. Shift bracket
16. Ferrule
17. Auger clutch bracket
18. Drive pulley support
19. Frame cover
20. Auger spring

smooth, paved surfaces and toward the highest setting (skid shoes lowered) when operating over rough surfaces. Be sure skid shoes are adjusted equally.

SHIFT ROD

On all models, disconnect shift control rod (6—Fig. M40) from shift lever (5) and place both the transmission and shift lever in neutral position. Turn shift control rod in or out of ferrule (14) at bottom of rod so upper end can be inserted in hole in shift lever. Secure rod in lever with cotter pin.

DRIVE CLUTCH

Drive spring (8—Fig. M40) at lower end of drive control rod (4) should be loose when drive control lever (1) is released. If not, unhook spring (8) from drive clutch bracket (11). Turn adjusting nut (9), located inside spring (8), so hook of spring is aligned with center of hole in drive clutch bracket then rehook spring in bracket. The clutch should engage when lever is held against handle grip and move the snowthrower without belt slippage.

AUGER CLUTCH

To check adjustment, remove belt cover and hold auger clutch lever against handle grip. The outer end of the slot in the end of auger brake linkage (25—Fig. M41) should be against the spacer on the auger belt idler as shown in Fig. M42. If not, disconnect auger control rod ferrule (Fig. M42) from auger clutch bracket and turn ferrule until proper adjustment is obtained when ferrule is connected to auger clutch bracket.

TRACK DRIVE

Drive track tension on Models 552 and 588 is properly adjusted when the track can be lifted approximately 1/2 inch (13 mm) at midpoint between track rollers. Adjustment is provided by adjuster cam (21—Fig. M45) on each side of left and right track assembly. If necessary to adjust track tension, loosen the cap screws retaining adjuster cams (21) on each side of track. Turn the adjuster cams equally with a screwdriver placed between tab on adjuster cam and heavy washer, then tighten the cap screws.

OVERHAUL

ENGINE

Engine make and model numbers are listed at beginning of this section. Re-

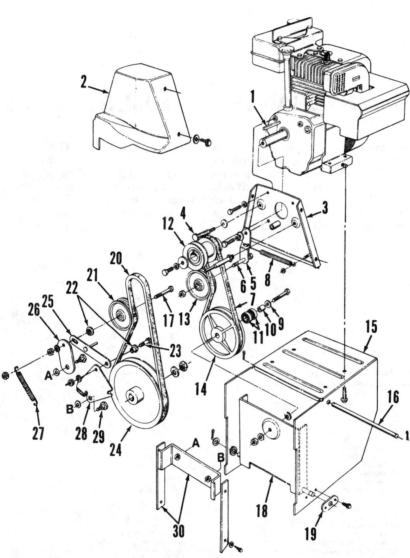

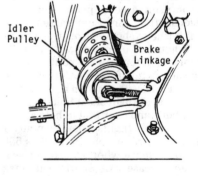

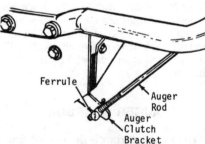

Fig. M41—Exploded view of drive belt and auger belt drive components and engine/frame assembly. Back side of drive pulley (14) has drive surface for friction wheel; depressing drive control lever moves pulley against friction wheel (38—Fig. M44).

1. Engine pulley key	8. Drive idler spring	17. Auger idler bolt	24. Auger pulley
2. Belt cover	9. Belleville washer	18. Drive pulley support	25. Brake linkage
3. Engine plate	10. Spacer	19. Support axle	26. Auger idler bracket
4. Shoulder bolt belt guide	11. Ball bearings	bracket	27. Auger idler spring
5. Drive clutch idler bracket	12. Engine pulley	20. Auger belt	28. Brake bracket
	13. Drive belt idler	21. Auger belt idler	29. Shoulder bolt
6. Idler bolt	14. Drive pulley	22. Shoulder spacers	30. Blower housing parts
7. Drive belt	15. Main frame	23. Brake pivot bolt	
	16. Sliding bracket rod		

Fig. M42—End of slot in brake linkage (25—Fig. M41) should be against spacer on the auger belt idler when auger control lever is depressed against snowthrower handle. Refer to text for adjustment procedure.

fer to Tecumseh 4-stroke engine section for engine overhaul information.

DRIVE BELTS

Remove the deflector chute crank at chute and remove upper belt cover. Remove the two shoulder bolt belt guides at engine crankshaft pulley. On track drive Models 552 and 588, place weight transfer lever in "Packed Snow" position. Remove the top bolts attaching auger housing to snowthrower frame assembly and loosen but do not remove the two bottom bolts. Lift up on auger belt to pull auger housing off of snowthrower frame and separate auger housing from the main unit.

To remove auger drive belt, remove the four shoulder bolt belt guides at auger pulley. The auger belt can now be removed from the bottom (auger) pulley. To remove traction drive belt, disconnect drive idler spring (8—Fig. M41) at engine plate and remove belt from engine pulley and bottom drive pulley. Install new belt(s) and reassemble by reversing disassembly procedure. Be sure pin on end of auger clutch bracket engages slot in auger brake bracket (see Fig. M43) before installing auger housing mounting bolts. Adjust auger and drive clutches as outlined in preceding paragraphs.

FRICTION WHEEL

To renew friction wheel, proceed as follows: Move shift lever to reverse position. Tip snowthrower up so it rests on front of auger housing and remove bottom cover, then remove friction wheel (38—Fig. M44) from wheel adapter (40). Place new wheel with cup side away from adapter and install retaining bolts finger tight. Turn the wheel to be sure it does not wobble, then securely tighten retaining bolts. Replace bottom cover.

Fig. M44—Exploded view of drive components; also see Fig. M41. Notch in shift linkage bracket (15) engages pin on sliding bracket (44). Sliding bracket is supported by sliding bracket rod (16). On all models except Model 586, only one flat washer (34) is used at right end of axle shaft (53).

6. Shift rod
14. Ferrule
15. Shift linkage bracket
16. Sliding bracket rod
31. Belleville washer
32. Flat washer
33. Flange bearing
33A. Flange bearing
34. Flat washer(s)
35. Spacer
36. Sprocket
37. Hi-pro key
38. Friction wheel
39. Hex shaft
40. Friction wheel adapter
41. Snap ring
42. Flat washer
43. Bearing
44. Sliding bracket
45. Flat washer
46. Roller chain
47. Sprocket
48. Snap ring
49. Sprocket
50. Flat washer
51. Bronze bearing
52. Flat washer
53. Sprocket & axle assy. (except 586)
54. Sprocket (Model 586)
54A. Axle (Model 586)
55. Roller chain
56. Shoulder bolt
57. Spacer (Model 586)
58. Flat washer (Model 586)
59. Wheel assy. (550 & 586)
60. Sleeve bearing (550 & 586)
61. Snap ring
62. Klick pin
63. Wheel assy. (440 & 450)

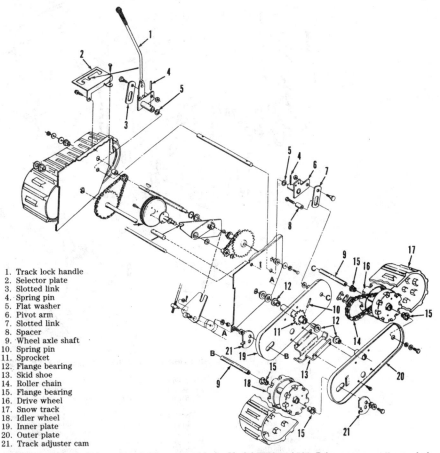

1. Track lock handle
2. Selector plate
3. Slotted link
4. Spring pin
5. Flat washer
6. Pivot arm
7. Slotted link
8. Spacer
9. Wheel axle shaft
10. Spring pin
11. Sprocket
12. Flange bearing
13. Skid shoe
14. Roller chain
15. Flange bearing
16. Drive wheel
17. Snow track
18. Idler wheel
19. Inner plate
20. Outer plate
21. Track adjuster cam

Fig. M45—Exploded view of track drive assembly for Models 552 and 588. Drive components are similar to wheel drive models shown in Fig. M44. Track lock handle (1) changes auger down force and has three positions: "Transport," "Normal Snow" and "Packed Snow."

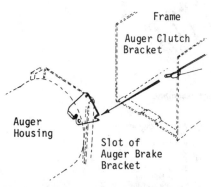

Fig. M43—Pin on end of auger clutch bracket must enter slot in auger brake bracket when reassembling snowthrower.

Frame

Auger Clutch Bracket

Auger Housing

Slot of Auger Brake Bracket

TRACK DRIVE

To renew drive track or chain inside track assembly on track drive Models 552 and 588, block up snowthrower frame so track assemblies are off the ground. Loosen the cap screws retaining track adjuster cams (21—Fig. M45) at each side of assembly and turn adjuster cams so all track tension is removed. Roll the track off of track rollers and side supports. Remove roller chain (14) by disconnecting master link. Reassemble by reversing disassembly procedure, making sure to install track with cleats pointing toward front of snowthrower. Adjust track tension as outlined in adjustment paragraph.

AUGER GEARBOX

To remove auger gearbox, follow procedure outlined in preceding paragraph for removal of auger drive belt. Then, remove auger drive pulley (7—Fig. M46) from rear of gearbox shaft and unbolt auger bearing housings (37) at each end of auger housing. Remove the gearbox and auger flight assembly from front of auger housing, then remove impeller (18) from input shaft and auger flights from driveshaft. Note the position of washers and spacers as you are disassembling the unit.

To disassemble gearbox, refer to exploded view in Fig. M46 and remove the screws retaining upper gear housing half (19). Note position of thrust washers and spacers as you disassemble gearbox. Clean and inspect all parts and renew any worn or damaged parts. Lubricate all parts prior to reassembly and pack gearbox with 10 ounces (295 mL) of Shell Alvania Grease EPR00.

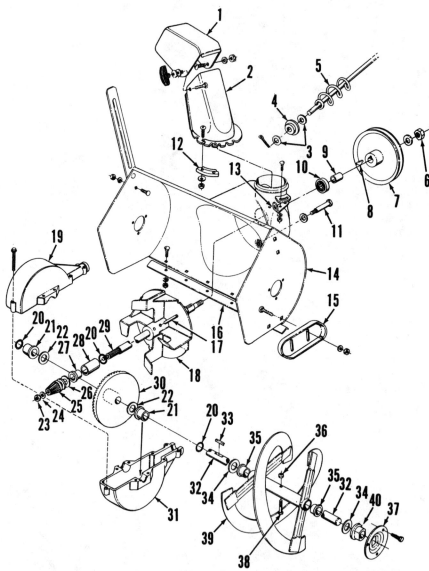

Fig. M46—Exploded view of auger housing, gearbox and auger flight assembly.

1. Upper chute	14. Auger housing	22. Flat washer
2. Lower chute	15. Skid shoes	23. Hex nut
3. Flat washers	16. Scraper bar (shave plate)	24. Flat washer
4. Bushing	17. Spring pin	25. Pinion gear
5. Chute crank	18. Impeller (blower fan)	26. Flat washer(s)
6. Pulley nut	19. Upper gear housing half	27. Flange bearing
7. Auger drive pulley	20. "O" ring	28. Sleeve bearing
8. Hi-pro key	21. Plastic flange bearing	29. Impeller (blower) axle
9. Spacer		30. Bevel gear
10. Ball bearing		31. Lower gear housing half
11. Shoulder bolt		
12. Chute flange keeper		
13. Chute crank bracket		

32. Auger flight (spiral) axle	
33. Key (Hi-pro or square)	
34. Flat washer	
35. Flange bearing	
36. Shear bolt nut	
37. Bearing housing	
38. Shear bolt	
39. Auger flight (spiral)	
40. Flange bearing	

MTD

Model	Engine Make	Engine Model	Self-Propelled?	No. of Stages	Scoop Width
590	Tecumseh	HMSK80	Yes	2	26 in. (660 mm)

LUBRICATION

ENGINE

Refer to Tecumseh 4-stroke engine section for engine lubrication information. Recommended fuel is regular or unleaded regular gasoline.

DRIVE MECHANISM

Lubricate drive chains after every 25 hours of operation or at least once each season using a good-quality chain oil. Auger shaft bushings and traction drive shaft bushings should be oiled after every 25 hours of operation with SAE 30 motor oil. Lubricate all control pivot and slide points to ensure smooth operation. Once each season or when necessary to replace auger flight shear bolts, remove shear bolts and squirt oil in hole and at ends of auger flights, then turn auger flight to distribute oil. Keep weight transfer moving parts lubricated with a clinging grease such as ''Lubriplate.''

AUGER GEARBOX

The auger gearbox is factory lubricated and should not require lubrication unless disassembled for overhaul. Recommended lubricant is Shell Alvania grease EPR00.

TRANSMISSION

The gear type transmission is factory lubricated and should not require lubrication unless disassembled for overhaul. Refer to Peerless transmission section of this manual for information.

MAINTENANCE

Inspect belts and pulleys and renew belts if excessively worn, frayed or burned. Renew pulleys if belt grooves are worn or damaged or if bearings in idler pulleys are loose or rough. Check drive chains and sprockets and renew if excessively worn or damaged.

ADJUSTMENTS

AUGER HEIGHT

Auger height can be adjusted by loosening skid shoe mounting bolts and repositioning the skid shoes. Auger height should be at lowest setting (skid shoes raised) when operating over

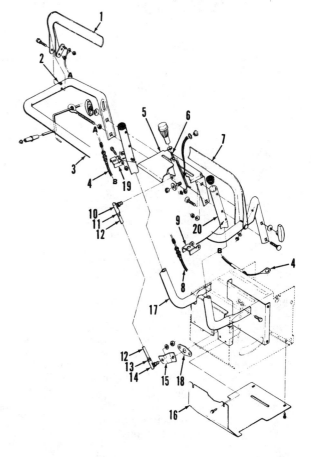

Fig. M50—Exploded view of handle assembly and snowthrower controls.

1. Drive clutch lever
2. Rubber bumper
3. Upper handle
4. Drive clutch cable
5. Control panel
6. Gearshift lever
7. Auger clutch lever
8. Auger clutch cable
9. Cable bracket
10. Ball joint assy.
11. Jam nut
12. Shift rod
13. Jam nut
14. Ball joint assy.
15. Shift arm
16. Bottom cover
17. Lower handle
18. Shift support
19. Cable bracket

smooth, paved surfaces and toward the highest setting (skid shoes lowered) when operating over rough surfaces. Be sure skid shoes are adjusted equally.

SHIFT ROD

Snowthrower should roll easily with shift lever in neutral position, and the wheels should lock up with shift lever in reverse. If not, loosen jam nuts (11 & 13—Fig. M50) (bottom nut is left-hand thread) at each end of shift rod (12). Move transmission shift arm (15) to find neutral position and turn shift rod to bring gearshift lever (6) in line with neutral "N" position on console.

CLUTCH CABLES

There should be a small amount of slack in the clutch control cables with clutch levers fully forward against bumpers on handles. To check adjustment, place a nickel on top of the rubber bumper at either handle and release the clutch lever. All slack in the cable should be removed without compressing the rubber bumper. If not, loosen and adjust jam nuts on cable housing at cable bracket (9 & 19—Fig. M50) to obtain correct adjustment, then tighten jam nuts.

OVERHAUL

ENGINE

Refer to Tecumseh 4-stroke engine section for engine overhaul information.

DRIVE BELTS

Remove the deflector chute crank at chute and remove upper belt cover. Loosen the upper left-hand belt guide and remove the right-hand belt guide at engine crankshaft pulley and roll belt off of pulley. Unhook the auger control cable from pin on auger brake link (7—Fig. M52) after removing hairpin clip from brake link pin. Place weight transfer lever in "packed snow" position. Remove the top bolts attaching auger housing to snowthrower frame assembly and loosen but do not remove the two bottom bolts. Lift up on auger belt to pull auger housing off of snowthrower frame and separate auger housing from the main unit.

To remove auger drive belt, remove shoulder bolt belt guides (21—Fig. M52) at auger pulley. Roll belt off of bottom (auger) pulley. Remove the second hairpin clip from brake link pin, push idler

pulley to right and work belt out between link pin and idler.

To remove traction drive belt, disconnect auger cable (16—Fig. M51) extension spring at idler arm (7) and unhook return spring (11) from arm. Loosen the two set screws (3 & 5) in engine PTO (camshaft) pulley (4) and remove pulley with belt. Remove transmission pulley (14) with belt.

Install new belt(s) and reassemble by reversing disassembly procedure. Adjust auger and drive clutches as outlined in preceding paragraphs.

TRANSMISSION

Model 590 is equipped with a Peerless Model 7002D031 transmission with six forward speeds and one reverse speed. Refer to Peerless transmission section in this manual for transmission overhaul information.

AUGER GEARBOX

To remove auger gearbox, follow procedure outlined in a preceding para-

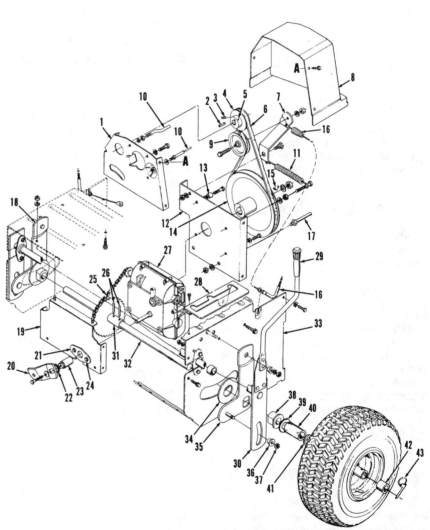

Fig. M51—Exploded view of drive mechanism. Traction drive is powered from pulley (4) on engine auxiliary PTO shaft.

1. Engine plate	13. Spacer	24. Hex head screw
2. Engine pulley key	14. Transmission pulley	25. Roller chain
3. Set screw	15. Key	26. Spring pins
4. Auxiliary PTO pulley	16. Auger cable & spring	27. Transmission assy.
5. Set screw	17. Belt guard bolt	28. Weight transfer plate
6. Drive belt	18. Side arm assy.	29. Handle grip
7. Idler arm	19. Back plate	30. Weight transfer handle
8. Belt cover	20. Shift arm	31. Sprocket & hub assy.
9. Flat idler	21. Shift support	32. Wheel axle
10. Belt guard bolts	22. Flat washer	33. Main frame assy.
11. Return spring	23. Shift arm extension	34. Slide washer
12. Transmission support plate		35. Side arm assy.
		36. Shoulder spacer
		37. Locknut
		38. Flange bearing
		39. Flat washer
		40. Spacer
		41. Flat washer
		42. Sleeve bearing
		43. Klick pin

graph for removal of auger drive belt. Then, remove auger drive pulley (16—Fig. M52) from rear of gearbox shaft and unbolt auger bearing supports (48) at each end of auger housing. Remove the gearbox and auger flight assembly from front of housing, then remove impeller (37) from input shaft and auger flights from driveshaft. Note the position of washers and spacers as you are disassembling the unit.

To disassemble gearbox, refer to exploded view in Fig. M52 and remove the screws retaining upper gear housing half (26). Note positions of thrust washers and spacers as you disassemble gearbox. Clean and inspect all parts and renew any worn or damaged parts. Lubricate all parts prior to reassembly and pack gearbox with 10 ounces (295 mL) of Shell Alvania Grease EPR00.

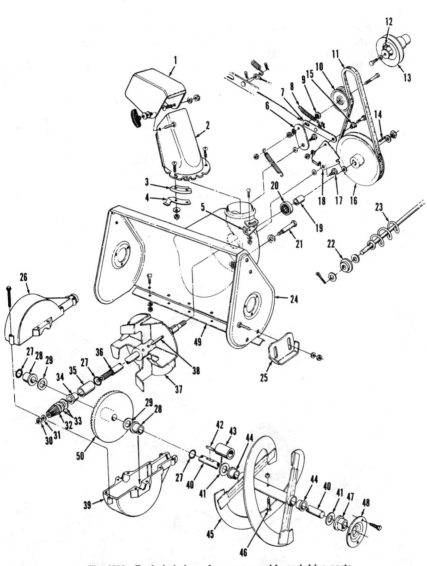

Fig. M52—Exploded view of auger assembly and drive parts.

1. Upper chute
2. Lower chute
3. Chute flange keeper
4. Chute stop
5. Chute crank bracket
6. Idler bracket
7. Brake link
8. Auger brake spring
9. Shoulder spacer
10. Flat idler
11. Auger drive belt
12. Square key
13. Engine pulley
14. Hi-pro key
15. Shoulder spacer
16. Auger drive pulley
17. Shoulder bolt
18. Brake bracket
19. Spacer
20. Ball bearing
21. Shoulder bolt belt guide
22. Bushing
23. Chute crank rod
24. Auger (blower) housing
25. Skid shoe
26. Upper gear housing half
27. "O" ring
28. Plastic flange bushing
29. Flat washer
30. Locknut
31. Flat washer
32. Pinion gear
33. Flat washers
34. Flange bearing
35. Sleeve bearing
36. Impeller (blower) axle
37. Impeller (blower fan)
38. Spring pin
39. Lower gear housing half
40. Auger (spiral) axle
41. Flat washer
42. Square key
43. Spacer
44. Flange bearing
45. Auger flight (spiral assy.)
46. Shear bolt
47. Flange bearing
48. Bearing support
49. Scraper bar (shave plate)
50. Bevel gear

MONTGOMERY WARD

MONTGOMERY WARD
Montgomery Ward Plaza
Chicago, IL 60671

Model	Engine Make	Engine Model	Self-Propelled?	No. of Stages	Scoop Width
TMO-35262A	Tecumseh	AH600	No	1	21 in. (533 mm)
TMO-35272B	B&S	62030	No	1	21 in. (533 mm)
TMO-35284A	B&S	62030	No	1	21 in. (533 mm)

1. Fuel tank cap
2. Upper cover
   (cowling)
3. Throttle lever (5 hp
   only)
4. Throttle cable (5 hp
   only)
5. Auger clutch cable
6. Chute crank
7. Belt tension spring
8. Engine support
   bracket
9. Engine pulley half
10. Flat washer
11. Idler bracket &
    brake
12. Spacer
13. Flat washer
14. Belt cover
15. Drive belt
16. Drive pulley
17. Bearing retainer cup
18. Ball bearing
19. Auger (blower)
    housing
20. Scraper bar (shave
    plate)
21. Bearing & collar kit
22. Rubber auger flights
    (spirals)
23. Rubber paddles
24. Auger assy.
25. Idler pulley

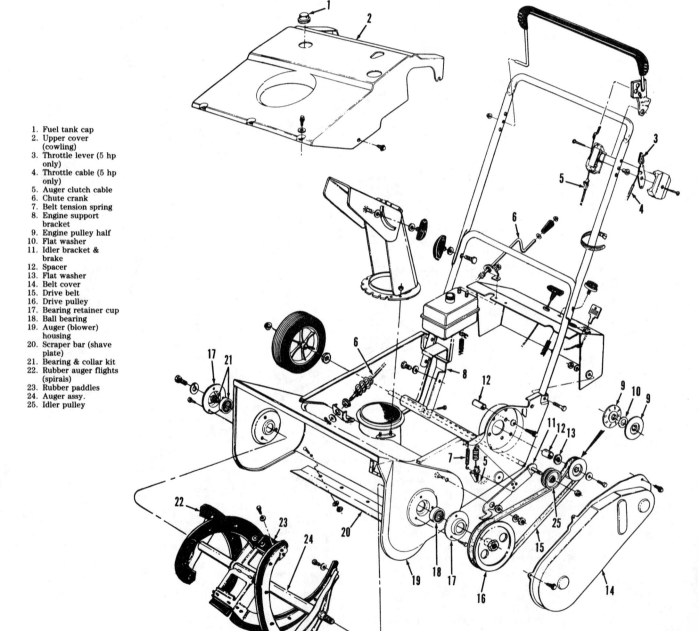

Fig. MW1—Typical exploded view of snowthrower assembly.

LUBRICATION

ENGINE

The engines used on these models are 2-stroke type which are lubricated by mixing oil with fuel. Recommended fuel:oil ratio is 32:1 when using a good-quality oil designed for 2-stroke engines. Mix with regular or low-lead gasoline. Refer to Briggs & Stratton or Tecumseh 2-stroke engine sections for additional information.

DRIVE MECHANISM

Lubricate control lever pivot points and cable periodically to ensure smooth operation.

MAINTENANCE

Inspect the drive belt and pulleys; renew belt if frayed, cracked, excessively worn or otherwise damaged. Renew pulleys if excessively worn or damaged. Inspect the rubber auger flights and renew if excessively worn or otherwise damaged.

ADJUSTMENTS

There are three holes in the belt idler bracket for belt tension adjustment. To adjust, remove the belt cover and refer to Fig. MW1. If belt slips (auger seems to slow down or hesitate at constant engine speed), connect the spring on end of clutch cable to next higher hole in idler bracket. If belt tension is not sufficient with spring in highest hole in bracket, it will be necessary to install a new idler belt.

OVERHAUL

ENGINE

Engine make and model are listed at front of this section. Refer to appropriate Briggs & Stratton or Tecumseh 2-stroke engine section for engine overhaul information.

DRIVE MECHANISM

Refer to exploded view of drive mechanism in Fig. MW1. Overhaul procedure is evident after inspection of unit and reference to appropriate exploded view.

MONTGOMERY WARD

Model	Engine Make	Engine Model	Self-Propelled?	No. of Stages	Scoop Width
TMO-35155A	Tecumseh	HSK50	Yes	2	24 in. (610 mm)
TMO-35156A	Tecumseh	HMSK80	Yes	2	26 in. (660 mm)
TMO-35275B	B&S	130202	Yes	2	24 in. (610 mm)
TMO-35275C	B&S	130202	Yes	2	24 in. (610 mm)
TMO-35288A	B&S	130202	Yes	2	24 in. (610 mm)

LUBRICATION

ENGINE

Refer to appropriate Briggs & Stratton or Tecumseh 4-stroke engine section for engine lubrication information. Recommended fuel is leaded or unleaded regular gasoline.

DRIVE MECHANISM

Lubricate drive chains, including chains in track drives, after every 25 hours of operation or at least once each season using a good-quality chain oil. Auger shaft bushings and traction drive shaft bushings should be oiled after every 25 hours of operation with SAE 30 motor oil. Lubricate all control pivot and slide points for smooth operation. Once each season or when necessary to replace auger flight shear bolts, remove shear bolts and squirt oil in hole and at ends of auger flights, then turn auger flight to distribute oil. Remove wheels once each season and grease axle shafts. On Model TMO-35288A, keep weight transfer linkage points greased with a clinging lubricant such as "Lubriplate."

AUGER GEARBOX

The auger gearbox is factory lubricated and should not require further lubrication unless disassembled for overhaul. Recommended lubricant is Shell Alvania grease EPR100.

MAINTENANCE

Inspect belts and pulleys and renew belts if excessively worn, frayed or burned. Renew pulleys if belt grooves are worn or damaged, or if bearings in idler pulleys are loose or rough. Check drive chains and sprockets and renew if excessively worn or damaged. On Model TMO-35288A, check track adjustment.

ADJUSTMENTS

AUGER HEIGHT

Auger height can be adjusted by loosening skid shoe mounting bolts and repositioning the skid shoes. Auger height should be at lowest setting (skid shoes raised) when operating over smooth, paved surfaces and toward the highest setting (skid shoes lowered) when operating over rough surfaces. Be sure skid shoes are adjusted equally.

Fig. MW20—Exploded view of handle assembly and controls.

1. Drive control lever
2. Auger control lever
3. Auger control rod
4. Drive control rod
5. Shift control lever
6. Shift control rod
7. Chute crank
8. Drive clutch spring
9. Adjusting nut
10. Shoulder bolt
11. Drive clutch bracket
12. Drive clutch rod
13. Drive spring
14. Ferrule
15. Shift bracket
16. Ferrule
17. Auger clutch bracket
18. Drive pulley support
19. Frame cover
20. Auger spring

SHIFT ROD

On all models, disconnect shift control rod (6—Fig. MW20) from shift control lever (5) and place both the transmission and shift lever in neutral position. Turn shift control rod in or out of ferrule (14) at bottom of rod so upper end can be inserted in hole in shift lever. Secure rod in lever with cotter pin.

DRIVE CLUTCH

With drive clutch control lever (1—Fig. MW20) released, drive spring (8) at bottom of drive control rod (4) should be loose. If not, unhook spring from drive clutch bracket (11) and turn adjusting

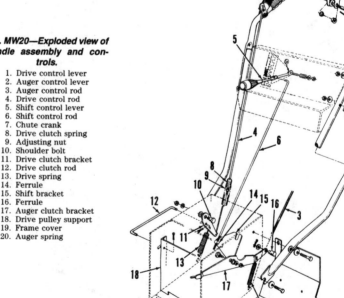

nut (9) located inside spring so hook of spring is aligned with center of hole in drive clutch bracket then rehook spring in bracket. With drive control lever (1) held against handle, the clutch should engage and move the snowthrower without belt slippage.

AUGER CLUTCH

To check adjustment, remove belt cover and hold auger control lever (2—Fig. MW20) against handle. The outer end of the slot in the end of auger brake linkage (25—Fig. MW21) should be against the spacer on the auger belt idler as shown in Fig. MW22. If not, disconnect auger control rod ferrule (Fig. MW22) from auger clutch bracket and turn ferrule until proper adjustment is obtained with ferrule reconnected to auger clutch bracket.

TRACK DRIVE

Drive track tension is properly adjusted when the track can be lifted approximately 1/2 inch (13 mm) at midpoint between track rollers. Adjustment is provided by adjuster cam (21—Fig. MW25) on each side of left and right track assembly. If necessary to adjust track tension, loosen the cap screws retaining the adjuster cams on each side of track. Turn the adjuster cams equally with a screwdriver placed between tab on adjuster cam and heavy washer, then tighten the cap screws.

OVERHAUL

ENGINE

Engine make and model numbers are listed at beginning of this section. Refer to Briggs & Stratton or Tecumseh 4-stroke engine section for engine overhaul information.

DRIVE BELTS

Remove the deflector chute crank at chute and remove upper belt cover. Remove the two shoulder bolt belt guides at engine crankshaft pulley. On track drive Model TMO-35288A, place weight transfer lever in "Packed Snow" position. Remove the top bolts attaching auger housing to snowthrower frame assembly and loosen but do not remove the two bottom bolts. Lift up on auger belt to pull auger housing off of snowthrower frame and separate auger housing from the main unit.

To remove auger drive belt, remove the four shoulder bolt belt guides at au-

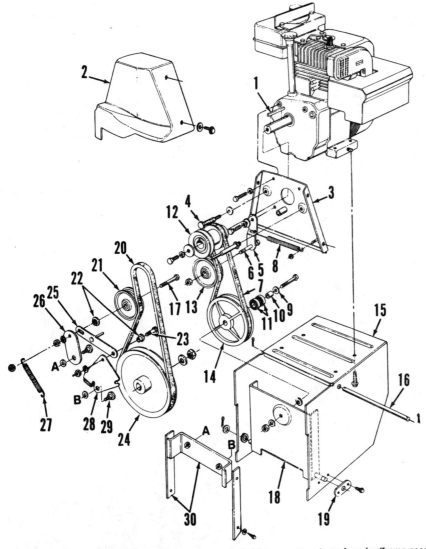

Fig. MW21—Exploded view of drive belt and auger belt drive components and engine/frame assembly. Back side of drive pulley (14) has drive surface for friction wheel; depressing drive control lever moves pulley against friction wheel (38—Fig. MW24).

1. Engine pulley key	8. Drive idler spring	16. Sliding bracket rod	23. Brake pivot bolt
2. Belt cover	9. Belleville washer	17. Auger idler bolt	24. Auger pulley
3. Engine plate	10. Spacer	18. Drive pulley support	25. Brake linkage
4. Shoulder bolt belt guide	11. Ball bearings	19. Support axle bracket	26. Auger idler bracket
5. Drive clutch idler bracket	12. Engine pulley	20. Auger belt	27. Auger idler spring
6. Idler bolt	13. Drive belt idler	21. Auger belt idler	28. Brake bracket
7. Drive belt	14. Drive pulley	22. Shoulder spacers	29. Shoulder bolt
	15. Main frame		30. Blower housing parts

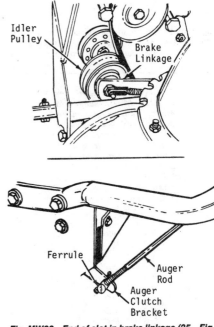

Fig. MW22—End of slot in brake linkage (25—Fig. MW21) should be against spacer on the auger belt idler when auger control lever is depressed against snowthrower handle. Refer to text for adjustment procedure.

ger pulley. The auger belt can now be removed from the bottom pulley. To remove traction drive belt, disconnect drive idler spring (8—Fig. MW21) at engine plate and remove belt from engine pulley and bottom drive pulley. Install new belt(s) and reassemble by reversing disassembly procedure. Be sure pin on end of auger clutch bracket engages slot in auger brake bracket as shown in Fig. MW23 when reassembling snowthrower. Adjust auger and drive clutches as outlined in preceding paragraphs.

FRICTION WHEEL

To renew friction disc, proceed as follows: Move shift lever to reverse position. Tip snowthrower up so it rests on front of auger housing and remove bottom cover, then remove friction wheel (38—Fig. MW24) from wheel adapter (40). Place new wheel with cup side away from adapter and install retaining bolts finger tight. Turn the wheel to be sure it does not wobble, then securely tighten retaining bolts. Replace bottom cover.

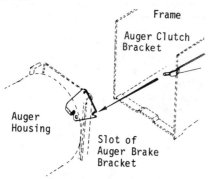

Fig. MW23—Pin on end of auger clutch bracket must engage slot in auger brake bracket when reassembling snowthrower.

Fig. MW25—Exploded view of track drive assembly for Model TMO-35288A. Drive components are similar to wheel drive models shown in Fig. MW24. Track lock handle changes auger down force and has three positions: "Transport," "Normal Snow" and "Packed Snow."

1. Track lock handle
2. Selector plate
3. Slotted link
4. Spring pin
5. Flat washer
6. Pivot arm
7. Slotted link
8. Spacer
9. Wheel axle shaft
10. Spring pin
11. Sprocket
12. Flange bearing
13. Skid shoe
14. Roller chain
15. Flange bearing
16. Drive wheel
17. Snow track
18. Idler wheel
19. Inner plate
20. Outer plate
21. Track adjuster cam

Fig. MW24—Exploded view of drive components; also see Fig. MW21. Notch in shift linkage bracket (15) engages pin on sliding bracket (44). Sliding bracket is supported by sliding bracket rod (16). On all models except Model TMO-35156A, only one flat washer (34) is used at right end of axle shaft (53).

6. Shift rod
14. Ferrule
15. Shift linkage bracket
16. Sliding bracket rod
31. Belleville washer
32. Flat washer
33. Flange bearing
33A. Flange bearing
34. Flat washer(s)
35. Spacer
36. Sprocket
37. Hi-pro key
38. Friction wheel
39. Hex shaft
40. Friction wheel adapter
41. Snap ring
42. Flat washer
43. Bearing
44. Sliding bracket
45. Flat washer
46. Roller chain
47. Sprocket
48. Snap ring
49. Sprocket
50. Flat washer
51. Bronze bearing
52. Flat washer
53. Sprocket & axle assy. (except TMO-35156A)
54. Sprocket (Model TMO-35156A)
54A. Axle (Model TMO-35156A)
55. Roller chain
56. Shoulder bolt
57. Spacer (Model TMO-35156A)
58. Flat washer (Model TMO-35156A)

59. Wheel assy. (TMO-35155A & TMO-35156A)
60. Sleeve bearing (TMO-35155A & TMO-35156A)
61. Snap ring
62. Klick pin
63. Wheel assy. (TMO-35275B & TMO-35275C)

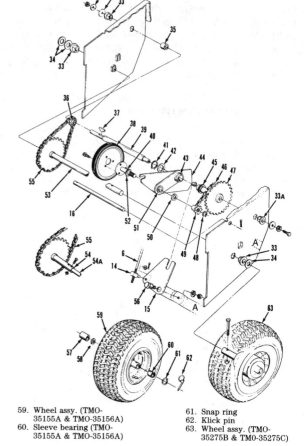

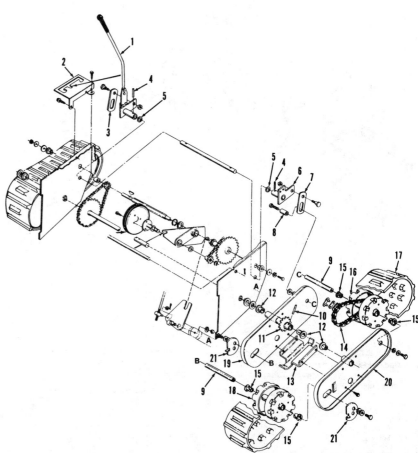

TRACK DRIVE

To renew drive track or chain inside track assembly on track drive Model TMO-35288A, block up snowthrower frame so track assemblies are off the ground. Loosen the cap screws retaining track adjuster cams (21—Fig. MW25) at each side of assembly and turn adjusters so all track tension is removed. Roll the track off of track rollers and side supports. Remove roller chain (14) by disconnecting master link. Reassemble by reversing disassembly procedure, making sure to install track with cleats pointing toward front of snowthrower. Adjust track tension as outlined in adjustment paragraph.

AUGER GEARBOX

To remove auger gearbox, follow procedure outlined in preceding paragraph for removal of auger drive belt. Then, remove auger drive pulley (7—Fig. MW26) and ball bearing (10) from rear of impeller shaft (29). Unbolt the auger bearing housings (37) at each end of auger housing. Remove the gearbox and auger flight assembly from front of auger housing, then remove impeller from input shaft and auger flights from driveshaft. Note the position of washers and spacers as you are disassembling the unit.

To disassemble gearbox, refer to exploded view in Fig. MW25 and remove the screws retaining upper gear housing half (19). Note positions of thrust washers and spacers as you disassemble gearbox. Clean and inspect all parts and renew any worn or damaged parts. Lubricate all parts prior to reassembly and pack gearbox with 10 ounces (295 mL) of Shell Alvania Grease EPR00.

1. Upper chute
2. Lower chute
3. Flat washers
4. Bushing
5. Chute crank
6. Pulley nut
7. Auger drive pulley
8. Hi-pro key
9. Spacer
10. Ball bearing
11. Shoulder bolt
12. Chute flange keeper
13. Chute crank bracket
14. Auger housing
15. Skid shoes
16. Scraper bar (shave plate)
17. Spring pin
18. Impeller (blower fan)
19. Upper gear housing half
20. "O" ring
21. Plastic flange bearing
22. Flat washer
23. Hex nut
24. Flat washer
25. Pinion gear
26. Flat washer(s)
27. Flange bearing
28. Sleeve bearing
29. Impeller (blower) axle
30. Bevel gear
31. Lower gear housing half
32. Auger flight (spiral) axle
33. Key (Hi-pro or square)
34. Flat washer
35. Flange bearing
36. Shear bolt nut
37. Bearing housing
38. Shear bolt
39. Auger flight (spiral)
40. Flange bearing

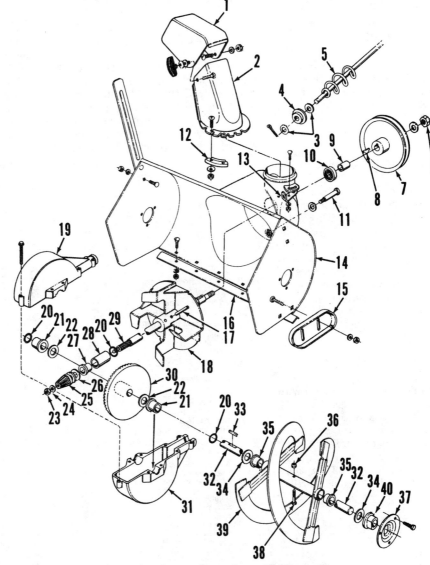

Fig. MW26—Exploded view of auger housing, gearbox and auger flight assembly.

MONTGOMERY WARD

Model	Engine Make	Engine Model	Self-Propelled?	No. of Stages	Scoop Width
TMO-35268B	B&S	190402	Yes	2	26 in. (660 mm)
TMO-35268C	B&S	190402	Yes	2	26 in. (660 mm)
TMO-35157A	Tecumseh	HMSK80	Yes	2	28 in. (711 mm)

LUBRICATION

ENGINE

Refer to appropriate Briggs & Stratton or Tecumseh 4-stroke engine section for engine lubrication information. Recommended fuel is leaded or unleaded regular gasoline.

DRIVE MECHANISM

Lubricate drive chains and sprockets at least once each season or after every 25 hours of operation with a good-quality chain oil. Auger shaft bushings, traction drive shaft bushings and axle shaft bushings should be oiled at least once each season or after every 25 hours operation. Remove wheels and grease axle shaft at least once each season. Lubricate control rod and lever pivot points to ensure smooth operation.

AUGER GEARBOX

The auger gearbox is lubricated with Shell Alvania grease EPROO. Remove filler plug (30—Fig. MW33) and check lubricant level using a wire as a dipstick. Read lubricant level on wire. Add lubricant through plug opening to bring gearbox level up to about 1/2 full. Capacity is 4 ounces (120 mL), do not overfill.

MAINTENANCE

Inspect drive belts and pulleys and renew belts if cracked, frayed, burned or otherwise damaged. Renew pulleys if belt grooves are excessively worn or damaged or if idler pulley bearings are loose or rough.

ADJUSTMENTS

AUGER HEIGHT

Auger height can be adjusted by loosening skid shoe mounting bolts and repositioning the skid shoes. Auger should be in lowest setting (skid shoes raised) when operating over smooth paved surfaces or raised (skid shoes lowered) when operating over rough surfaces.

Fig. MW30—Exploded view of handle and controls. Models produced in 1986 also had a throttle control assembly mounted in handle panel (20).

1. Shoulder nut
2. Shoulder bolt
3. Plastic bushing
4. Drive clutch lever
5. Handle grip
6. Drive clutch rod
7. Bumper
8. Upper L.H. handle
9. Chute crank
10. Lower L.H. handle
11. Main frame cover
12. Drive clutch spring
13. Adjusting nut
14. Drive clutch bracket
15. Shift linkage bracket
16. Auger clutch bracket
17. Auger clutch spring
18. Adjusting nut
19. Auger clutch spring
20. Handle panel
21. Shift rod
22. Auger clutch rod
23. Lower R.H. handle
24. Shifting lever
25. Upper R.H. handle
26. Auger clutch lever
27. Shift lever spring

CLUTCH RODS

With auger clutch lever (26—Fig. MW30) released against bumper (7), lower end of auger clutch spring (19) should fit loosely in the hole at rear of auger clutch bracket (16). If not, disconnect spring and slide spring upward on auger clutch rod (22) to expose adjusting nut (18) and turn nut up or down so bottom of hook in spring will be even with middle of hole in bracket when lifting clutch rod up to take slack out of rod

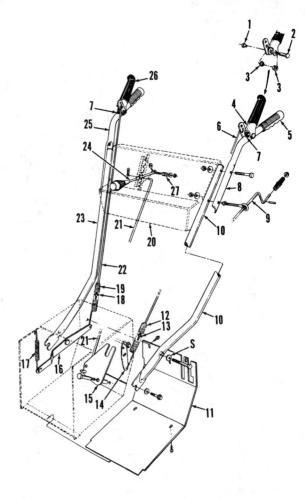

and spring. There must not be any tension on clutch rod spring with the clutch lever in disengaged position. Repeat adjustment procedure for drive clutch rod (6).

SHIFT ROD

Move shift lever to neutral position on handle panel; sliding bracket assembly (13—Fig. MW32) should be aligned with slot (S—Fig. MW30) in main frame cover (11). Also, with the drive clutch lever (4) in engaged position, the snowthrower should roll back and forth with shift lever in neutral. If necessary to adjust shift rod, disconnect shift rod (21) from shift lever (24) at control console. Move shift lever to neutral position and move shift rod until sliding bracket is aligned with slot in main frame cover. If cover is removed, sliding bracket should be aligned on center of neutral stop bolt (19—Fig. MW32). Thread shift rod in or out of ferrule (41) in shift bracket (40) at bottom end of rod until top end aligns with hole in shift lever. Reconnect shift rod and check adjustment.

Shift lever spring (27—Fig. MW30) should be tight enough to hold lever in handle panel detent notches, yet allow lever to move without binding on detents.

quired to separate the auger housing from main frame. With someone holding handles, push in on auger idler pulley to lift brake pad (BR) out of pulley groove, then lift auger housing from main frame. At this point, either the auger belt or drive belt, or both, can be renewed.

To remove auger belt after separating front and rear units, remove the five shoulder bolts (5—Fig. MW33) and Belleville washers from auger housing and remove belt from pulley. An alternate method is to remove the cap screw holding auger drive pulley to impeller shaft and remove the pulley and belt; be careful not to lose the drive key and Belleville washer. Install new belt by reversing removal procedure. If auger pulley was removed, be sure key is in place on shaft and install Belleville washer with cup side toward pulley. Tighten pulley retaining cap screw securely.

To remove traction drive belt, unhook drive idler spring (16—Fig. MW31) from engine plate, then remove the belt from drive pulley. Remove shoulder bolt (4), spacer (5) and Belleville washer at right side of engine pulley and remove the belt. Install new belt by reversing removal procedure.

Be sure that idler pulley and/or brake pad are not behind the drive pulley

when assembling auger housing to main frame.

DRIVE MECHANISM

To overhaul drive mechanism, first remove drive belt as outlined in a preceding paragraph, then remove engine assembly and traction drive pulley (19—Fig. MW31). Further disassembly procedure is evident after inspection of unit and reference to Fig. MW32.

Be sure not to over tighten the hex shaft self-aligning bearing (51—Fig. MW32) as hex shaft (50) needs to flex in the bearing for free movement of sliding bracket (13). If adjusting auger clutch rod (37) with main frame cover removed, forward point of sliding bracket (13) should be centered on head of neutral stop bolt (19) when shift lever on handle panel is in neutral position.

FRICTION WHEEL

To remove rubber faced friction wheel, first tip snowthrower forward onto front of auger housing and remove the cover from main frame. Friction wheel (5—Fig. MW32) can then be unbolted from friction wheel adapter (4) and removed from the end of hex shaft. Install new friction wheel with cup side

OVERHAUL

ENGINE

Engine make and model numbers are listed at beginning of this section. Refer to Tecumseh 4-stroke engine section for engine overhaul information.

DRIVE BELTS

To renew drive belts, it is first necessary to split the snowthrower between the auger housing and main frame assembly as follows: Disconnect the deflector chute crank at the chute end and remove drive belt cover. Remove shoulder bolt (17—Fig. MW31) and Belleville washer at left side of engine pulley. Remove shoulder bolt (4) and spacer (5) at right hand side of engine pulley, being careful not to lose the Belleville (spring cup) washer located between idler bracket and engine plate. Slip auger drive belt off engine pulley and loosely install shoulder bolt, roller and Belleville washer at right side of engine pulley. Remove the top cap screws retaining auger housing flanges (24) to main frame (21). Two people are re-

Fig. MW31—Exploded view of belt drive system. Auger pulley (1) is also shown in Fig. MW33 and traction drive pulley (19) mounts on drive plate spindle (35—Fig. MW32).

1. Auger pulley
2. Auger belt
3. Auger belt idler
4. Shoulder bolt
5. Spacer
6. Auger idler bracket
7. Auger clutch rod
8. Engine pulley
9. Traction drive belt
10. Belt cover
11. Square key
12. Auger brake linkage
13. Traction drive idler
14. Shoulder bolt
15. Drive idler bracket
16. Drive idler spring
17. Shoulder bolt
18. Engine spacer plate
19. Traction drive pulley
20. Engine bracket assy.
21. Main (motor mount) frame
22. Brake bracket assy.
23. Shoulder bolt
24. Auger housing flanges

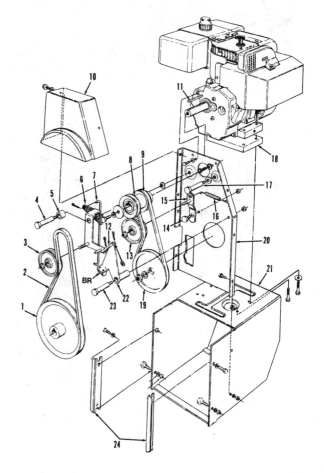

facing toward adapter and install retaining lockwashers and nuts finger tight. Turn the wheel to be sure it does not wobble (cocked on hub), then securely tighten nuts.

AUGER SHEAR BOLTS

The auger flights are driven through shear bolts (37—Fig. MW33) which will shear when a solid obstruction is encountered to prevent damage to snowthrower. If this occurs, be sure to replace the bolts with factory replacement parts; substitute bolts may not shear and cause extensive damage. Before installing new bolts, oil the auger flight bushings and be sure flights spin freely on the auger shaft.

AUGER GEARBOX

To remove auger gearbox, first remove auger belt as outlined in a preceding paragraph. If auger pulley (1—Fig. MW33) was not removed, remove it at this time. Remove outer bearing retainer (3), then remove set screws and unlock bearing (4) from gearbox input shaft. Unbolt auger shaft bearing flanges (39) at each end of auger housing, remove flange bearings (38) and thrust washers (34) and withdraw the gearbox with impeller (10) and auger flights (36). Remove shear bolts (37), auger flights, inner thrust washers and spacers (33) from auger shaft. Remove the carriage bolt and spring pins (9), then pull impeller from shaft. Remove the bolts holding the gearbox halves (19 and 31) together and disassemble gearbox. Clean and inspect all parts and renew any excessively worn or damaged parts. Worm gear (27) can be removed from hub (28).

Reassemble by reversing disassembly procedure and lubricate gearbox using 4 ounces (120 mL) of Shell Alvania Grease EPROO.

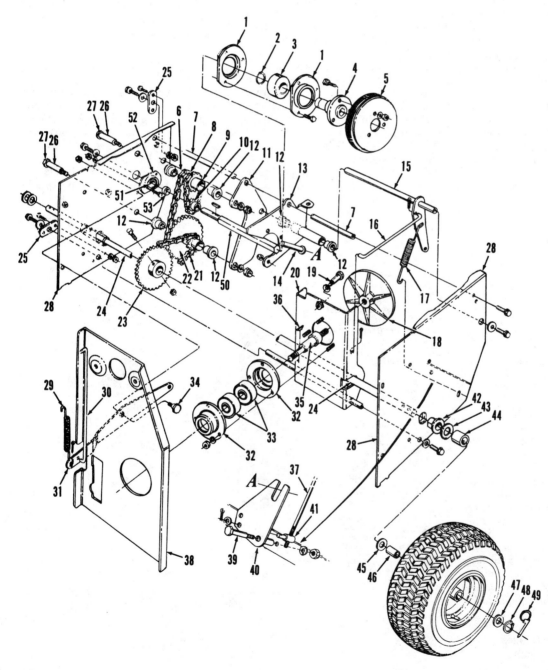

Fig. MW32—Exploded view of friction drive and chain speed reduction drive components.

1. Bearing housing
2. Snap ring
3. Ball bearing
4. Friction wheel adapter
5. Friction wheel
6. Roller chain
7. Sliding support axle
8. Roller chain
9. Double sprocket
10. Hex shaft sprocket
11. Chain support bracket
12. Flange bearing
13. Sliding bracket assy.
14. Pin (part of 13)
15. Drive clutch bracket
16. Drive clutch rod
17. Drive clutch spring
18. Aluminum drive plate
19. Neutral stop bolt
20. Drive pivot plate
21. Roller chain
22. Double sprocket
23. Axle sprocket
24. Axle shaft
25. Pivot shaft bracket
26. Sprocket hub tube
27. Sprocket bolt
28. Main frame assy.
29. Auger clutch spring
30. Auger clutch rod
31. Auger clutch lever
32. Bearing housing
33. Ball bearings
34. Shoulder bolt
35. Drive plate spindle
36. Hi-pro key
37. Auger clutch rod
38. Engine bracket assy.
39. Shoulder bolt
40. Shift bracket
41. Ferrule
42. Axle flange bearing
43. Spacer washer
44. Spacer
45. Wheel spacer washer
46. Wheel bushing
47. Wheel spacer washer
48. Snap ring
49. Klick pin
50. Hex shaft
51. Bearing
52. Bearing housing
53. Spacer

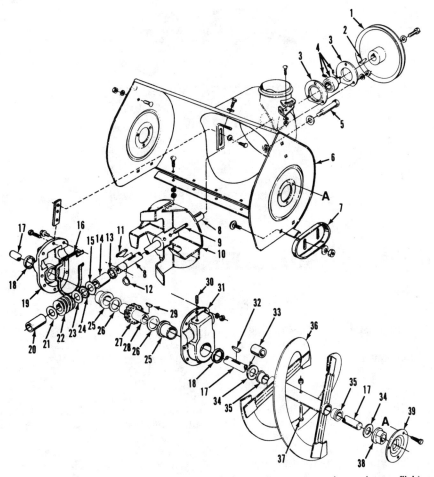

Fig. MW33—Exploded view of auger housing, impeller, auger gearbox and auger flights.

1. Auger pulley	11. Hi-pro key	21. Flat washer	30. Filler plug
2. Hi-pro key	12. Snap ring	22. Worm gear	31. Housing half
3. Bearing retainer	13. Oil seal	23. Flat washer	32. Woodruff key
4. Self-aligning bearing	14. Sleeve bearing	24. Thrust bearing	33. Spacer
5. Shoulder bolt	15. Flat washer	25. Flange bearing	34. Thrust washer
6. Auger housing	16. Gasket	26. Flat washer	35. Plastic bushing
7. Skid shoe	17. Auger shaft	27. Worm gear	36. Auger flight
8. Impeller shaft	18. Oil seal	28. Gear hub	37. Shear bolt
9. Spring pin	19. Housing half	29. Woodruff key	38. Flange bearing
10. Impeller	20. Sleeve bushing		39. Bearing flange

NOMA

NOMA OUTDOOR PRODUCTS, INC.
210 American Dr., P.O. Box 7000
Jackson, TN 38308-7000

Model	Engine Make	Engine Model	Self-Propelled?	No. of Stages	Scoop Width
420T	Tecumseh	HS40	Yes	2	20 in. (508 mm)

LUBRICATION

ENGINE

Refer to Tecumseh 4-stroke engine section for engine lubrication information. Recommended fuel is regular or unleaded regular gasoline.

DRIVE MECHANISM

Recommended lubrication period is after every 10 hours of operation and before storing unit at end of season. Oil the drive chains and sprockets with SAE 10W-30 oil. The hex shaft assembly in the variable speed friction drive requires no lubrication except to wipe the hex shaft and gears with SAE 10W-30 motor oil at end of season to prevent rusting. The hex shaft bearings are of the sealed type. Be careful not to get any grease or oil on the disc drive plate or friction wheel; clean these parts thoroughly if oil or grease comes in contact with these parts.

Lubricate all moving part pivot points with SAE 10W-30 oil. Lubricate weight transfer plate and pivot points with clinging type grease such as "Lubriplate."

The auger flights do not have a grease fitting; before storage at end of season, remove the auger shear bolts and squirt oil into shear bolt holes. Turn the auger on shaft to distribute the oil throughout length of auger shaft.

AUGER GEARBOX

The auger gearbox is factory lubricated for life and unless grease leaks out, lubrication should not be necessary unless gearbox is being overhauled. Some models have grease plug (33—Fig. N4) that can be removed. Use a wire as a dipstick to check for presence of grease.

ADJUSTMENTS

AUGER HEIGHT

Lower the adjusting skids to raise auger scraper bar on rough surfaces. On paved surfaces, the skids can be raised to bring the auger assembly down. Keep in mind that the scraper bar below the auger should be 1/8 inch (3.2 mm) above sidewalk or other paved area to be cleaned. This clearance can be obtained by placing spacers (the extra shear pins stored under snowthrower handle can be used for spacers) under the scraper bar, then adjusting the skids so they contact paved surface. To compensate for wear, the scraper bar position is adjustable, and the bar can be reversed or replaced. Be sure the bar is adjusted parallel with the working surface, and for proper clearance, above skid shoes.

DRIVE BELTS

The traction drive belt is spring loaded and does not require adjustment; check control cable adjustment or replace belt if it is slipping. The auger drive belt can be adjusted after remov-

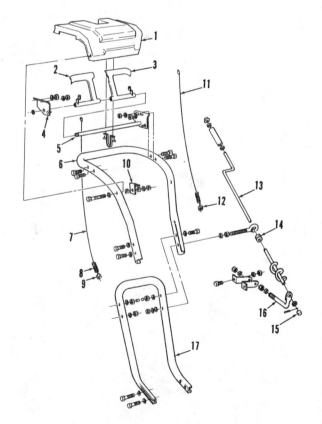

Fig. N1—Exploded view of control and handle assembly.

1. Control panel
2. Auger clutch lever
3. Drive clutch lever
4. Lever rod bracket
5. Rod & bracket assy.
6. Upper handle
7. Auger cable
8. Auger clutch spring
9. Locknut
10. Panel bracket
11. Drive clutch cable & spring
12. Locknut
13. Chute crank assy.
14. Grommet
15. Plastic cap
16. Chute crank bracket
17. Lower handle

ing belt cover, then proceed as follows: Loosen nut on auger belt idler (16—Fig. N3) and move idler in slot of idler arm (14) toward belt about 1/8 inch (3 mm). Tighten nut, engage auger drive clutch lever and check belt adjustment; belt should deflect about 1/2 inch (12.7 mm) with moderate pressure. Reset idler pulley until proper belt tension is obtained with drive engaged. The belt guides at engine pulley should be adjusted to clear belt 3/32 inch (2.4 mm) with auger drive clutch engaged.

CONTROL CABLES

Refer to Fig. N1 and with auger cable (7) "z" fitting disconnected from auger clutch lever (2), move lever forward to contact plastic bumper and pull upward on cable; the center of the "z" fitting should be between the top and the center of the hole in the lever. If not, slide cable down through auger clutch spring (8) and turn lock nut (9) to obtain proper adjustment. Then, pull cable up through spring and reconnect "z" fitting on cable in hole of lever. Repeat the procedure for drive clutch cable (11) adjustment.

FRICTION WHEEL

Remove bottom panel from snow-thrower. Place shifter lever in first (Position 1) gear and check position of friction wheel on disc drive plate. Distance from left outer side of disc drive plate to right side of friction wheel should be 2-3/4 inches (70 mm) as shown in Fig. N2. If necessary to adjust position of friction wheel, loosen bolts retaining upper speed shift control lever (29—Fig. N3) to shift lever adapter (32). Then move friction wheel to proper position and tighten lever bolts.

TRACK DRIVE

To check track tension, pull up gently on center of track at midpoint between track rollers. Distance between track and top of track support frame should then be no more than 1-1/4

inches (32 mm). To adjust track, loosen jam nuts (3—Fig. N5) on adjusting bolt (11) at rear end of track frame (4) and tighten or loosen adjusting bolt as necessary. Adjust opposite track in same manner. Be sure adjusting bolt jam nuts are securely tightened.

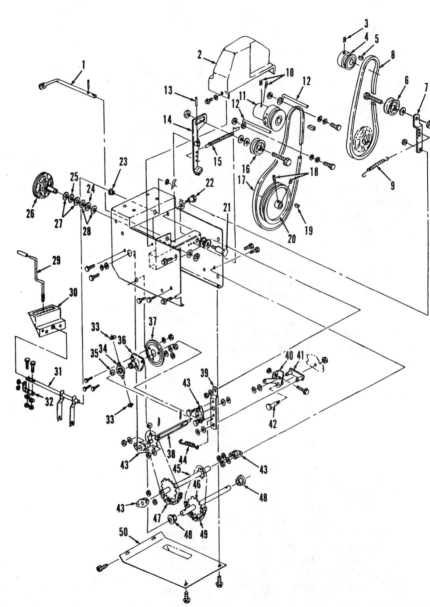

Fig. N3—Exploded view of drive mechanism. The traction drive is from engine PTO pulley (4) and the auger is driven from engine crankshaft pulley (11). Drive disc (26) and shaft are separate items on some units.

1. Auger clutch lever
2. Belt cover
3. Set screw
4. Engine PTO pulley
5. Woodruff key
6. Drive idler pulley
7. Drive idler arm
8. Drive belt
9. Idler arm spring
10. Set screws
11. Engine crankshaft pulley
12. Belt guide
13. Roll pin
14. Auger brake & idler arm
15. Auger brake arm spring
16. Auger belt idler
17. Auger belt
18. Set screws
19. Woodruff key
20. Auger drive pulley
21. Shoulder bolt
22. Shoulder bolt
23. Plastic bushing
24. Thrust washer
25. Needle thrust bearing
26. Drive disc
27. Thrust washers
28. Needle thrust bearings
29. Shift lever
30. Shift bracket
31. Shift yoke
32. Shift lever/yoke adapter
33. Flat washers
34. Trunion bearing
35. Snap ring
36. Friction wheel hub
37. Friction wheel
38. Hex shaft
39. Drive clutch lever
40. Drive lever assy.
41. Drive lever link
42. Shoulder bolt
43. Self-centering bearing
44. Return spring
45. Jackshaft assy.
46. Wheel axle & sprocket
47. Roller chain
48. Axle bearing
49. Roller chain
50. Bottom cover panel

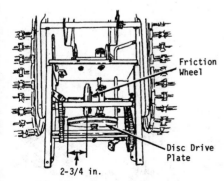

Fig. N2—View showing measurement to check friction wheel adjustment.

Friction Wheel

Disc Drive Plate

2-3/4 in.

OVERHAUL

ENGINE

Engine make and model numbers are listed at beginning of section. Refer to Tecumseh engine section for engine overhaul.

AUGER DRIVE BELT

To remove auger drive belt, remove belt cover (2—Fig. N3), loosen the right-hand belt guide (12) and move it away from drive pulley. Loosen auger belt idler (16) and move it away from the belt. Engage auger drive clutch lever to move brake away from belt and slip the belt out from between auger pulley and brake. Remove the top two bolts securing auger housing to motor mount frame. Place weight transfer system pedal in transport position (latched in top notch). Slightly loosen bottom two bolts attaching auger housing to motor mount frame. Separate auger housing and motor mount frame by hinging on bottom two bolts. Remove belt from engine auger drive pulley. Install new belt by reversing removal procedure. Place weight transfer system in lowest notch in bracket to pivot auger housing and motor mount frame back into position. Adjust belt as outlined in preceding paragraph.

TRACTION DRIVE BELT

To remove traction drive belt, first remove upper belt cover, pull traction idler pulley back and slip belt past idler pulley, then remove belt from engine pulley (It may be necessary to loosen and remove engine auger drive pulley for clearance to remove belt). Place speed selector in sixth gear position and remove belt from between rubber drive disc and the combination drive pulley and disc drive plate. Install new belt by reversing removal procedure. If engine auger drive pulley was removed, be sure auger drive pulleys are aligned and drive key is flush with end of engine shaft before tightening engine pulley set screw. Adjust belts as outlined in belt adjustment paragraph.

AUGER SHEAR BOLT

The augers are driven from the auger shaft through special bolts (37—Fig. N4) that are designed to break to protect the machine if an object becomes lodged in the auger housing. Only original equipment replacement shear bolts should be used for safety reasons. Before installing new bolt, squirt oil between auger flight and auger shaft and turn auger flight to distribute oil throughout the full length of shaft.

AUGER ASSEMBLY

To remove auger, first remove auger drive belt as outlined in previous paragraph. Remove auger drive pulley and

Fig. N4—Exploded view of auger housing, gearbox assembly and augers.

1. Deflector chute
2. Hand guard (not all models)
3. Chute extension
4. Mounting flange
5. Flange retainer
6. Impeller shaft bearing
7. Auger housing
8. Left side plate
9. Right side plate
10. Auger shaft bearing
11. Skid shoe
12. Scraper bar
13. Oil seal
14. Bushing
15. Thrust washers
16. Needle bearing
17. Roll pin
18. Thrust collar
19. Impeller shaft
20. Right case half
21. Oil seal
22. Bushing
23. Thrust washer
24. Woodruff key
25. Worm gear
26. Bushing
27. Gasket
28. Auger shaft
29. Thrust washer
30. Bushing
31. Oil seal
32. Left case half
33. Grease plug
34. Impeller bolts
35. Impeller assy.
36. Locknuts
37. Auger shear bolts
38. Left auger assy.
39. Right auger assy.

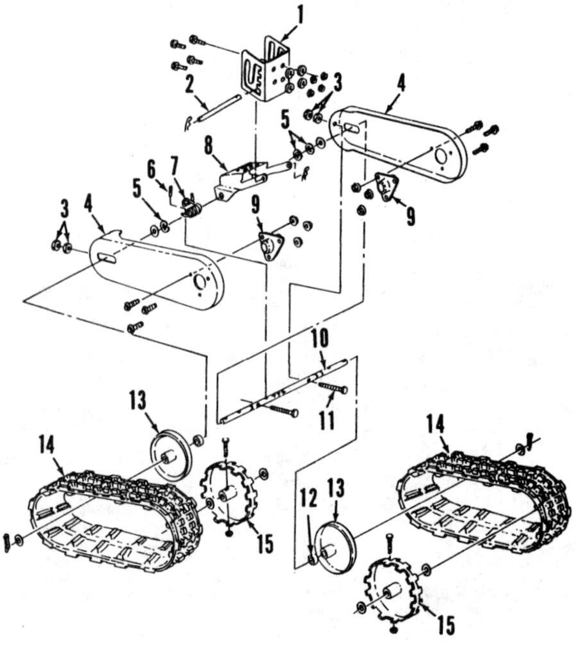

impeller shaft bearing (6—Fig. N4) from rear of housing, unbolt auger shaft bearings (10) at each side of housing and remove the auger and gearbox assembly. Remove the auger drive shear bolts (37) and slide the auger flights from auger shaft. Remove impeller bolts (34) and pull impeller from shaft. Remove the bolts holding the case halves (20 & 32) together and disassemble gearbox. Clean and inspect all parts and renew any worn or damaged parts. Coat the

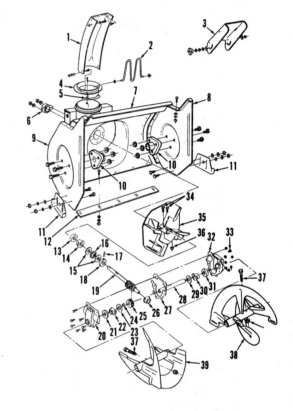

Fig. N5—Exploded view of track drive assembly. Track drive wheels (15) mount on axle shaft (46—Fig. N3). Foot pedal (8) operates weight transfer system.

1. Mounting bracket
2. Pivot pin
3. Jam nuts
4. Track frame
5. Snap ("E") rings
6. Roll pin
7. Foot pedal spring
8. Foot pedal
9. Frame support bearing
10. Idler axle shaft
11. Track adjusting bolt
12. Spacer
13. Track idler wheel
14. Rubber track
15. Track drive wheel

parts with clean grease and pack the gearbox with approximately 3 ounces (85 grams) grease on assembly. Some recommended greases are Benalene #372, Shell Darina 1, Texaco Thermatex EP1, and Mobiltem 78.

FRICTION DRIVE WHEEL

To replace friction wheel, proceed as follows: Stand snowthrower up on au-ger housing end and remove bottom cover panel (50—Fig. N3). Unbolt bearings (43) from main frame at each end of hex shaft (38). Slide the hex shaft with friction wheel, trunnion bearing and end bearings out of snowthrower. Be careful not to lose flat washers (33) from studs on trunnion bearing (34). Unbolt friction wheel (37) from hub and install new wheel with cup side away from hub and reassemble by reversing removal procedure. Check friction wheel adjustment as outlined in preced-ing paragraph before reinstalling bottom panel.

TRACK ASSEMBLY

Refer to exploded view of track assembly in Fig. N5. Disassembly and overhaul is evident after inspection of unit and reference to exploded view. Readjust track as outlined in previous paragraph.

NOMA

Model	Engine Make	Engine Model	Self-Propelled?	No. of Stages	Scoop Width
523	Tecumseh	HS50	Yes	2	23 in. (584 mm)
825	Tecumseh	HS80	Yes	2	25 in. (635 mm)
825T	Tecumseh	HS80	Yes	2	25 in. (635 mm)

LUBRICATION

ENGINE

Refer to Tecumseh engine section for engine lubrication information. Recommended fuel is leaded or unleaded regular gasoline.

DRIVE MECHANISM

Recommended lubrication period is after every 10 hours of operation and before storing unit at end of season. Oil the drive chains and sprockets with SAE 10W-30 oil. The hex shaft assembly in the variable speed friction drive requires no lubrication except to wipe the hex shaft and gears with SAE 10W-30 motor oil at end of season to prevent rusting. The hex shaft bearings are of the sealed type. Be careful not to get any grease or oil on the disc drive plate or friction wheel; clean these parts thoroughly if oil or grease comes in contact.

Lubricate all moving part pivot points with SAE 10W-30 oil. On track drive model, lubricate weight transfer plate and pivot points with clinging type grease such as "Lubriplate."

Using a hand-operated grease gun, lubricate the auger shaft grease fittings after every 10 hours of operation.

AUGER GEARBOX

Once each season, remove the grease level plug (17—Fig. N15) and, if grease is not visible, use a wire as a dipstick to check for grease in gearbox. Recommended lubricants are Benalene #372, Shell Darina 1, Texaco Thermatex EP1 and Mobiltem 78.

ADJUSTMENTS

AUGER HEIGHT

Lower the adjusting skids to raise auger scraper bar on rough surfaces.

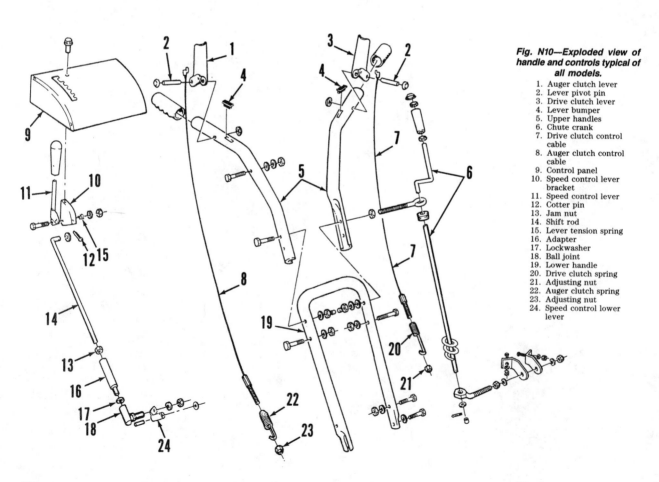

Fig. N10—Exploded view of handle and controls typical of all models.

1. Auger clutch lever
2. Lever pivot pin
3. Drive clutch lever
4. Lever bumper
5. Upper handles
6. Chute crank
7. Drive clutch control cable
8. Auger clutch control cable
9. Control panel
10. Speed control lever bracket
11. Speed control lever
12. Cotter pin
13. Jam nut
14. Shift rod
15. Lever tension spring
16. Adapter
17. Lockwasher
18. Ball joint
19. Lower handle
20. Drive clutch spring
21. Adjusting nut
22. Auger clutch spring
23. Adjusting nut
24. Speed control lower lever

NOTE: On track drive models, the weight transfer system will not operate with skid plates adjusted for maximum ground clearance.

On paved surfaces, the skids can be raised to bring the auger assembly down. Keep in mind that the scraper bar below the auger should be 1/8 inch (3.2 mm) above sidewalk or other paved area to be cleaned. This clearance can be obtained by placing spacers under the scraper bar, then adjusting the skids so they contact paved surface. To compensate for wear, the scraper bar position is adjustable, and the bar can be reversed or renewed. Be sure the bar is adjusted parallel with the working surface and for proper clearance, above skid shoes.

DRIVE BELTS

The traction drive belt is spring loaded and does not require adjustment; check control cable adjustment or renew belt if it is slipping. The auger drive belt can be adjusted after removing belt cover, then proceed as follows: Loosen nut on idler pulley (18—Fig. N11) and move idler in idler arm slot toward belt

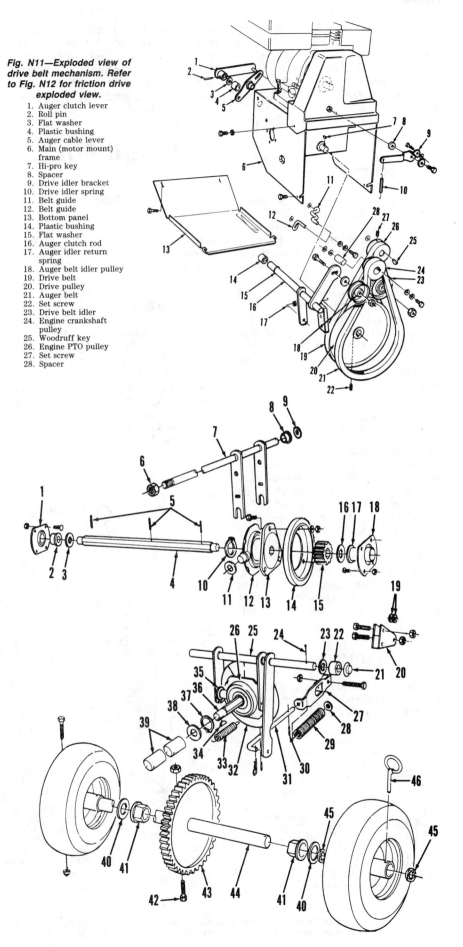

Fig. N11—Exploded view of drive belt mechanism. Refer to Fig. N12 for friction drive exploded view.

1. Auger clutch lever
2. Roll pin
3. Flat washer
4. Plastic bushing
5. Auger cable lever
6. Main (motor mount) frame
7. Hi-pro key
8. Spacer
9. Drive idler bracket
10. Drive idler spring
11. Belt guide
12. Belt guide
13. Bottom panel
14. Plastic bushing
15. Flat washer
16. Auger clutch rod
17. Auger idler return spring
18. Auger belt idler pulley
19. Drive belt
20. Drive pulley
21. Auger belt
22. Set screw
23. Drive belt idler
24. Engine crankshaft pulley
25. Woodruff key
26. Engine PTO pulley
27. Set screw
28. Spacer

Fig. N12—Exploded view of disc drive mechanism.

1. Bearing plate
2. Bearing
3. Flat washer
4. Hex shaft
5. Roll pins
6. Locknut
7. Speed control yoke
8. Plastic bushing
9. Washer
10. Snap ring
11. Washer
12. Trunnion bearing
13. Friction wheel hub
14. Friction wheel
15. Pinion gear
16. Flat washer
17. Bearing
18. Bearing plate
19. Spacers
20. Traction pivot
21. Push nut
22. Bushing
23. Flat washer
24. Cotter pin
25. Traction shaft
26. Trunnion bearing
27. Traction spring bracket
28. Flat washer
29. Traction spring
30. Spring pin
31. Traction rod
32. Drive disc
33. Return spring
34. Hi-pro key
35. Washer
36. Drive disc shaft
37. Snap ring
38. Thrust bearing
39. Needle bearings
40. Washer
41. Axle bearing
42. Bolt
43. Driven gear
44. Axle
45. Grip rings
46. Klick pin

about 1/8 inch (3 mm). Tighten nut, engage auger drive clutch lever and check belt adjustment; belt should deflect about 1/2 inch (12.7 mm) with moderate pressure. Reset idler pulley until proper belt tension is obtained with drive engaged. Belt guides (11 & 12) should be adjusted to clear belt 3/32 inch (2.4 mm) with auger drive clutch engaged.

CONTROL CABLES

Refer to Fig. N10 and with drive clutch control cable (7) "z" fitting disconnected from lever (3), move lever forward to contact plastic bumper (4); the center of the "z" fitting should be between the top and the center of the hole in the lever. If not, slide cable down through spring (20), hold square part of cable end from turning and rotate adjusting nut (21) to obtain correct cable length. Then, pull cable up through spring and reconnect "z" fitting on cable to hole in lever. Repeat adjustment procedure for auger clutch control cable (8).

FRICTION WHEEL

Remove bottom panel from snowthrower. Place speed control lever in first gear and check position of friction wheel on disc drive plate. The right side of friction wheel should be 3-3/8 inches (86 mm) from outer edge of disc drive plate as shown in Fig. N13. If adjustment is necessary, loosen jam nut (13—Fig. N10) on speed select rod. Remove ball joint (18) from speed control lower lever (24) and turn adapter (16) to obtain correct friction wheel position. Reinstall ball joint, tighten jam nut and reinstall bottom panel.

TRACKS

Refer to Fig. N16 and loosen cap screw (13) on each side of track assembly. Rotate adjusting cam (14) equally on each side of assembly so deflected distance between top of track plate and inside of track is not greater than 2 inches (50.8 mm). Uneven adjustment of adjusting cams will result in a twist in the rubber track. Repeat adjustment for opposite track assembly.

OVERHAUL

ENGINE

Engine make and model numbers are listed at beginning of section. Refer to Tecumseh engine section for engine overhaul.

AUGER DRIVE BELT

To remove auger drive belt, remove belt drive cover and loosen belt guides (11 & 12—Fig. N11) and move them away from engine crankshaft pulley (24). Loosen auger belt idler pulley (18) and move it away from the belt. Engage auger drive clutch lever to move brake away from belt and slip the belt off auger pulley. Remove belt from engine drive pulley. Reverse removal procedure to install new belt and adjust belt and belt guides as outlined in a preceding paragraph.

TRACTION DRIVE BELT

To remove traction drive belt, remove belt drive cover and loosen the left-hand belt guide (11—Fig. N11) and move guide away from engine PTO pulley. Pull spring loaded traction idler pulley back and slip belt past idler pulley. Remove belt from engine pulley and up between the auger and traction drive pulleys. Reverse removal procedure to install new belt and adjust belt and belt guide as outlined in a preceding paragraph.

AUGER SHEAR BOLT

The augers are driven via the auger shaft through special shear bolts (16—Fig. N14) that are designed to shear to protect the machine if an object becomes lodged in the auger housing. Only original equipment replacement shear bolts should be used for safety reasons. When replacing a shear bolt, be sure to install new shear bolt spacer (17) and,

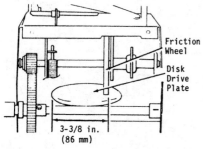

Fig. N13—With speed control lever in first gear, friction wheel should be positioned as shown.

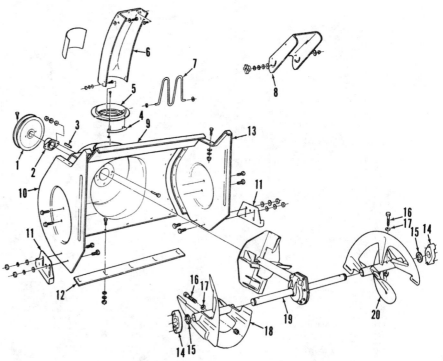

Fig. N14—Exploded view of auger housing assembly.

1. Auger drive pulley	
2. Impeller shaft bearing	
3. Square key	7. Hand guard
4. Chute retainer	8. Upper chute extension
5. Chute flange	9. Center auger housing
6. Bottom chute extension	10. R.H. side plate

11. Skid shoe	16. Shear bolt
12. Scraper bar	17. Shear bolt spacer
13. L.H. side plate	18. R.H. auger flight
14. Auger shaft bushing	19. Gearbox assy.
15. Spacer washer	20. L.H. auger flight

using a hand operated grease gun, lubricate auger shaft grease fittings.

AUGER ASSEMBLY

To remove auger, remove auger drive belt, disconnect the chute crank from

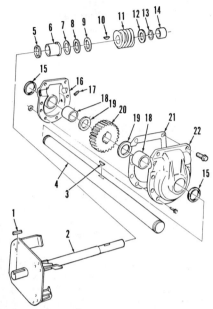

Fig. N15—Exploded view of auger gearbox. Impeller is an integral part of impeller shaft (2).

1. Square key
2. Impeller & shaft assy.
3. Woodruff key
4. Auger shaft
5. Square cut "O" ring
6. Impeller shaft bushing
7. Thrust bearing race
8. Needle thrust bearing
9. Thrust bearing race
10. Woodruff key
11. Worm gear
12. Thrust washer
13. Snap ring
14. Impeller shaft bushing
15. Oil seals
16. L.II. housing half
17. Grease plug
18. Auger shaft bushing
19. Thrust washer
20. Gear
21. Gasket
22. R.II. housing half

auger housing and separate the auger housing from main (motor mount) frame. Remove auger drive pulley (1—Fig. N14) and impeller shaft bearing (2) from rear of housing. Unbolt auger shaft bushings (14) at each side of housing and remove the assembly. Remove the auger drive shear bolts (16) and remove the auger flights from auger shaft.

With auger flights removed, refer to Fig. N15 for exploded view of auger drive gearbox. To disassemble, remove the cap screws holding gearbox housing halves (16 & 22) and separate the unit. Remove impeller shaft bushing (14), snap ring (13) and thrust washer (12) from impeller shaft. Press shaft from worm gear (11) and remove needle thrust bearing (8) and races (7 & 9), rear impeller shaft bushing (6) and "O" ring (5) from impeller shaft. Clean and inspect all parts and renew any excessively worn or damaged parts. To install new auger shaft bushings (18), press bushings into case from inside and flush with inside of casting. Install new seals (15) in gearbox halves.

Complete the reassembly by reversing disassembly procedure. During reassembly, coat all parts with grease and pack gearbox with a maximum of 3-1/4 ounces (96 mL) using one of the following recommended greases: Benalene #372, Shell Darina 1, Texaco Thermatex EP1 or Mobiltem 78.

FRICTION DRIVE

To renew friction wheel, proceed as follows: Stand snowthrower up on auger housing end and remove bottom panel (13—Fig. N11). On Model 825T, disconnect track connecting rod (2—Fig. N16) at right track plate (1) and rotate right track assembly parallel to ground to gain clearance to hex shaft end bearing. Remove the three bolts securing friction wheel (14—Fig. N12) to hub (13) and move speed control lever to first speed position. Loosen but do not remove the four nuts securing right hex shaft bearing plate (1). Move speed control lever to sixth speed position. Place tape over the four bearing plate bolts inside main (motor mount) frame and remove the four bearing plate retaining nuts. Remove bearing plate and slide hex shaft to right until friction wheel can be removed from end of hex shaft. Install replacement friction wheel retaining bolts loosely, then complete reassembly in reverse order of disassembly. Check friction wheel adjustment as outlined in a preceding paragraph.

TRACK ASSEMBLY

Refer to exploded view of track assembly in Fig. N16. Disassembly and overhaul is evident after inspection of unit and reference to exploded view. Readjust track as outlined in a previous paragraph.

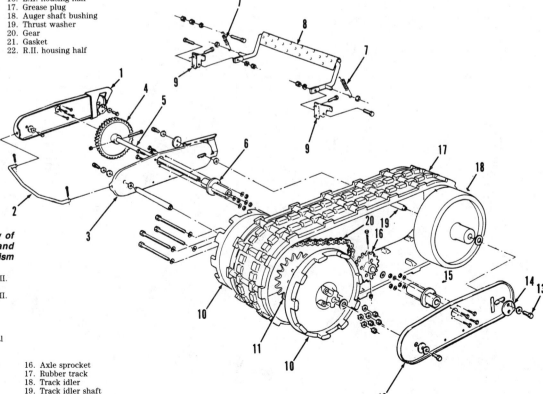

Fig. N16—Exploded view of left track assembly and weight transfer mechanism for Model 825T.

1. Track plate, inner R.II.
2. Connecting rod
3. Track plate, inner L.II.
4. Axle drive gear
5. Axle shaft
6. Inner axle hub
7. Pedal spring
8. Weight transfer pedal
9. Pedal bracket
10. Track drive wheel
11. Trace drive sprocket
12. Track plae, out L.II
13. Cap screw
14. Adjusting cam
15. Outer axle hub
16. Axle sprocket
17. Rubber track
18. Track idler
19. Track idler shaft
20. Track drive chain

NOMA

Model	Engine Make	Engine Model	Self-Propelled?	No. of Stages	Scoop Width
1033	Tecumseh	HMSK100	Yes	2	33 in. (838 mm)

LUBRICATION

ENGINE

Refer to Tecumseh engine section for engine lubrication information. Recommended fuel is regular or low-lead gasoline.

DRIVE MECHANISM

Recommended lubrication period is after every 10 hours of operation and before storing unit at end of season. Oil the drive chains and sprockets with SAE 10W-30 oil. Using a hand grease gun, lubricate grease fitting (4—Fig. N22) on the spindle assembly and grease fittings (23) on speed control bracket (13) once each season. Be careful not to get any grease or oil on the friction wheel drive plate or friction wheel; clean these parts thoroughly if oil or grease comes in contact. Remove the drive wheels and lubricate axle shafts with any automotive type grease at least once a year and prior to storage. Lubricate all moving part pivot points with SAE 10W-30 oil. Using a hand operated grease gun, lubricate the auger flight grease fittings after every 10 hours of operation and each time an auger shear bolt is replaced.

AUGER GEARBOX

Once each season, remove grease plug (18—Fig. N26) and, if grease is not visible, use a wire as a dipstick to check for grease in gearbox. If gearbox is dry, fill to plug hole level with a recommended lubricant such as Benalene #372, Shell Darina 1, Texaco Thermatex EP1 and Mobiltem 78.

ADJUSTMENTS

AUGER HEIGHT

Lower the adjusting skids to raise auger scraper bar on rough surfaces. On paved surfaces, the skids can be raised to bring the auger assembly down. Keep in mind that the scraper bar below the auger should be 1/8 inch (3 mm) above

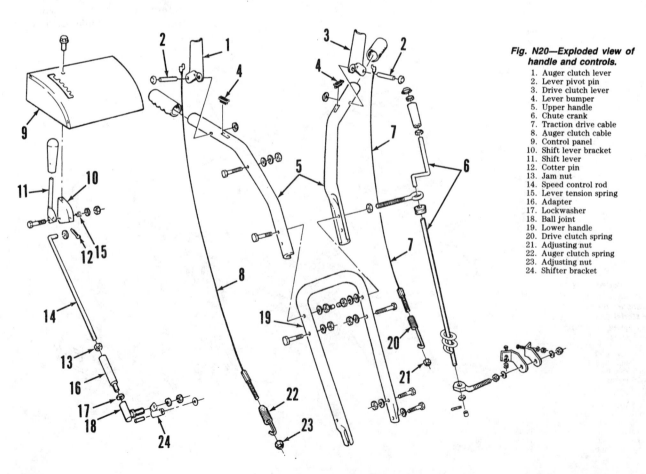

Fig. N20—Exploded view of handle and controls.

1. Auger clutch lever
2. Lever pivot pin
3. Drive clutch lever
4. Lever bumper
5. Upper handle
6. Chute crank
7. Traction drive cable
8. Auger clutch cable
9. Control panel
10. Shift lever bracket
11. Shift lever
12. Cotter pin
13. Jam nut
14. Speed control rod
15. Lever tension spring
16. Adapter
17. Lockwasher
18. Ball joint
19. Lower handle
20. Drive clutch spring
21. Adjusting nut
22. Auger clutch spring
23. Adjusting nut
24. Shifter bracket

sidewalk or other paved area to be cleaned. This clearance can be obtained by placing spacers under the scraper bar, then adjusting the skids so they contact paved surface.

To compensate for wear, the scraper bar position is adjustable, and the bar can be reversed for longer wear or replaced when both edges are worn. Be sure the bar is adjusted parallel with the working surface and for proper clearance above skid shoes.

DRIVE BELTS

The traction drive belt is spring loaded and does not require adjustment; replace belt if it is slipping. The auger drive belt can be adjusted after removing belt cover, then proceed as follows: Loosen nut on auger idler pulley (17—Fig. N21) and move idler in slot of bracket (16) toward belt about 1/8 inch (3 mm). Tighten nut, engage auger drive clutch lever and check belt adjustment; belt should deflect about 1/2 inch (12.7 mm) with moderate pressure. Reset idler pulley until proper belt tension is obtained with drive engaged.

CONTROL CABLES

Refer to Fig. N20 and with "z" fitting on upper end of auger clutch cable (8) disconnected from lever (1), move lever forward to contact plastic bumper (4); the center of the "z" fitting should be between the top and the center of the hole in the lever when pulling slack up out of cable. If not, push cable through auger clutch spring (22) and adjust nut (23) on lower end of cable to align "z" fitting on cable with hole in lever. Then, pull cable up through spring and reconnect "z" fitting on cable to hole in lever. Repeat this procedure to adjust traction drive cable (7).

FRICTION WHEEL

Remove rear cover panel (30—Fig. N21) from main (motor mount) frame. Place speed control lever in first (1) speed position and check position of

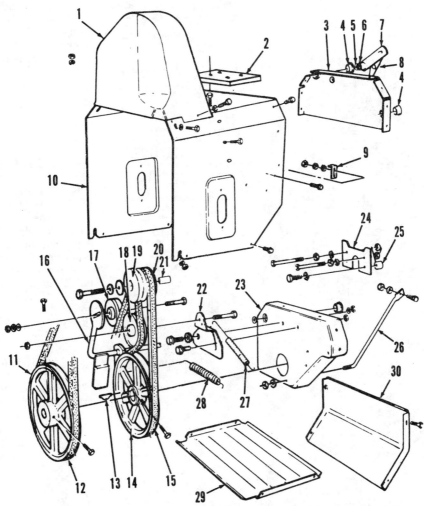

Fig. N21—Exploded view of belt drive mechanism. Refer to Fig. N22 for friction wheel drive components.

1. Belt cover		
2. Engine riser plate	10. Main (motor mount)	
3. Rear main frame	frame	
plate	11. Auger pulley	17. Auger idler pulley	23. Drive spindle plate
4. Plastic bushings	12. Auger belt	18. Drive idler pulley	24. Belt guide plate
5. Flat washer	13. Drive pulley key	19. Auger drive pulley	25. Spacer
6. Roll pin	14. Traction drive	20. Traction drive	26. Plate support rod
7. Auger clutch lever	pulley	pulley	27. Drive idler spring
8. Clutch cable lever	15. Traction drive belt	21. Engine pulley	28. Auger brake spring
9. Cable adjustment	16. Auger idler bracket	spacer	29. Bottom cover panel
bracket	& brake	22. Drive idler bracket	30. Rear cover panel

friction wheel on friction wheel drive plate as shown in Fig. N23. The left side of friction wheel should be 2-5/8 inches (66.7 mm) from left outer edge of friction wheel drive plate. If necessary to adjust position of friction wheel, refer to Fig. N20 and loosen jam nut (13) on speed control rod (14). Remove ball joint (18) from shifter bracket (24) and turn adapter (16) to obtain correct friction wheel position with ball joint inserted in bracket hole. Reinstall ball joint, tighten jam nut and reinstall rear cover panel.

OVERHAUL

ENGINE

Engine make and model numbers are listed at beginning of section. Refer to Tecumseh engine section for engine overhaul.

DRIVE BELTS

To remove auger belt, remove upper (belt) cover and remove auger drive en-

gine pulley and belt from engine. Disconnect chute crank mechanism at chute. Remove the two top bolts and loosen the bottom two bolts holding auger housing to engine mount frame. Separate auger housing from engine mount by pivoting on bottom bolts. Bend belt retainer tabs (Fig. N24) away from auger pulley only as necessary to remove belt. Install new belt on auger pulley and while holding it tightly in pulley groove, bend retainer tabs back into position leaving 1/16 to 1/8 inch (1.5-3 mm) clearance between tabs and belt. Pull up on belt, engage auger clutch and swing engine mount frame back into position. Replace two top housing bolts and tighten the two lower bolts. Install auger drive belt on engine pulley and reinstall pulley on engine shaft. Adjust auger belt as outlined in a preceding paragraph.

To remove traction drive belt, remove auger belt as in preceding paragraph. Pull spring loaded traction idler pulley back and slip belt past idler pulley and off of engine pulley. It may be necessary to remove auger brake assembly to remove belt from traction drive pulley. Install new belt and reassemble in reverse of disassembly procedure. Adjust belts as outlined in a preceding paragraph.

AUGER SHEAR BOLT

The augers are driven by the auger shaft via shear bolts (15—Fig. N25) that are designed to shear to protect the machine if an object becomes lodged in the auger housing. Only original equipment replacement shear bolts should be used for safety reasons. When replacing shear bolt(s), be sure to install new shear bolt spacer (14) and, using a hand operated grease gun, lubricate auger shaft grease fittings.

AUGER ASSEMBLY

To remove auger, first remove auger drive belt as previously outlined and re-

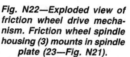

Fig. N22—Exploded view of friction wheel drive mechanism. Friction wheel spindle housing (3) mounts in spindle plate (23—Fig. N21).

1. Seal
2. Needle roller bearing
3. Spindle housing
4. Grease fitting
5. Thrust bearing races
6. Needle thrust bearing
7. Friction wheel drive plate & spindle
8. Thrust washer
9. Snap ring
10. Hex shaft drive sprocket
11. Roller chain
12. Hex shaft sliding bearing
13. Speed control bracket
14. Speed control yoke
15. Snap ring
16. Thrust washer
17. Friction wheel hub
18. Friction wheel
19. Hex shaft end bearing
20. Square key
21. Gear & shaft assy.
22. Bearing
23. Grease fitting
24. Push on end cap
25. Flat washer
26. Bushing
27. Hex shaft end bearing
28. Snap ring
29. Hex shaft
30. Axle shaft
31. Axle drive chain
32. Axle bearing bracket
33. Thrust washer
34. Grip type retainer
35. Klick pin
36. Bolt
37. Yoke drive
38. Yoke return spring
39. Cotter pin

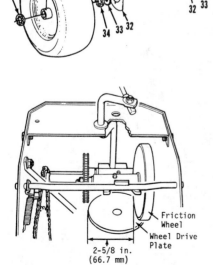

Fig. N23—With shift (speed control) lever in first speed position, friction wheel position should be as shown.

Friction Wheel

Wheel Drive Plate

2-5/8 in. (66.7 mm)

Fig. N24—Bend auger belt retainer tabs back to remove auger belt. On reassembly, bend tabs to clear belt as shown.

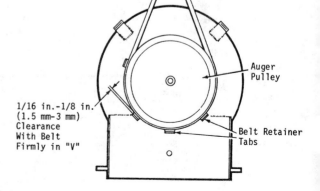

1/16 in.-1/8 in. (1.5 mm-3 mm) Clearance With Belt Firmly in "V"

Auger Pulley

Belt Retainer Tabs

fer to Fig. N25. Remove auger drive pulley (1) and impeller shaft bearing (2) from rear of housing. Unbolt auger shaft bearings (12) at each side of housing and remove the assembly. Remove the auger drive shear bolts (15) and pull the auger flights from auger shaft.

With auger flights removed, refer to Fig. N26 for exploded view of auger drive gearbox. To disassemble, remove the cap screws holding gearbox housing halves (17 & 23) and separate the unit. Remove impeller shaft bushing (15), snap ring (14) and thrust washer (13) from impeller shaft. Press shaft from worm gear (12) and remove needle thrust bearing (9) and races (8 & 10), impeller shaft bearing (7) and "O" ring (6) from impeller shaft. If necessary, impeller fan (2) can be removed from the impeller shaft. Clean and inspect all parts and renew any excessively worn or damaged parts. To install auger shaft bushings (19), press bushings into case from inside until flush with inside of casting. Install new seals (16) in gearbox halves.

When reassembling, coat all parts with grease and pack gearbox with a maximum of 3-1/4 ounces (92 grams) of one of the following greases: "Benalene #372, Shell Darina 1, Texaco Thermatex EP1, or Mobiltem 78.

FRICTION DRIVE

To replace friction wheel, stand snowthrower up on auger housing end and remove rear panel and bottom panel. Hold hex shaft from turning with a 11/16 inch open-end wrench and remove the nut and washers holding friction wheel (18—Fig. N22) and hub (17) assembly on shaft. Remove the friction wheel and hub assembly, install new friction wheel (cup side away from hub) and reassemble by reversing disassembly procedure. Be sure that square key (20) is in place in friction wheel hub and that thrust washer (16) is completely on the shoulder of the shaft, not trapped between the shoulder and threaded part of shaft.

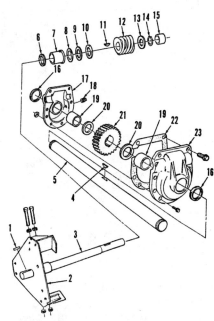

Fig. N26—Exploded view of auger gearbox and impeller assembly.

1. Square key
2. Impeller fan
3. Impeller shaft
4. Woodruff key
5. Auger shaft
6. Square cut "O" ring
7. Impeller shaft bearing
8. Thrust bearing race
9. Needle thrust bearing
10. Thrust bearing race
11. Key
12. Worm gear
13. Thrust washer
14. Snap ring
15. Impeller shaft bearing
16. Auger shaft seal
17. L.H. housing half
18. Grease plug
19. Auger shaft bushing
20. Thrust washer
21. Auger shaft gear
22. Gasket
23. R.H. housing half

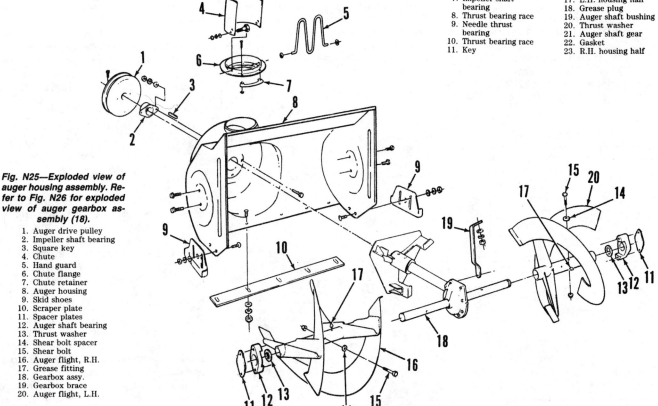

Fig. N25—Exploded view of auger housing assembly. Refer to Fig. N26 for exploded view of auger gearbox assembly (18).

1. Auger drive pulley
2. Impeller shaft bearing
3. Square key
4. Chute
5. Hand guard
6. Chute flange
7. Chute retainer
8. Auger housing
9. Skid shoes
10. Scraper plate
11. Spacer plates
12. Auger shaft bearing
13. Thrust washer
14. Shear bolt spacer
15. Shear bolt
16. Auger flight, R.H.
17. Grease fitting
18. Gearbox assy.
19. Gearbox brace
20. Auger flight, L.H.

J.C. PENNEY

J.C. PENNEY CO., INC.
11800 West Burleigh Street
Milwaukee, Wisconsin 53201

Model	Engine Make	Engine Model	Self-Propelled?	No. of Stages	Scoop Width
A420TRAC	Tecumseh	HS40	Yes	2	20 in. (508 mm)

LUBRICATION

ENGINE

Refer to Tecumseh 4-stroke engine section for engine lubrication information. Recommended fuel is regular or unleaded regular gasoline.

DRIVE MECHANISM

Recommended lubrication period is after every 10 hours of operation and before storing unit at end of season. Oil the drive chains and sprockets with SAE 10W-30 oil. The hex shaft assembly in the variable speed friction drive requires no lubrication except to wipe the hex shaft and gears with SAE 10W-30 motor oil at end of season to prevent rusting. The hex shaft bearings are of the sealed type. Be careful not to get any grease or oil on the disc drive plate or friction wheel; clean these parts thoroughly if oil or grease comes in contact with these parts.

Lubricate all moving part pivot points with SAE 10W-30 oil. Lubricate weight transfer plate and pivot points with clinging type grease such as "Lubriplate."

The auger flights do not have a grease fitting; before storage at end of season, remove the auger shear bolts and squirt oil into shear bolt holes. Turn the auger to distribute the oil throughout length of auger shaft.

AUGER GEARBOX

The auger gearbox is factory lubricated for life and unless grease leaks out, lubrication should not be necessary unless gearbox is being overhauled.

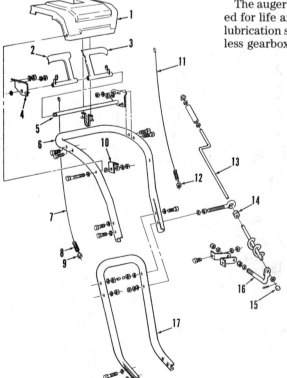

Fig. P1—Exploded view of control and handle assembly.
1. Control panel
2. Auger clutch lever
3. Drive clutch lever
4. Lever rod bracket
5. Rod & bracket assy.
6. Upper handle
7. Auger cable
8. Auger clutch spring
9. Locknut
10. Panel bracket
11. Drive clutch cable & spring
12. Locknut
13. Chute crank assy.
14. Grommet
15. Plastic cap
16. Chute crank bracket
17. Lower handle

ADJUSTMENTS

AUGER HEIGHT

Lower the adjusting skids to raise auger scraper bar on rough surfaces. On paved surfaces, the skids can be raised to bring the auger assembly down. Keep in mind that the scraper bar below the auger should be 1/8 inch (3.2 mm) above sidewalk or other paved area to be cleaned. This clearance can be obtained by placing spacers (the extra shear pins stored under snowthrower handle can be used for spacers) under the scraper bar, then adjusting the skids so they contact paved surface. To compensate for wear, the scraper bar position is adjustable, and the bar can be reversed or replaced. Be sure the bar is adjusted parallel with the working surface, and for proper clearance, above skid shoes.

DRIVE BELTS

The traction drive belt is spring loaded and does not require adjustment; check control cable adjustment or replace belt if it is slipping. The auger drive belt can be adjusted after removing belt cover, then proceed as follows: Loosen nut on auger belt idler (16—Fig. P3) and move idler in slot of idler arm (14) toward belt about 1/8 inch (3 mm). Tighten nut, engage auger drive clutch lever and check belt adjustment; belt should deflect about 1/2 inch (12.7 mm) with moderate pressure. Reset idler pulley until proper belt tension is obtained with drive engaged. The belt guides at engine pulley should be adjusted to clear belt 3/32 inch (2.4 mm) with auger drive clutch engaged.

CONTROL CABLES

Refer to Fig. P1 and with auger cable (7) "z" fitting disconnected from auger clutch lever (2), move lever forward to contact plastic bumper and pull up on cable; the center of the "z" fitting should be between the top and the center of the hole in the lever. If not, slide cable down through auger clutch spring (8) and turn lock nut (9) to obtain proper adjustment. Then, pull cable up through spring and reconnect "z" fitting on cable in hole of lever. Repeat the procedure for drive clutch cable (11) adjustment.

FRICTION WHEEL

Remove bottom panel from snowthrower. Place shifter lever in first (Position 1) gear and check position of friction wheel on disc drive plate. Distance from left outer side of disc drive plate to right side of friction wheel should be 2-3/4 inches (70 mm) as shown in Fig. P2. If necessary to adjust position of friction wheel, loosen bolts retaining upper speed shift lever (29—Fig. P3) to shift lever adapter (32). Then move friction wheel to proper position and retighten lever bolts.

TRACK DRIVE

To check track tension, pull up gently on center of track at midpoint between track rollers. Distance between track and top of track support frame should then be no more than 1-1/4 inches (32 mm). To adjust track, loosen jam nuts (3—Fig. P5) on adjusting bolt (11) at rear of track frame (4) and tighten or loosen adjusting bolt as necessary. Adjust opposite track in same manner. Be sure adjusting bolt jam nuts are securely tightened.

OVERHAUL

ENGINE

Engine make and model numbers are listed at beginning of section. Refer to

Tecumseh engine section for engine overhaul.

AUGER DRIVE BELT

To remove auger drive belt, remove belt cover (2—Fig. P3), loosen the right-hand belt guide (12) and move guide away from drive pulley. Loosen auger belt idler (16) and move it away from the belt. Engage auger drive clutch lever to move brake away from belt and slip the belt out from between pulley and brake. Remove the top two bolts securing auger housing to motor mount frame. Place weight transfer system in transport position (latched in top notch). Slightly loosen bottom two bolts attaching auger housing to motor mount

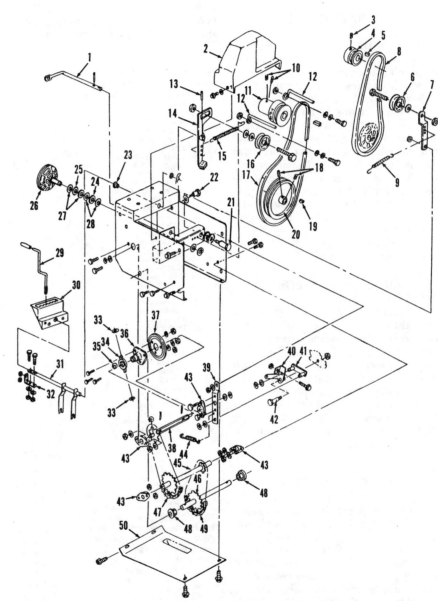

Fig. P3—Exploded view of drive mechanism. The traction drive is from engine PTO pulley (4) and the auger is driven from engine crankshaft pulley (11). Drive disc (26) and shaft are separate items on some models.

1. Auger clutch lever		27. Thrust washers	39. Drive clutch lever
2. Belt cover	15. Auger brake arm	28. Needle thrust	40. Drive lever assy.
3. Set screw	spring	bearings	41. Drive lever link
4. Engine PTO pulley	16. Auger belt idler	29. Shift lever	42. Shoulder bolt
5. Woodruff key	17. Auger belt	30. Shift bracket	43. Self-centering
6. Drive idler pulley	18. Set screws	31. Shift yoke	bearing
7. Drive idler arm	19. Woodruff key	32. Shift lever/yoke	44. Return spring
8. Drive belt	20. Auger drive pulley	adapter	45. Jackshaft assy.
9. Idler arm spring	21. Shoulder bolt	33. Flat washers	46. Wheel axle &
10. Set screws	22. Shoulder bolt	34. Trunion bearing	sprocket
11. Engine crankshaft	23. Plastic bushing	35. Snap ring	47. Roller chain
pulley	24. Thrust washer	36. Friction wheel hub	48. Axle bearing
12. Belt guide	25. Needle thrust	37. Friction wheel	49. Roller chain
13. Roll pin	bearing	38. Hex shaft	50. Bottom cover panel
14. Auger brake & idler	26. Drive disc		
arm			

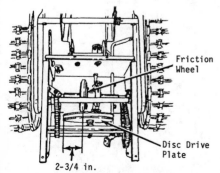

Fig. P2—View showing measurement to check friction wheel adjustment.

Friction Wheel

Disc Drive Plate

2-3/4 in.

frame. Separate auger housing and motor mount frame by hinging on bottom two bolts. Remove belt from engine auger drive pulley. Install new belt by reversing removal procedure. Place weight transfer system in lowest notch in bracket to pivot auger housing and motor mount frame back into position. Adjust belt as outlined in preceding paragraph.

TRACTION DRIVE BELT

To remove traction drive belt, first remove upper belt cover, pull traction idler pulley back and slip belt past idler pulley, then remove belt from engine pulley (It may be necessary to loosen and remove engine auger drive pulley for clearance to remove belt). Place speed selector in sixth gear position and remove belt from between rubber drive disc and drive disc plate. Install new belt by reversing removal procedure. If engine auger drive pulley was removed, be sure auger drive pulleys are aligned and drive key is flush with end of engine shaft before tightening engine pulley set screw. Adjust belts as outlined in belt adjustment paragraph.

AUGER SHEAR BOLT

The augers are driven from the auger shaft through special bolts (37—Fig. P4) that are designed to shear to protect the machine if an object becomes lodged in the auger housing. Only original equipment replacement shear bolts should be used for safety reasons. Before installing new bolt, squirt oil between auger flight and auger shaft and turn auger flight to distribute oil throughout the full length of shaft.

AUGER ASSEMBLY

To remove auger, first remove auger drive belt as outlined in previous paragraph. Remove auger drive pulley and bearing (6—Fig. P4) from rear of housing, unbolt impeller shaft bearings (10) at each side of housing and remove the auger and gearbox assembly. Remove the auger drive shear bolts (37) and slide auger flights from auger shaft. Remove impeller bolts (34) and pull impeller from shaft. Remove the bolts holding case halves (20 & 32) together and disassemble gearbox. Clean and inspect all parts and renew any worn or damaged parts. Coat the parts with clean grease and pack gearbox with about 3 ounces (85 grams) of grease on assembly. Recommended greases are Benalene #372, Shell Darina 1, Texaco Thermatex EP1 or Mobiltem 78.

FRICTION DRIVE WHEEL

To replace friction wheel, proceed as follows: Stand snowthrower up on auger housing end and remove bottom cover panel (50—Fig. P3). Unbolt bearings (43) from engine frame at each end of hex shaft (38). Slide the hex shaft with friction wheel, trunnion bearing and end bearings out of snowthrower. Be careful not to lose flat washers (33) from studs on trunnion bearing (34). Unbolt friction wheel (37) from hub and install new wheel (cup side away from hub) by reversing removal procedure. Check friction wheel adjustment as outlined in preceding paragraph before reinstalling bottom panel.

TRACK ASSEMBLY

Refer to exploded view of track assembly in Fig. P5. Disassembly and overhaul is evident after inspection of unit and reference to exploded view. Readjust track as outlined in previous paragraph.

Fig. P4—Exploded view of auger housing, gearbox assembly and augers.

1. Deflector chute
2. Hand guard (not all models)
3. Chute extension
4. Mounting flange
5. Flange retainer
6. Impeller shaft bearing
7. Auger housing
8. Left side plate
9. Right side plate
10. Auger shaft bearing
11. Skid shoe
12. Scraper bar
13. Oil seal
14. Bushing
15. Thrust washers
16. Needle bearing
17. Roll pin
18. Thrust collar
19. Impeller shaft
20. Right case half
21. Oil seal
22. Bushing
23. Thrust washer
24. Woodruff key
25. Worm gear
26. Bushing
27. Gasket
28. Auger shaft
29. Thrust washer
30. Bushing
31. Oil seal
32. Left case half
33. Grease plug
34. Impeller bolts
35. Impeller assy.
36. Locknuts
37. Auger shear bolts
38. Left auger assy.
39. Right auger assy.

Fig. P5—Exploded view of track drive assembly. Track drive wheels (15) mount on axle shaft (46—Fig. P3). Foot pedal (8) operates weight transfer system.

1. Mounting bracket
2. Pivot pin
3. Jam nuts
4. Track frame
5. Snap ("E") rings
6. Roll pin
7. Foot pedal spring
8. Foot pedal
9. Frame support bearing
10. Idler axle shaft
11. Track adjusting bolt
12. Spacer
13. Track idler wheel
14. Rubber track
15. Track drive wheel

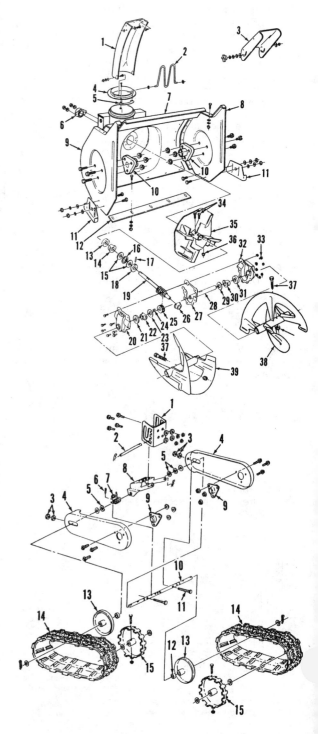

J.C. PENNEY

Model	Engine Make	Engine Model	Self-Propelled	No. of Stages	Scoop Width
4923	Tecumseh	HS80	Yes	2	25 in. (635 mm)
4923A	Tecumseh	HS80	Yes	2	25 in. (635 mm)

LUBRICATION

ENGINE

Refer to Tecumseh engine section for engine lubrication information. Recommended fuel is regular or low-lead gasoline.

DRIVE MECHANISM

Recommended lubrication period is after every 10 hours of operation and before storing unit at end of season. Oil the drive chains and sprockets with SAE 10W-30 oil. The hex shaft assembly in the variable speed friction drive requires no lubrication except to wipe the hex shaft and gears with SAE 10W-30 motor oil at end of season to prevent rusting. The hex shaft bearings are of the sealed type. Be careful not to get any grease or oil on the disc drive plate or friction wheel; clean these parts thoroughly if oil or grease comes in contact.

Lubricate all moving part pivot points with SAE 10W-30 oil. Using a hand-operated grease gun, lubricate the auger shaft zerk fittings after every 10 hours of operation.

AUGER GEARBOX

The auger gearbox is factory lubricated and unless oil should leak out, lubrication should not be necessary unless gearbox is being overhauled. Once each season, remove plug (17—Fig. P15) and, if grease is not visible, use a wire as a dipstick to check for grease in gearbox. Recommended lubricants are Benalene #372, Shell Darina 1, Texaco Thermatex EP1 and Mobiltem 78.

ADJUSTMENTS

AUGER HEIGHT

Lower the adjusting skids to raise auger scraper bar on rough surfaces. On paved surfaces, the skids can be raised to bring the auger assembly down. Keep in mind that the scraper bar below the

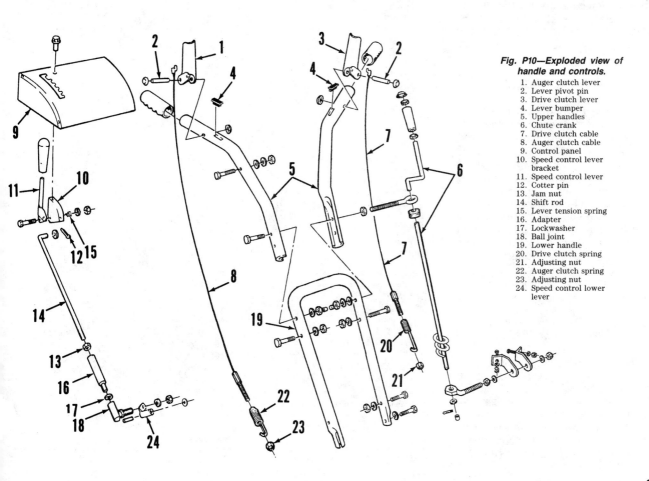

Fig. P10—Exploded view of handle and controls.

1. Auger clutch lever
2. Lever pivot pin
3. Drive clutch lever
4. Lever bumper
5. Upper handles
6. Chute crank
7. Drive clutch cable
8. Auger clutch cable
9. Control panel
10. Speed control lever bracket
11. Speed control lever
12. Cotter pin
13. Jam nut
14. Shift rod
15. Lever tension spring
16. Adapter
17. Lockwasher
18. Ball joint
19. Lower handle
20. Drive clutch spring
21. Adjusting nut
22. Auger clutch spring
23. Adjusting nut
24. Speed control lower lever

auger should be 1/8 inch (3.2 mm) above sidewalk or other paved area to be cleaned. This clearance can be obtained by placing spacers under the scraper bar, then adjusting the skids so they contact paved surface.

NOTE: Weight transfer system will not operate with auger housing raised to highest position (skid shoes in lowest setting).

To compensate for wear, the scraper bar position is adjustable, and the bar can be reversed for longer wear or renewed when both edges are worn. Be sure the bar is adjusted parallel with the working surface and for proper clearance, above skid shoes.

DRIVE BELTS

The traction drive belt is spring loaded and does not require adjustment; renew belt if it is slipping. The auger drive belt can be adjusted after removing belt cover, then proceed as follows: Loosen nut on idler pulley (18—Fig. P11) and move idler in idler arm slot toward belt about 1/8 inch (3 mm). Tighten nut, engage auger drive clutch lever and check belt adjustment; belt should deflect about 1/2 inch (12.7 mm) with moderate pressure. Reset idler pulley until proper belt tension is obtained with drive engaged. Belt guides (11 & 12) should be adjusted to clear belt 3/32 inch (2.4 mm) with auger drive clutch engaged.

CONTROL CABLES

Refer to Fig. P10 and with drive clutch cable (7) "z" fitting disconnected from drive clutch lever (3), move lever forward to contact plastic bumper (4); the center of the "z" fitting should be between the top and the center of the hole in the lever. If not, slide cable down through drive clutch spring (20), hold square part of cable end from turning and rotate adjusting nut (21) to obtain correct cable length. Then, pull cable up through spring and reconnect "z" fitting on cable to hole in lever. Repeat adjustment procedure for auger clutch cable (8).

FRICTION WHEEL

Remove bottom panel from snowthrower. Place speed control lever in first gear and check position of friction wheel on disc drive plate. The right side of friction wheel should be 3-3/8 inches (86 mm) from outer edge of disc drive plate as shown in Fig. P13. If adjustment is necessary, loosen jam nut (13—Fig.

P10) on speed select rod. Remove ball joint (18) from speed control lower lever (24) and turn adapter (16) to obtain correct friction wheel position with ball joint inserted in speed control lower lever. Reinstall ball joint, tighten jam nut and reinstall bottom panel.

Fig. P11—Exploded view of drive belt mechanism. Refer to Fig. P12 for friction drive exploded view.
1. Auger clutch lever
2. Roll pin
3. Flat washer
4. Plastic bushing
5. Auger cable lever
6. Main (motor mount) frame
7. Hi-pro key
8. Spacer
9. Drive idler bracket
10. Drive idler spring
11. Belt guide
12. Belt guide
13. Bottom panel
14. Plastic bushing
15. Flat washer
16. Auger clutch rod
17. Auger idler return spring
18. Auger belt idler pulley
19. Drive belt
20. Drive pulley
21. Auger belt
22. Set screw
23. Drive belt idler
24. Engine crankshaft pulley
25. Woodruff key
26. Engine PTO pulley
27. Set screw
28. Spacer

Fig. P12—Exploded view of disc drive mechanism.
1. Bearing plate
2. Bearing
3. Flat washer
4. Hex shaft
5. Roll pins
6. Locknut
7. Speed control yoke
8. Plastic bushing
9. Washer
10. Snap ring
11. Washer
12. Trunnion bearing
13. Friction wheel hub
14. Friction wheel
15. Pinion gear
16. Flat washer
17. Bearing
18. Bearing plate
19. Spacers
20. Traction pivot
21. Push nut
22. Bushing
23. Flat washer
24. Cotter pin
25. Traction shaft
26. Trunnion bearing
27. Traction spring bracket
28. Flat washer
29. Traction spring
30. Spring pin
31. Traction rod
32. Drive disc
33. Return spring
34. Hi-pro key
35. Washer
36. Drive disc shaft
37. Snap ring
38. Thrust bearing
39. Needle bearings

TRACKS

Refer to Fig. P16 and loosen cap screws (13) on each side of track assembly. Rotate adjusting cam (14) equally on each side of assembly so deflected distance between top of track plate (12)

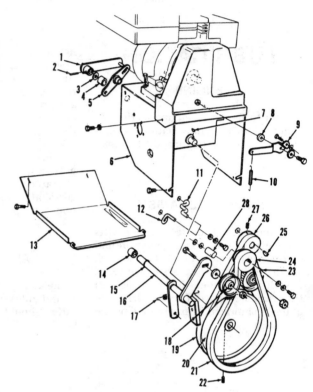

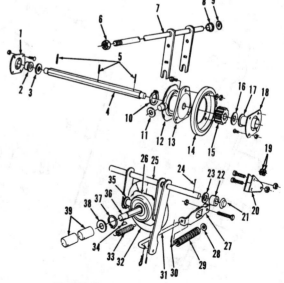

and inside of track is not greater than 2 inches (50.8 mm). Uneven adjustment of adjusting cams will result in a twist in the rubber track. Repeat adjustment for opposite track assembly.

OVERHAUL

ENGINE

Engine make and model numbers are listed at beginning of section. Refer to Tecumseh engine section for engine overhaul.

AUGER DRIVE BELT

To remove auger drive belt, remove belt drive cover and loosen belt guides (11 & 12—Fig. P11) and move them away from engine crankshaft pulley (24). Loosen auger belt idler pulley (18) and move it away from the belt. Engage auger drive clutch lever to move brake away from belt and slip the belt off auger pulley. Remove belt from engine drive pulley. Install new belt by reversing removal procedure. Adjust belt and belt guides as outlined in a previous paragraph.

TRACTION DRIVE BELT

To remove traction drive belt, remove belt cover and loosen the left-hand belt guide (11—Fig. P11) and move guide away from engine PTO pulley. Pull spring loaded traction idler pulley back and slip belt past idler pulley. Remove belt from engine pulley and up between the auger and traction drive pulleys. Install new belt by reversing removal procedure. Adjust belt and belt guide as outlined in a preceding paragraph.

AUGER SHEAR BOLT

The augers are driven via the auger shaft through special shear bolts (16—Fig. P14) that are designed to shear to

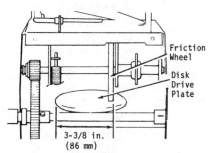

Fig. P13—With speed control lever in first gear, friction wheel should be positioned as shown.

protect the machine if an object becomes lodged in the auger housing. Only original equipment replacement shear bolts should be used for safety reasons. When replacing shear bolt(s), be sure to install new shear bolt spacer (17) and, using a hand-operated grease gun, lubricate auger shaft grease fittings.

AUGER ASSEMBLY

To remove auger, remove auger drive belt, disconnect the chute crank from auger housing and separate auger housing from main (motor mount) frame. Remove auger drive pulley (1—Fig. P14) and impeller shaft bearing (2) from rear of housing. Unbolt auger shaft bushings (14) at each side of housing and remove the assembly. Remove the auger drive shear bolts (16) and remove the auger flights from auger shaft.

With auger flights removed, refer to Fig. P15 for exploded view of auger drive gearbox. To disassemble, remove the cap screws holding the gearbox housing halves (16 & 22) and separate the unit. Remove impeller shaft bushing (14), snap ring (13) and thrust washer (12) from impeller shaft. Press shaft from worm gear (11) and remove needle thrust bearing (8) and races (7 & 9), rear impeller shaft bushing (6) and "O" ring (5) from impeller shaft. Clean and inspect all parts and renew any excessive-

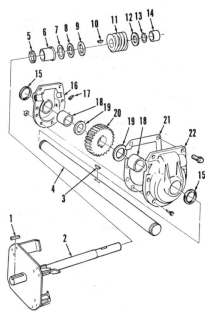

Fig. P15—Exploded view of auger gearbox. Impeller is an integral part of impeller shaft (2).

1. Square key	11. Worm gear
2. Impeller & shaft assy.	12. Thrust washer
3. Woodruff key	13. Snap ring
4. Auger shaft	14. Impeller shaft bushing
5. Square cut "O" ring	15. Oil seals
6. Impeller shaft bushing	16. L.H. housing half
7. Thrust bearing race	17. Grease plug
8. Needle thrust bearing	18. Auger shaft bushing
9. Thrust bearing race	19. Thrust washer
10. Woodruff key	20. Gear
	21. Gasket
	22. R.H. housing half

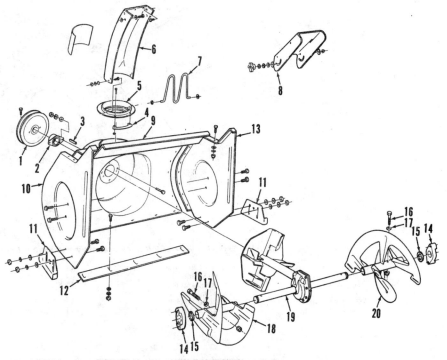

Fig. P14—Exploded view of auger housing assembly.

1. Auger drive pulley		
2. Impeller shaft bearing	7. Hand guard	11. Skid shoe	16. Shear bolt
3. Square key	8. Upper chute extension	12. Scraper bar	17. Shear bolt spacer
4. Chute retainer	9. Center auger housing	13. L.H. side plate	18. R.H. auger flight
5. Chute flange	10. R.H. side plate	14. Auger shaft bushing	19. Gearbox assy.
6. Bottom chute extension		15. Spacer washer	20. L.H. auger flight

ly worn or damaged parts. To install auger shaft bushings (18), press bushings into case from inside and flush with inside of casting. Install new seals (15) in gearbox halves.

When reassembling, coat all parts with grease and pack gearbox with a total of 3-1/4 ounces (96 mL) of one of the following greases: Benalene #372, Shell Darina 1, Texaco Thermatex EP1, or Mobiltem 78.

FRICTION DRIVE

To renew friction wheel, proceed as follows: Stand snowthrower up on au-ger housing end and remove bottom panel (13—Fig. P11). Disconnect track connecting rod (2—Fig. P16) at right track plate (1) and rotate right track assembly parallel to ground to gain clearance to hex shaft end bearing. Remove the three bolts securing friction wheel (14—Fig. P12) to hub (13) and move speed control lever to first speed position. Loosen but do not remove nuts securing right hex shaft bearing plate (1). Move speed control lever to sixth speed position. Place tape over the four hex shaft bearing plate bolts inside motor mount and remove the previously loosened retaining nuts. Remove bearing plate and slide hex shaft to right un-til friction wheel can be removed. Install replacement friction wheel retaining bolts loosely, then complete reassembly in reverse order of disassembly. Check friction wheel adjustment as outlined in a previous paragraph.

TRACK ASSEMBLY

Refer to exploded view of track assembly in Fig. P16. Disassembly and overhaul is evident after inspection of unit and reference to exploded view. Readjust track after servicing as outlined in a previous paragraph.

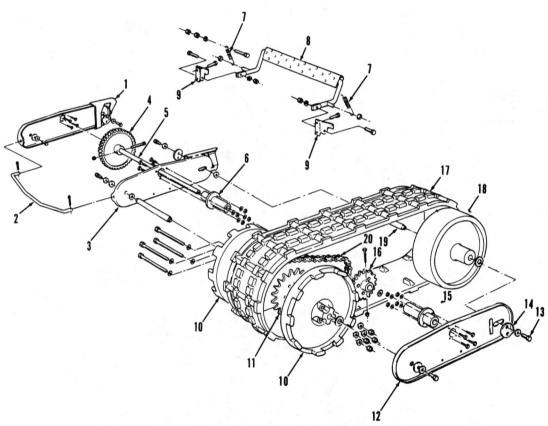

Fig. P16—Exploded view of left track assembly and weight transfer mechanism.

1. Track plate, inner R.H.
2. Connecting rod
3. Track plate, inner L.H.
4. Axle drive gear
5. Axle shaft
6. Inner axle hub
7. Pedal spring
8. Weight transfer pedal
9. Pedal bracket
10. Track drive wheel
11. Track drive sprocket
12. Track plate, outer L.H.
13. Cap screw
14. Adjusting cam
15. Outer axle hub
16. Axle sprocket
17. Rubber track
18. Track idler
19. Track idler shaft
20. Track drive chain

J.C. PENNEY

Model	Engine Make	Engine Model	Self-Propelled?	No. of Stages	Scoop Width
4924	Tecumseh	HM100	Yes	2	32 in. (813 mm)
4924A	Tecumseh	HM100	Yes	2	32 in. (813 mm)

LUBRICATION

ENGINE

Refer to Tecumseh engine section for engine lubrication information. Recommended fuel is regular or low-lead gasoline.

DRIVE MECHANISM

Recommended lubrication period is after every 10 hours of operation and before storing unit at end of season. Oil the drive chains and sprockets with SAE 10W-30 oil. Using a hand grease gun, lubricate grease fitting (4—Fig. P22) on the spindle assembly and grease fittings (23) on speed control bracket (13) once each season. Be careful not to get any grease or oil on the friction wheel drive plate or friction wheel; clean these parts thoroughly if oil or grease comes in contact. Remove the drive wheels and lubricate axle shafts with any automotive type grease at least once a year and prior to storage. Lubricate all moving part pivot points with SAE 10W-30 oil. Using a hand operated grease gun, lubricate the auger flight grease fittings after every 10 hours of operation and each time an auger shear bolt is replaced.

AUGER GEARBOX

Once each season, remove grease plug (18—Fig. P26) and, if grease is not visible, use a wire as a dipstick to check for grease in gearbox. Fill to plug hole level with a recommended lubricants such as Benalene #372, Shell Darina 1, Texaco Thermatex EP1 or Mobiltem 78.

ADJUSTMENTS

AUGER HEIGHT

Lower the adjusting skids to raise auger scraper bar on rough surfaces. On paved surfaces, the skids can be raised to bring the auger assembly down. Keep in mind that the scraper bar below the auger should be 1/8 inch (3 mm) above sidewalk or other paved area to be cleaned. This clearance can be obtained

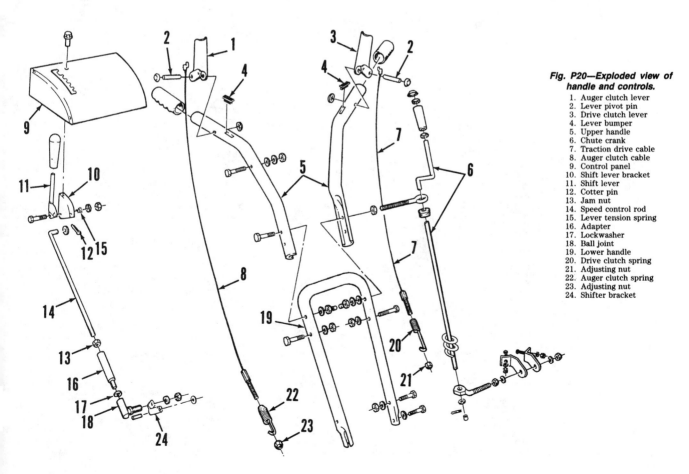

Fig. P20—Exploded view of handle and controls.
1. Auger clutch lever
2. Lever pivot pin
3. Drive clutch lever
4. Lever bumper
5. Upper handle
6. Chute crank
7. Traction drive cable
8. Auger clutch cable
9. Control panel
10. Shift lever bracket
11. Shift lever
12. Cotter pin
13. Jam nut
14. Speed control rod
15. Lever tension spring
16. Adapter
17. Lockwasher
18. Ball joint
19. Lower handle
20. Drive clutch spring
21. Adjusting nut
22. Auger clutch spring
23. Adjusting nut
24. Shifter bracket

by placing spacers under the scraper bar, then adjusting the skids so they contact paved surface.

To compensate for wear, the scraper bar position is adjustable, and the bar can be reversed for longer wear or replaced when both edges are worn. Be sure the bar is adjusted parallel with the working surface and for proper clearance above skid shoes.

DRIVE BELTS

The traction drive belt is spring loaded and does not require adjustment; replace belt if it is slipping. The auger drive belt can be adjusted after removing belt cover, then proceed as follows:

Loosen nut on auger idler pulley (17—Fig. P21) and move idler in slot of bracket (16) toward belt about 1/8 inch (3 mm). Tighten nut, engage auger drive clutch lever and check belt adjustment; belt should deflect about 1/2 inch (12.7 mm) with moderate pressure. Reset idler pulley until proper belt tension is obtained with drive engaged.

CONTROL CABLES

Refer to Fig. P20 and with "z" fitting on upper end of auger clutch cable (8) disconnected from lever (1), move lever forward to contact plastic bumper (4); the center of the "z" fitting should be between the top and the center of the

hole in the lever when pulling slack up out of cable. If not, push cable through auger clutch spring (22) and adjust nut (23) on lower end of cable to align "z" fitting on cable with hole in lever. Then, pull cable up through spring and reconnect "z" fitting on cable to hole in lever. Repeat this procedure to adjust traction drive cable (7).

FRICTION WHEEL

Remove rear cover panel (30—Fig. P21) from main (motor mount) frame. Place speed control lever in first speed position and check position of friction wheel on friction wheel drive plate as shown in Fig. P23. The left side of fric-

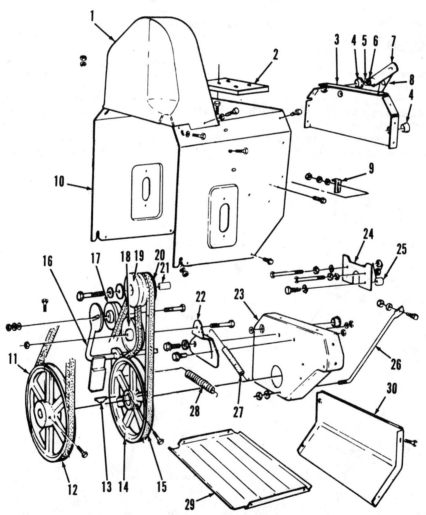

Fig. P21—Exploded view of belt drive mechanism. Refer to Fig. P22 for friction wheel drive components.

1. Belt cover
2. Engine riser plate
3. Rear main frame plate
4. Plastic bushings
5. Flat washer
6. Roll pin
7. Auger clutch lever
8. Clutch cable lever
9. Cable adjustment bracket
10. Main (motor mount) frame
11. Auger pulley
12. Auger belt
13. Drive pulley key
14. Traction drive pulley
15. Traction drive belt
16. Auger idler bracket & brake
17. Auger idler pulley
18. Drive idler pulley
19. Auger drive pulley
20. Traction drive pulley
21. Engine pulley spacer
22. Drive idler bracket
23. Drive spindle plate
24. Belt guide plate
25. Spacer
26. Plate support rod
27. Drive idler spring
28. Auger brake spring
29. Bottom cover panel
30. Rear cover panel

tion wheel should be 2-5/8 inches (66.7 mm) from left outer edge of friction wheel drive plate. If necessary to adjust position of friction wheel, refer to Fig. P20 and loosen jam nut (13) on speed control rod (14). Remove ball joint (18) from shifter bracket (24) and turn adapter (16) to obtain correct friction wheel position with ball joint inserted in bracket hole. Reinstall ball joint, tighten jam nut and reinstall rear cover panel.

OVERHAUL

ENGINE

Engine make and model numbers are listed at beginning of section. Refer to Tecumseh engine section for engine overhaul.

DRIVE BELTS

To remove auger belt, remove upper (belt) cover and remove auger drive en-

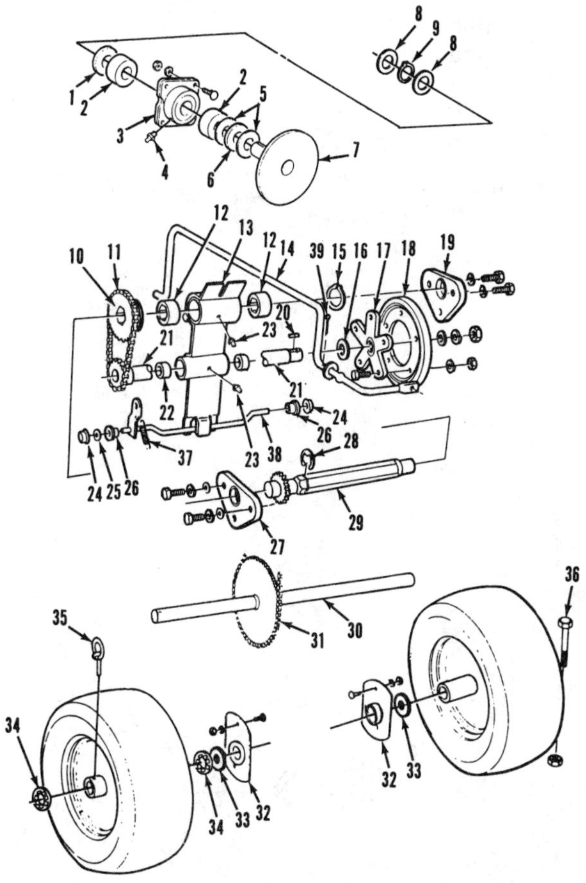

Fig. P22—Exploded view of friction wheel drive mechanism. Friction wheel spindle housing (3) mounts in spindle plate (23—Fig. P21).

1. Seal
2. Needle roller bearing
3. Spindle housing
4. Grease fitting
5. Thrust bearing races
6. Needle thrust bearing
7. Friction wheel drive plate & spindle
8. Thrust washer
9. Snap ring
10. Hex shaft drive sprocket
11. Roller chain
12. Hex shaft sliding bearing
13. Speed control bracket
14. Speed control yoke
15. Snap ring
16. Thrust washer
17. Friction wheel hub
18. Friction wheel
19. Hex shaft end bearing
20. Square key
21. Gear & shaft assy.
22. Bearing
23. Grease fitting
24. Push on end cap
25. Flat washer
26. Bushing
27. Hex shaft end bearing
28. Snap ring
29. Hex shaft
30. Axle shaft
31. Axle drive chain
32. Axle bearing bracket
33. Thrust washer
34. Grip type retainer
35. Klick pin
36. Bolt
37. Yoke drive
38. Yoke return spring
39. Cotter pin

gine pulley and belt from engine. Disconnect chute crank mechanism at chute. Remove the two top bolts and loosen the bottom two bolts holding auger housing to main (motor mount) frame. Separate auger housing from engine mount by hinging on bottom bolts. Bend belt retainer tabs (Fig. P24) away from auger pulley only as necessary to remove belt. Install new belt on auger pulley and while holding it tightly in pulley groove, bend retainer tabs back into position leaving 1/16 to 1/8 inch (1.5-3 mm) clearance between tabs and belt. Pull up on belt, engage auger clutch and swing engine mount frame back into position. Replace two top housing bolts and tighten the two lower bolts. Install auger drive belt on engine pulley and reinstall pulley on engine shaft. Adjust auger belt as outlined in a preceding paragraph.

To remove traction drive belt, remove auger belt as in preceding paragraph. Pull spring loaded traction idler pulley back and slip belt past idler pulley and off of engine pulley. It may be necessary to remove auger brake assembly to remove belt from traction drive pulley. Install new belt and reassemble in reverse of disassembly procedure. Adjust belts as outlined in a preceding paragraph.

AUGER SHEAR BOLT

The augers are driven by the auger shaft via shear bolts (15—Fig. P25) that are designed to shear to protect the machine if an object becomes lodged in the auger housing. Only original equipment replacement shear bolts should be used for safety reasons. When replacing shear bolt(s), be sure to install new shear bolt spacer (14) and, using a hand operated grease gun, lubricate auger shaft grease fittings.

AUGER ASSEMBLY

To remove auger, first remove auger drive belt as previously outlined and re-

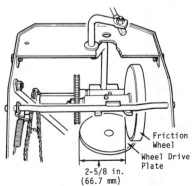

Fig. P23—With shift (speed control) lever in first speed position, friction wheel position should be as shown.

Fig. P24—Bend auger belt retainer tabs back to remove auger belt. On reassembly, bend tabs to clear belt as shown.

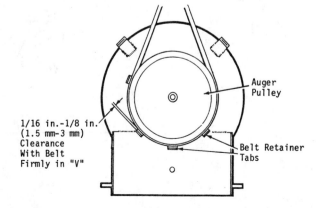

fer to Fig. P25. Remove auger drive pulley (1) and impeller shaft bearing (2) from rear of housing. Unbolt auger shaft bearings (12) at each side of housing and remove the assembly. Remove the auger drive shear bolts (15) and pull the auger flights from auger shaft.

With auger flights removed, refer to Fig. P26 for exploded view of auger drive gearbox. To disassemble, remove the cap screws holding gearbox housing halves (17 & 23) and separate the unit. Remove impeller shaft bushing (15), snap ring (14) and thrust washer (13) from impeller shaft. Press shaft from worm gear (12) and remove needle thrust bearing (9) and races (8 & 10), impeller shaft bearing (7) and "O" ring (6) from impeller shaft. If necessary, impeller fan (2) can be removed from impeller shaft. Clean and inspect all parts and renew any excessively worn or damaged parts. To install new auger shaft bushings (19), press bushings into case from inside until flush with inside of casting. Install new seals (16) in gearbox halves.

When reassembling, coat all parts with grease and pack gearbox with a maximum of 3-1/4 ounces (92 grams) of a lubricant such as Benalene #372, Shell Darina 1, Texaco Thermatex EP1 or Mobiltem 78.

FRICTION DRIVE

To replace friction wheel, stand snowthrower up on auger housing end and remove rear panel and bottom panel. Hold hex shaft from turning with a 11/16-inch open-end wrench and remove the nut and washers holding friction wheel (18—Fig. P22) and hub (17) assembly on shaft. Remove the friction wheel and hub assembly, install new friction wheel (cup side away from hub) and reassemble by reversing disassembly procedure. Be sure that square key (20) is in place in friction wheel hub and that thrust washer (16) is completely on the shoulder of the shaft, not trapped between the shoulder and threaded part of shaft.

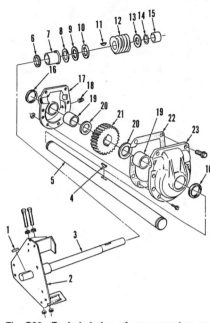

Fig. P26—Exploded view of auger gearbox and impeller assembly.

1. Square key
2. Impeller fan
3. Impeller shaft
4. Woodruff key
5. Auger shaft
6. Square cut "O" ring
7. Impeller shaft bearing
8. Thrust bearing race
9. Needle thrust bearing
10. Thrust bearing race
11. Key
12. Worm gear
13. Thrust washer
14. Snap ring
15. Impeller shaft bearing
16. Auger shaft seal
17. L.H. housing half
18. Grease plug
19. Auger shaft bushing
20. Thrust washer
21. Auger shaft gear
22. Gasket
23. R.H. housing half

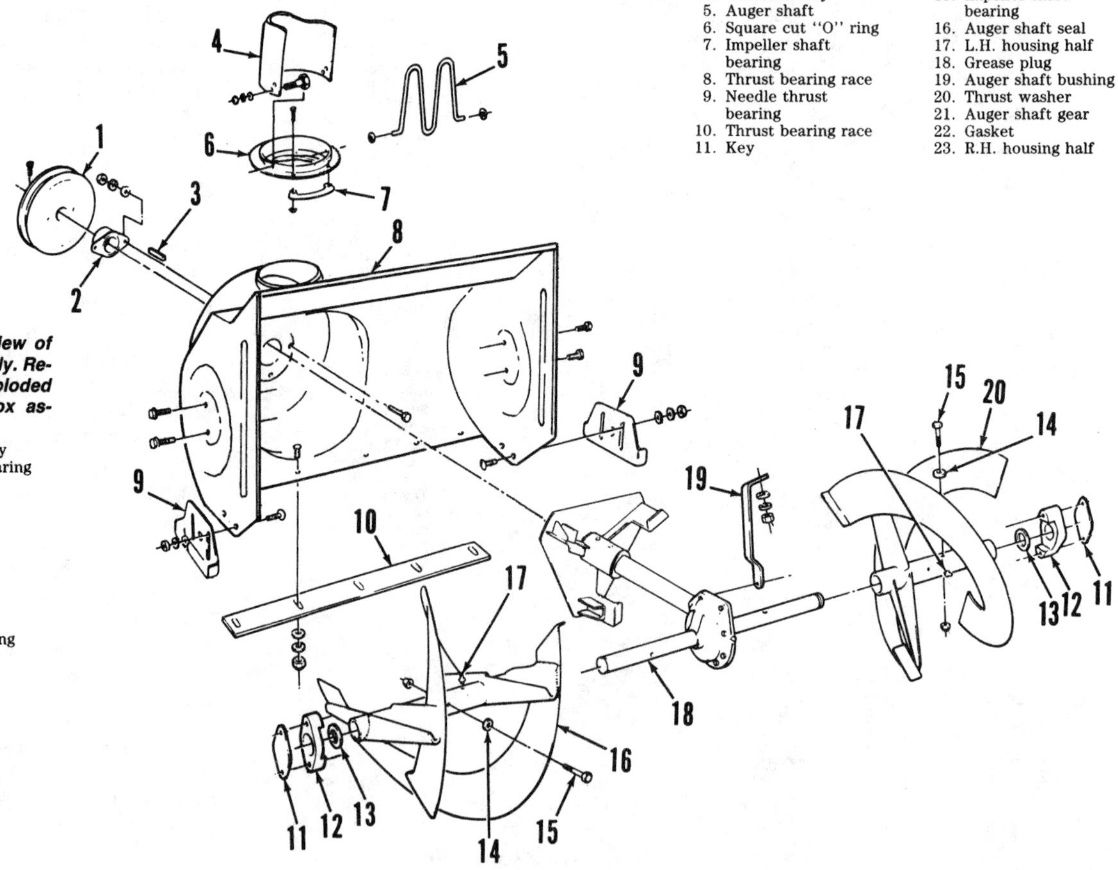

Fig. P25—Exploded view of auger housing assembly. Refer to Fig. P26 for exploded view of auger gearbox assembly (18).

1. Auger drive pulley
2. Impeller shaft bearing
3. Square key
4. Chute
5. Hand guard
6. Chute flange
7. Chute retainer
8. Auger housing
9. Skid shoes
10. Scraper plate
11. Spacer plates
12. Auger shaft bearing
13. Thrust washer
14. Shear bolt spacer
15. Shear bolt
16. Auger flight, R.H.
17. Grease fitting
18. Gearbox assy.
19. Gearbox brace
20. Auger flight, L.H.

J.C. PENNEY

Model	Engine Make	Engine Model	Self-Propelled?	No. of Stages	Scoop Width
4050	Tecumseh	HSSK40	Yes	2	20 in. (508 mm)
4050A	Tecumseh	HSSK40	Yes	2	20 in. (508 mm)
4049	Tecumseh	HSSK50	Yes	2	24 in. (610 mm)

LUBRICATION

ENGINE

Refer to Tecumseh 4-stroke engine section for engine lubrication information. Recommended fuel is regular or unleaded regular gasoline.

DRIVE MECHANISM

Lubricate drive chains after every 25 hours of operation or at least once each season using a good-quality chain oil. Auger shaft bushings and traction drive shaft bushings should be oiled after every 25 hours of operation with SAE 30 motor oil. Lubricate all control pivot and slide points for smooth operation. Once each season or when necessary to replace auger flight shear bolts, remove shear bolts and squirt oil in hole and at ends of auger flights, then turn auger flight to distribute oil. On Model 4049, remove wheels once each season and grease axle shafts.

AUGER GEARBOX

The auger gearbox is factory lubricated and should not require further lubrication unless disassembled for overhaul.

MAINTENANCE

Inspect belts and pulleys and renew belts if excessively worn, frayed or burned. Renew pulleys if belt grooves are worn or damaged, or if bearings in idler pulleys are loose or rough. Check drive chains and sprockets and renew if excessively worn or damaged.

ADJUSTMENTS

AUGER HEIGHT

Auger height can be adjusted by loosening skid shoe mounting bolts and repositioning the skid shoes. Auger height should be at lowest setting (skid shoes raised) when operating over smooth, paved surfaces and toward the highest setting (skid shoes lowered) when operating over rough surfaces. Be sure skid shoes are adjusted equally.

SHIFT ROD

Disconnect shift rod (6—Fig. P30) from shift control lever (5) and place both the transmission and shift lever in neutral position. Turn shift control rod in or out of ferrule (14) at bottom of rod so upper end can be inserted in hole in shift lever. Secure rod in lever with cotter pin.

DRIVE CLUTCH

Drive clutch spring (8—Fig. P30) at lower end of drive control rod (4) should be loose with drive control lever (1) released. If not, unhook spring from drive clutch bracket (11). Turn adjusting nut (9) inside drive clutch spring (8) so hook of spring is aligned with center of hole in drive clutch bracket then rehook spring in bracket. With drive control lever held against handle grip, the drive clutch should engage and move the snowthrower without belt slippage.

AUGER CLUTCH

To check adjustment, remove belt cover and hold auger control lever (2—Fig P30) against handle. The outer end of the slot in the end of auger brake linkage (25—Fig. P31) should be against the spacer on the auger belt idler as shown in Fig. P32. If not, disconnect auger control rod ferrule (Fig. P32) from auger clutch bracket and turn ferrule until proper adjustment is obtained with ferrule reconnected to auger clutch bracket.

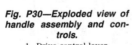

Fig. P30—Exploded view of handle assembly and controls.

1. Drive control lever
2. Auger control lever
3. Auger control rod
4. Drive control rod
5. Shift control lever
6. Shift control rod
7. Chute crank
8. Drive clutch spring
9. Adjusting nut
10. Shoulder bolt
11. Drive clutch bracket
12. Drive clutch rod
13. Drive spring
14. Ferrule
15. Shift bracket
16. Ferrule
17. Auger clutch bracket
18. Drive pulley support
19. Frame cover
20. Auger spring

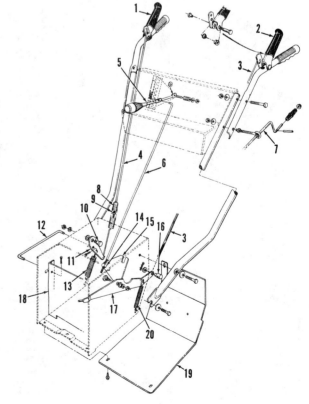

OVERHAUL

ENGINE

Engine make and model numbers are listed at beginning of this section. Refer to Tecumseh 4-stroke engine section for engine overhaul information.

DRIVE BELTS

Disconnect chute crank (5—Fig. P35) at chute and remove upper belt cover. Remove the two shoulder bolt belt guides at engine crankshaft pulley. Re-

move the top bolts attaching auger housing to snowthrower frame assembly and loosen but do not remove the two bottom bolts. Lift up on auger belt to pull auger housing off of snowthrower frame and separate auger housing from the main unit.

To remove auger drive belt, remove the four shoulder bolt belt guides at auger pulley. The auger belt can now be removed from the bottom pulley. To remove traction drive belt, disconnect drive idler spring (8—Fig. P31) at engine plate and remove belt from engine pulley and bottom drive pulley. Install new belt(s) and reassemble by reversing disassembly procedure. Be sure pin on

front end of auger clutch bracket engages slot in auger brake bracket as shown in Fig. P33 before installing auger housing mounting bolts. Adjust auger and drive clutches as outlined in preceding paragraphs.

FRICTION WHEEL

To renew friction disc, proceed as follows: Move shift lever to reverse position. Tip snowthrower up so it rests on front of auger housing and remove bottom cover, then remove friction wheel (38—Fig. P34) from wheel adapter (40). Place new wheel with cup side away from adapter and install retaining bolts finger tight. Turn the wheel to be sure it does not wobble, then securely tighten retaining bolts. Replace bottom cover.

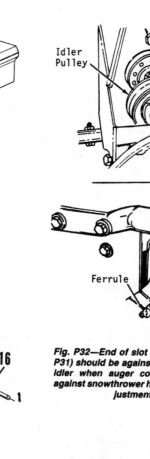

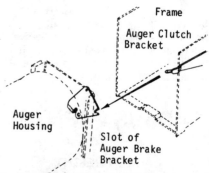

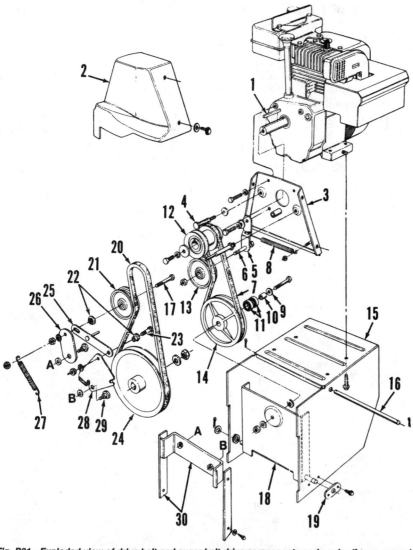

Fig. P31—Exploded view of drive belt and auger belt drive components and engine/frame assembly. Back side of drive pulley (14) has drive surface for friction wheel; depressing drive control lever moves pulley against friction wheel (38—Fig. P34).

1. Engine pulley key
2. Belt cover
3. Engine plate
4. Shoulder bolt belt guide
5. Drive clutch idler bracket
6. Idler bolt
7. Drive belt
8. Drive idler spring
9. Belleville washer
10. Spacer
11. Ball bearings
12. Engine pulley
13. Drive belt idler
14. Drive pulley
15. Main frame
16. Sliding bracket rod
17. Auger idler bolt
18. Drive pulley support
19. Support axle bracket
20. Auger belt
21. Auger belt idler
22. Shoulder spacers
23. Brake pivot bolt
24. Auger pulley
25. Brake linkage
26. Auger idler bracket
27. Auger idler spring
28. Brake bracket
29. Shoulder bolt
30. Blower housing parts

Fig. P32—End of slot in brake linkage (25—Fig. P31) should be against spacer on the auger belt idler when auger control lever is depressed against snowthrower handle. Refer to text for adjustment procedure.

Fig. P33—Pin on end of auger clutch bracket must engage slot in auger brake bracket when reassembling snowthrower.

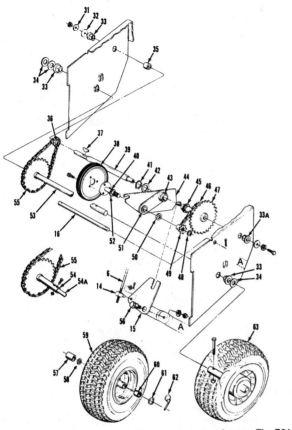

6. Shift rod
14. Ferrule
15. Shift linkage bracket
16. Sliding bracket rod
31. Belleville washer
32. Flat washer
33. Flange bearing
33A. Flange bearing
34. Flat washer(s)
35. Spacer
36. Sprocket
37. Hi-pro key
38. Friction wheel
39. Hex shaft
40. Friction wheel adapter
41. Snap ring
42. Flat washer
43. Bearing
44. Sliding bracket
45. Flat washer
46. Roller chain
47. Sprocket
48. Snap ring
49. Sprocket
50. Flat washer
51. Bronze bearing
52. Flat washer
53. Sprocket & axle assy. (Models 4050 & 4050A)
54. Sprocket (Model 4049)
55. Roller chain
56. Shoulder bolt
57. Spacer (Model 4049)
58. Flat washer (Model 4049)
59. Wheel assy. (Model 4049)
60. Sleeve bearing (Model 4049)
61. Snap ring
62. Klick pin
63. Wheel assy. (Models 4050 & 4050A)

AUGER GEARBOX

To remove auger gearbox, follow procedure outlined in preceding paragraph for removal of auger drive belt. Then, remove auger drive pulley (7—Fig. P35) from rear of impeller shaft (29) and unbolt auger bearing housings (37) at each end of auger housing. Remove the gearbox and auger flight assembly from front of housing, then remove impeller (18) from input shaft and auger flights from driveshaft. Note the position of washers and spacers as you are disassembling the unit.

To disassemble gearbox, refer to exploded view in Fig. P35 and remove the screws retaining upper gear housing half (19). Note thrust washer and spacer positions as you disassemble gearbox. Clean and inspect all parts and renew any worn or damaged parts. Lubricate all parts prior to reassembly and pack gearbox with 10 ounces (295 mL) of Shell Alvania Grease EPR00.

Fig. P34—Exploded view of drive components; also see Fig. P31. Notch in shift linkage bracket (15) engages pin on sliding bracket (44). Sliding bracket is supported by sliding bracket rod (16).

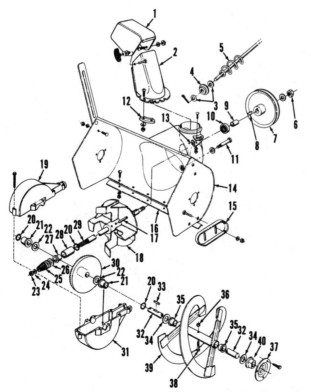

1. Upper chute
2. Lower chute
3. Flat washers
4. Bushing
5. Chute crank
6. Pulley nut
7. Auger drive pulley
8. Hi-pro key
9. Spacer
10. Ball bearing
11. Shoulder bolt
12. Chute flange keeper
13. Chute crank bracket
14. Auger housing
15. Skid shoes
16. Scraper bar (shave plate)
17. Spring pin
18. Impeller (blower fan)
19. Upper gear housing half
20. "O" ring
21. Plastic flange bearing
22. Flat washer
23. Hex nut
24. Flat washer
25. Pinion gear
26. Flat washer(s)
27. Flange bearing
28. Sleeve bearing
29. Impeller (blower) axle
30. Bevel gear
31. Lower gear housing half
32. Auger flight (spiral) axle
33. Key (Hi-pro or square)
34. Flat washer
35. Flange bearing
36. Shear bolt nut
37. Bearing housing
38. Shear bolt
39. Auger flight (spiral)
40. Flange bearing

Fig. P35—Exploded view of auger housing, gearbox and auger flight assembly.

J.C. PENNEY

Model	Engine Make	Engine Model	Self-Propelled?	No. of Stages	Scoop Width
4048	Tecumseh	HMSK80	Yes	2	28 in. (711 mm)
4048A	Tecumseh	HMSK80	Yes	2	28 in. (711 mm)
4051	Tecumseh	HMSK100	Yes	2	28 in. (711 mm)

LUBRICATION

ENGINE

Refer to Tecumseh 4-stroke engine section for engine lubrication information. Recommended fuel is regular or unleaded regular gasoline.

DRIVE MECHANISM

Lubricate drive chains and sprockets at least once each season or after every 25 hours of operation with a SAE 30 engine oil. Auger shaft bushings, traction drive shaft bushings and axle shaft bushings should be oiled with SAE 30 engine oil at least once each season or after every 25 hours of operation. Remove wheels and grease axle shaft with multipurpose automotive grease at least once each season. Lubricate control rod and lever pivot points to ensure smooth operation.

AUGER GEARBOX

The auger gearbox is lubricated with Shell Alvania grease EPROO. Remove filler plug (30—Fig. P43) and check lubricant level using a wire as a dipstick. Read lubricant level on wire. Add lubricant through plug opening to bring gearbox level up to about 1/2 full. Capacity is 4 ounces (120 mL), do not overfill.

MAINTENANCE

Inspect drive belts and pulleys and renew belts if cracked, frayed, burned or otherwise damaged. Renew pulleys if belt grooves are excessively worn or damaged or if idler pulley bearings are loose or rough.

ADJUSTMENTS

AUGER HEIGHT

Auger height can be adjusted by loosening skid shoe mounting bolts and repositioning the skid shoes. Auger should be in lowest setting (skid shoes raised) when operating over smooth paved surfaces or raised (skid shoes lowered) when operating over rough surfaces.

CLUTCH RODS

With auger clutch lever (26—Fig. P40) released against bumper (7), lower end of auger clutch spring (19) should fit loosely in the hole at rear of auger clutch bracket (16). If not, disconnect spring and slide spring upward on auger clutch rod (22) to expose adjusting nut (18) and turn nut up or down so bottom of hook in spring will be even with middle of hole in bracket when lifting clutch rod up to take slack out of rod

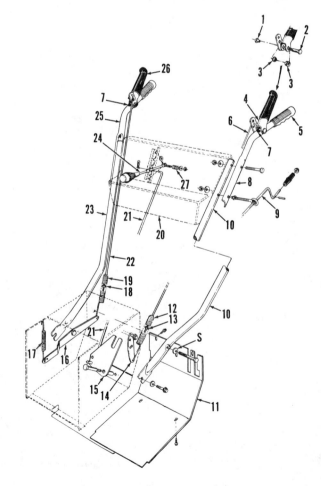

Fig. P40—Exploded view of handle and controls.

1. Shoulder nut
2. Shoulder bolt
3. Plastic bushing
4. Drive clutch lever
5. Handle grip
6. Drive clutch rod
7. Bumper
8. Upper L.H. handle
9. Chute crank
10. Lower L.H. handle
11. Main frame cover
12. Drive clutch spring
13. Adjusting nut
14. Drive clutch bracket
15. Shift linkage bracket
16. Auger clutch bracket
17. Auger clutch spring
18. Adjusting nut
19. Auger clutch spring
20. Handle panel
21. Shift rod
22. Auger clutch rod
23. Lower R.H. handle
24. Shift lever
25. Upper R.H. handle
26. Auger clutch lever
27. Shift lever spring

and spring. There must not be any tension on clutch rod spring with the clutch lever in disengaged position. Repeat adjustment procedure for drive clutch rod (6).

SHIFT ROD

Move shift lever to neutral position on handle panel; sliding bracket assembly (13—Fig. P42) should be aligned with slot (S—Fig. P40) in main frame cover (11). Also, with the drive clutch lever (4) in engaged position, the snowthrower should roll back and forth with shift lever in neutral. If necessary to adjust shift rod, disconnect shift rod (21) from shift lever (24) at control console. Move shift lever to neutral position and move shift rod until edge of sliding bracket is visible through slot (S) in main frame cover. If cover is removed, sliding bracket should be aligned on center of neutral stop bolt (19—Fig. P42). Thread shift rod in or out of ferrule (41) in shift bracket (40) at bottom end of rod until top end aligns with hole in shift lever. Reconnect shift rod and check adjustment.

Shift lever spring (27—Fig. P40) should be tight enough to hold lever in handle panel detent notches, yet allow lever to move without binding on detents.

main frame (21). Two people are required to separate the auger housing from main frame. With someone holding handles, push in on auger idler pulley to lift brake pad (BR) out of pulley groove, then lift auger housing from main frame. At this point, either the auger belt or drive belt, or both, can be renewed.

To remove auger belt after separating front and rear units, remove the five shoulder bolts (5—Fig. P43) and Belleville washers from auger housing and remove belt from pulley. An alternate method is to remove the cap screw holding auger drive pulley to impeller shaft and remove the pulley and belt; be careful not to lose the drive key and Belleville washer. Install new belt by reversing removal procedure. If auger pulley was removed, be sure key is in place on shaft and install Belleville washer with cup side toward pulley. Tighten pulley retaining cap screw securely.

To remove traction drive belt, unhook drive idler spring (16—Fig. P41) from engine plate, then remove the belt from drive pulley. Remove shoulder bolt (4), spacer (5) and Belleville washer at right side of engine pulley and remove the belt. Install new belt by reversing removal procedure.

Be sure that idler pulley and/or brake pad (BR) are not behind the drive pulley when assembling auger housing to main frame.

DRIVE MECHANISM

To overhaul drive mechanism, first remove drive belt as outlined in a preceding paragraph, then remove engine assembly and traction drive pulley (19—Fig. P41). Further disassembly procedure is evident after inspection of unit and reference to Fig. P42. Be sure not to over tighten the hex shaft self-aligning bearing (51—Fig. P42) as hex shaft (50) needs to flex in the bearing for free movement of sliding bracket (13). If adjusting auger clutch rod (37) with main frame cover removed, forward point of sliding bracket (13) should be centered on head of neutral stop bolt (19) when shift lever on handle panel is in neutral position.

FRICTION WHEEL

To remove rubber faced friction wheel, first tip snowthrower forward onto front of auger housing and remove the cover from main frame. Friction

OVERHAUL

ENGINE

Engine make and model numbers are listed at beginning of this section. Refer to Tecumseh 4-stroke engine section for engine overhaul information.

DRIVE BELTS

To renew drive belts, it is first necessary to split the snowthrower between the auger housing and main frame assembly as follows: Disconnect the deflector chute crank at the chute end and remove drive belt cover. Remove shoulder bolt (17—Fig. P41) and Belleville washer at left side of engine pulley. Remove shoulder bolt (4) and spacer (5) at right hand side of engine pulley, being careful not to lose the Belleville (spring cup) washer located between idler bracket and engine plate. Slip auger drive belt off engine pulley and loosely install shoulder bolt, roller and Belleville washer at right side of engine pulley. Remove the top cap screws retaining auger housing flanges (24) to

Fig. P41—Exploded view of belt drive system. Auger pulley (1) is also shown in Fig. P43 and traction drive pulley (19) mounts on drive plate spindle (35—Fig. P42).

1. Auger pulley
2. Auger belt
3. Auger belt idler
4. Shoulder bolt
5. Spacer
6. Auger idler bracket
7. Auger clutch rod
8. Engine pulley
9. Traction drive belt
10. Belt cover
11. Square key
12. Auger brake linkage
13. Traction drive idler
14. Shoulder bolt
15. Drive idler bracket
16. Drive idler spring
17. Shoulder bolt
18. Engine spacer plate
19. Traction drive pulley
20. Engine bracket assy.
21. Main (motor mount) frame
22. Brake bracket assy.
23. Shoulder bolt
24. Auger housing flanges

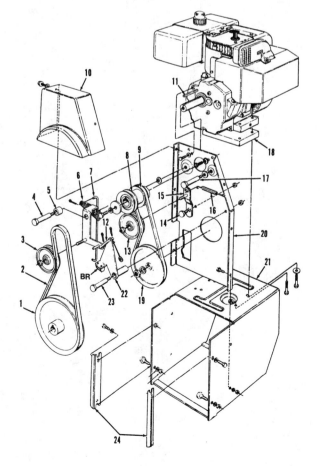

wheel (5—Fig. P42) can then be unbolted from friction wheel adapter (4) and removed from the end of hex shaft. Install new friction wheel with cup side facing toward adapter and install retaining lockwashers and nuts finger tight. Turn the wheel to be sure it does not wobble (cocked on hub), then securely tighten nuts.

AUGER SHEAR BOLTS

The auger flights are driven through special bolts (37—Fig. P43) which are designed to shear when a solid obstruction is encountered to prevent damage to snowthrower. If this occurs, be sure to replace the bolts with factory replacement parts; substitute bolts may not shear and cause extensive damage. Before installing new bolts, oil the auger flight bushings and be sure flights spin freely on the auger shaft.

AUGER GEARBOX

To remove auger gearbox, first remove auger belt as outlined in a preceding paragraph. If auger pulley (1—Fig. P43) was not removed, remove it at this time. Remove outer bearing retainer (3), then remove set screws and unlock bearing (4) from gearbox input shaft. Unbolt auger shaft bearing flanges (39) at each end of auger housing, remove flange bearings (38) and thrust washers (34) and withdraw the gearbox with impeller (10) and auger flights (36). Remove shear bolts (37), auger flights, inner thrust washers and spacers (33) from auger shaft. Remove the carriage bolt and spring pins (9), then pull impeller from shaft. Remove the bolts holding the gearbox halves (19 and 31) together and disassemble gearbox. Clean and inspect all parts and renew any excessively worn or damaged parts. Worm gear (27) can be removed from hub (28).

Reassemble by reversing disassembly procedure and lubricate gearbox using 4 ounces (120 mL) of Shell Alvania Grease EPROO.

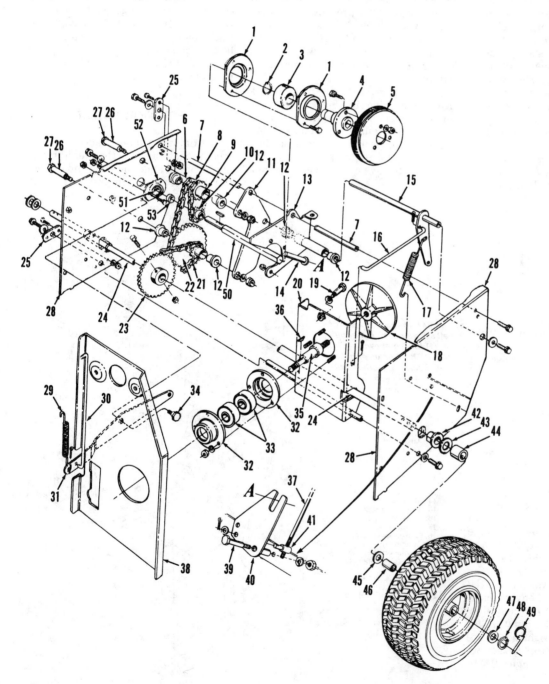

Fig. P42—Exploded view of friction drive and chain speed reduction drive components.

1. Bearing housing
2. Snap ring
3. Ball bearing
4. Friction wheel adapter
5. Friction wheel
6. Roller chain
7. Sliding support axle
8. Roller chain
9. Double sprocket
10. Hex shaft sprocket
11. Chain support bracket
12. Flange bearing
13. Sliding bracket assy.
14. Pin (part of 13)
15. Drive clutch bracket
16. Drive clutch rod
17. Drive clutch spring
18. Aluminum drive plate
19. Neutral stop bolt
20. Drive pivot plate
21. Roller chain
22. Double sprocket
23. Axle sprocket
24. Axle shaft
25. Pivot shaft bracket
26. Sprocket hub tube
27. Sprocket bolt
28. Main frame assy.
29. Auger clutch spring
30. Auger clutch rod
31. Auger clutch lever
32. Bearing housing
33. Ball bearings
34. Shoulder bolt
35. Drive plate spindle
36. Hi-pro key
37. Auger clutch rod
38. Engine bracket assy.
39. Shoulder bolt
40. Shift bracket
41. Ferrule
42. Axle flange bearing
43. Spacer washer
44. Spacer
45. Wheel spacer washer
46. Wheel bushing
47. Wheel spacer washer
48. Snap ring
49. Klick pin
50. Hex shaft
51. Bearing
52. Bearing housing
53. Spacer

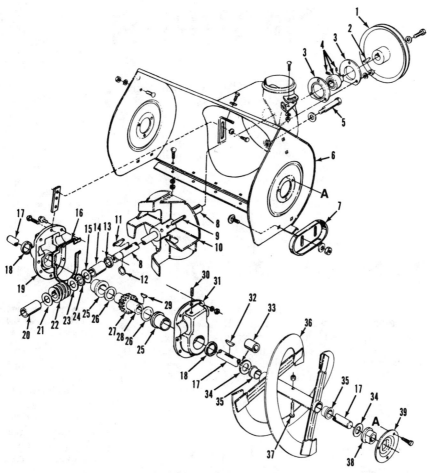

Fig. P43—Exploded view of auger housing, impeller, auger gearbox and auger flights.

1. Auger pulley
2. Hi-pro key
3. Bearing retainer
4. Self-aligning bearing
5. Shoulder bolt
6. Auger housing
7. Skid shoe
8. Impeller shaft
9. Spring pin
10. Impeller
11. Hi-pro key
12. Snap ring
13. Oil seal
14. Sleeve bearing
15. Flat washer
16. Gasket
17. Auger shaft
18. Oil seal
19. Housing half
20. Sleeve bushing
21. Flat washer
22. Worm gear
23. Flat washer
24. Thrust bearing
25. Flange bearing
26. Flat washer
27. Worm gear
28. Gear hub
29. Woodruff key
30. Filler plug
31. Housing half
32. Woodruff key
33. Spacer
34. Thrust washer
35. Plastic bushing
36. Auger flight
37. Shear bolt
38. Flange bearing
39. Bearing flange

SNAPPER

SNAPPER POWER EQUIPMENT
McDonough, GA 30253

Model	Engine Make	Engine Model	Self-Propelled?	No. of Stages	Scoop Width
3200	*	*	No	1	20 in. (508 mm)
3201	*	*	No	1	20 in. (508 mm)

*Briggs & Stratton 62032 or Tecumseh AH600

LUBRICATION

ENGINE

Recommended fuel is regular or unleaded gasoline mixed with a good quality oil designed for two-stroke engines. Spefied fuel:oil mix ratio is 32:1. Be sure fuel and oil are mixed well.

MAINTENANCE

Inspect drive belt (16–Fig. S2) and renew if frayed, cracked, burnt or otherwise damaged. Renew belt pulleys (4 and 12) if grooves are excessively worn or damaged. Inspect impeller (21) and impeller flighting (22 and 23) for damage which may reduce efficiency of the unit. Renewal may be necessary if damage is severe.

ADJUSTMENTS

AUGER CLUTCH

Pull clutch control lever (1–Fig. S1) up against handle (3). There should be no slack in clutch cable and the cable spring should be extended 3/8 to ½ inch (9.5-12.7 mm). Turn adjuster nut on cable until proper tension is achieved. Cable should be slack when clutch control lever is released.

DRIVE BELT

Refer to Fig. S3. If clutch does not release completely when lever is down, it may be necessary to readjust belt guide (4). Loosen screws and adjust for a clearance of 1/16 inch (1.6 mm) between belt and lower guide when clutch lever is in up position.

Note that auger pulley (17–Fig. S2) is threaded on auger (21) with left-hand threads. To remove pulley, hold auger and turn pulley clockwise.

OVERHAUL

ENGINE

Engine make and model numbers are listed at beginning of section. Refer to Briggs & Stratton or Tecumseh engine section for overhaul procedures.

BELT

Refer to Fig. S3. Remove belt guard and turn auger pulley (7) clockwise while holding auger. This will loosen belt enough to allow removal and replacement of belt. At this time, inspect idler pulley bearing and renew as needed. With new belt installed, check for proper clutch adjustment and readjustment as needed. Refer to adjustment section for procedure.

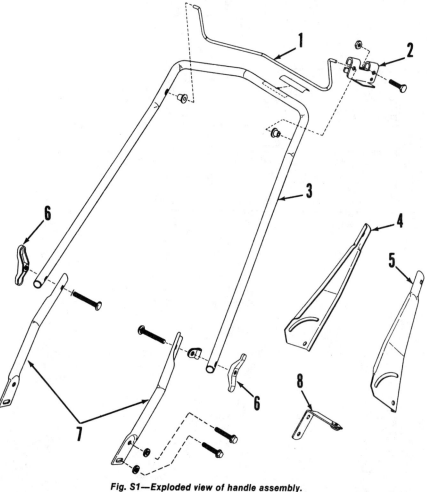

Fig. S1—Exploded view of handle assembly.

1. Control lever	3. Upper handle assy.	5. Brace, L.H.	7. Lower handle assy.
2. Clip	4. Brace, R.H.	6. Wing nut	8. Bracket

1. Cable assy.
2. Collector assy.
3. Engine brace
4. Brake spring
5. Woodruff key
6. Engine pulley
7. Set screw
8. Belt guide
9. Idler spring
10. Idler pulley
11. Idler arm (3200)
12. Idler arm (3201)
13. Spacer bearing (3201)
14. Idler brake spring
    (3201)
15. Cover
16. Belt
17. Auger pulley
18. Bearing retainer
19. Bearing
20. Scraper blade
21. Auger
22. Flighting
23. Flighting

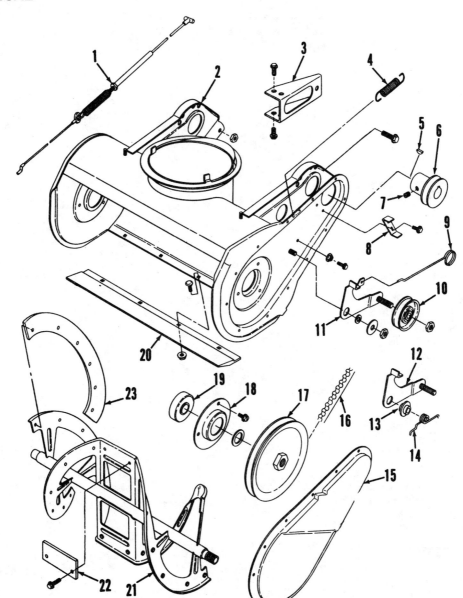

Fig. S2—Exploded view of snowthrower.

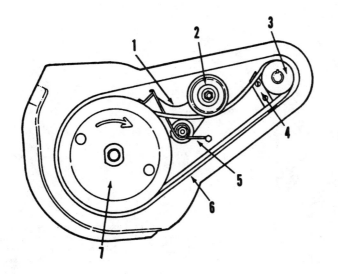

Fig. S3—View of belt and pulley system.

1. Idler arm
2. Idler pulley
3. Drive pulley
4. Belt guide
5. Idler brake spring
6. Auger drive belt
7. Auger pulley

SNAPPER

Model	Engine Make	Engine Model	Self-Propelled?	No. of Stages	Scoop Width
4220	Tecumseh	HSSK40	Yes	2	22 in. (559 mm)
5240	Tecumseh	HSSK50	Yes	2	24 in. (610 mm)

LUBRICATION

ENGINE

Refer to Tecumseh engine section for lubrication requirements. Recommended fuel is regular or unleaded gasoline.

AUGER ASSEMBLY

AUGER. At the end of each season, remove shear bolts and lubricate the shaft through the two fittings on the auger shaft shown in Fig. S4. Spin auger on shaft to distribute grease, then reinstall shear bolts.

AUGER BEARINGS. Once each season, lubricate bearings through auger grease fittings shown in Fig. S4 with two shots of grease.

GEARBOX. Once each season, lubricate gearbox grease fitting with five shots of grease. Recommended grease is Snapper "00" or equivalent.

CHUTE ASSEMBLY

Once each season, spread a light coat of grease on the chute worm gear/ring gear assembly. Oil the chute crank pivot point between chute and collector.

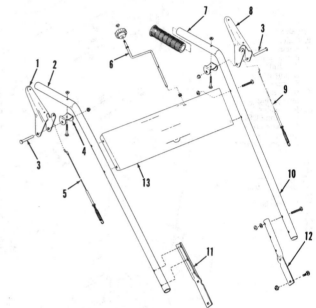

1. Auger clutch lever
2. Right upper handle
3. Pin
4. Bracket
5. Auger clutch cable
6. Chute crank
7. Left upper handle
8. Traction clutch lever
9. Traction clutch cable
10. Left upper handle
11. Right lower handle
12. Left lower handle
13. Center brace

Fig. S5—Exploded view of handle and control assembly.

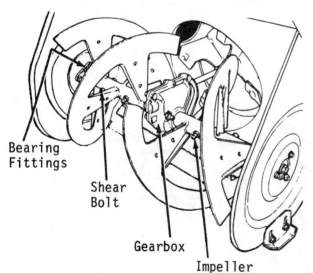

Bearing Fittings

Shear Bolt

Gearbox

Impeller Grease Fitting

Fig. S4—View of lubrication points.

WHEELS

Once each season, remove wheels and apply a light coat of grease to the axle shafts.

MAINTENANCE

AUGERS. Augers are secured to the auger shaft with shear pins which will fracture when the auger contacts an obstruction. Use only original equipment type shear pins.

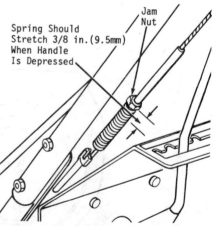

Jam Nut

Spring Should Stretch 3/8 in. (9.5mm) When Handle Is Depressed

Fig. S6—View showing proper adjustment of traction or auger clutch cable. Spring should stretch 3/8 inch (9.5 mm) when control handle is depressed.

BELTS. Inspect drive belts and belt pulley grooves frequently. Renew belt(s) if frayed, cracked, burnt or otherwise damaged.

HEIGHT

To adjust auger height, loosen skid (21 – Fig. S8) mounting bolts and reposition skids. Increase height when operating snowthrower over an irregular surface.

BELT TENSION

TRACTION BELT. Traction drive belt (2 – Fig. S7) tension is automatically maintained by idler pulley spring (13). No adjustment is needed.

AUGER BELT. Auger drive belt (3 – Fig. S7) tension is adjusted by moving idler pulley (20) in or out as needed. Belt guide should be positioned approximately 1/16 inch (1.6 mm) away from auger belt.

CLUTCH CABLES

Refer to Fig. S6. When either cable is properly adjusted, the cable spring should be stretched out 3/8 inch (9.5 mm) when clutch lever is depressed. To adjust, loosen jam nut and rotate adjuster sleeve until adjustment is correct, then retighten jam nut.

CHUTE CRANK

Chute worm gear (18 – Fig. S8) may be moved in or out as needed to adjust for easy cranking. Loosen the adjusting nut and jam nut which holds the worm gear bracket (20) to collector housing (16). Retighten both nuts when adjustment is completed.

OVERHAUL

ENGINE

Engine make and model numbers are listed at the beginning of this section. Refer to Tecumseh engine section for overhaul procedure.

BELTS

To replace belts, remove upper belt cover (1 – Fig. S7) and lower cover (32 – Fig. S9). Loosen belt guide (8 – Fig. S7) and slide away from belt. Loosen traction idler pulley and bracket (18 and 20) and slide idler pulley assembly away from belt. Roll traction belt (2) off engine pulley (5). Roll auger belt (3) off engine pulley (7) in a similar manner. Tip snowthrower on left side and remove

Fig. S7—Exploded view of pulley and belt system.

1. Cover
2. Traction belt
3. Auger belt
4. Woodruff key
5. Engine pulley (traction)
6. Key
7. Engine pulley (auger)
8. Belt guide
9. Plate
10. Belt finger
11. Carriage bolt
12. Idler bracket
13. Spring
14. Idler pulley
15. Frame
16. Brake rod
17. Cotter key
18. Brake arm
19. Break tube
20. Idler pulley

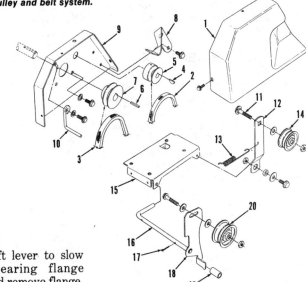

right wheel. Move shift lever to slow position. Remove bearing flange (41 – Fig. S9) screws and remove flange. Remove hex shaft (36), gear (38) and bearing (40) as an assembly. Tilt chain case (Fig. S10) sideways and remove. Depress auger clutch lever, remove auger belt first and then remove the traction drive belt.

Install new belts by reversing disassembly procedure. Adjust auger belt idler for approximately ½ inch (12.7 mm) of travel.

DRIVE DISC AND CHAIN CASE

Tip snowthrower on left side and remove right wheel. Move shift lever to

"slow" position. Remove bearing flange (41 – Fig. S9) screws and remove flange. Remove hex shaft (36), gear (38) and bearing (40) as an assembly. Tilt chain case sideways and remove. Disassembly is evident after inspection and referral to exploded view in Fig. S10. Inspect drive disc ring (2), chain (15), sprockets (14 and 10) and bearings (12) for wear

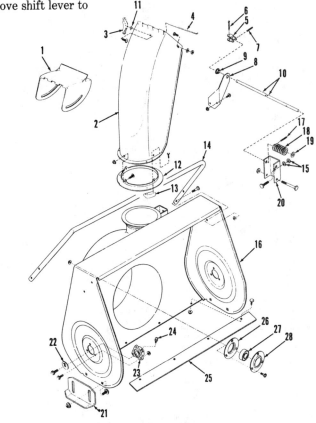

1. Deflector
2. Chute
3. Wing nut
4. Hinge pin
5. Connector
6. Cotter pin
7. Roll pin
8. Bracket
9. Bushing
10. Worm shaft
11. Washer
12. Ring
13. Retainer
14. Drift cutter
15. Nut
16. Collector
17. Roll pin
18. Worm gear
19. Washer
20. Bracket
21. Skid
22. Washer
23. Bearing
24. Grease fitting
25. Blade
26. Housing
27. Bearing
28. Housing

Fig. S8—Exploded view of collector and chute assembly.

or damage and renew as needed. Coat chain and sprockets with Snapper "OO" grease or equivalent during reassembly.

AUGER GEARBOX

Disassembly of auger assembly requires the separation of the auger/collector assembly from the traction unit. Unbolt retainers (13 – Fig. S8) and remove discharge chute (2). Remove belt guard (1 – Fig. S7) and lower cover (32 – Fig. S9). Remove auger and traction drive belts as outlined in BELTS section. Unbolt chute worm gear bracket from blower housing. Remove the two bolts between the blower chute and the belt cavity and carefully separate the collector and traction units.

Remove drive plate (22 – Fig. S11) and pulley (24) from auger shaft. Remove auger drive shaft bearing and housing assembly (26, 27 and 28 – Fig. S8) and unbolt the left and right auger shaft bearings (23). Remove impeller, gearbox

and augers as an assembly. Refer to Fig. S11 for an exploded view. Drive out roll pins (1) and remove impeller (3). Remove bearings from ends of auger shaft and remove shear bolts. Remove augers noting that tabs face inside toward the gearbox. Remove bolts and split gearcase halves (12 and 14). Inspect worm shaft (10), worm gear (19), thrust washers (18) and bearings (5 and 20) for excessive wear or damage. Renew as needed. Be careful not to damage the

"O" ring (13) during disassembly. Reassembly is reverse of disassembly procedure. Refill gearbox with 5 ounce (148 mL) of Snapper "OO" grease or equivalent.

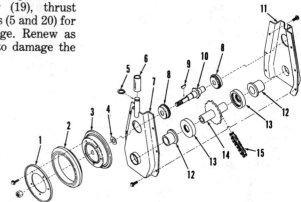

1. Plate
2. Driven ring
3. Hub
4. Washer
5. Snap ring
6. Roller
7. Case
8. Bearing
9. Woodruff key
10. Shaft & sprocket
11. Cover
12. Bearing
13. Spacer
14. Hub & sprocket
15. Chain

Fig. S10—Exploded view of chain case assembly.

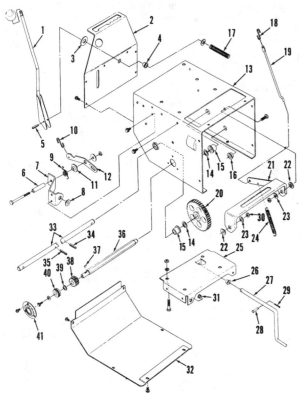

Fig. S9—Exploded view of frame and traction drive system.

1. Shift lever	15. Bushing	
2. Rear frame cover	16. Bearing	29. Cotter pin
3. Thrust washer	17. Shift spring	30. Nut
4. Bearing	18. "S" hook	31. Bushing
5. Roll pin	19. Clutch rod	32. Bottom cover
6. Spacer	20. Gear	33. Axle shaft
7. Bracket	21. Clutch bracket	34. Roll pin
8. Bushing	22. Bushing	35. Roll pin
9. Spring	23. Washer	36. Hex shaft
10. "S" hook	24. Spring	37. Woodruff key
11. Bushing	25. Engine support	38. Pinion gear
12. Lever	26. Bearing	39. Shim
13. Frame	27. Shift rod	40. Bearing
14. Washer	28. Pin	41. Bearing retainer

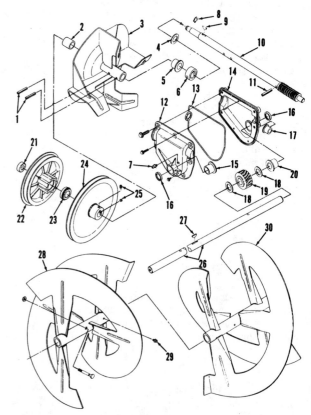

Fig. S11—Exploded view of impeller, auger and gearbox assembly.

1. Roll pins	11. Roll pin	21. Bearing
2. Spacer	12. R.H. gearcase cover	22. Drive plate
3. Impeller	13. "O" ring	23. Bearing
4. Thrust washer	14. L.H. gearcase cover	24. Pulley
5. Bearing	15. Bearing	25. Set screws
6. Collar	16. Seal	26. Auger shaft
7. Grease fitting	17. Bushing	27. Woodruff key
8. Snap ring	18. Thrust washer	28. R.H. auger
9. Woodruff key	19. Worm gear	29. Grease fitting
10. Worm shaft	20. Bearing	30. L.H. auger

SNAPPER

Model	Engine Make	Engine Model	Self-Propelled?	No. of Stages	Scoop Width
5230	Tecumseh	HSK50	Yes	2	23 in. (584 mm)
5241	*	*	Yes	2	23 in. (584 mm)
8230	B&S	190412	Yes	2	23 in. (584 mm)
8241	B&S	190402	Yes	2	24 in. (610 mm)
8242	B&S	190402	Yes	2	24 in. (610 mm)
8260	B&S	190412	Yes	2	26 in. (660 mm)
8261	B&S	190402	Yes	2	26 in. (660 mm)
8262	B&S	190402	Yes	2	26 in. (660 mm)
10300S	Tecumseh	HMSK100	Yes	2	30 in. (762 mm)
10301	Tecumseh	HMSK100	Yes	2	30 in. (762 mm)
10302	Tecumseh	HMSK100	Yes	2	30 in. (762 mm)

*B&S Model 130202 or Tecumseh Model HSK50

LUBRICATION

ENGINE

Refer to Briggs & Stratton or Tecumseh engine section for lubrication requirements. Recommended fuel is regular or unleaded gasoline.

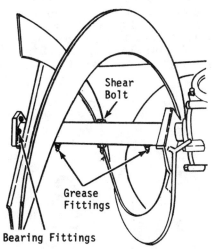

Fig. S12—View of lubrication points for auger assembly.

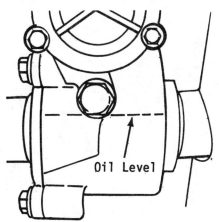

Fig. S13—Fill auger gearbox to oil level shown.

AUGER ASSEMBLY

AUGER. At the end of each season, remove shear bolts and lubricate the shaft through the four fittings on the auger shaft shown in Fig. S12. Spin auger on shaft to distribute grease, then reinstall shear bolts.

AUGER BEARINGS. Once each season, lubricate auger grease fittings with two shots of grease.

GEARBOX. Refer to Fig. S13. After every 25 hours of operation, check gearbox oil level as shown. Specified lubricant is a good grade of engine oil.

CHUTE ASSEMBLY

Once each season, spread a light coat of grease on the chute worm gear ring gear assembly. Oil the chute crank pivot point between chute and collector.

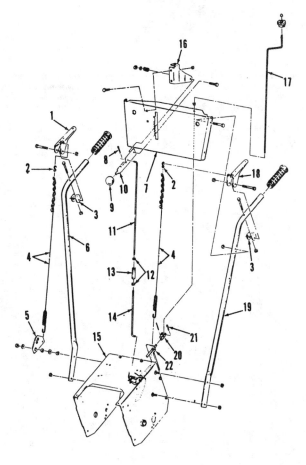

Fig. S14—View of handle and control assembly.

1. Auger clutch lever
2. S-hook
3. Bracket
4. Cable assy.
5. Lever
6. Right handlebar
7. Center brace
8. Cotter pin
9. Knob
10. Shift lever
11. Upper shift rod
12. Jam nuts
13. Connector
14. Lower shift rod
15. Frame
16. Bracket
17. Chute crank
18. Traction clutch lever
19. Left handlebar
20. U-joint
21. Cotter pin
22. Bracket

WHEELS

Once each season, remove wheels and apply a light coat of grease to the axle shafts.

MAINTENANCE

AUGERS. Augers are secured to the auger shaft with shear pins which will

fracture when the auger contacts an obstruction. Use only original equipment type shear pins.

BELTS. Inspect drive belts and belt pulley grooves frequently. Renew belt(s) if frayed, cracked, burnt or otherwise damaged.

ADJUSTMENTS

AUGER HEIGHT

To adjust auger height, loosen skid (23—Fig. S18) mounting bolts and reposition skids. Increase height when operating snowthrower over an irregular surface.

BELT TENSION

TRACTION BELT. Traction drive belt (16—Fig. S17) tension is automatically maintained by idler pulley spring. No adjustment is required.

AUGER BELT. Auger drive belt (15—Fig. S17) tension is adjusted by moving idler pulley (21) in or out as needed. Belt guide should be positioned approximately 1/16 inch (1.6 mm) away from auger belt.

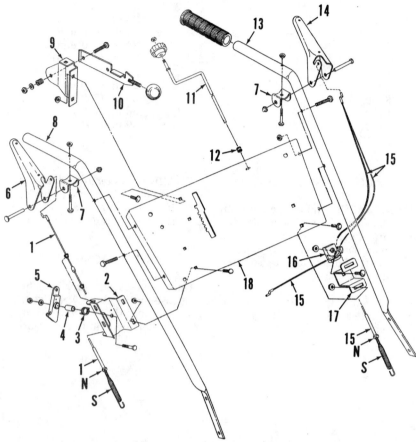

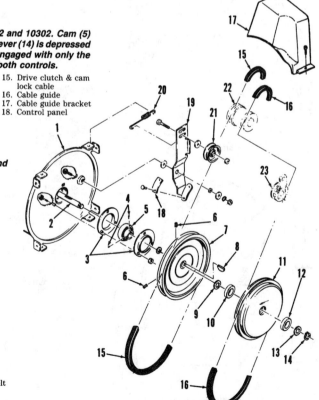

Fig. S15—Exploded view of handle and control assembly for Models 8242, 8262 and 10302. Cam (5) locks the auger control cable (1) in engaged position whenever the drive control lever (14) is depressed while auger control lever (6) is engaged. Both the auger and drive will remain engaged with only the drive control lever depressed; releasing the drive control disengages both controls.

1. Auger control cable	6. Auger control lever	11. Chute crank	15. Drive clutch & cam lock cable
2. Cam lock bracket	7. Lever bracket	12. Chute crank bushing	16. Cable guide
3. Cam lock spring	8. R.H. handlebar	13. L.H. handlebar	17. Cable guide bracket
4. Spacer	9. Shift lever bracket	14. Drive control lever	18. Control panel
5. Cam lock	10. Shift lever		

Fig. S17—View of belt and pulley system.

1. Collector housing
2. Impeller shaft
3. Bearing retainer
4. Set screw
5. Bearing
6. Set screw
7. Auger/impeller drive pulley
8. Woodruff key
9. Thrust washer
10. Bearing
11. Traction drive plate/pulley
12. Bearing
13. Washer
14. Snap ring
15. Auger/impeller drive belt
16. Traction drive belt
17. Belt cover
18. Brake pad
19. Bracket
20. Spring
21. Auger/impeller idler belt
22. Engine pulley
23. Traction idler pulley

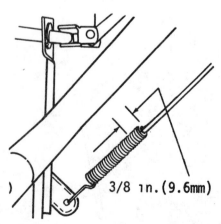

3/8 in. (9.6mm)

Fig. S16—View of clutch cable spring. Spring should stretch 3/8 inch (9.5 mm) when control lever is depressed.

CLUTCH CABLES

Refer to Fig. S16. When either cable properly adjusted, the cable spring should be stretched 3/8 inch (9.6 mm) when clutch lever is depressed.

All Models Except 8242, 8262 And 10302

To adjust, shorten or lengthen cable by repositioning S-hooks (2—Fig. S14) to a different cable (4) chain link.

Models 8242, 8262 And 10302

The adjustment on control cables is made by loosening cable jam nuts (N—Fig. S15), disconnecting lower end of spring (S) from control arm and turning spring on control cable end.

CHUTE CRANK

Chute worm gear (17—Fig. S18) may be moved in or out as needed to adjust for easy cranking. Loosen the adjusting nut and jam nut which holds the worm gear bracket (20) to the collector housing (9). Retighten bolt nuts when adjustment is completed.

OVERHAUL

ENGINE

Engine make and model numbers are listed at the beginning of this section. Refer to Briggs & Stratton or Tecumseh engine section for overhaul procedures.

BELTS

To remove belts, remove upper belt cover (17—Fig. S17) and lower cover (15—Fig. S19). Loosen traction idler pulley and bracket (21 and 19—Fig. S17) and slide idler pulley assembly away from belt. Unhook traction belt idler pulley spring from frame. Roll both belts off engine pulley (22). Unbolt retainers (8—Fig. S18) and remove discharge chute (5). Unbolt and disconnect worm gear (17) and bracket (20) assembly from collector (9). Remove collector-to-traction assembly bolts on top and sides between discharge chute and belt cavity. Separate collector and traction assembly being careful not to let traction assembly fall backwards on handles.

With collector and traction units split, slide belts off auger drive pulley (7—Fig.

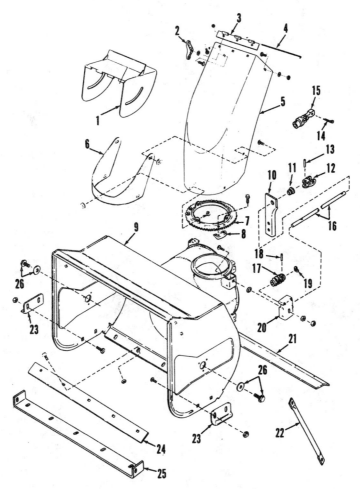

Fig. S18—Exploded view of collector assembly.

1. Deflector	9. Collector housing	(30 in. only)
2. Wing nut	10. Bracket	22. Brace (30 in. only)
3. Hinge	11. Bushing	23. Skid shoe
4. Hinge pin	12. Connector	24. Scraper blade (24 &
5. Chute	13. Roll pin	16. Worm shaft	26 in.)
6. Brace	14. Cotter pin	17. Worm gear	25. Scraper blade (30 in.)
7. Ring	15. U-joint	18. Roll pin	26. Bolt & washer
8. Retainer		19. Wave washer	(auger)
		20. Bracket	
		21. Stiffner bar	

S17) and traction drive pulley (11). Renew as needed. Reassemble by reversing removal procedure and readjust auger idler pulley as outlined in belt adjustment section.

DRIVE DISC AND CHAIN CASE

REMOVE AND REINSTALL. Tip snowthrower up on its collector housing and remove the lower cover (15—Fig. S19). Remove the left side wheel.

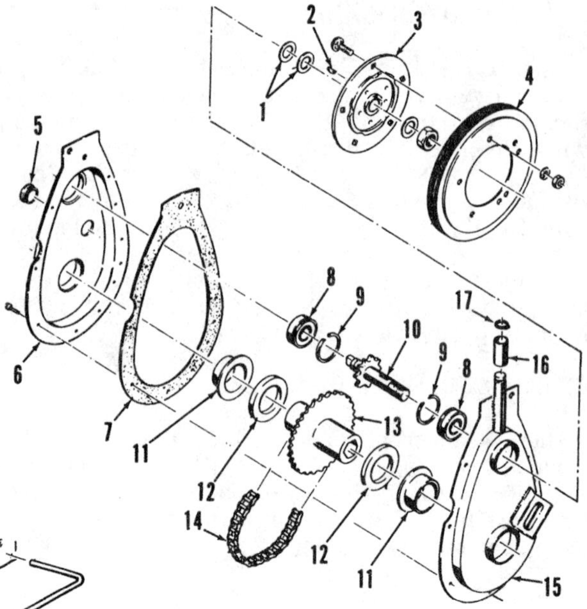

Fig. S20—Exploded view of chaincase.

1. Washers
2. Woodruff key
3. Hub
4. Drive disc
5. Plug
6. Cover
7. Gasket
8. Bearing
9. Retainer
10. Shaft & sprocket
11. Bushing
12. Thrust washer
13. Hub & sprocket
14. Chain
15. Case
16. Roller
17. Snap ring

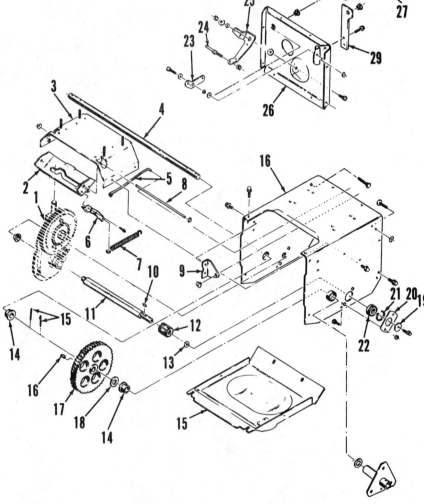

Unbolt and remove bearing retainer (20). Pull hex shaft and gear assembly (11 and 12) out of frame. Tilt chain case and remove from frame.

Reassembly is reverse of disassembly procedure.

OVERHAUL. Refer to Fig. S20 for an exploded view. Remove plug (7) and drain lubricant. Remove the seven retaining screws and pry case halves (6 and 15) apart. Inspect internal components for excessive wear and damage and renew as needed. During reassembly, renew gasket (7). Reassembly is reverse of disassembly procedure. Inspect drive disc (4) for excessive wear and renew if needed.

AUGER GEARBOX

REMOVE AND REINSTALL. Split snowthrower as outlined in belt section. Remove snap ring (14—Fig. S17) and washer(s) (13). Slide drive plate (11) off end of shaft (2). Remove auger pulley (15). Loosen set screws (4) on bearing (5). Remove auger end bearing (8—Fig. S21) bolts and slide augers (1 and 5), gearbox (2) and impeller (4) out of collector housing as an assembly.

Inspect bearings (5—Fig. S17) and (8—Fig. S21) for smooth operation. Renew if needed. Reassemble by reversing disassembly procedure.

OVERHAUL. Remove roll pins (3—Fig. S21) and slide impeller (4) off impeller shaft. Remove shear bolts (6) and

Fig. S19—Exploded view of traction drive and frame assembly.

1. Chain case
2. Bracket
3. Brace
4. Axle shaft
5. Push rod
6. Bracket
7. Spring
8. Cross shaft
9. Bracket
10. Woodruff key
11. Hex shaft
12. Pinion gear
13. Washer
14. Bushing
15. Roll pin
16. Woodruff key
17. Gear, axle shaft
18. Washer
19. Washer
20. Retainer
21. Retainer
22. Bearing
23. Bellcrank
24. Stud
25. Bracket
26. Rear cover
27. Rod
28. Bracket
29. Bracket

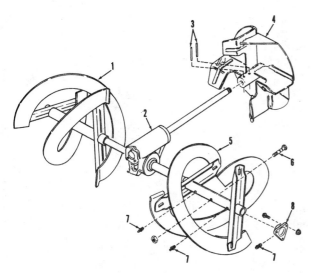

Fig. S21—Exploded view of auger and impeller assembly.

1. Right auger
2. Gearbox
3. Roll pin
4. Impeller
5. Left auger
6. Shear bolt
7. Grease fitting
8. Bearing

components for breakage and excessive wear and renew as needed. Renew "O" rings (5 & 12) and seals (3) during reassembly. Reassemble by reversing disassembly procedure. Refill gearbox as outlined in LUBRICATION section.

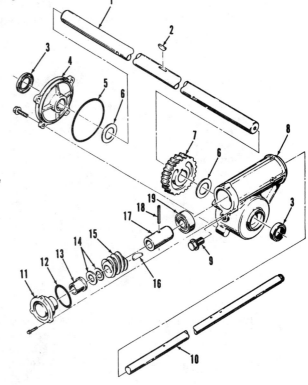

Fig. S22—Exploded view of auger gearbox.

1. Auger shaft
2. Woodruff key
3. Seal
4. Cover
5. "O" ring
6. Thrust washer
7. Gear
8. Case
9. Plug
10. Impeller shaft
11. Cap
12. "O" ring
13. Bushing
14. Thrust washers
15. Worm gear
16. Woodruff key
17. Spacer
18. Roll pin
19. Bearing

slide augers (1 & 5) off auger shaft. Refer to Fig. S22 for exploded view of auger gearbox. Remove drain plug (9) and drain oil completely from unit. Remove the three end cap (11) retaining screws and tap cap out of housing with a soft hammer. Remove impeller shaft (10) assembly by turning shaft enough to roll worm gear (15) out of mesh, then pull shaft assembly out through front of case (8). Remove the four case cover (4) retaining screws and slide cover out over auger shaft (1). Pull auger shaft and gear (7) assembly out of housing. Inspect all

TORO

TORO COMPANY
8111 Lyndale Ave. South
Minneapolis, Minnesota 55420

Model	Engine Make	Engine Model	Self-Propelled?	No. of Stages	Scoop Width
Snow Pup	Tecumseh	AH520	No	1	14 in. (356 mm)
Snow Husky	Tecumseh	AH520	No	1	21 in. (533 mm)
Snow Pup	Tecumseh	AH520	No	1	21 in. (533 mm)
Snow Master	Tecumseh	AH520	No	1	14 in. (356 mm)
Snow Master	Tecumseh	AH520	No	1	20 in. (508 mm)
S-140	Tecumseh	AH520	No	1	14 in. (356 mm)
S-200	Tecumseh	AH520	No	1	20 in. (508 mm)
S-620	Tecumseh	AH600	No	1	20 in. (508 mm)
CR 20E	Tecumseh	AH600	No	1	20 in. (508 mm)

LUBRICATION

ENGINE

Refer to following information for engine fuel and lubrication requirements. All models must use fuel mixed with 2-stroke engine oil as engines are 2-stroke type. Recommended fuel is regular or low-lead gasoline. Recommended fuel:oil mixture ratio is 32:1 for all 20-inch models, for 14-inch Model S-140, and for 14-inch Snow Pup and Snow Master models manufactured after 1975 (model numbers 31405 and 38014). All 21-inch models and 14-inch models manufactured in 1975 and earlier require a 16:1 fuel:oil mixture ratio as these engines use machined bushings in crankshaft and connecting rod areas.

DRIVE

All 20-inch models have a belt drive. On early (1975 and 1976) 20-inch Snow Master, left auger shaft bearing is an oil impregnated bushing. On all later 20-inch models, bearings are factory lubricated for life of bearing (A ball bearing kit was made available for the early units.). The belt drive idler bracket pivot bolt should be lubricated with SAE 30 oil after every 5 hours of operation.

A chain drive is used on all 14-inch and 21-inch models. The drive bearings on these models are a factory lubricated and sealed type, and should not require further lubrication for life of bearing. The drive chain should be lubricated with a good-quality chain oil at least once each season.

MAINTENANCE

On 20-inch models, periodically inspect impeller drive belt. Renew belt if frayed, cracked, excessively worn or otherwise damaged. Drive belt can be removed after removing drive belt cover and lifting spring loaded idler away from belt. With belt removed, check condition of idler and auger shaft bearings; renew if rough or loose.

On 14-inch and 21-inch models, inspect drive chain and sprockets and renew if excessively worn or damaged.

ADJUSTMENTS

IMPELLER CLEARANCE

On 14-inch models manufactured from 1978 to 1980, the scraper is a 2-piece unit with the plastic edge available separately. The scraper assembly is adjustable. Impeller-to-scraper clearance should be 1/16 to 1/8 inch (1.6-3.2 mm) and is adjustable by loosening the brackets retaining the scraper assembly at each end of impeller housing. Move the scraper assembly until correct adjustment is obtained then retightening the bolts retaining the scraper brackets.

On all other models, the scraper is a one-piece unit and bolts directly to the bottom of the housing. No adjustment is necessary.

DRIVE CHAIN

On 14 and 21-inch models, drive chain deflection should not be greater than 1/8 inch (3.2 mm). Adjustment is made by loosening engine mounting bolts and, if so equipped, electric starter mounting bolts. Move the engine and starter assembly in the slotted mounting holes and re-tighten all bolts. Recheck chain tension and be sure that engine and impeller sprockets are aligned.

DRIVE BELT

On 20-inch models, the drive belt idler is spring loaded. Other than renewing idler spring if spring is weak or broken, there is no adjustment of drive belt.

OVERHAUL

ENGINE

Refer to appropriate Tecumseh engine section for engine overhaul information.

TORO

Model	Engine Make	Engine Model	Self-Propelled?	No. of Stages	Scoop Width
CCR-1000	Tecumseh	AH600	No	1	20 in. (508 mm)
CCR-2000	Toro*	47PK9	No	1	20 in. (508 mm)

*Service data not available for this engine at time of publication.

LUBRICATION

ENGINE

Refer to appropriate Tecumseh 2-stroke engine section for engine lubrication requirements on Model CCR-1000.

DRIVE

All bearings are factory lubricated for life of bearing. The belt drive idler bracket pivot bolt should be lubricated with SAE 30 oil after every 5 hours of operation.

MAINTENANCE

Periodically inspect impeller drive belt. Renew belt if frayed, cracked, excessively worn or otherwise damaged. Drive belt can be removed after removing drive belt cover and moving spring loaded brake away from belt.

ADJUSTMENTS

CLUTCH CABLE

If drive belt slips with clutch handle in fully engaged position, it will be necessary to increase clutch control cable spring pressure. Slide the spring cover, located on cable between control bail on handle and drive mechanism, upward to expose adjuster. Hook ''Z'' end of cable in hole of adjuster closer to clutch spring. If clutch cable spring tension is too great, impeller may turn with clutch lever released. Spring on idler bracket should pull brake tab of bracket (Model CCR-1000) or brake arm (Model CCR-2000) against belt when clutch lever is released.

OVERHAUL

ENGINE

Refer to appropriate Tecumseh 2-stroke engine section for overhaul procedures on Model CCR-1000. Service information on Toro engine was not available at time of publication. A NGK BPMR4A spark plug is recommended for the Toro engine.

TORO

Model	Engine Make	Engine Model	Self-Propelled?	No. of Stages	Scoop Width
524	Tecumseh	H50	Yes	2	24 in. (610 mm)
624*	Tecumseh	H60	Yes	2	24 in. (610 mm)
724	Tecumseh	H70	Yes	2	24 in. (610 mm)
824*	Tecumseh	HM80	Yes	2	24 in. (610 mm)

*None gear drive ''Power Shift'' models.

LUBRICATION

ENGINE

Refer to Tecumseh engine section for engine lubrication requirements.

DRIVE MECHANISM

Periodically lubricate chain, gears, hex shaft, bushings and linkage. Do not allow oil on drive belts or friction wheel and disc. Lubricate control linkage for smooth operation.

AUGER GEARBOX

To check auger gearbox oil level, remove pipe plug (P—Fig. TR1) in front of gearbox. Oil should be level with bottom of plug hole with unit on level ground. Fill gearbox with SAE 90EP oil. Approximate capacity is 3 ounces (89 mL) of oil.

MAINTENANCE

Inspect drive belts and pulley belt grooves. Renew belt if frayed, cracked, burnt or otherwise damaged. Inspect drive chains and sprockets and renew if excessively worn or damaged. Engine throttle control should move freely without binding.

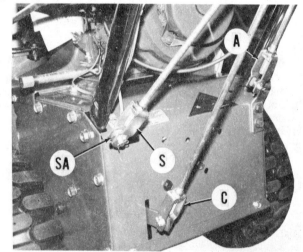

Fig. TR2—View of control rod lower ends. Refer to text for adjustment.

Fig. TR1—Remove plug (P) to check auger gearbox oil level.

Fig. TR3—View of control panel. Refer to Fig. TR5 for view of underside.

ADJUSTMENTS

SHIFT CONTROL LEVER

To adjust shift control lever detach shift clevis (S—Fig. TR2) from shift arm then position shift control lever (L—Fig. TR3) so lever is against tip of projection between "N" and "1" slots on control panel as shown in Fig. TR3. Move shift arm (SA—Fig. TR2) fully to left, rotate clevis (S) so holes in clevis and arm align and reattach clevis with pin.

FRICTION WHEEL

Friction wheel position must be adjusted to provide neutral. Move shift control lever to either neutral slot, remove bottom panel and measure distances between friction wheel (FW—Fig. TR4) and friction disc (FD) and driven pulley (DP). Distances should be equal. Obtain equal distances by detaching, rotating and reattaching clevis (C—Fig. TR2).

AUGER DRIVE BELT TENSION

To adjust auger drive belt tension rotate clevis (A—Fig. TR2). Turn clevis out on rod to increase belt tension or in on rod to decrease belt tension. Engage auger and check belt tension. Insufficient belt tension will cause slippage while excessive belt tension will cause premature failure.

HEIGHT

Auger height is adjusted by repositioning skid shoes on both sides of auger frame. Normal setting is 1/8 inch (3.2 mm) between auger and level surface. Auger height should be increased when working over an irregular surface.

OVERHAUL

ENGINE

Engine make and models are listed at the front of this section. Refer to Tecumseh engine section for engine overhaul.

AUGER GEARBOX

Overhaul information for auger gearbox was not available at time of publication.

Fig. TR4—View of friction drive.

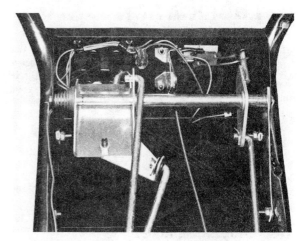

Fig. TR5—View of control panel underside. Refer to Fig. TR3 for top view of control panel.

Fig. TR6—View of drive mechanism.

TORO

Model	Engine Make	Engine Model	Self-Propelled?	No. of Stages	Scoop Width
826	B&S	19000	Yes	2	26 in. (660 mm)
832	B&S	19000	Yes	2	32 in. (813 mm)
1032	B&S	25000	Yes	2	32 in. (813 mm)

LUBRICATION

ENGINE

Refer to Briggs and Stratton engine section for engine lubrication requirements.

DRIVE MECHANISM

Periodically lubricate chain, gears, hex shaft, bushings and linkage. Do not allow oil on drive belts or friction wheel and disc. Lubricate control linkage for smooth operation.

AUGER GEARBOX

On models so equipped, an auger gearbox is located between the augers. To check auger gearbox oil level, remove pipe plug (P—Fig. TR10) in front of gearbox. Oil should be level with bottom of plug hole with unit on level ground. Fill gearbox with SAE 90 EP oil. Approximate capacity is 3 ounces (89 mL) of oil.

MAINTENANCE

Inspect drive belts and pulley belt grooves. Renew belt if frayed, cracked, burnt or otherwise damaged. Inspect drive chains and sprockets and renew if excessively worn or damaged. Engine throttle control should move freely without binding.

ADJUSTMENTS

AUGER HEIGHT

Auger height is adjusted by repositioning skid shoes on both sides of auger frame. Normal setting is 1/8 inch (3.2 mm) between auger and level surface. Auger height should be increased when working over an irregular surface.

FRICTION WHEEL

If friction wheel does not disengage from friction disc in neutral or does not contact friction wheel when in gear to drive snowthrower, then friction wheel must be adjusted. Remove rear panel then place shift control lever in neutral

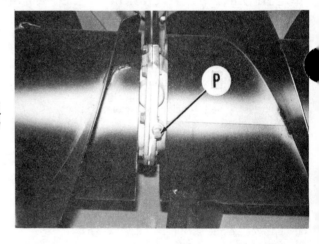

Fig. TR10—Remove plug (P) to fill or drain auger gearbox and to check oil level.

Fig. TR11—View of drive mechanism.

and measure gap between friction wheel (FW—Fig. TR11) and friction disc (FD). Gap should be 3/32 inch (2.4 mm). To obtain desired gap, remove clevis pin (P—Fig. TR12) and rotate swivel bracket (B). Reattach control rod to swivel bracket and check adjustment. Note that clevis pin (P) must be installed with pin head on right side.

DRIVE SPEED

If drive speed in 1st gear requires adjustment or if it is difficult to move speed control lever into third gear slot, loosen bolts securing swivel mount (M—Fig. TR12) and reposition mount. Tighten bolts and check adjustment.

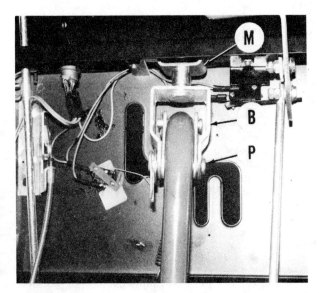

Fig. TR12—View of control panel underside.

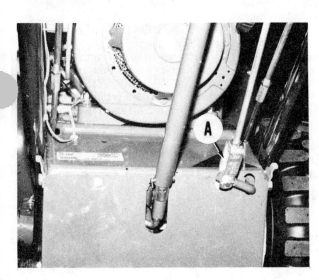

Fig. TR13—Rotate clevis (A) to adjust auger belt tension.

AUGER DRIVE BELT TENSION

To adjust auger drive belt tension rotate clevis (A—Fig. TR13). Turn clevis out on rod to increase belt tension or in on rod to decrease belt tension. Engage auger and check belt tension. Insufficient belt tension will cause slippage while excessive belt tension will cause premature failure.

OVERHAUL

ENGINE

Engine make and model numbers are listed at the front of this section. Refer to Briggs & Stratton engine section for engine overhaul.

AUGER GEARBOX

Overhaul information for auger gearbox was not available at time of publication.

Fig. TR14—View of dog type wheel clutch.

TROY-BILT

GARDEN WAY MANUFACTURING CO.
102nd Street & 9th Ave.
Troy, NY 12180

Model	Engine Make	Engine Model	Self-Propelled?	No. of Stages	Scoop Width
5210R	Tecumseh	HSSK50	Yes	2	22 in. (559 mm)

LUBRICATION

ENGINE

Refer to Tecumseh 4-stroke engine section for engine lubrication information. Recommended fuel is unleaded or leaded regular grade gasoline.

DRIVE MECHANISM

Auger gearbox oil level should be maintained level with plug opening (23—Fig. TB6) in gearbox with SAE 90 gear lubricant. After each use, lubricate all handlebar controls, lever pivot points and auger shaft bearings with SAE 30 motor oil; silicone spray may be used on controls instead of motor oil. Lubricate traction drive chain after every 50 hours of operation with a light coat of oil. After every 50 hours of operation, remove drive wheels and apply a light coat of multipurpose grease on axle shaft bearings. Lubricate chute worm gear and

toothed part of chute flange with multipurpose grease. The gear type transmission is lubricated at the factory and does not require lubrication except during overhaul; refer to Peerless transmission section for information.

MAINTENANCE

Inspect belts and pulleys and renew belts if frayed, cracked, burned or excessively worn. Check pulley grooves for excessive wear or damage and renew if defect is found. Check idler pulley bearings and renew idler pulley assembly if bearings are loose or rough. Inspect traction drive chain and sprockets and renew if excessively worn or damaged.

The auger flights are driven by the auger gearbox shaft via auger shear bolts (18—Fig. TB5) which may shear if the auger encounters a solid obstruction. If this happens, turn auger so holes in auger tube and gearbox shaft are aligned. Drive out remainder of old bolt and install a new factory replacement shear bolt.

ADJUSTMENTS

AUGER HEIGHT

The auger height may be adjusted by loosening skid shoe mounting bolts and

repositioning the skid shoes. Auger should be in lowest position (skid shoes raised on housing) when operating over smooth surfaces, and in highest position when operating over rough surfaces.

SHIFT CONTROL

The shift lever is mounted directly to the transmission shift rod and fork assembly and adjustment is not required.

DRIVE CABLE

To check drive cable adjustment, first remove belt cover at front of engine. Refer to Fig. TB1 and hold the idler pulley against belt so belt is tight. Loosen idler stop screw nut (J) and adjust drive idler stop (I) so end of stop is 3/8 inch (9.5 mm) away from main frame, then tighten nut. With stop adjusted, all slack should be out of drive cable with control lever released. If cable is too tight or loose, adjust cable housing (K—Fig. TB2) position at bracket on handlebar

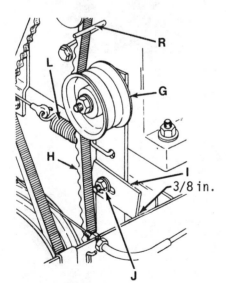

Fig. TB1—View showing adjustment of drive idler stop.
G. Drive belt idler
H. Traction drive belt
I. Drive idler stop
J. Stop screw nut
R. Belt guide

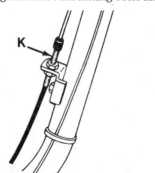

Fig. TB2—Cables (K) for auger and drive controls are adjusted at bracket on snowthrower handle. Auger cable bracket is on right handle and drive cable on left handle.

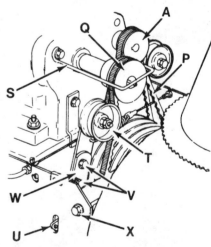

Fig. TB3—View showing adjustment of auger drive brake arm.
A. Engine PTO pulley
P. Auger drive belt
Q. Engine crankshaft pulley
S. Belt guide
T. Auger belt idler
U. Brake arm adjustment indicator
V. Brake arm screws
W. Brake arm
X. Auger housing bolt

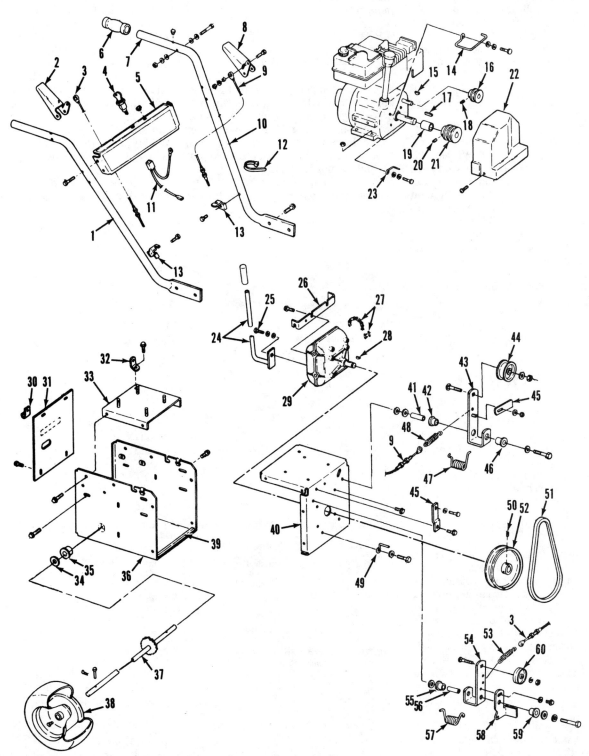

Fig. TB4—Exploded view showing handles, controls and drive system. Letters following callouts indicate same parts shown in Figs. TB1, TB2 or TB3.

1. Right handle
2. Auger control lever
3. Auger control cable (K)
4. Ignition switch
5. Console panel
6. Handle grips
7. Lever bumpers Drive control lever
8. Drive control cable (K)
10. Left handle
11. Ignition cable assy.
12. Wiring tie strap

13. Cable brackets
14. Belt guide (S)
15. Woodruff key
16. Engine PTO pulley (A)
17. Square key
18. Set screw
19. Spacer
20. Set screw
21. Engine pulley (Q)
22. Belt cover
23. Belt guide (R)
24. Shift lever

25. Shift lever bolt
26. Reinforcement bracket
27. Drive chain & connector
28. Woodruff key
29. Peerless transmission
30. Auger cable clamp
31. Rear cover
32. Auger crank support
33. Motor mount plate

34. Flat washer
35. Axle flange bushing
36. Main frame
37. Drive axle assy.
38. Drive wheel
39. Rubber pad
40. Frame cover
41. Spacer
42. Bushing
43. Drive idler arm
44. Drive belt idler (G)
45. Drive idler stop (I)
46. Bushing

47. Torsion spring
48. Drive cable spring
49. Belt guide
50. Set screw
51. Drive belt (H)
52. Transmission pulley
53. Auger cable spring
54. Auger idler bracket
55. Bushing
56. Spacer
57. Torsion spring
58. Auger brake (W)
59. Bushing
60. Auger belt idler (T)

to obtain desired cable tension. Then, with spark plug removed from engine, pull engine recoil starter while watching engine drive pulley. The belt should not move as engine is turned; if it does, decrease the gap between idler stop and frame below the distance previously set.

AUGER CABLE

To check auger control cable adjustment, remove belt cover and push auger belt idler against belt. While holding idler in this position, check to see that hole in brake arm adjustment indicator (U—Fig. TB3) is flush with main frame. If not, loosen the two screws (V) and move brake arm (W) sideways as necessary, then tighten the screws and recheck position of brake arm hole. When this adjustment is completed, check auger control cable at control lever; all slack should be out of cable with control lever released. If not, adjust position of control cable housing (K—Fig. TB2) at bracket on handlebar.

OVERHAUL

ENGINE

Engine make and model numbers are listed at beginning of this section. Refer to Tecumseh 4-stroke engine section for engine overhaul information.

DRIVE BELTS

To remove either drive belt, remove belt cover, unbolt chute worm gear bracket and remove the worm gear assembly. Remove the two auger housing bolts (X—Fig. TB3), one at each side of frame. Pivot rear half of snowthrower back and rest handlebars on solid support. Loosen bolt holding belt guide (S) and remove auger drive belt (P) from engine crankshaft pulley (Q). Loosen the two belt guides located on back of auger housing below the auger pulley, turn guides out of the way and remove belt from auger pulley.

The traction drive belt can be removed after removing auger belt from engine pulley. Loosen belt guide (R—Fig. TB1) on engine and the belt guides on front of main frame chassis below the transmission pulley. Take the drive belt off of transmission pulley, then remove it from engine auxiliary PTO pulley.

Reassemble by reversing disassembly procedure and adjust drive and auger controls as outlined in adjustment sections. When remounting chute worm

gear, position gear mounting bracket so gear turns freely after tightening mounting bolt.

AUGER GEARBOX

To remove auger gearbox, first remove auger belt and drive belt as outlined in preceding paragraph. Completely separate auger housing from snowthrower frame. Remove set screw (1—Fig. TB5) from auger pulley (2), then remove auger pulley and Woodruff key (13) from rear end of impeller shaft. It may be necessary to heat auger pulley hub as

thread locking compound is used on installation. Remove impeller shaft bearing retainer (3), shims (5) and ball bearing (4). Remove auger bearings (20) from ends of auger shaft and bearing supports (6) from housing, then withdraw auger and blower fan assembly from housing. Remove shear bolts (18) and slide augers from auger shaft. Remove roll pin (14) from impeller and remove impeller from rear of auger gearbox input shaft. When reinstalling auger assembly, apply thread locking compound when placing auger pulley on impeller shaft.

To disassemble gearbox, remove cover (1—Fig. TB6) and withdraw auger

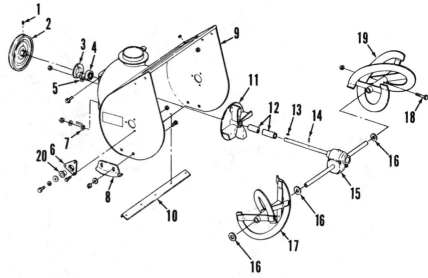

Fig. TB5—Exploded view of auger housing assembly. Exploded view of auger gearbox is shown in Fig. TB6.

1. Set screw	6. Bearing support	11. Impeller	16. Special washer
2. Auger pulley	7. Belt guide	12. Sleeve	17. R.H. auger
3. Bearing retainer	8. Skid shoes	13. Woodruff key	18. Auger shear bolt
4. Ball bearing	9. Auger housing	14. Roll pin	19. L.H. auger
5. Shim	10. Scraper blade	15. Gearbox assy.	20. Auger bearing

Fig. TB6—Exploded view of auger gearbox assembly.

1. Gearbox cover
2. Cover gasket
3. Oil seal
4. Flange bushing
5. Thrust bearing race
6. Worm gear
7. Gearbox case
8. Auger shaft
9. Woodruff key
10. Oil seal
11. Needle bearing
12. Thrust bearing race
13. Thrust bearing
14. Spacer
15. Impeller shaft & gear
16. Spacer
17. Thrust washer
18. Thrust race
19. Bearing spacer/retainer
20. Needle bearing
21. Internal snap ring
22. Expansion plug
23. Oil plug

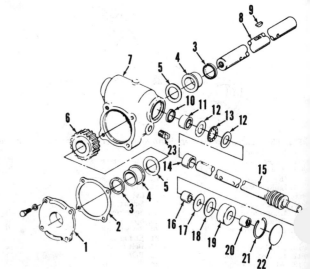

shaft and worm gear (6) as an assembly. If necessary to renew auger shaft (8) or gear, press shaft from gear and remove Woodruff key (9). Remove expansion plug (22) and internal snap ring (21) from front of gearbox. Impeller shaft (15) with spacer (19), needle bearing (20), thrust race (18), thrust washer (17) and spacer (16) can then be removed out front opening of gearbox case. Removal of flange bushings (4), needle bearings (11 and 20) and seals (3 and 10) can now be accomplished. When installing new needle bearings, press only on lettered end of bearing cage. On reassembly, refill gearbox to level plug opening with SAE 90 gear lubricant.

TRANSMISSION

With transmission removed, refer to Peerless transmission section of this manual for overhaul information.

TROY-BILT

Model	Engine Make	Engine Model	Self-Propelled?	No. of Stages	Scoop Width
826R	Tecumseh	HMSK80	Yes	2	26 in. (660 mm)

LUBRICATION

ENGINE

Refer to Tecumseh 4-stroke engine section for engine lubrication information.

DRIVE MECHANISM

After each use, use clean engine oil or silicone spray to lubricate handlebar control lever pivot points; be careful not to get oil or spray on handles. Apply a few drops of oil on the pivot points of engine control lever and speed control lever. After every 25 hours of operation, check oil level at plug (23—Fig. TB18) in auger gearbox and add SAE 90 weight gear oil as necessary to bring oil level to bottom of plug opening. Use a hand-operated grease gun filled with multipurpose grease to lubricate the four auger flight grease fittings. Clean any dirt and old grease from the chute worm gear and tooth portion of chute flange and apply new multipurpose grease. Remove left drive wheel and grease axle shaft. Remove bottom panel of snowthrower, then lubricate hex fork shaft with multipurpose grease and apply light motor oil on drive chains.

MAINTENANCE

Inspect drive belts and renew if frayed, cracked, burned or excessively worn. Renew belt pulleys if pulley groove is worn or damaged. Idler pulleys should be renewed if bearing is rough or loose. Inspect drive chains and sprockets and renew if worn or if chains are stretched.

Augers are driven from auger shaft by shear bolts (19—Fig. TB17) which will fracture if a solid obstruction is encountered. If shear bolt fails, drive out any remaining part of bolt and install new factory replacement bolt. Before installing new bolt, use a hand-operated grease gun to lubricate auger shaft grease fittings and be sure auger turns

freely on shaft. Do not use substitute shear bolts.

Recommended tire air pressure is 8-12 psi (56-82 kPa). Be sure both tires are inflated to same air pressure.

ADJUSTMENTS

AUGER HEIGHT

Both the skid shoes and scraper plate are adjustable. With auger height at

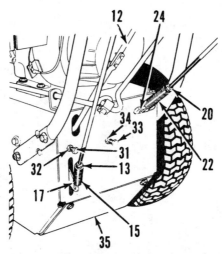

Fig. TB10—View showing control linkage adjusting points at rear of snowthrower.

12. Speed control rod
13. Jam nuts
15. Drive control spring
17. Drive control arm
20. Jam nuts
22. Auger clutch
    control rod spring
24. Auger control arm
31. Adjusting bolt
32. Locknut
33. Adjusting bolt
34. Locknut
35. Main frame cover

lowest setting (skid shoes raised), scraper bar should be adjusted to height of 1/8 inch (3.2 mm) from flat, smooth surface. The skid shoes should be lowered (auger housing raised) when operating over rough, irregular surface.

SPEED CONTROL ROD

To adjust speed control rod, place control lever in fifth speed position on console, then loosen locknut (34—Fig. TB10) and back nut off as far as possible toward head of adjusting bolt (33). Turn locating screw in until locknut contacts main frame housing. Disconnect control rod pivot (10—Fig. TB11) from speed control lever and place lever in No. 1 position. While holding control rod up, the control rod pivot should just fit into hole in lever. If not, loosen jam nut (11) and turn control rod pivot up or down on speed control rod (28) until it will fit into lever when pulling up on rod. Reinstall control rod pivot into lever and tighten control rod jam nut. Unscrew adjusting bolt (33—Fig. TB10) out of housing approximately 1/2 inch (12.7 mm) and tighten locknut (34). There should be no interference to movement of the speed control lever throughout its range of travel; if interference is noted, check drive disc clearance as outlined in following paragraph.

Fig. TB11—Bottom view of control panel showing speed control rod adjustment.

10. Control rod pivot
11. Jam nut
28. Speed control rod

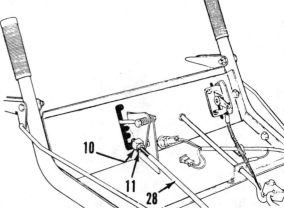

Information Courtesy of Garden Way Incorporated, Troy, New York

DRIVE DISC CLEARANCE

Remove main frame cover (35—Fig. TB10) from main frame to provide access to drive disc. When drive clutch lever is released, there should be from 0.060 to 0.125 inch (1.5-3.2 mm) clearance between rubber friction drive wheel (FW—Fig. TB13) and friction disc (FD). If adjustment is necessary, loosen locknut (32—Fig. TB10) and turn adjusting bolt (31) to obtain correct clearance, then tighten locknut.

DRIVE BELT TENSION

To check adjustment, refer to Fig. TB10 and measure length of drive control spring (15) with drive clutch lever released, then measure length of spring again with lever held against handlebar grip. The difference in measurements should be 1/2 inch (12.7 mm). If not, un-

hook spring from drive control arm (17); it may be necessary to remove hand lever pivot pin to gain slack necessary to unhook spring. Loosen jam nuts (13) and turn the nut inside spring up to increase spring measurement or down to decrease measurement. Each 1-1/4 turn of

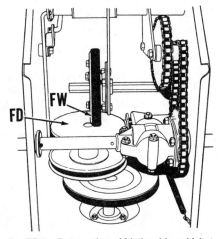

Fig. TB13—Bottom view of friction drive with bottom cover removed. Clearance between friction wheel (FW) and friction disc (FD) should be 0.060-0.125 inch (1.5-3.2 mm).

nut will equal approximately 1/16 inch (1.6 mm). Tighten jam nuts, reconnect spring to drive control arm and recheck adjustment.

AUGER BELT TENSION

To check adjustment, measure length of auger clutch control rod spring (22—Fig. TB10) with auger control lever released. Then, hold lever against handlebar grip and measure spring in extended length. The difference between the two measurements should be 1-9/16 inches (40.3 mm). If not, unhook spring from auger control arm (24); it may be necessary to remove handlebar lever pivot bolt to provide slack needed to unhook the spring. Loosen jam nuts (20) and turn hex nut inside spring up the control rod to increase spring extension or down to decrease measurement. One turn of the nut will change spring extension about 1/16 inch (1.6 mm). After adjusting, tighten jam nuts, then reconnect spring to lever and recheck adjustment.

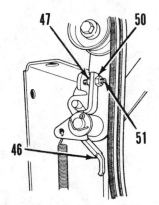

Fig. TB12—View showing adjustment of auger brake arm.

46. Brake arm
47. Jam nut
50. Jam nut
51. Adjusting screw

1. Throttle assy.	27. Idler arm bushing
2. Console panel	28. Lower speed control rod
3. Handle grip	29. Right handle
4. Drive control lever	30. Main frame back plate
5. Drive control rod	31. Adjusting bolt
6. Lever bumper	32. Locknut
7. Left handle	33. Adjusting bolt
8. Speed control lever	34. Locknut
9. Lever friction spring	35. Main frame cover
10. Speed control rod pivot	36. Main frame
11. Jam nut	37. Drive pin
12. Speed control rod	38. Auger control arm
13. Jam nuts	39. Plastic tube
14. Lockwasher	40. Cross member
15. Drive control spring	41. Spring hook
16. Adjusting nut	42. Drive idler spring
17. Drive control arm	43. Drive belt idler
18. Auger control lever	44. Drive idler arm
19. Auger control rod	45. Auger idler arm
20. Jam nuts	46. Auger brake arm
21. Lockwasher	47. Jam nut
22. Auger clutch control rod spring	48. Auger idler spring
23. Adjusting nut	49. Spring hook
24. Auger control arm	50. Jam nut
25. Flat washer	51. Adjusting screw
26. Spring washer	52. Auger belt idler

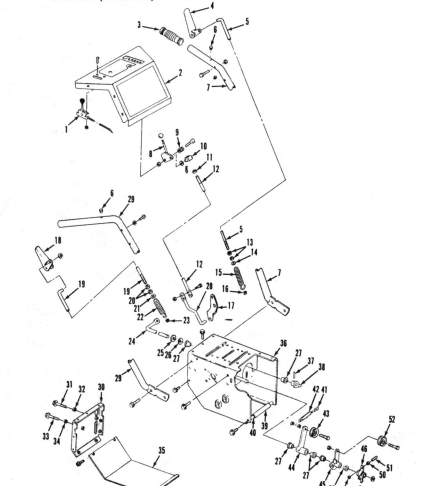

Fig. TB14—Exploded view showing handles, control linkage and main frame assembly.

AUGER BRAKE ARM

To check brake arm adjustment, first remove belt cover at front of engine. Hold auger clutch control lever against handlebar grip and measure gap between brake arm (46—Fig. TB12) and auger drive belt. The gap should measure between 7/16 and 1/2 inch (11.1-12.7 mm). If gap is not correct, loosen jam nuts (47 and 50) and position screw (51) to obtain desired gap, then tighten jam nuts. The adjusting screw can be held with an Allen wrench while turning jam nuts.

LIMITED SLIP DIFFERENTIAL

Remove dust cap (54—Fig. TB16) from right end of drive axle and loosen jam nuts (55). Remove outer nut, turn inner jam nut in finger tight and then using a wrench, tighten inner jam nut 1-1/4 additional turns. Hold inner jam nut in this position with a wrench, then install and tighten outer jam nut.

OVERHAUL

ENGINE

Engine make and model number is listed at beginning of this section. Refer to Tecumseh 4-stroke engine section for engine overhaul information.

TRACTION DRIVE BELT

Remove belt cover at front of engine and remove the belt guide (G—Fig. TB15) from engine. Take auger drive belt off of engine pulley. Release tension

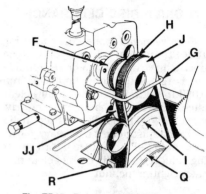

Fig. TB15—References for belt removal.

F. Friction drive belt
G. Belt guide
H. Auger drive belt
J. Engine pulley
JJ. Drive belt idler
Q. Auger drive pulley
R. Auger brake arm

1. Set screws
2. Engine pulley
3. Drive belt
4. Belt guide
5. Square key
6. Disc drive plate
7. Woodruff key
8. Drive disc spindle
9. Tube
10. Bearing retainer
11. Ball bearing
12. Bearing retainer
13. Spacer
14. Drive disc pivot plate
15. Link
16. Speed control arm
17. Link
18. Snap ring
19. Link
20. Speed control rod
21. Speed control arm
22. Control arm pivot bushing
23. Axle shaft, L.H.
24. Drive pin
25. Drive pin
26. Snap ring
27. Nylon bushing
28. Support shaft
29. Roller chain
30. Support plate
31. Flat washer
32. Compression spring
33. Spacer
34. Sprocket bushings
35. Double sprocket
36. Double sprocket
37. Roller chain
38. Sliding plate
39. Snap ring
40. Control arm bushing
41. Drive control arm
42. Nylon bushing
43. Bearing retainer
44. Ball bearing
45. Hex shaft sprocket
46. Woodruff key
47. Hex shaft
48. Snap ring
49. Bearing retainers
50. Thrust bearing
51. Friction wheel hub
52. Friction wheel
53. Differential locking shaft
54. Dust cap
55. Jam nuts
56. Belleville washers
57. D-hole washer
58. Friction disc
59. Nylon bushing
60. Axle tube, R.H.
61. Flange bushing
62. Flat washer
63. Differential sprocket
64. Roller chain
65. Differential assy.
66. Drive pulley

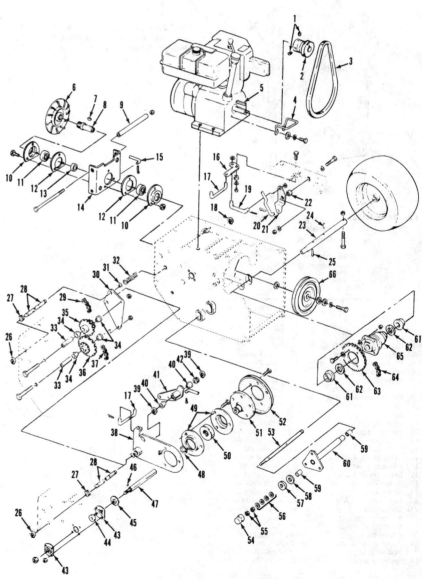

Fig. TB16—Exploded view of disc drive mechanism.

from drive belt idler (JJ) and take drive belt off of lower (driven) pulley. Then move lower part of belt up the gap between the auger and drive pulleys and remove belt from engine pulley. Install new belt by reversing removal procedure and check drive belt adjustment as outlined in adjustment paragraph.

AUGER DRIVE BELT

Remove belt cover at front of engine and remove belt guide (G—Fig. TB15) from engine. Take auger drive belt off of engine pulley, then remove it from lower pulley; it may be necessary to pull auger brake arm (R) outward to have room for removing belt. Pull belt up through the gap between the auger and drive pulleys. Install new belt by reversing removal procedure and check auger belt adjustment as outlined in adjustment paragraph.

FRICTION WHEEL

To renew rubber faced friction wheel, first tip snowthrower forward onto front of auger housing and be sure it is securely braced in this position. Remove lower cover from main frame and refer to Fig. TB13. Hold the hex shaft from turning with correct size open-end wrench, then remove the cap screws retaining wheel to adapter. The friction wheel can then be removed from over the end of the hex shaft. Install new wheel with cup side toward adapter.

DRIVE MECHANISM

Overhaul of the disc drive mechanism should be evident after inspection of unit and reference to Fig. TB13 and Fig. TB16. Friction wheel thrust bearing (50—Fig. TB16) should be installed on hub (51) using a locking compound. Refer to adjustment paragraph on reassembly.

AUGER GEARBOX

To remove auger gearbox, first remove auger belt and drive belt as outlined in preceding paragraphs. Completely separate auger housing from snowthrower frame. Remove set screw (2—Fig. TB17) from auger drive pulley (1), then remove auger pulley and Woodruff key (12) from rear end of impeller shaft. Remove impeller shaft bearing retainer (4) and ball bearing (3). It may be necessary to heat bearing as locking compound is used on installation. Remove auger shaft bushings (5) from ends of auger shaft and bushing supports (6) from housing, then

withdraw auger and blower fan assembly from housing. Remove shear bolts (19) and slide augers from auger shaft. Remove roll pin (13) from impeller and remove impeller from rear of auger gearbox input shaft. When reinstalling auger assembly, apply locking compound when placing rear ball bearing (3) on impeller shaft.

To disassemble gearbox, remove cover (1—Fig. TB18) and withdraw auger shaft and worm gear (6) as an assembly. If necessary to renew auger shaft (8) or gear, press shaft from gear and remove Woodruff key (9). Remove expansion plug (22) and internal snap ring (21) from front of gearbox. The impeller shaft (15) with spacer (19), needle bearing (20), thrust race (18), thrust washer (17) and spacer (16) can then be removed out front opening of gearbox case.

Removal of flange bushings (4), needle bearings (11 & 20) and seals (3 & 10) can now be accomplished. When installing new needle bearings, press only on lettered end of bearing cage. On reassembly, refill gearbox to level plug opening with SAE 90 gear lubricant.

DIFFERENTIAL

Although the limited slip differential (65—Fig. TB16) can be disassembled, the unit is not serviceable. If worn or damaged, differential must be renewed as a complete assembly. If disassembled for inspection and all parts are reusable, clean the parts thoroughly, apply grease on all parts and pack differential with 1-1/2 ounces (44.4 mL) of multipurpose grease.

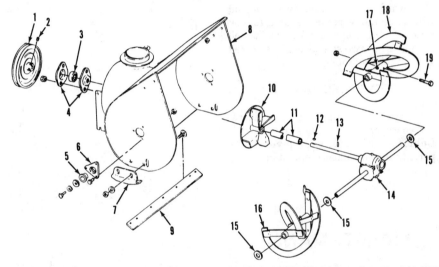

Fig. TB17—Exploded view of auger housing assembly. Refer to Fig. TB18 for gearbox assembly (14) exploded view.

1. Auger drive pulley
2. Set screw
3. Ball bearing
4. Bearing retainers
5. Auger shaft bushing
6. Bushing support
7. Skid shoe
8. Auger housing
9. Scraper bar
10. Impeller
11. Sleeve
12. Woodruff key
13. Roll pin
14. Gearbox assy.
15. Flat washer
16. Auger, R.H.
17. Grease fitting
18. Auger, L.H.
19. Shear bolt

Fig. TB18—Exploded view of auger gearbox assembly.

1. Gearbox cover
2. Cover gasket
3. Oil seal
4. Flange bushing
5. Thrust bearing race
6. Worm gear
7. Gearbox case
8. Auger shaft
9. Woodruff key
10. Oil seal
11. Needle bearing
12. Thrust bearing race
13. Thrust bearing
14. Spacer
15. Impeller shaft & gear
16. Spacer
17. Thrust washer
18. Thrust race
19. Bearing spacer/retainer
20. Needle bearing
21. Internal snap ring
22. Expansion plug
23. Oil plug

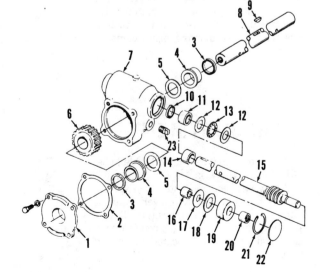

WHITE

WHITE OUTDOOR PRODUCTS
P.O. Box 8875
Cleveland, OH 44136

Model	Engine Make	Engine Model	Self-Propelled?	No. of Stages	Scoop Width
300	Tecumseh	AH520	No	1	20 in. (508 mm)

LUBRICATION

ENGINE

Refer to Tecumseh engine section for engine lubrication requirements.

The engine is a two-stroke type engine which is lubricated by mixing oil with fuel. White recommends a fuel:oil ratio of 32:1 using a good quality oil designed for two-stroke engines mixed with regular gasoline.

DRIVE MECHANISM

Lubricate shafts and bushings as required to prevent binding and galling. Lubricate control lever pivot point periodically for smooth operation.

MAINTENANCE

Periodically inspect drive belt and pulleys. Renew belt if frayed, cracked, excessively worn or otherwise damaged.

The auger is secured to drive pulley shaft by a shear pin which will fracture when auger contacts an obstruction. Renew shear pin with a suitable replacement. Do not use ordinary fasteners which may damage drive components.

ADJUSTMENT

The clutch rod (15 – Fig. W1) should be adjusted so there is 1/8 inch (3.2 mm) end play between rod and clutch spring (16) with auger clutch lever against handlebar in disengaged position. To adjust end play, detach clutch rod (15) and turn rod in nut (17) located in spring (16).

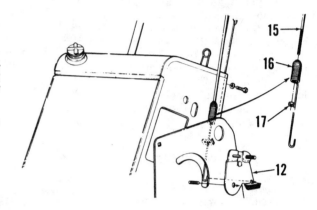

Fig. W1—View of auger control clutch rod and spring. Refer to text for adjustment.

12. Idler arm & brake
15. Auger clutch control rod
16. Spring
17. Nut

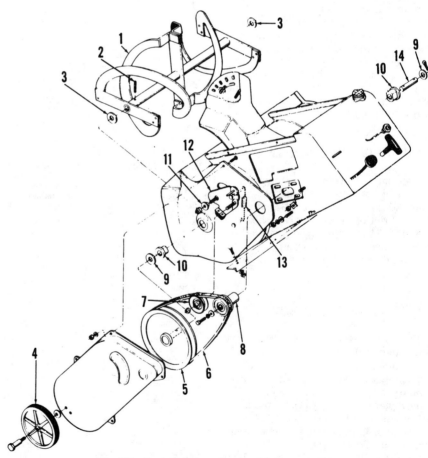

Fig. W2—Exploded view of auger drive components.

1. Auger	5. Auger pulley	8. Engine pulley	11. Spacer
2. Pin	6. Belt	9. Washer	12. Idler arm & brake
3. Washer	7. Idler pulley	10. Bushing	13. Spring
4. Guide wheel			14. Shaft

OVERHAUL

ENGINE

Engine make and model are listed in front of section. Refer to Tecumseh section for engine overhaul.

DRIVE MECHANISM

Overhaul of drive components is evident after inspection of unit and referral to Fig. W2.

WHITE

Model	Engine Make	Engine Model	Self-Propelled?	No. of Stages	Scoop Width
400	Tecumseh	HS50	Yes	1	21 in. (533 mm)

LUBRICATION

ENGINE

Refer to Tecumseh engine section for engine lubrication requirements.

DRIVE MECHANISM

Lubricate drive chains at least once each season with good quality chain oil. Lubricate control cables and control rod and lever pivot points for smooth operation.

MAINTENANCE

Inspect chain and sprockets and renew if excessively worn. Inspect belts and pulleys and renew if frayed, cracked, excessively worn or otherwise damaged. Inspect control cables and renew if frayed or binding.

Recommended tire pressure is 7-10 psi (48-69 kPa).

ADJUSTMENTS

HEIGHT

Auger height may be adjusted by loosening skid shoe mounting bolts and repositioning skid shoe. Skid shoe should be located in lowest setting when operating over smooth surfaces and in highest setting when operating over rough, irregular surfaces.

SHIFT LINKAGE

The shift linkage must be properly adjusted so shift hub (17—Fig. W11) correctly engages dogs on forward sprocket (18) and reverse gear (16). To adjust shift linkage, detach shift rod (4—Fig. W10) from shift arm (5). Roll snowthrower back and forth while moving shift arm (5) until neutral is found. Position shift control lever in neutral then rotate shift rod (4) so rod end will engage shift arm (5) hole and attach rod. Check adjustment by engaging forward and reverse and being sure unit rolls freely in neutral.

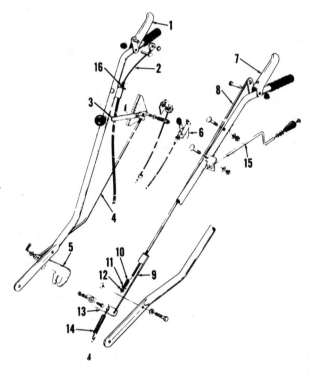

Fig. W10—Drawing of control assembly.

1. Drive clutch control lever
2. Clutch cable
3. Shift control lever
4. Shift rod
5. Shift arm
6. Throttle assy.
7. Auger clutch control lever
8. Auger clutch rod
9. Auger clutch rod
10. Spring
11. Washer
12. Locknut
13. Lever
14. Spring
15. Discharge chute crank
16. Nut

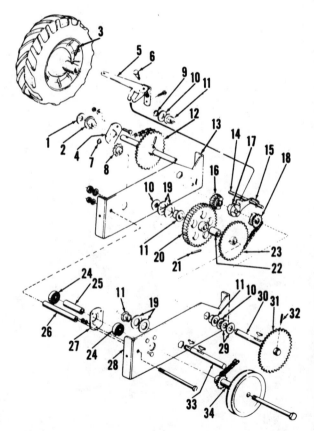

Fig. W11—Exploded view of drive assembly.

1. Shim
2. Bushing
3. Pin
4. Bushing plate
5. Shift arm
6. Pivot bolt
7. Snap ring
8. Sprocket
9. Snap ring
10. Washer
11. Bushing
12. Sprocket
13. Bracket
14. Shift shaft
15. Shift fork
16. Reverse gear
17. Shift hub
18. Forward sprocket
19. Washers
20. Gear
21. Pin
22. Spacer
23. Sprocket
24. Bearing
25. Spacer
26. Spacer
27. Bearing retainer
28. Bracket
29. Washers
30. Shaft
31. Sprocket
32. Pin
33. Shaft
34. Sprocket & pulley assy.

DRIVE CLUTCH

The drive clutch is adjusted by turning adjusting nut (16—Fig. W10) on clutch cable housing. Rotate adjusting nut (16) so clutch control lever just contacts handlebar grip with lever in engaged position.

AUGER CLUTCH

The auger clutch is adjusted by turning adjusting rod nut (12—Fig. W10). Turn nut so there is approximately 0.015 inch (0.38 mm) between compressed coils of spring (10) when auger clutch control lever is in engaged position.

AUGER DRIVE CHAIN

Auger drive chain tension is adjusted by loosening nut (37—Fig. W12) and repositioning chain idler (31) to remove chain slack.

AXLE SPROCKET

Wheel axle sprocket (12—Fig. W11) must be aligned with drive sprocket (8). Check alignment by laying a straightedge across faces of sprockets. Wheel axle sprocket position is determined by shims (1) between wheel hub and bushing (2). Add or delete shims as required to align sprockets. In some cases it may be necessary to grind material from wheel hub and trim flange of bushing.

OVERHAUL

ENGINE

Engine make and model number are listed at front of this section. Refer to Tecumseh engine section for overhaul procedure.

DRIVE BELTS

To remove drive belts, remove cover plates on front and right side of main frame then tip unit forward and remove bottom panel. Unscrew shoulder bolt adjacent to lower pulleys which serves as belt guide and remove belts. Reverse removal procedure to install belts.

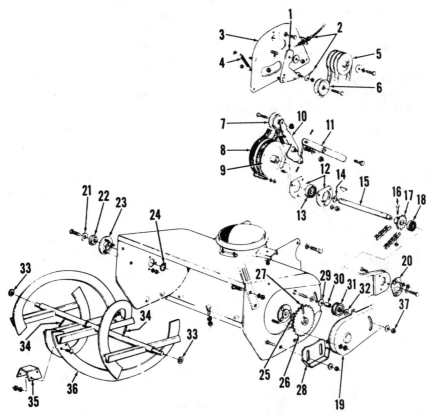

W12—Exploded view of auger drive and drive belt assemblies.

1. Drive idler arm	11. Auger clutch arm	20. Bearing retainer	29. Washer
2. Drive clutch cable	12. Bearing retainer	21. Washer	30. Spacer
3. Engine plate	13. Bearing	22. Bushing	31. Idler
4. Spring	14. Snap ring	23. Bushing retainer	32. Belleville washer
5. Engine pulley	15. Shaft	24. Snap ring	33. Washer
6. Drive idler pulley	16. Bolt	25. Washer	34. Spring pin
7. Auger idler pulley	17. Sprocket	26. Sprocket	35. Plate
8. Drive pulley	18. Bearing	27. Pin	36. Augers
9. Auger pulley	19. Chain cover	28. Skid shoe	37. Nut
10. Auger idler arm			

WHITE

Model	Engine Make	Engine Model	Self-Propelled?	No. of Stages	Scoop Width
500	Tecumseh	HS50	Yes	2	24 in. (610 mm)
510	Tecumseh	HS50	Yes	2	24 in. (610 mm)
800	Tecumseh	HM80	Yes	2	26 in. (660 mm)
810	Tecumseh	HM100	Yes	2	26 in. (660 mm)
1000	Tecumseh	HM100	Yes	2	33 in. (838 mm)

LUBRICATION

ENGINE

Refer to Tecumseh engine section for engine lubrication requirements.

Recommended fuel is regular or low-lead gasoline.

DRIVE MECHANISM

Lubricate drive chains at least once each season with good quality chain oil. Auger shaft and axle bushings should be oiled and axle should be greased at least once each season or after 25 hours of operation. Lubricate control rod and lever pivot points for smooth operation.

AUGER GEARBOX

The auger gearbox is lubricated with either SAE 90 oil or Plastilube. On models equipped with a plug (20—Fig. W24), fill gearbox with SAE 90 until oil level reaches plug opening. If gearbox is not equipped with plug (20), then gearbox is lubricated with Plastilube during assembly and does not normally require new lubricant.

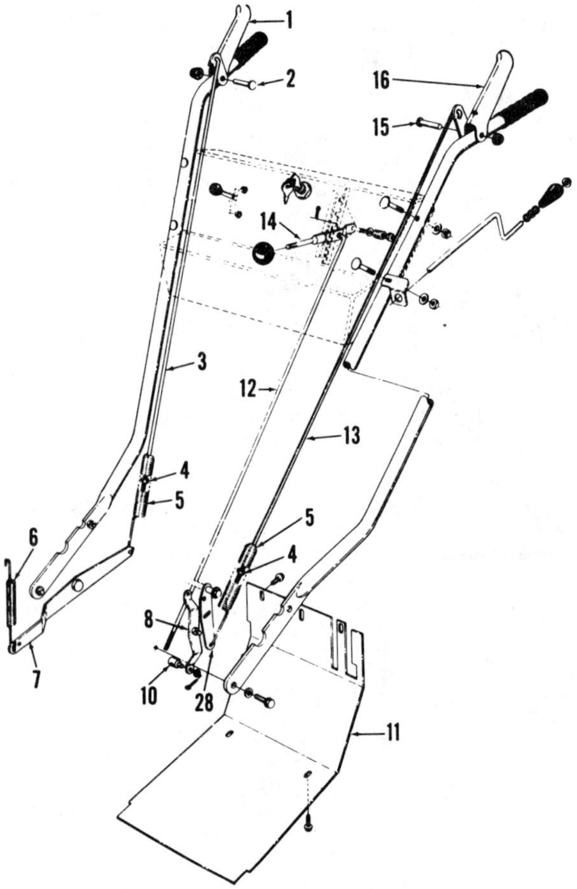

MAINTENANCE

Inspect chain and sprockets and renew if excessively worn. Inspect belts and pulleys and renew belts if frayed, cracked, burnt or otherwise damaged. Check throttle cable for binding and stiff operation.

Recommended tire pressure is 7-10 psi (48-69 kPa).

ADJUSTMENTS

HEIGHT

Auger height may be adjusted by loosening skid shoe mounting bolts and repositioning skid shoe. Skid shoes should be located in lowest setting when operating over smooth surfaces and in highest setting when operating over rough, irregular surfaces.

SHIFT CONTROL ROD

Detach shift control rod (12—Fig. W20) from shift control lever (14). Note locating slot in lower frame cover as shown in Fig. W21, move shift rod so shift plate (27—Fig. W23) is centered in slot. Position shift control lever in neutral detent then rotate shift control rod in ferrule at bottom of rod so upper rod end will fit easily into shift control

Fig. W20—Drawing of control assembly.

1. Auger clutch control lever
2. Pin
3. Auger clutch control rod
4. Nut
5. Spring
6. Spring
7. Lever
8. Lever
10. Pivot nut
11. Lower panel
12. Shift rod
13. Drive clutch control rod
14. Shift control lever
15. Pivot pin
16. Drive clutch control lever

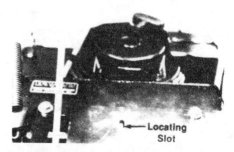

Fig. W21—View showing location of locating slot used for shift control adjustment on later models.

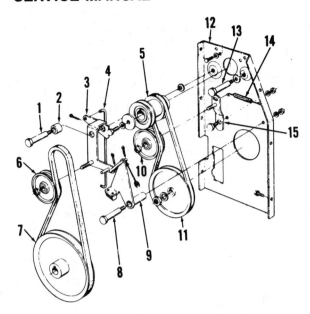

Fig. W22—Exploded view of pulley system.

1. Pivot bolt
2. Spacer
3. Auger idler bracket
4. Auger clutch rod
5. Engine pulley
6. Auger idler pulley
7. Auger pulley
8. Pivot bolt
9. Auger pulley brake
10. Drive idler pulley
11. Drive pulley
12. Engine plate
13. Belt guide bolt
14. Spring
15. Drive idler arm

lever. Attach rod then check adjustment by noting if unit will roll forward and backward with shift control lever in neutral.

CLUTCH RODS

Adjustment of auger and drive clutch control rods is accomplished by turning nut at spring end of rod. There should be a small amount of end play between rod and spring (5—Fig. W20) with clutch disengaged so idler will relieve pressure against drive belt. With clutch engaged, drive belt should tighten around pulleys without slippage.

OVERHAUL

ENGINE

Engine make and model numbers are listed at front of this section. Refer to Tecumseh engine section for engine overhaul procedure.

AUGER GEARBOX

Auger gearbox may be serviced after removing complete auger assembly from frame then detaching impeller and augers. Overhaul of gearbox is evident after referral to Fig. W24 and inspection of unit. Note thickness of washers during disassembly and reassembly.

All gearboxes may be filled with 4 ounces (118.3 mL) of Plastilube or other similar light grease while models equipped with plug (20–Fig. W24) may be lubricated with oil as outlined in LUBRICATION section. Models without plug (20) must be lubricated with grease.

DRIVE MECHANISM

Refer to Figs. W22 and W23 for

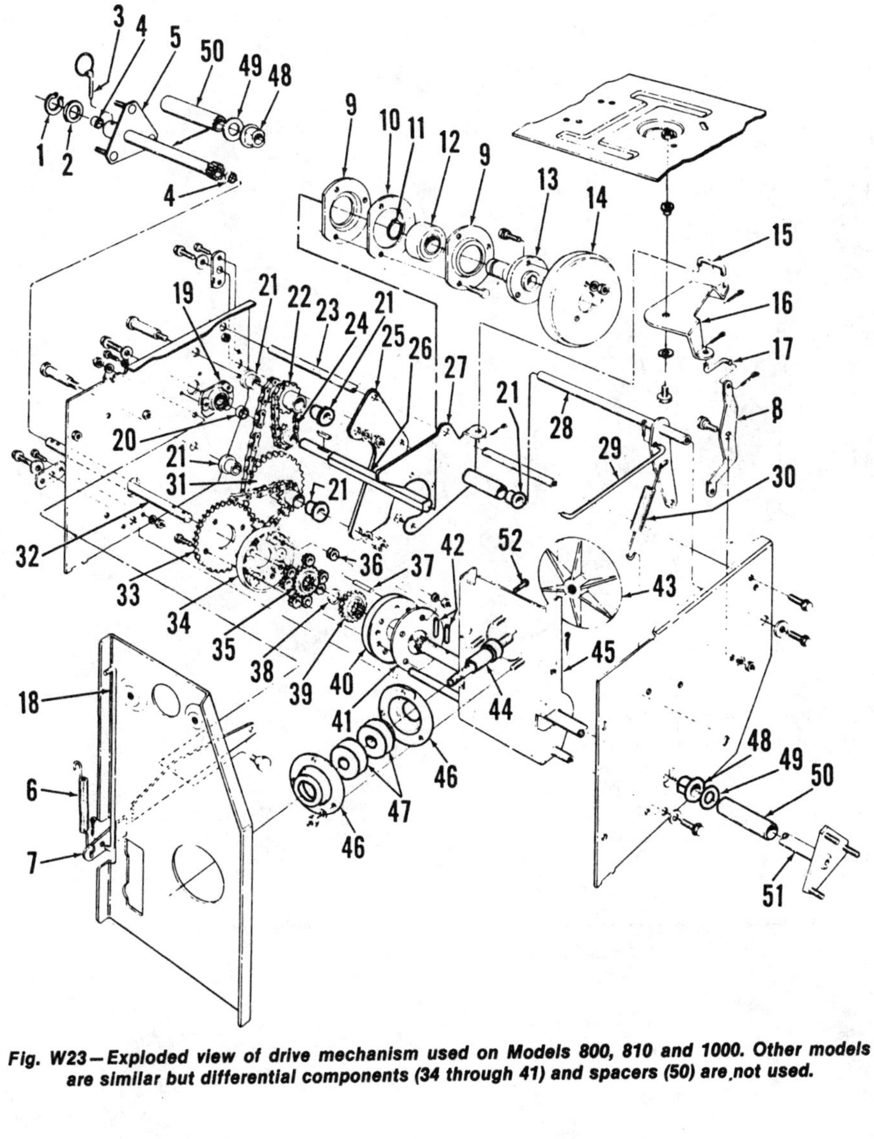

Fig. W23—Exploded view of drive mechanism used on Models 800, 810 and 1000. Other models are similar but differential components (34 through 41) and spacers (50) are not used.

1. Snap ring	28. Shaft
2. Washer	29. Link
3. Lock pin	30. Spring
4. Bushing	31. Sprocket
5. Right axle	32. Differential lock shaft
6. Spring	33. Sprocket
7. Lever	34. Gear housing
8. Lever	35. Sun gear
9. Bearing retainer	36. Planetary gear
10. Spacer	37. Shaft
11. Snap ring	38. Spacer
12. Bearing	39. Gear
13. Friction hub	40. Gear housing
14. Friction wheel	41. Plate
15. Link	42. Pins
16. Bellcrank	43. Friction disc
17. Link	44. Shaft
18. Auger clutch rod	45. Drive support
19. Bushing plate	46. Bearing housing
20. Bushing	47. Bearings
21. Bushing	48. Bushing
22. Sprocket	49. Washer
23. Shaft	50. Spacer
24. Sprocket	51. Left axle
25. Support	52. Shoulder bolt
26. Hex shaft	
27. Shift plate	

exploded view of drive components. A differential is used on Models 800 and 1000. On all models a lock pin may be used to lock wheel and axle together so both wheels are driven. The lock pin is located in right wheel of Models 800 and 1000 and in left wheel of Model 500. Do not overtighten locknuts securing bearing retaining plates (9). Bearing (12) is self-centering and must be able to move in retaining plates or

bearing will bind on friction hub (13). Install friction wheel (14) with cupped side towards hub (13). Be sure auger idler bracket (3—Fig. W22) has sufficient travel to operate properly. If not, grind approximately 1/16 inch (1.6 mm) from end of spacer (2) and recheck movement.

DRIVE BELTS

To remove drive belts, disconnect

chute linkage and remove belt cover. Unscrew shoulder bolts—a roller is mounted on right side bolt—adjacent to engine pulley. Lift up on auger idler pulley (6—Fig. W22) so brake disengages auger drive pulley then lift up auger frame slightly and separate auger and main frames. Unscrew shoulder bolts which serve as belt guides and remove drive belts. Reverse removal procedure to install belts.

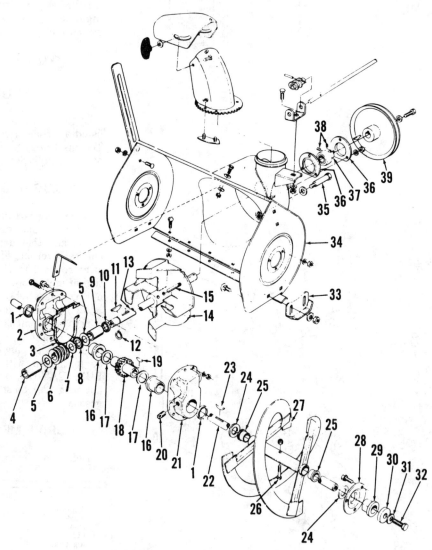

Fig. W24—Exploded view of auger assembly.

1. Seal	9. Bushing	17. Washer	25. Bushing
2. Gearbox half R.H.	10. Seal	18. Gear	26. Bolt
3. Gasket	11. Key	19. Key	27. Auger
4. Bushing	12. Snap ring	20. Plug	28. Bushing retainer
5. Washer	13. Drive shaft	21. Gearbox half L.H.	29. Bushing
6. Worm gear	14. Impeller	22. Auger shaft	30. Washer
7. Washer	15. Pin	23. Key	31. Belleville washer
8. Thrust bearing	16. Bushing	24. Washer	32. Screw

WHITE

Model	Engine Make	Engine Model	Self-Propelled?	No. of Stages	Scoop Width
350 (180)*	Tecumseh	AH600	No	1	21 in. (533 mm)
350 (190)	Tecumseh	HS50	No	1	21 in. (533 mm)

*Model 350 (181) is same as 350 (180) except it is equipped with 120 Volt AC starter.

LUBRICATION

ENGINE

Model 180-181

The engine is a two-stroke type and a fuel:oil mix ratio of 32:1 is recommended. Mix SAE 30 or 40 outboard or two-stroke oil with regular or unleaded gasoline.

Model 190

Refer to Tecumseh engine section for lubrication requirements.

AUGER

Auger and belt idler bearings are sealed and no lubrication is required.

MAINTENANCE

Periodically inspect drive belt and pulleys. Renew belt if frayed, cracked, excessively worn or otherwise damaged. Inspect flighting on auger and renew if chipped or excessively worn.

ADJUSTMENT

BELT

Refer to Fig. W26 and remove belt cover. Move spring (3) end to next highest hole (A, B or C) to increase belt tension. Reinstall cover.

OVERHAUL

ENGINE

Engine make and model numbers are listed at the front of this section. Refer to Tecumseh engine section for overhaul procedures.

BELT

To renew belt (18–Fig. W25), remove cover (14) and disconnect springs (25 and 26). Lift up on idler (16), bracket (24) and slide belt off pulleys (15 and 19). Reverse procedure to reinstall.

AUGER ASSEMBLY

Refer to Fig. W25 for an exploded view of auger and collector assembly. Remove belt as outlined in BELT section. Thread auger pulley (19) off end of auger (23). Unbolt and remove bearing retainers (20). Slide bearings (21) off end of auger and remove auger from collector housing (22). Inspect bearings for smooth operation and renew if needed. To reassemble, reverse disassembly procedure.

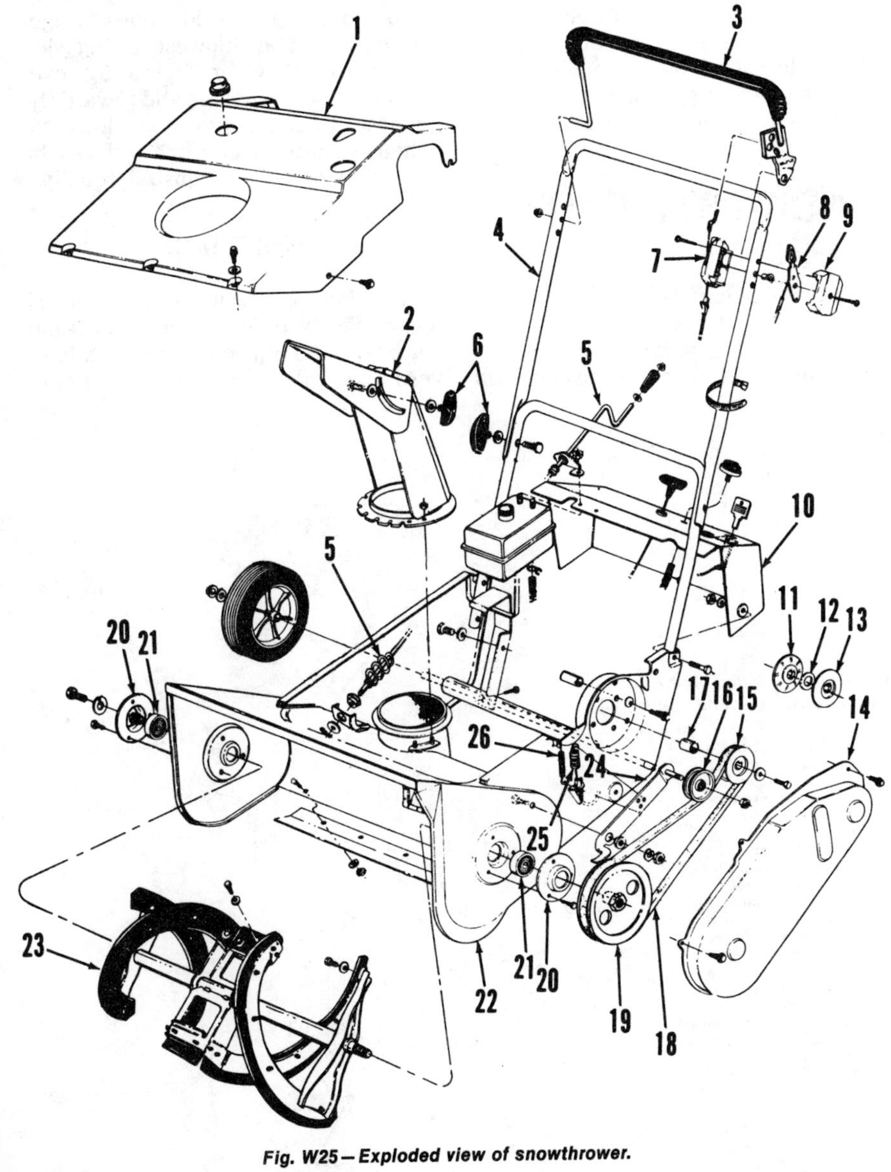

Fig. W25 — Exploded view of snowthrower.

1. Cover	8. Lever	15. Pulley	21. Bearing
2. Chute	9. Bracket	16. Idler pulley	22. Collector housing
3. Clutch handle	10. Rear cover	17. Spacer	23. Auger
4. Handle	11. Pulley half	18. Belt	24. Idler arm
5. Chute crank	12. Washer	19. Auger pulley	25. Spring
6. Wing nut	13. Pulley half	20. Retainer	26. Spring
7. Bracket	14. Cover		

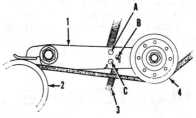

Fig. W26—View of belt and pulley system showing idler bracket (1), auger pulley (2), spring (3), idler pulley (4) and high, middle and low spring position holes (A, B and C).

WHITE

Model	Engine Make	Engine Model	Self-Propelled?	No. of Stages	Scoop Width
410	Tecumseh	HSSK50	Yes	2	20 in. (508 mm)
510	Tecumseh	HSSK50	Yes	2	24 in. (610 mm)
520	Tecumseh	HSSK50	Yes	2	24 in. (610 mm)

LUBRICATION

ENGINE

Refer to Tecumseh 4-stroke engine section for engine lubrication information. Recommended fuel is regular or unleaded regular gasoline.

DRIVE MECHANISM

Lubricate drive chains after every 25 hours of operation or at least once each season using a good-quality chain oil. Auger shaft bushings and traction drive shaft bushings should be oiled after every 25 hours of operation with SAE 30 motor oil. Lubricate all control pivot and slide points for smooth operation. Once each season or when necessary to replace auger flight shear bolts, remove shear bolts and squirt oil in hole and at ends of auger flights, then turn auger flight to distribute oil. On wheel drive Models 410 and 510, remove wheels once each season and grease axle shafts. On track drive Model 520, keep weight transfer linkage points greased with a clinging lubricant such as "Lubriplate."

AUGER GEARBOX

The auger gearbox is factory lubricated and should not require further lubrication unless disassembled for overhaul. Recommended lubricant is 10 ounces (295 mL) of Shell Alvania grease EPR00.

MAINTENANCE

Inspect belts and pulleys and renew belts if excessively worn, frayed or burned. Renew pulleys if belt grooves are worn or damaged, or if bearings in idler pulleys are loose or rough. Check drive chains and sprockets and renew if excessively worn or damaged.

ADJUSTMENTS

AUGER HEIGHT

Auger height can be adjusted by loosening skid shoe mounting bolts and repositioning the skid shoes. Auger height should be at lowest setting (skid shoes raised) when operating over smooth, paved surfaces and toward the highest setting (skid shoes lowered) when operating over rough surfaces. Be sure skid shoes are adjusted equally.

SHIFT ROD

On all models, disconnect shift control rod (6—Fig. W30) from shift lever (5) and place both the transmission and shift lever in neutral position. Turn shift con-

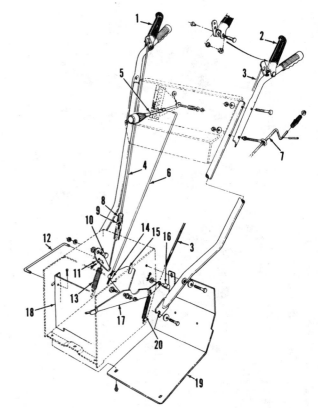

Fig. W30—Exploded view of handle assembly and controls.

1. Drive control lever
2. Auger control lever
3. Auger control rod
4. Drive control rod
5. Shift control lever
6. Shift control rod
7. Chute crank
8. Drive clutch spring
9. Adjusting nut
10. Shoulder bolt
11. Drive clutch bracket
12. Drive clutch rod
13. Drive spring
14. Ferrule
15. Shift bracket
16. Ferrule
17. Auger clutch bracket
18. Drive pulley support
19. Frame cover
20. Auger spring

trol rod in or out of ferrule (14) at bottom of rod until upper end can be inserted in hole in shift lever. Secure rod in lever with cotter pin.

DRIVE CLUTCH

On all models, with drive control lever (1—Fig. W30) released, drive spring (8) at bottom of drive control rod (4) should be loose. If not, unhook drive spring (8) from drive clutch bracket (11). Turn adjusting nut (9), located inside

spring, so hook of spring is aligned with center of hole in drive clutch bracket. Hook spring into bracket; the spring should be loose. When drive clutch lever (1) is held against handle grip, the drive clutch should engage and move the snowthrower without belt slippage.

AUGER CLUTCH

To check adjustment, remove belt cover and hold auger control lever (2—Fig. W30) against handle grip. The outer end

of the slot in the end of auger brake linkage should be against the spacer on the auger belt idler as shown in Fig. W32. If end of slot is not against idler spacer, disconnect auger control rod ferrule (Fig. W32) from auger clutch bracket and turn ferrule on control rod until proper adjustment is obtained when ferrule is reconnected.

TRACK DRIVE

Drive track tension on Model 520 is properly adjusted when the track can

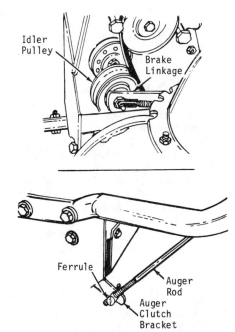

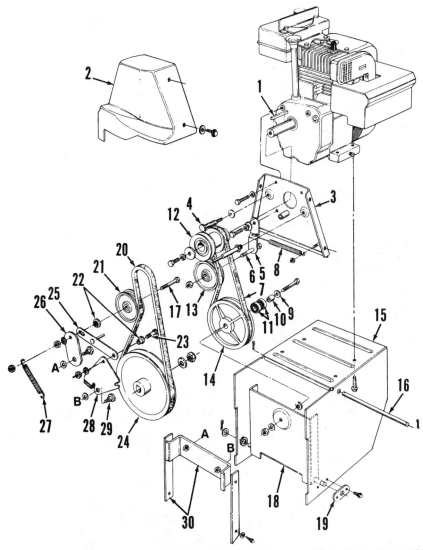

Fig. W31—Exploded view of drive belt and auger belt drive components and engine/frame assembly. Back side of drive pulley (14) has drive surface for friction wheel; depressing drive control lever moves pulley against friction wheel (38—Fig. W34).

1. Engine pulley key	8. Drive idler spring	16. Sliding bracket rod
2. Belt cover	9. Belleville washer	17. Auger idler bolt
3. Engine plate	10. Spacer	18. Drive pulley support
4. Shoulder bolt belt guide	11. Ball bearings	19. Support axle bracket
5. Drive clutch idler bracket	12. Engine pulley	20. Auger belt
6. Idler bolt	13. Drive belt idler	21. Auger belt idler
7. Drive belt	14. Drive pulley	22. Shoulder spacers
	15. Main frame	

23. Brake pivot bolt
24. Auger pulley
25. Brake linkage
26. Auger idler bracket
27. Auger idler spring
28. Brake bracket
29. Shoulder bolt
30. Blower housing parts

Fig. W32—End of slot in brake linkage (25—Fig. W31) should be against spacer on the auger belt idler when auger control lever is depressed against snowthrower handle. Refer to text for adjustment procedure.

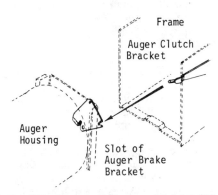

Fig. W33—Pin on end of auger clutch bracket must engage slot in auger brake bracket when reassembling snowthrower.

be lifted approximately 1/2 inch (13 mm) at midpoint between track rollers. Adjustment is provided by adjuster cam (21—Fig. W35) on each side of left and right track assembly. If necessary to adjust track tension, loosen the cap screws retaining adjuster cams (21) on each side of track. Turn the adjuster cams equally with a screwdriver placed between tab on adjuster cam and heavy washer, then tighten the cap screws.

OVERHAUL

ENGINE

Engine make and model numbers are listed at beginning of this section. Refer to Tecumseh 4-stroke engine sections for engine overhaul information.

DRIVE BELTS

Remove the deflector chute crank at chute and remove upper belt cover. Remove the two shoulder bolt belt guides at engine pulley. On track drive Model 520, place weight transfer lever in "Packed Snow" position. Remove the top bolts attaching auger housing to snowthrower frame assembly and loosen but do not remove the two bottom bolts. Lift up on auger belt to pull auger housing off of snowthrower frame and separate auger housing from the main unit.

To remove auger drive belt, remove the four shoulder bolt belt guides at auger pulley. The auger belt can now be removed from the bottom pulley. To remove traction drive belt, disconnect drive idler spring (8—Fig. W31) at engine plate and remove belt from engine pulley and bottom drive pulley. Install new belt(s) and reassemble by reversing disassembly procedure. Be sure pin on end of auger clutch bracket engages slot in auger brake bracket (see Fig. W33) before installing auger housing mounting bolts. Adjust auger and drive clutches as outlined in preceding paragraphs.

FRICTION WHEEL

To renew friction wheel, proceed as follows: Move shift lever to reverse position. Tip snowthrower up so it rests on front of auger housing and remove bottom cover, then remove friction wheel (38—Fig. W34) from wheel adapter (40). Place new wheel with cup side away from adapter and install retaining bolts finger tight. Turn the wheel to be sure it does not wobble, then securely tight-

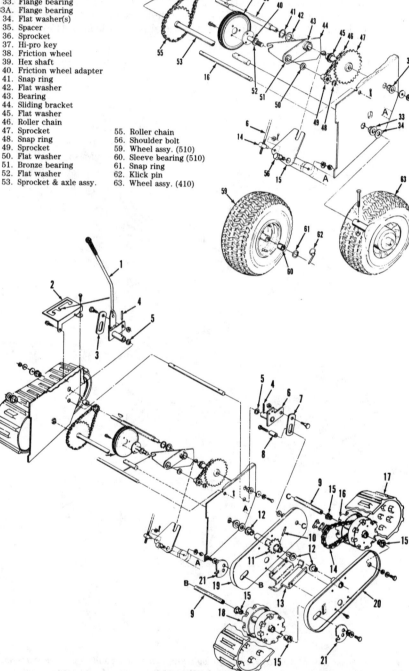

Fig. W34—Exploded view of drive components; also see Fig. W31. Notch in shift linkage bracket (15) engages pin on sliding bracket (44). Sliding bracket is supported by shaft (16). On all models, only one flat washer (34) is used at right end of axle shaft (53).

6. Shift rod
14. Ferrule
15. Shift linkage bracket
16. Sliding bracket rod
31. Belleville washer
32. Flat washer
33. Flange bearing
33A. Flange bearing
34. Flat washer(s)
35. Spacer
36. Sprocket
37. Hi-pro key
38. Friction wheel
39. Hex shaft
40. Friction wheel adapter
41. Snap ring
42. Flat washer
43. Bearing
44. Sliding bracket
45. Flat washer
46. Roller chain
47. Sprocket
48. Snap ring
49. Sprocket
50. Flat washer
51. Bronze bearing
52. Flat washer
53. Sprocket & axle assy.
55. Roller chain
56. Shoulder bolt
59. Wheel assy. (510)
60. Sleeve bearing (510)
61. Snap ring
62. Klick pin
63. Wheel assy. (410)

Fig. W35—Exploded view of track drive assembly for Model 520. Drive components are similar to wheel drive models shown in Fig. W34. Track lock handle changes auger down force and has three positions: "Transport," "Normal Snow" and "Packed Snow."

1. Track lock handle
2. Selector plate
3. Slotted link
4. Spring pin
5. Flat washer
6. Pivot arm
7. Slotted link
8. Spacer
9. Wheel axle shaft
10. Spring pin
11. Sprocket
12. Flange bearing
13. Skid shoe
14. Roller chain
15. Flange bearing
16. Drive wheel
17. Snow track
18. Idler wheel
19. Inner plate
20. Outer plate
21. Track adjuster cam

en retaining bolts. Replace bottom cover.

TRACK DRIVE

To renew drive track or chain inside track assembly on Model 520, block up snowthrower frame so track assemblies are off the ground. Loosen the cap screws retaining track adjuster cams (21—Fig. W35) at each side of assembly and turn adjusters so all track tension is removed. Roll the track off of track rollers and side supports. Remove roller chain (14) by disconnecting master

link. Reassemble by reversing disassembly procedure, making sure to install track with cleats pointing toward front of snowthrower. Adjust track tension as outlined in adjustment paragraph.

AUGER GEARBOX

To remove auger gearbox, follow procedure outlined in preceding paragraph for removal of auger drive belt. Then, remove auger drive pulley (7—Fig. W36) from rear of gearbox shaft and unbolt auger bearing housings (37) at each end of auger housing. Remove the gear-

box and auger flight assembly from front of auger housing, then remove impeller (18) from input shaft and auger flights from driveshaft. Note the position of washers and spacers as you are disassembling the unit.

To disassemble gearbox, refer to exploded view in Fig. W36 and remove the screws retaining upper gear housing half (19). Note thrust washer and spacer positions as you disassemble gearbox. Clean and inspect all parts and renew any worn or damaged parts. Lubricate all parts prior to reassembly and pack gearbox with 10 ounces (295 mL) of Shell Alvania Grease EPR00.

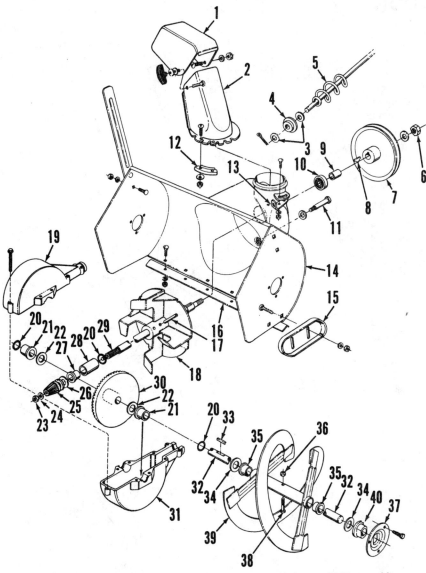

Fig. W36—Exploded view of auger housing, gearbox and auger flight assembly.

1. Upper chute	14. Auger housing	22. Flat washer	32. Auger flight (spiral) axle
2. Lower chute	15. Skid shoes	23. Hex nut	33. Key (Hi-pro or square)
3. Flat washers	16. Scraper bar (shave plate)	24. Flat washer	34. Flat washer
4. Bushing	17. Spring pin	25. Pinion gear	35. Flange bearing
5. Chute crank	18. Impeller (blower fan)	26. Flat washer(s)	36. Shear bolt nut
6. Pulley nut	19. Upper gear housing half	27. Flange bearing	37. Bearing housing
7. Auger drive pulley	20. "O" ring	28. Sleeve bearing	38. Shear bolt
8. Hi-pro key	21. Plastic flange bearing	29. Impeller (blower) axle	39. Auger flight (spiral)
9. Spacer		30. Bevel gear	40. Flange bearing
10. Ball bearing		31. Lower gear housing half	
11. Shoulder bolt			
12. Chute flange keeper			
13. Chute crank bracket			

YARD-MAN

MTD PRODUCTS, INC.
P.O. Box 36900
Cleveland, Ohio 44136

Model	Engine Make	Engine Model	Self-Propelled?	No. of Stages	Scoop Width
31100	Tecumseh	AH520	No	1	20 in. (508 mm)

LUBRICATION

ENGINE

Refer to Tecumseh engine section for engine lubrication requirements.

The engine is a two-stroke type engine which is lubricated by mixing oil with fuel. Yard-man recommends a fuel:oil ratio of 32:1 using a good quality oil designed for two-stroke engines mixed with regular gasoline.

DRIVE MECHANISM

Lubricate shafts and bushings as required to prevent binding and galling. Lubricate control lever pivot point periodically for smooth operation.

On early models, periodically lubricate clutch bushing (5—Fig. Y4) with oil. Do not allow oil on clutch shoes or drum as clutch slippage will result.

MAINTENANCE

On early models inspect drive chain and sprockets. On later models inspect drive belt and pulleys and renew belt if frayed, cracked, excessively worn or otherwise damaged.

On later models the auger is secured to drive pulley shaft by a shear pin which will fracture when auger contacts an obstruction. Renew shear pin with a suitable replacement. Do not use ordinary fasteners which may damage drive components.

ADJUSTMENTS

CHAIN

On early models a chain is used to drive auger. Remove chain guard and top panel then loosen engine mount bolts and locating bolt on right side of engine. Adjust engine location so chain will deflect ½ inch (12.7 mm) with thumb pressure. Retighten engine mounting bolts.

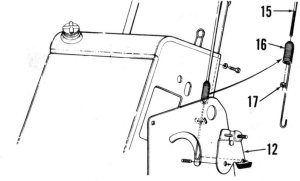

Fig. Y1—View of auger control clutch rod and spring on later models. Refer to text for adjustment.

12. Idler arm & brake
15. Auger clutch control rod
16. Spring
17. Nut

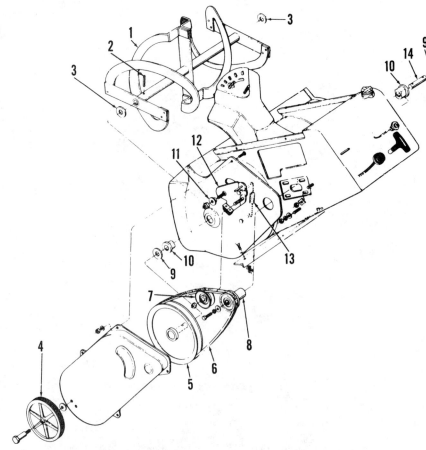

Fig. Y2—Exploded view of auger drive components on later models.

1. Auger	5. Auger pulley	8. Engine pulley	11. Spacer
2. Pin	6. Belt	9. Washer	12. Idler arm & brake
3. Washer	7. Idler pulley	10. Bushing	13. Spring
4. Guide wheel			14. Shaft

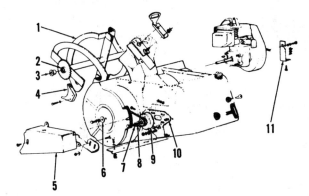

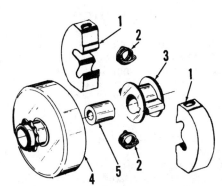

Fig. Y3—Exploded view of auger drive components on early models. Refer to Fig. Y4 for clutch components.

1. Auger
2. Auger shaft
3. Bushing
4. Chain cleaner
5. Chain guard
6. Washer
7. Spring washer
8. Washer
9. Clutch
10. Engine plate
11. Engine bracket

Fig. Y4—Exploded view of centrifugal clutch used on early models.

1. Clutch shoes
2. Springs
3. Hub
4. Clutch drum
5. Bushing

CLUTCH ROD

On later models the clutch rod (15—Fig. Y1) should be adjusted so there is 1/8 inch (3.2 mm) end play between rod and clutch spring (16) with auger clutch lever against handlebar in disengaged position. To adjust end play, detach clutch rod (15) and turn rod in nut (17) located in spring (16).

OVERHAUL

ENGINE

Engine make and model are listed in front of section. Refer to Tecumseh section for engine overhaul.

DRIVE MECHANISM

Overhaul of drive mechanism is evident after inspection of unit and referral to Fig. Y2 or Y3.

CENTRIFUGAL CLUTCH

Early models are equipped with a centrifugal clutch as shown in Fig. Y4. The clutch may be removed after removing chain guard, disconnecting chain and unscrewing retaining screw. Inspect components for excessive wear and damage. Clutch shoes (1) are available only as a set. Bushing (5) should be lubricated with a light oil. Check clutch drum for warpage or out-of-round due to overheating.

YARD-MAN

Model	Engine Make	Engine Model	Self-Propelled?	No. of Stages	Scoop Width
31220	Tecumseh	HS35	Yes	1	20 in. (508mm)
31230	Tecumseh	HS50	Yes	1	20 in. (508 mm)
31430	Tecumseh	HS50	Yes	1	21 in. (533 mm)

LUBRICATION

ENGINE

Refer to Tecumseh engine section for engine lubrication requirements.

DRIVE MECHANISM

Lubricate drive chains at least once each season with good quality chain oil. Lubricate control cables and control rod and lever pivot points for smooth operation.

MAINTENANCE

Inspect chain and sprockets and renew if excessively worn. Inspect belts and pulleys and renew belts if frayed, cracked, excessively worn or otherwise damaged. Inspect control cables and renew if frayed or binding.

Recommended tire pressure is 7-10 psi (48-69 kPa).

ADJUSTMENTS

HEIGHT

Auger height may be adjusted by loosening skid shoe mounting bolts and repositioning skid shoe. Skid shoes should be located in lowest setting when operating over smooth surface and in highest setting when operating over rough, irregular surfaces.

SHIFT LINKAGE

The shift linkage must be properly adjusted so shift hub (17—Fig. Y12) correctly engages dogs on forward sprocket (18) and reverse gear (16). To adjust shift linkage, detach shift rod (4—Fig. Y10) from shift arm (5). Roll snowthrower back and forth while moving shift arm (5) until neutral is found. Position shift control lever in neutral then rotate shift rod (4) so rod end will engage shift arm (5) hole and attach rod. Check adjustment by en-

gaging forward and reverse and being sure unit rolls freely in neutral.

DRIVE CLUTCH

The drive clutch is adjusted by turning adjusting nut (16—Fig. Y10) on clutch cable housing. Rotate adjusting nut (16) so clutch control lever just contacts handlebar grip with lever in engaged position.

AUGER CLUTCH

The auger clutch is adjusted by turning adjusting rod nut (12—Fig. Y10). Turn nut so there is approximately 0.015 inch (0.38 mm) between compressed coils of spring (10) when auger clutch control lever is in engaged position.

AUGER DRIVE CHAIN

Auger drive chain tension on Model 31220 is adjusted by loosening screws in each end of auger shaft then turning adjusting cam on left side of auger frame as shown in Fig. Y11. Place a suitable tool through hole in auger shaft to secure shaft when loosening or tightening end screws.

Auger drive chain tension on Model 31430 is adjusted by loosening nut (37—Fig. Y13) and repositioning chain idler (31) to remove chain slack.

AXLE SPROCKET

Wheel axle sprocket (12—Fig. Y12) must be aligned with drive sprocket

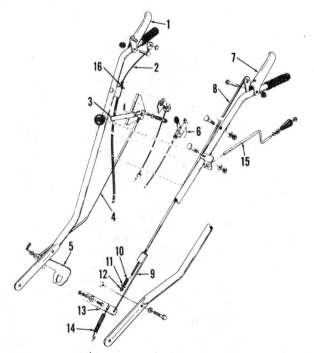

Fig. Y10—Drawing of control assembly.

1. Drive clutch control lever
2. Clutch cable
3. Shift control lever
4. Shift rod
5. Shift arm
6. Throttle assy.
7. Auger clutch control lever
8. Auger clutch rod
9. Auger clutch rod
10. Spring
11. Washer
12. Locknut
13. Lever
14. Spring
15. Discharge chute crank
16. Nut

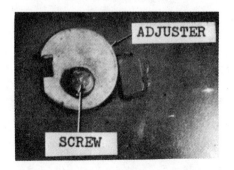

Fig. Y11—View showing chain adjuster cam and auger shaft screw on Model 31220.

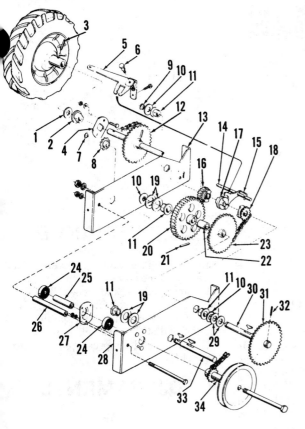

Fig. Y12—Exploded view of drive assembly.

1. Shim
2. Bushing
3. Pin
4. Bushing plate
5. Shift arm
6. Pivot bolt
7. Snap ring
8. Sprocket
9. Snap ring
10. Washer
11. Bushing
12. Sprocket
13. Bracket
14. Shift shaft
15. Shift fork
16. Reverse gear
17. Shift hub
18. Forward sprocket
19. Washers
20. Gear
21. Pin
22. Spacer
23. Sprocket
24. Bearing
25. Spacer
26. Spacer
27. Bearing retainer
28. Bracket
29. Washers
30. Shaft
31. Sprocket
32. Pin
33. Shaft
34. Sprocket & pulley assy.

(8). Check alignment by laying a straight edge across faces of sprockets. Wheel axle sprocket position is determined by shims (1) between wheel hub and bushing (2). Add or delete shims as required to align sprockets. In some cases it may be necessary to grind material from wheel hub and trim flange of bushing.

OVERHAUL

ENGINE

Engine make and model numbers are listed in front of this section. Refer to Tecumseh engine section for overhaul procedure.

DRIVE BELTS

To remove drive belts, remove cover plates on front and right side of main frame then tip unit forward and remove bottom panel. Unscrew shoulder bolt adjacent to lower pulleys which serves as belt guide and remove belts. Reverse removal procedure to install belts.

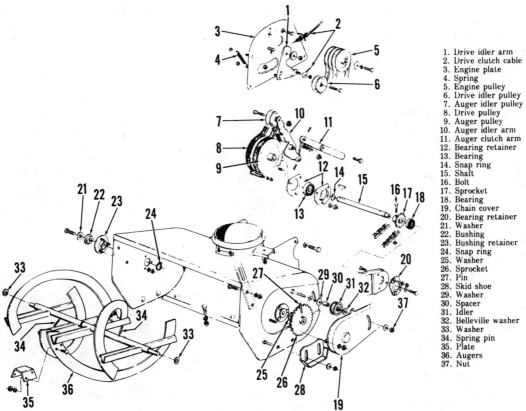

1. Drive idler arm
2. Drive clutch cable
3. Engine plate
4. Spring
5. Engine pulley
6. Drive idler pulley
7. Auger idler pulley
8. Drive pulley
9. Auger pulley
10. Auger idler arm
11. Auger clutch arm
12. Bearing retainer
13. Bearing
14. Snap ring
15. Shaft
16. Bolt
17. Sprocket
18. Bearing
19. Chain cover
20. Bearing retainer
21. Washer
22. Bushing
23. Bushing retainer
24. Snap ring
25. Washer
26. Sprocket
27. Pin
28. Skid shoe
29. Washer
30. Spacer
31. Idler
32. Belleville washer
33. Washer
34. Spring pin
35. Plate
36. Augers
37. Nut

Fig.—Y13—Exploded view of auger drive and drive belt assemblies used on Model 31430. Model 31220 is similar.

YARD-MAN

Model	Engine Make	Engine Model	Self-Propelled?	No. of Stages	Scoop Width
31550	Tecumseh	HS50	Yes	2	24 in. (610 mm)
31600	Tecumseh	HS50	Yes	2	24 in. (610 mm)
31650	Tecumseh	HS70	Yes	2	24 in. (610 mm)
31750	Tecumseh	HS70	Yes	2	26 in. (660 mm)
31800	Tecumseh	HM80	Yes	2	26 in. (660 mm)
31850	Tecumseh	HM80	Yes	2	26 in. (660 mm)
31950	Tecumseh	HM80	Yes	2	33 in. (838 mm)
31960	Tecumseh	HM100	Yes	2	33 in. (838 mm)

LUBRICATION

ENGINE

Refer to Tecumseh engine section for engine lubrication requirements.

Recommended fuel is regular or low-lead gasoline.

DRIVE MECHANISM

Lubricate drive chains at least once each season with good quality chain oil. Auger shaft and axle bushings should be oiled and axle should be greased at least once each season or after 25 hours of operation. Lubricate control rod and lever pivot points for smooth operation.

AUGER GEARBOX

The auger gearbox is lubricated with either SAE 90 oil or Plastilube. On models equipped with a plug (20—Fig. Y24), fill gearbox with SAE 90 until oil level reaches plug opening. If gearbox is not equipped with plug (20), then gearbox is lubricated with Plastilube during assembly and does not normally require new lubricant.

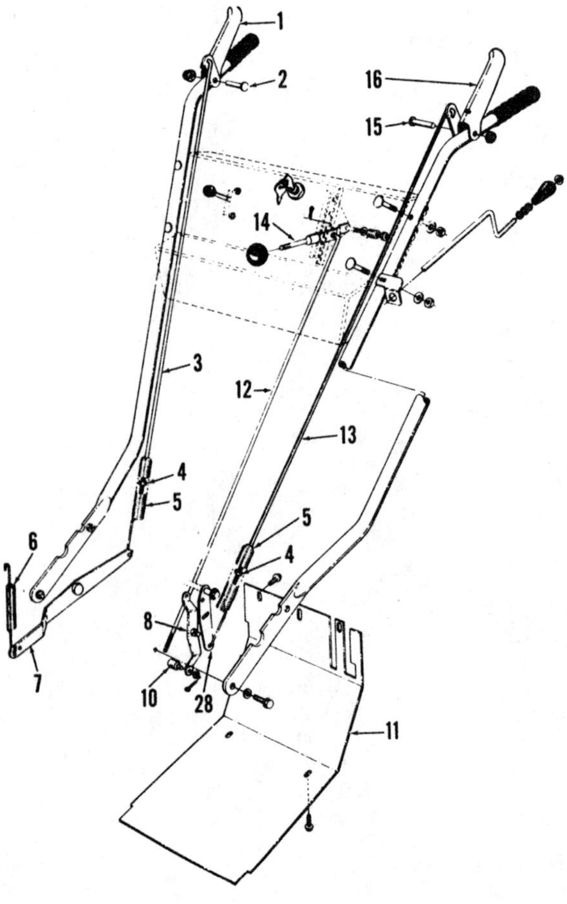

MAINTENANCE

Inspect chain and sprockets and renew if excessively worn. Inspect belts and pulleys and renew belts if frayed, cracked, burnt or otherwise damaged. Check throttle cable for binding and stiff operation.

Recommended tire pressure is 7-10 psi (48-69 kPa).

ADJUSTMENTS

HEIGHT

Auger height may be adjusted by loosening skid shoe mounting bolts and repositioning skid shoe. Skid shoes should be located in lowest setting when operating over smooth surfaces and in highest setting when operating over rough, irregular surfaces.

SHIFT CONTROL ROD

Detach shift control rod (12—Fig. Y20) from shift control lever (14). On later models with locating slot in lower frame cover as shown in Fig. Y21, move shift rod so shift plate (27—Fig. Y23) is centered in slot. On early models without locating slot, remove lower frame cover and move shift control rod so friction wheel is

Fig. Y20—Drawing of control assembly.

1. Auger clutch control lever
2. Pin
3. Auger clutch control rod
4. Nut
5. Spring
6. Spring
7. Lever
8. Lever
10. Pivot nut
11. Lower panel
12. Shift rod
13. Drive clutch control rod
14. Shift control lever
15. Pivot pin
16. Drive clutch control lever

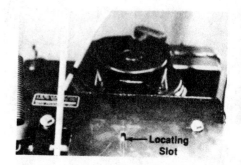

Fig. Y21—View showing location of locating slot used for shift control adjustment on later models.

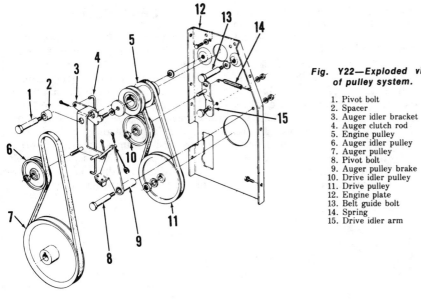

Fig. Y22—Exploded view of pulley system.

1. Pivot bolt
2. Spacer
3. Auger idler bracket
4. Auger clutch rod
5. Engine pulley
6. Auger idler pulley
7. Auger pulley
8. Pivot bolt
9. Auger pulley brake
10. Drive idler pulley
11. Drive pulley
12. Engine plate
13. Belt guide bolt
14. Spring
15. Drive idler arm

centered on neutral stop bracket (N—Fig. Y23). On all models, position shift control lever in neutral detent then rotate shift control rod in ferrule at bottom of rod so upper rod end will fit easily into shift control lever. Attach rod then check adjustment by noting if unit will roll forward and backward with shift control lever in neutral.

CLUTCH RODS

Adjustment of auger and drive clutch control rods is accomplished by turning nut at spring end of rod. There should be a small amount of end play between rod and spring (5—Fig. Y20) with clutch disengaged so idler will relieve pressure against drive belt. With clutch engaged, drive belt should tighten around pulleys without slippage.

OVERHAUL

ENGINE

Engine make and model numbers are listed at front of this section. Refer to Tecumseh engine section for engine overhaul procedure.

AUGER GEARBOX

Auger gearbox may be serviced after removing complete auger assembly from frame then detaching impeller and augers. Overhaul of gearbox is evident after referral to Figs. Y24 & Y24A and inspection of unit. Note thickness of washers during disassembly and reassembly.

All gearboxes may be filled with 4 ounces of Plastilube grease while models equipped with plug (20) may be

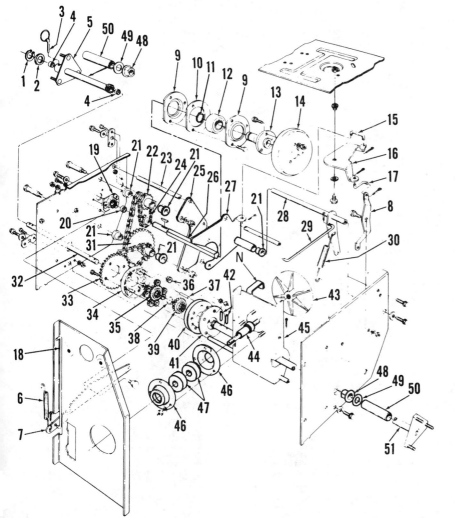

Fig. Y23— Exploded view of drive mechanism used on Models 31850, 31950 and 31960. Models 31600, 31650, 31750 and 31800 are similar but different components (34 through 41) and spacers (50) are not used.

1. Snap ring	27. Shift plate
2. Washer	28. Shaft
3. Lock pin	29. Link
4. Bushing	30. Spring
5. Right axle	31. Sprocket
6. Spring	32. Differential lock
7. Lever	shaft
8. Lever	33. Sprocket
9. Bearing retainer	34. Gear housing
10. Spacer	35. Sun gear
11. Snap ring	36. Planetary gear
12. Bearing	37. Shaft
13. Friction hub	38. Spacer
14. Friction wheel	39. Gear
15. Link	40. Gear housing
16. Bellcrank	41. Plate
17. Link	42. Pins
18. Auger clutch rod	43. Friction disc
19. Bushing plate	44. Shaft
20. Bushing	45. Drive support
21. Bushing	46. Bearing housing
22. Sprocket	47. Bearings
23. Shaft	48. Bushing
24. Sprocket	49. Washer
25. Support	50. Spacer
26. Hex shaft	51. Left axle

lubricated with oil as outlined in LUBRICATION section. Models without plug (20) must be lubricated with grease.

DRIVE MECHANISM

Refer to Figs. Y22, Y23 and Y24B for exploded views of drive components. A differential is used on Models 31850, 31950 and 31960. On all models a lock pin may be used to lock wheel and axle together so both wheels are driven.

The lock pin is located in right wheel of Models 31850, 31950 and 31960 and in left wheel of all other models. Do not overtighten locknuts securing bearing retaining plates (9). Bearing (12) is self-centering and must be able to move in retaining plates or bearing will bind on friction hub (13). Install friction wheel (14) with cupped side towards hub (13). Be sure auger idler bracket (3—Fig. Y22) has sufficient travel to operate properly. If not, grind approximately 1/16-inch (1.6 mm) from end of spacer (2) and recheck movement.

DRIVE BELTS

To remove drive belts, disconnect chute linkage and remove belt cover. Unscrew shoulder bolts adjacent to engine pulley while noting roller mounted on right side bolt. Lift up on auger idler pulley (6—Fig. Y22) so brake disengages auger drive pulley then lift up auger frame slightly and separate auger and main frames. Unscrew shoulder bolts which serve as belt guides and remove drive belts. Reverse removal procedure to install belts.

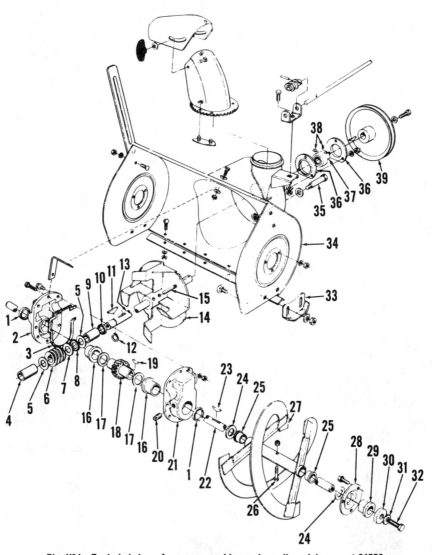

Fig. Y24—Exploded view of auger assembly used on all models except 31550.

1. Seal	9. Bushing	17. Washer	25. Bushing
2. Gearbox half R.H.	10. Seal	18. Gear	26. Bolt
3. Gasket	11. Key	19. Key	27. Auger
4. Bushing	12. Snap ring	20. Plug	28. Bushing retainer
5. Washer	13. Drive shaft	21. Gearbox half L.H.	29. Bushing
6. Worm gear	14. Impeller	22. Auger shaft	30. Washer
7. Washer	15. Pin	23. Key	31. Belleville washer
8. Thrust bearing	16. Bushing	24. Washer	32. Screw

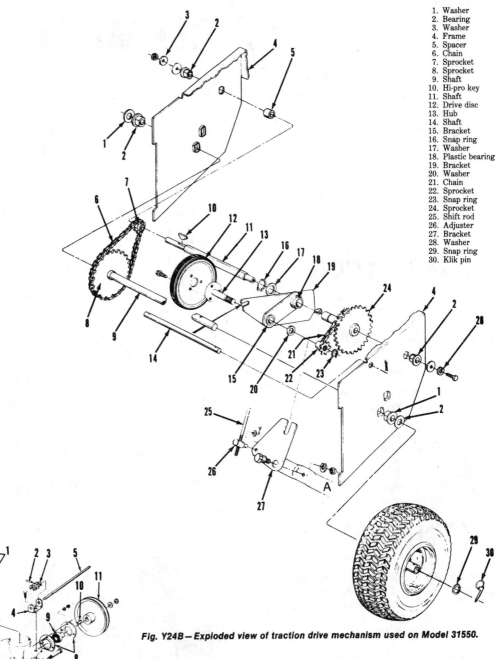

1. Washer
2. Bearing
3. Washer
4. Frame
5. Spacer
6. Chain
7. Sprocket
8. Sprocket
9. Shaft
10. Hi-pro key
11. Shaft
12. Drive disc
13. Hub
14. Shaft
15. Bracket
16. Snap ring
17. Washer
18. Plastic bearing
19. Bracket
20. Washer
21. Chain
22. Sprocket
23. Snap ring
24. Sprocket
25. Shift rod
26. Adjuster
27. Bracket
28. Washer
29. Snap ring
30. Klik pin

Fig. Y24B — Exploded view of traction drive mechanism used on Model 31550.

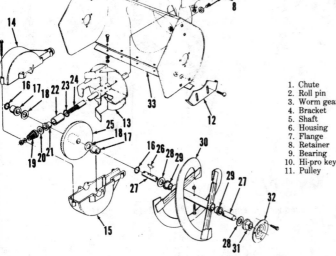

1. Chute	12. Skid
2. Roll pin	13. Impeller
3. Worm gear	14. Upper case half
4. Bracket	15. Lower case half
5. Shaft	16. "O" ring
6. Housing	17. Plastic bearing
7. Flange	18. Washer
8. Retainer	19. Pinion gear
9. Bearing	20. Washer
10. Hi-pro key	21. Bearing
11. Pulley	22. Sleeve
	23. "O" ring
	24. Shaft
	25. Gear
	26. Woodruff key
	27. Auger shaft
	28. Washer
	29. Bearing
	30. Auger
	31. Bearing
	32. Retainer
	33. Shave plate

Fig. Y24A — Exploded view of auger assembly used on Model 31550.

YARD-MAN

Model	Engine Make	Engine Model	Self-Propelled?	No. of Stages	Scoop Width
31150	Tecumseh	AH600	No	1	21 in. (533 mm)
31160	Tecumseh	HS50	No	1	21 in. (533 mm)
31180	Tecumseh	AH600	No	1	21 in. (533 mm)
31190	Tecumseh	HS50	No	1	21 in. (533 mm)

LUBRICATION

ENGINE

Models 31150 and 31180

The engine is a two-stroke type and a fuel:oil mix ratio of 32:1 is recommended. Mix SAE 30 or 40 outboard or two-stroke oil with regular or unleaded gasoline.

Models 31160 and 31190

Refer to Tecumseh engine section for lubrication requirements. Recommended fuel is regular or unleaded gasoline.

MAINTENANCE

Periodically inspect drive belt and pulleys. Renew belt if frayed, cracked, excessively worn or otherwise damaged.

Inspect flighting on auger and renew if chipped or excessively worn.

ADJUSTMENT

BELT

Refer to Fig. Y32 and remove belt cover. Move spring (3) end to next highest hole (A, B or C) to increase belt tension. Reinstall cover.

OVERHAUL

ENGINE

Engine make and model numbers are listed at the front of this section. Refer to Tecumseh engine section for overhaul procedures.

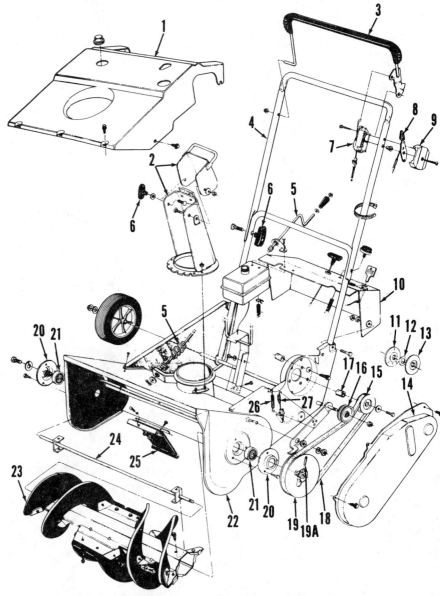

Fig. Y30 — Exploded view of snowthrower Models 31150 and 31160.

1. Cover
2. Chute
3. Clutch handle
4. Handle
5. Chute crank
6. Wing nut
7. Bracket
8. Lever
9. Bracket
10. Rear cover
11. Pulley half
12. Washer
13. Pulley half
14. Cover
15. Pulley
16. Idler pulley
17. Spacer
18. Belt
19. Auger pulley
19A. Roll pin
20. Retainer
21. Bearing
22. Collector housing
23. Auger
24. Auger shaft
25. Shave plate & flap

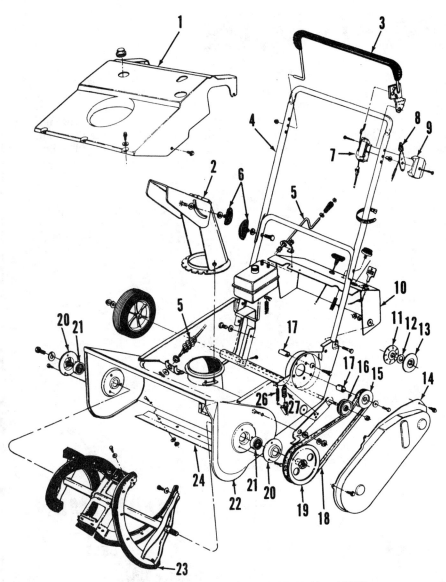

Fig. Y32 — View of belt and pulley system showing idler bracket (1), auger pulley (2), springs (3 and 5), idler pulley (4) and high, middle and low spring position holes (A, B and C).

BELT

To renew belt (18 – Figs. Y30 and Y31), remove cover (14) and disconnect springs (26 and 27). Lift up on idler (16) bracket and slide belt off pulleys (15 and 19). Reverse procedure to reinstall.

AUGER ASSEMBLY

Models 31150 and 31160

Refer to Fig. Y30 for an exploded view of auger and collector assembly. Remove belt as outlined in BELT section. Remove roll pin (19A) and slide pulley (19) off end of auger shaft (24). Unbolt and remove bearing retainers (20). Slide bearings (21) off end of auger shaft and remove auger assembly (23 and 24) from collector housing (22). Inspect bearings for smooth operation and renew as needed. To reassemble, reverse disassembly procedure.

Models 31180 and 31190

Refer to Fig. Y31 for an exploded view of auger and collector assembly. Remove belt as outlined in BELT section. Thread auger pulley (19) off end of auger (23). Unbolt and remove bearing retainers (20). Slide bearings (21) off end of auger. Remove auger assembly from collector housing (22). Inspect bearings for smooth operation and renew as needed. To reassemble, reverse disassembly procedure.

Fig. Y31 — Exploded view of snowthrower Models 31180 and 31190.

1. Cover	7. Bracket	13. Pulley half
2. Chute	8. Lever	14. Cover
3. Clutch handle	9. Bracket	15. Pulley
4. Handle	10. Rear cover	16. Idler pulley
5. Chute crank	11. Pulley half	17. Spacer
6. Wing nut	12. Washer	18. Belt

19. Auger pulley	
20. Retainer	
21. Bearing	
22. Collector housing	
23. Auger	

YARD-MAN

Model	Engine Make	Engine Model	Self-Propelled?	No. of Stages	Scoop Width
31340	Tecumseh	HS40	Yes	2	22 in. (559 mm)
31350	Tecumseh	H50	Yes	2	24 in. (610 mm)
31380	Tecumseh	HM80	Yes	2	28 in. (711 mm)
31960	Tecumseh	HM100	Yes	2	33 in. (838 mm)

LUBRICATION

ENGINE

Refer to Tecumseh engine section for lubrication requirements. Recommended fuel is regular or unleaded gasoline.

DRIVE MECHANISM

Once each season, lubricate all chains, bearings, gears and shift linkage with engine oil.

MAINTENANCE

Periodically inspect drive belts and pulleys. Renew belts if frayed, cracked, excessively worn or otherwise damaged.

1. Spring
2. Idler bracket
3. Brake linkage
4. Spacer
5. Brake bracket
6. Auger pulley
7. Idler pulley
8. Auger belt
9. Engine pulley
10. Idler pulley
11. Traction pulley
12. Bearings
13. Cotter pin
14. Idler bracket
15. Plate
16. Spacer
17. Spring
18. Frame
19. Shaft
20. Bracket

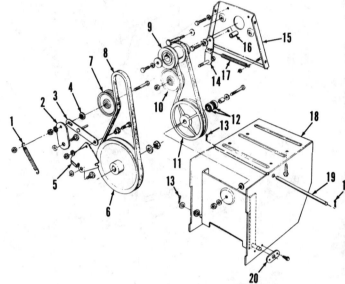

Fig. Y41—Exploded view of belt and pulley system.

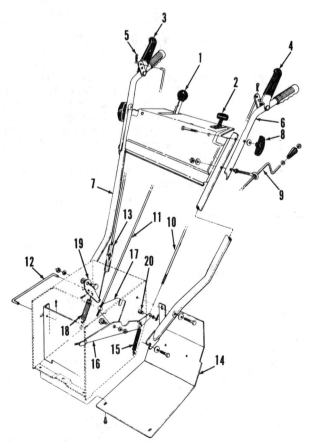

1. Shift lever
2. Throttle lever
3. Traction clutch lever
4. Auger clutch lever
5. Pin
6. Handle, upper
7. Handle, lower
8. Wing nut
9. Chute crank
10. Cable
11. Shift rod
12. Clutch rod
13. Extension spring
14. Frame cover
15. Spring
16. Bracket
17. Bracket
18. Spring
19. Bracket
20. Ferrule

Fig. Y40—Exploded view of handle and control assembly.

ADJUSTMENT

BELT

Traction Drive

Unhook spring (13–Fig. Y40) from drive bracket (19). With the clutch lever (3) in the raised position and the spring not stretched, spring hook should line up with center hole of drive bracket. Use adjuster nut on cable to change length as needed.

Auger Drive

Disconnect ferrule (20–Fig. Y40) from bracket (16). With clutch lever (4) in the raised position, align ferrule with the hole in the bracket. Thread ferrule in or out on shaft (10) to achieve proper adjustment.

OVERHAUL

ENGINE

Engine make and model numbers are listed at the front of this section. Refer to Tecumseh engine section for overhaul procedures.

BELT

All Belts

Disconnect cotter pin and remove chute crank assembly (30 – Fig. Y43) from collector housing. Remove skid (23) from right hand side. Unbolt and remove right hand side belt cover (22). Unbolt and remove collector housing rear belt cover (24). Unbolt and remove the upper plastic belt cover. Remove the top collector housing-to-frame bolts and loosen the bottom bolts. Remove the two engine pulley belt guards and slide auger belt off engine pulley. Separate the collector housing from the frame assembly. Disconnect idler spring (17). Remove center bolt from auger pulleys (19 and

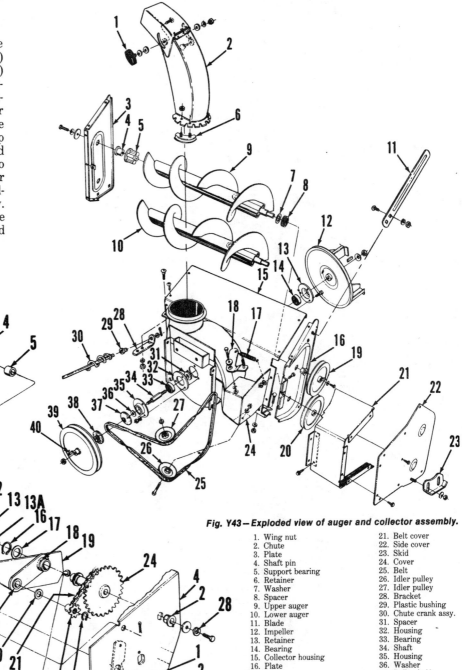

Fig. Y43 – Exploded view of auger and collector assembly.

1. Wing nut	21. Belt cover
2. Chute	22. Side cover
3. Plate	23. Skid
4. Shaft pin	24. Cover
5. Support bearing	25. Belt
6. Retainer	26. Idler pulley
7. Washer	27. Idler pulley
8. Spacer	28. Bracket
9. Upper auger	29. Plastic bushing
10. Lower auger	30. Chute crank assy.
11. Blade	31. Spacer
12. Impeller	32. Housing
13. Retainer	33. Bearing
14. Bearing	34. Shaft
15. Collector housing	35. Housing
16. Plate	36. Washer
17. Spring	37. Pulley half
18. Idler bracket	38. Pulley half
19. Upper pulley	39. Pulley
20. Lower pulley	40. Square key

20) and slide pulleys out enough to remove belt.

Install new belt and reassemble by reversing disassembly procedure.

DRIVE DISC

Move shift lever to reverse position and tip snowthrower on end. Remove bottom cover (14 – Fig. Y40). Unbolt drive disc (12 – Fig. Y42) from hub (13) and remove. Install the new drive disc so that the cupped side is opposite the hub. Reinstall bottom cover.

1. Washer		
2. Bearing	12. Drive disc	
3. Washer	13. Hub	
4. Frame	13A. Washer	
5. Spacer	14. Shaft	21. Chain
6. Chain	15. Bracket	22. Sprocket
7. Sprocket	16. Snap ring	23. Snap ring
8. Sprocket	17. Washer	24. Sprocket
9. Shaft	18. Plastic bearing	25. Shift rod
10. Hi-pro key	19. Bracket	26. Adjuster
11. Shaft	20. Washer	27. Bracket
		28. Washer

Fig. Y42 – Exploded view of traction drive mechanism.

FUNDAMENTALS SECTION

ENGINE FUNDAMENTALS

OPERATING PRINCIPLES

The one, two or four cylinder engines used to power riding lawn mowers, garden tractors, pumps, generators, welders, mixers, windrowers, hay balers and many other items of power equipment in use today are basically similar. All are technically known as "Internal Combustion Reciprocating Engines."

The source of power is heat formed by the burning of a combustible mixture of petroleum products and air. In a reciprocating engine, this burning takes place in a closed cylinder containing a piston. Expansion resulting from the heat of combustion applies pressure on the piston to turn a shaft by means of a crank and connecting rod.

The fuel-air mixture may be ignited by means of an electric spark (Otto Cycle Engine) or by heat formed from compression of air in the engine cylinder (Diesel Cycle Engine). The complete series of events which must take place in order for the engine to run occurs in two revolutions of the crankshaft (four strokes of the piston in cylinder) and is referred to as a "Four-Stroke Cycle Engine."

OTTO CYCLE. In a spark ignited engine, a series of five events is required in order for the engine to provide power. This series of events is called the "Cycle" (or "Work Cycle") and is repeated in each cylinder of the engine as long as work is being done. This series of events which comprise the "Cycle" is as follows:

1. The mixture of fuel and air is pushed into the cylinder by atmospheric pressure when the pressure within the engine cylinder is reduced by the piston moving downward in the cylinder.

2. The mixture of fuel and air is compressed by the piston moving upward in the cylinder.

3. The compressed fuel-air mixture is ignited by a timed electric spark.

4. The burning fuel-air mixture expands, forcing the piston downward in the cylinder thus converting the chemical energy generated by combustion into mechanical power.

5. The gaseous products formed by the burned fuel-air mixture are exhausted from the cylinder so that a new "Cycle" can begin.

The above described five events which comprise the work cycle of an engine are commonly referred to as (1), INTAKE;

(2), COMPRESSION; (3), IGNITION; (4), EXPANSION (POWER); and (5), EXHAUST.

DIESEL CYCLE. The Diesel Cycle differs from the Otto Cycle in that air alone is drawn into the cylinder during the intake period. The air is heated from being compressed by the piston moving upward in the cylinder, then a finely atomized charge of fuel is injected into the cylinder where it mixes with the air and is ignited by the heat of the compressed air. In order to create sufficient heat to ignite the injected fuel, an engine operating on the Diesel Cycle must compress the air to a much greater degree than an engine operating on the Otto Cycle where the fuel-air mixture is ignited by an electric spark. The power and exhaust events of the Diesel Cycle are similar to the power and exhaust events of the Otto Cycle.

FOUR-STROKE CYCLE. In a four-stroke cycle engine operating on the Otto Cycle (spark ignition), the five events of the cycle take place in four strokes of the piston, or in two revolutions of the engine crankshaft. Thus, a power stroke occurs only on alternate

downward strokes of the piston.

In view "A" of Fig. 1−1, the piston is on the first downward stroke of the cycle. The mechanically operated intake valve has opened the intake port and, as the downward movement of the piston has reduced the air pressure in the cylinder to below atmospheric pressure, air is forced through the carburetor, where fuel is mixed with the air, and into the cylinder through the open intake port. The intake valve remains open and the fuel-air mixture continues to flow into the cylinder until the piston reaches the bottom of its downward stroke. As the piston starts on its first upward stroke, the mechanically operated intake valve closes and since the exhaust valve is closed, the fuel-air mixture is compressed as in view "B".

Just before the piston reaches the top of its first upward stroke, a spark at the spark plug electrodes ignites the compressed fuel-air mixture. As the engine crankshaft turns past top center, the burning fuel-air mixture expands rapidly and forces the piston downward on its power stroke as shown in view "C". As the piston reaches the bottom of the power stroke, the mechanically operated exhaust valve starts to open and as the

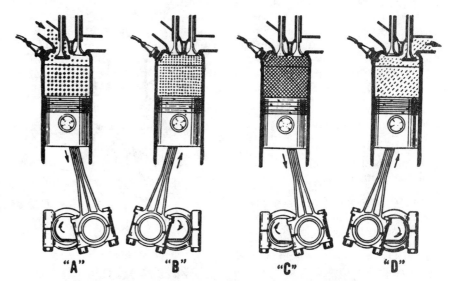

Fig. 1-1 — Schematic diagram of four-stroke cycle engine operating on the Otto (spark ignition) cycle. In view "A", piston is on first downward (intake) stroke and atmospheric pressure is forcing fuel-air mixture from carburetor into cylinder through the open intake valve. In view "B", both valves are closed and piston is on its first upward stroke compressing the fuel-air mixture in cylinder. In view "C", spark across electrodes of spark plug has ignited fuel-air mixture and heat of combustion rapidly expands the burning gaseous mixture forcing the piston on its second downward (expansion or power) stroke. In view "D", exhaust valve is open and piston on its second upward (exhaust) stroke forces the burned mixture from cylinder. A new cycle then starts as in view "A".

pressure of the burned fuel-air mixture is higher than atmospheric pressure, it starts to flow out the open exhaust port. As the engine crankshaft turns past bottom center, the exhaust valve is almost completely open and remains open during the upward stroke of the piston as shown in view "D". Upward movement of the piston pushes the remaining burned fuel-air mixture out of the exhaust port. Just before the piston reaches the top of its second upward or exhaust stroke, the intake valve opens and the exhaust valve closes. The cycle is completed as the crankshaft turns past top center and a new cycle begins as the

piston starts downward as shown in view "A".

In a four-stroke cycle engine operating on the Diesel Cycle, the sequence of events of the cycle is similar to that described for operation on the Otto Cycle, but with the following exceptions: On the intake stroke, air only is taken into the cylinder. On the compression stroke, the air is highly compressed which raises the temperature of the air. Just before the piston reaches the top dead center, fuel is injected into the cylinder and is ignited by the heated, compressed air. The remainder of the cycle is similar to that of the Otto Cycle.

Fig. 2-2, air cannot enter into the carburetor and pressure in the carburetor decreases greatly as the engine is turned at cranking speed. Fuel can then flow from the fuel nozzle. In manufacturing the carburetor choke plate or disc, a small hole or notch is cut in the plate so that some air can flow through the plate when it is in closed position to provide air for the starting fuel-air mixture. In some instances, after starting a cold engine, it is advantageous to leave the choke plate in a partly closed position as the restriction of air flow will decrease the air pressure in carburetor venturi, thus causing more fuel to flow from the nozzle, resulting in a richer fuel-air mixture. The choke plate or disc should be in fully open position for normal engine operation.

If, after the engine has been started, the throttle plate is in the wide-open position as shown by the solid line in Fig. 2-2, the engine can obtain enough fuel and air to run at dangerously high speeds. Thus, the throttle plate or disc must be partly closed as shown by the dotted lines to control engine speed. At no load, the engine requires very little air and fuel to run at its rated speed and the throttle must be moved on toward the closed position as shown by the dash lines. As more load is placed on the engine, more fuel and air are required for the engine to operate at its rated speed and the throttle must be moved closer to the wide open position as shown by the solid line. When the engine

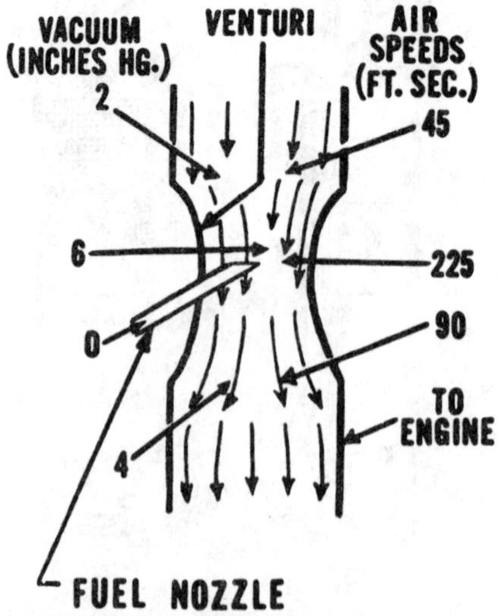

CARBURETOR FUNDAMENTALS

OPERATING PRINCIPLES

Function of the carburetor on a spark-ignition engine is to atomize the fuel and mix the atomized fuel in proper proportions with air flowing to the engine intake port or intake manifold. Carburetors used on engines that are to be operated at constant speeds and under even loads are of simple design since they only have to mix fuel and air in a relatively constant ratio. On engines operating at varying speeds and loads, the carburetors must be more complex because different fuel-air mixtures are required to meet the varying demands of the engine.

FUEL-AIR MIXTURE RATIO REQUIREMENTS. To meet the demands of an engine being operated at varying speeds and loads, the carburetor must mix fuel and air at different mixture ratios. Fuel-air mixture ratios required for different operating conditions are approximately as follows.

	Fuel	Air
Starting, cold weather	1 lb.	7 lbs.
Accelerating	1 lb.	9 lbs.
Idling (no load)	1 lb.	11 lbs.
Part open throttle	1 lb.	15 lbs.
Full load, open throttle	1 lb.	13 lbs.

BASIC DESIGN. Carburetor design is based on the venturi principle which simply means that a gas or liquid flowing through a necked-down section (venturi) in a passage undergoes an increase in velocity (speed) and a decrease in pressure as compared to the velocity and pressure in full size sections of the passage. The principle is illustrated in Fig. 2-1, which shows air passing through a carburetor venturi. The figures given for air speeds and vacuum are approximate for a typical wide-open

throttle operating condition. Due to low pressure (high vacuum) in the venturi, fuel is forced out through the fuel nozzle by the atmospheric pressure (0 vacuum) on the fuel; as fuel is emitted from the nozzle, it is atomized by the high velocity air flow and mixes with the air.

In Fig. 2-2, the carburetor choke plate and throttle plate are shown in relation to the venturi. Downward pointing arrows indicate air flow through the carburetor.

At cranking speeds, air flows through the carburetor venturi at a slow speed; thus, the pressure in the venturi does not usually decrease to the extent that atmospheric pressure on the fuel will force fuel from the nozzle. If the choke plate is closed as shown by dotted line in

Fig. 2-1 – Drawing illustrating the venturi principle upon which carburetor design is based. Figures at left are inches of mercury vacuum and those at right are air speeds in feet per second that are typical of conditions found in a carburetor operating at wide open throttle. Zero vacuum in fuel nozzle corresponds to atmospheric pressure.

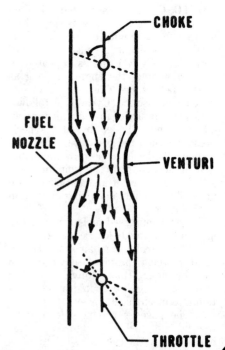

Fig. 2-2 – Drawing showing basic carburetor design. Text explains operation of the choke and throttle valves. In some carburetors, a primer pump may be used instead of the choke valve to provide fuel for the starting fuel-air mixture.

is required to develop maximum power or speed the throttle must be in the wide open position.

Although some carburetors may be as simple as the basic design just described, most engines require more complex design features to provide variable fuel-air mixture ratios for different operating conditions. These design features will be described in the following paragraph.

CARBURETOR TYPE

Carburetors are of the float type and are either of the downdraft or side draft design. The following paragraph describes the features and operating principles of the float type carburetor.

FLOAT TYPE CARBURETOR. The principle of float type carburetor operation is illustrated in Fig. 2–3. Fuel is delivered at inlet (I) by gravity with fuel tank placed above carburetor, or by a fuel lift pump when tank is located below carburetor inlet. Fuel flows into the open inlet valve (V) until fuel level (L) in bowl lifts float against fuel valve needle and closes the valve. As fuel is emitted from the nozzle (N) when engine is running, fuel level will drop, lowering the float and allowing valve to open so that fuel will enter the carburetor to meet the requirements of the engine.

In Fig. 2–4, a cut-away view of a well known make of float type carburetor is shown. Atmospheric pressure is maintained in fuel bowl through passage (20) which opens into carburetor air horn ahead of the choke plate (21). Fuel level is maintained at just below level of opening (O) in nozzle (22) by float (19) actuating inlet valve needle (8). Float

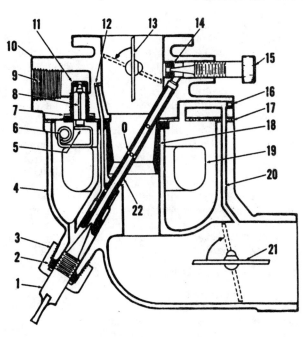

Fig. 2-4 – Cross-sectional drawing of float type carburetor used on some engines.

0. Orifice
1. Main fuel needle
2. Packing
3. Packing nut
4. Carburetor bowl
5. Float tang
6. Float hinge pin
7. Gasket
8. Inlet valve
9. Fuel inlet
10. Carburetor body
11. Inlet valve seat
12. Vent
13. Throttle plate
14. Idle orifice
15. Idle fuel needle
16. Plug
17. Gasket
18. Venturi
19. Float
20. Fuel bowl vent
21. Choke
22. Fuel nozzle

height can be adjusted by bending float tang (5).

When starting a cold engine, it is necessary to close the choke plate (21) as shown by dotted lines so as to lower the air pressure in carburetor venturi (18) as engine is cranked. Then, fuel will flow up through nozzle (22) and will be emitted from openings (O) in nozzle. When an engine is hot, it will start on a leaner fuel-air mixture than when cold and may start without the choke plate being closed.

When engine is running at slow idle speed (throttle plate nearly closed as indicated by dotted lines in Fig. 2–4), air pressure above the throttle plate is low and atmospheric pressure in fuel bowl forces fuel up through the nozzle and out through orifice in seat (14) where it mixes with air passing the throttle plate. The idle fuel mixture is adjustable by turning needle (15) in or out as required. Idle speed is adjustable by turning the throttle stop screw (not shown) in or out to control amount of air passing the throttle plate.

When throttle plate is opened to increase engine speed, velocity of air flow through venturi (18) increases, air pressure at venturi decreases and fuel will flow from openings (O) in nozzle instead of through orifice in idle seat (14). When engine is running at high speed, pressure in nozzle (22) is less than at vent (12) opening in carburetor throat above venturi. Thus, air will enter vent

and travel down the vent into the nozzle and mix with the fuel in the nozzle. This is referred to as air bleeding and is illustrated in Fig. 2–5.

Many different designs of float type carburetors will be found when servicing the different makes and models of engines. Reference should be made to the engine repair section of this manual for adjustment and overhaul specifications. Refer to carburetor servicing paragraphs in fundamentals section for service hints.

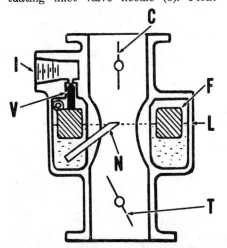

Fig. 2-3 – Drawing showing basic float type carburetor design. Fuel must be delivered under pressure either by gravity or by use of fuel pump, to the carburetor fuel inlet (I). Fuel level (L) operates float (F) to open and close inlet valve (V) to control amount of fuel entering carburetor. Also shown are the fuel nozzle (N), throttle (T) and choke (C).

Fig. 2-5 – Illustration of air bleed principle explained in text.

IGNITION SYSTEM FUNDAMENTALS

The ignition system provides a properly timed surge of extremely high voltage electrical energy which flows across the

spark plug electrode gap to create the ignition spark. Engines may be equipped with either a magneto or battery ignition system. A magneto ignition system generates electrical energy, intensifies (transforms) this electrical energy to the extremely high voltage required and delivers this electrical energy at the proper time for the ignition spark. In a battery ignition system, a storage battery is used as a source of electrical energy and the system transforms the relatively low electrical voltage from the battery into the high voltage required and delivers the high voltage at proper time for the ignition spark. Thus, the function of the two systems is somewhat similar except for the basic source of electrical energy. The fundamental operating principles of ignition systems are explained in the following paragraphs.

MAGNETISM AND ELECTRICITY

The fundamental principles upon which ignition systems are designed are presented in this section. As the study of magnetism and electricity is an entire scientific field, it is beyond the scope of this manual to fully explore these subjects. However, the following information will impart a working knowledge of basic principles which should be of value in servicing engines.

MAGNETISM. The effects of magnetism can be shown easily while the theory of magnetism is too complex to be presented here. The effects of magnetism were discovered many years ago when fragments of iron ore were found to attract each other and also attract other pieces of iron. Further, it was found that when suspended in air, one end of the iron ore fragment would always point in the directin of the North Star. The end of iron ore fragment pointing north was called the "north pole" and the opposite end the "south pole." By stroking a piece of steel with a "natural magnet," as these iron ore fragments were called, it was found that the magnetic properties of the natural magnet could be transferred or "induced" into the steel.

Steel which will retain magnetic properties for an extended period of time after being subjected to a strong magnetic field are called "permanent magnets;" iron or steel that loses such magnetic properties soon after being subjected to a magnetic field are called "temporary magnets." Soft iron will lose magnetic properties almost immediately after being removed from a magnetic field, and so is used where this property is desirable.

The area affected by a magnet is called a "field of force." The extent of this field of force is related to the strength of the magnet and can be determined by use of a compass. In practice, it is common to illustrate the field of force surrounding a magnet by lines as shown in Fig. 3–1 and the field of force is usually called "lines of force" or "flux." Actually, there are no "lines;" however, this is a convenient method of illustrating the presence of the invisible magnetic forces and if a certain magnetic force is defined as a "line of force," then all magnetic forces may be measured by comparison. The number of "lines of force" making up a strong magnetic field is enormous.

Most materials when placed in a magnetic field are not attracted by the magnet, do not change the magnitude or direction of the magnetic field, and so are called "non-magnetic materials." Materials such as iron, cobalt, nickel or their alloys, when placed in a magnetic field will concentrate the field of force and hence are magnetic conductors or "magnetic materials." There are no materials known in which magnetic fields will not penetrate and magnetic lines of force can be deflected only by magnetic materials or by another magnetic field.

Alnico, an alloy containing aluminum, nickel and cobalt, retains magnetic properties for a very long period of time after being subjected to a strong magnetic field and is extensively used as a permanent magnet. Soft iron, which loses magnetic properties quickly, is used to concentrate magnetic fields as in Fig. 3–1.

ELECTRICITY. Electricity, like magnetism, is an invisible physical force whose effects may be more readily explained than the theory of what electricity consists of. All of us are familiar with the property of electricity to produce light, heat and mechanical power. What must be explained for the purpose of understanding ignition system operation is the inter-relationship of magnetism and electricity and how the ignition spark is produced.

Electrical current may be defined as a flow of energy in a conductor which, in some ways, may be compared to flow of water in a pipe. For electricity to flow, there must be a pressure (voltage) and a complete circuit (closed path) through which the electrical energy may return, a comparison being a water pump and a pipe that receives water from the outlet (pressure) side of the pump and returns the water to the inlet side of the pump. An electrical circuit may be completed by electricity flowing through the earth (ground), or through the metal framework of an engine or other equipment ("grounded" or "ground" connections). Usually, air is an insulator through which electrical energy will not flow. However, if the force (voltage) becomes great, the resistance of air to the flow of electricity is broken down and a current will flow, releasing energy in the form of a spark. By high voltage electricity breaking down the resistance of the air gap between the spark plug electrodes, the ignition spark is formed.

ELECTRO-MAGNETIC INDUCTION. The principle of electro-magnetic induction is as follows: When a wire (conductor) is moved through a field of magnetic force so as to cut across the lines of force (flux), a potential voltage or electromotive force (emf) is induced in the wire. If the wire is a part of a completed electrical circuit, current will flow through the circuit as illustrated in Fig. 3–2. It should be noted that the movement of the wire through the lines of magnetic force is a relative motion; that is, if the lines of force of a moving magnetic field cut across a wire, this will also induce an emf to the wire.

The direction of an induced current is related to the direction of magnetic force and also to the direction of movement of the wire through the lines of force, or flux. The voltage of an induced

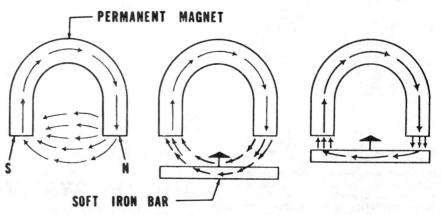

PERMANENT MAGNET

SOFT IRON BAR

Fig. 3-1 — In left view, field of force of permanent magnet is illustrated by arrows showing direction of magnetic force from north pole (N) to south pole (S). In center view, lines of magnetic force are being attracted by soft iron bar that is being moved into the magnetic field. In right view, the soft iron bar has been moved to the magnet and the field of magnetic force is concentrated within the bar.

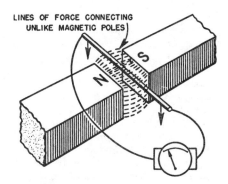

Fig. 3-2 – When a conductor is moved through a magnetic field so as to cut across lines of force, a potential voltage will be induced in the conductor. If the conductor is a part of a completed electrical circuit, current will flow through the circuit as indicated by the gage.

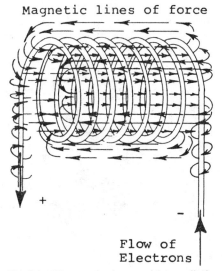

Fig. 3-4 – When a wire is wound in a coil, the magnetic force created by a current in the wire will tend to converge in a single strong magnetic field as illustrated. If the loops of the coil are wound closely together, there is little tendency for lines of force to surround individual loops of the coil.

current is related to the strength, or concentration of lines of force, of the magnetic field and to the rate of speed at which the wire is moved through the flux. If a length of wire is wound into a coil and a section of the coil is moved through magnetic lines of force, the voltage induced will be proportional to the number of turns of wire in the coil.

ELECTRICAL MAGNETIC

FIELDS. When a current is flowing in a wire, a magnetic field is present around the wire as illustrated in Fig. 3-3. The direction of lines of force of this magnetic field is related to the direction of current in the wire. This is known as the left hand rule and is stated as follows: If a wire carrying a current is grasped in the left hand with thumb pointing in direction electrons are moving, the curved fingers will point the direction of lines of magnetic force (flux) encircling the wire.

NOTE: The currently used electron theory explains the movement of electrons from negative to positive. Be sure to use the LEFT HAND RULE with the thumb pointing the direction electrons are moving (the positive end of a conductor).

If a current is flowing in a wire that is wound into a coil, the magnetic flux surrounding the wire converge to form a stronger magnetic field as shown in Fig. 3-4. If the coils of wire are very close together, there is little tendency for

magnetic flux to surround individual loops of the coil and a strong magnetic field will surround the entire coil. The strength of this field will vary with the current flowing through the coil.

STEP-UP TRANSFORMERS (IGNITION COILS). In both battery and magneto ignition systems, it is necessary to step-up, or transform, a relatively low primary voltage to the 15,000 to 20,000 volts required for the ignition spark. This is done by means of an ignition coil which utilizes the interrelationship of magnetism and electricity as explained in preceding paragraphs.

Basic ignition coil design is shown in Fig. 3-5. The coil consists of two separate coils of wire which are called the primary coil winding and the secondary coil winding, or simply the primary

winding and the secondary winding. The primary winding as indicated by the heavy, black line is of larger diameter wire and has a smaller number of turns when compared to the secondary winding indicated by the light line.

A current passing through the primary winding creates a magnetic field (as indicated by the "lines of force") and this field, concentrated by the soft iron core, surrounds both the primary and secondary windings. If the primary winding current is suddenly interrupted, the magnetic field will collapse and the lines of force will cut through the coil windings. The resulting induced voltage in the secondary winding is greater than the voltage of the current that was flowing in the primary winding and is related to the number of turns of wire in each winding. Thus:

Induced secondary voltage = primary voltage ×

$$\frac{\text{No. of turns in secondary winding}}{\text{No. of turns in primary winding}}$$

For example, if the primary winding of an ignition coil contained 100 turns of wire and the secondary winding contained 10,000 turns of wire, a current having an emf of 200 volts flowing in the primary winding, when suddenly interrupted, would result in and emf of:

$$200 \text{ Volts} \times \frac{10,000 \text{ turns of wire}}{100 \text{ turns of wire}}$$
$$= 20,000 \text{ volts}$$

SELF-INDUCTANCE. It should be noted the collapsing magnetic field resulting from interrupted current in the primary winding will also induce a current in the primary winding. This effect is termed "self-inductance." This self-induced current is such as to oppose any interruption of current in the primary winding, slowing the collapse of

Fig. 3-3 – A magnetic field surrounds a wire carrying an electrical current. The direction of magnetic force is indicated by the left hand rule; that is, if thumb of left hand points in direction that electrical current is flowing in conductor, fingers of the left hand will indicate direction of magnetic force.

Fig. 3-5 – Drawing showing principles of ignition coil operation. A current in primary winding will establish a magnetic field surrounding both the primary and secondary windings and the field will be concentrated by the iron core. When primary current is interrupted, the magnetic field will "collapse" and the lines of force will cut the coil windings inducing a very high voltage in the secondary winding.

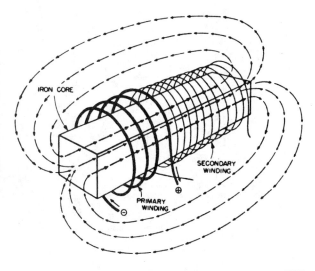

the magnetic field and reducing the efficiency of the coil. The self-induced primary current flowing across the slightly open breaker switch, or contact points, will damage the contact surfaces due to the resulting spark.

To momentarily absorb, then stop the flow of current across the contact points, a capacitor or, as commonly called, a condenser is connected in parallel with the contact points. A simple con-

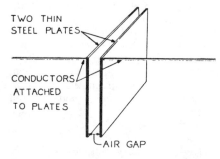

Fig. 3-6—Drawing showing construction of a simple condenser. Capacity of such a condenser to absorb current is limited due to the relatively small surface area. Also, there is a tendency for current to arc across the air gap. Refer to Fig. 3-9 for construction of typical ignition system condenser.

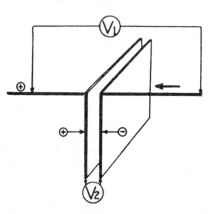

Fig. 3-7—A condenser in an electrical circuit will absorb electrons until an opposing voltage (V2) is built up across condenser plates which is equal to the voltage (V1) of the electrical current.

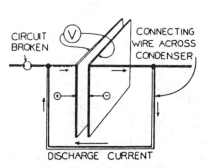

Fig. 3-8—When a circuit containing a condenser is interrupted (circuit broken), the condenser will retain a potential voltage (V). If a wire is connected across the condenser, a current will flow in reverse direction of charging current until condenser is discharged (voltage across condenser plates is zero).

denser is shown in Fig. 3–6; however, the capacity of such a condenser to absorb current (capacitance) is limited by

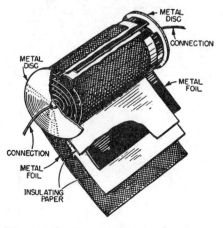

Fig. 3-9—Drawing showing construction of typical ignition system condenser. Two layers of metal foil, insulated from each other with paper, are rolled tightly together and a metal disc contacts each layer, or strip of foil. Usually, one disc is grounded through the condenser shell.

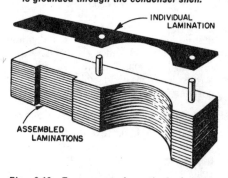

Fig. 3-10—To prevent formation of "eddy currents" within soft iron cores used to concentrate magnetic fields, core is assembled of plates or "laminations" that are insulated from each other. In a solid iron core, there is a tendency for counteracting magnetic forces to build up from stray currents induced in the core.

Fig. 3-11—Schematic diagram of typical battery ignition system used on single cylinder engine. On unit shown, breaker points are actuated by timer cam; on some units, points may be actuated by cam on engine camshaft. Refer to Fig. 3-12 for cut-away view of typical battery ignition coil. In view above, primary coil winding is shown as heavy black line (outside coil loops) and secondary winding is shown by lighter line (inside coil loops).

the small surface area of the plates. To increase capacity to absorb current, the condenser used in ignition systems is constructed as shown in Fig. 3–9.

EDDY CURRENTS. It has been found that when a solid soft iron bar is used as a core for an ignition coil, stray electrical currents are formed in the core. These stray, or "eddy currents" create opposing magnetic forces causing the core to become hot and also decrease efficiency of the coil. As a means of preventing excessive formation of eddy currents within the core, or other magnetic field carrying parts of a magneto, a laminated plate construction as shown in Fig. 3–10 is used instead of solid material. The plates, or laminations, are insulated from each other by a natural oxide coating formed on the plate surfaces or by coating the plates with varnish. The cores of some ignition coils are constructed of soft iron wire instead of plates and each wire is insulated by a varnish coating. This type of construction serves the same purpose as laminated plates.

BATTERY IGNITION SYSTEMS

Some engines are equipped with a battery ignition system. A schematic diagram of a typical battery ignition system for a single cylinder engine is shown in Fig. 3–11. Designs of battery to location of breaker points and method for actuating the points; however, all operate on the same basic principles.

BATTERY IGNITION SYSTEM PRINCIPLES. Refer to the schematic diagram in Fig. 3–11. When the timer cam is turned so the contact points are

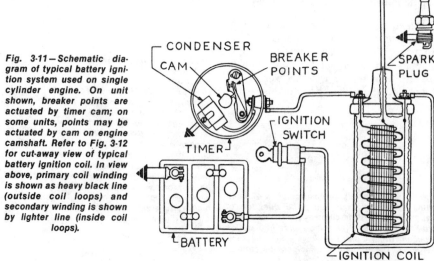

closed, a current is established on the primary circuit by the emf of the battery. This current flowing through the primary winding of the ignition coil establishes a magnetic field concentrated in the core laminations and surrounding the windings. A cut-away view of a typical ignition coil is shown in Fig. 3–12. At the proper time for the ignition spark, contact points are opened by the timer cam and primary ignition circuit is interrupted. The condenser, wired in parallel with breaker contact points between timer terminal and ground, absorbs self-induced current in the primary circuit for an instant and brings the flow of current to a quick, controlled stop. The magnetic field surrounding the coil rapidly cuts the primary and secondary windings creating an emf as high as 250 volts in the primary winding and up to 25,000 volts in the secondary winding. Current absorbed by the condenser is discharged as breaker points close, grounding the condenser lead wire.

Due to resistance of the primary winding, a certain period of time is required for maximum primary current flow after breaker contact points are closed. At high engine speeds, points remain closed for a smaller interval of time,

hence primary current does not build up to maximum and secondary voltage is somewhat less than at low engine speed. However, coil design is such that the minimum voltage available at high engine speed exceeds the normal maximum voltage required for ignition spark.

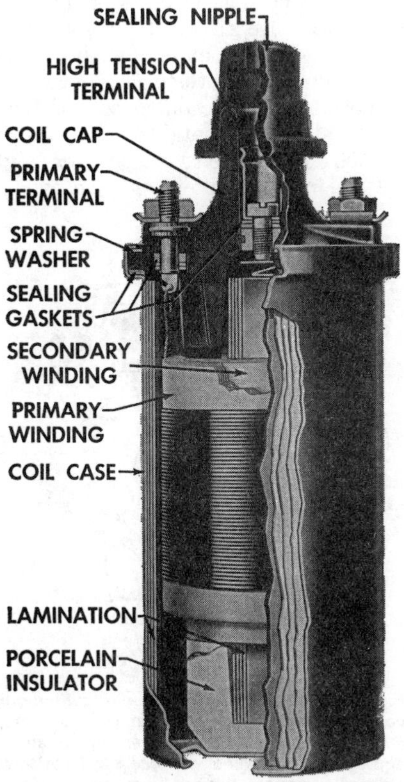

MAGNETO IGNITION SYSTEMS

By utilizing principles of magnetism and electricity as outlined in previous paragraphs, a magneto generates an electrical current of relatively low voltage, then transforms this voltage into the extremely high voltage necessary to produce ignition spark. This surge of

high voltage is timed to create the ignition spark and ignite the compressed fuel-air mixture in the engine cylinder at the proper time in the Otto cycle as described in the paragraphs on fundamentals of engine operation principles.

Two different types of magnetos are used on air-cooled engines and, for discussion in this section of the manual, will be classified as "flywheel type magnetos" and "self-contained unit type magnetos."

Flywheel Type Magnetos

The term "flywheel type magneto" is derived from the fact that the engine

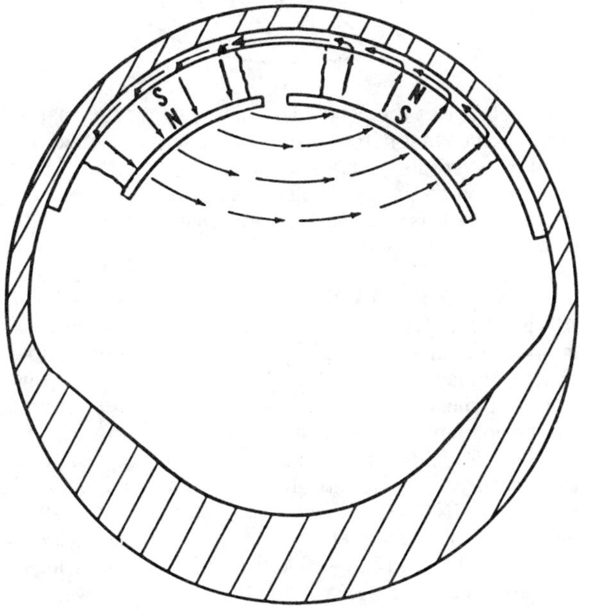

Fig. 3-13 — Cut-away view of typical engine flywheel used with flywheel magneto type ignition system. The permanent magnets are usually cast into the flywheel. For flywheel type magnetos having the ignition coil and core mounted to outside of flywheel, magnets would be flush with outer diameter of flywheel.

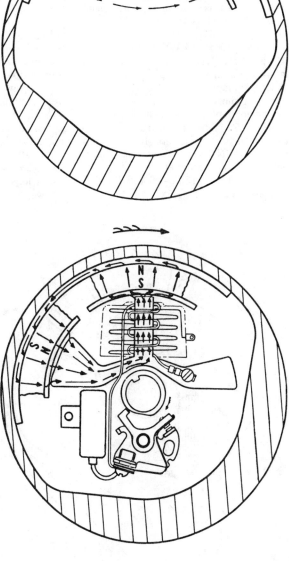

Fig. 3-12 — Cut-away view of typical battery ignition system coil. Primary winding consists of approximately 200-250 turns (loops) of heavier wire; secondary winding consists of several thousand turns of fine wire. Laminations concentrate magnetic lines of force and increase efficiency of coil.

Fig. 3-14 — View showing flywheel turned to a position so lines of force of the permanent magnets are concentrated in the left and center core legs and are interlocking the coil windings.

SEALING NIPPLE

HIGH TENSION TERMINAL

COIL CAP

PRIMARY TERMINAL

SPRING WASHER

SEALING GASKETS

SECONDARY WINDING

PRIMARY WINDING

COIL CASE

LAMINATION

PORCELAIN INSULATOR

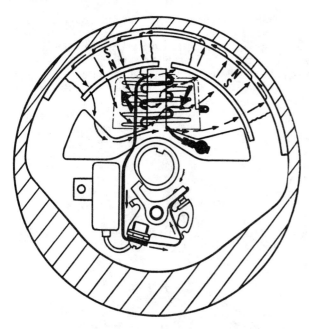

Fig. 3-15 — View showing flywheel turned to a position so lines of force of the permanent magnets are being withdrawn from the left and center core legs and are being attracted by the center and right core legs. While this event is happening, the lines of force are cutting up through the coil windings section between the left and center legs and are cutting down through the section between the right and center legs as indicated by the heavy black arrows. As the breaker points are now closed by the cam, a current is induced in the primary ignition circuit as the lines of force cut through the coil windings.

flywheel carries the permanent magnets and is the magneto rotor. In some similar systems, magneto rotor is mounted on engine crankshaft as is the flywheel, but is a part separate from flywheel.

FLYWHEEL MAGNETO OPERATING PRINCIPLES. In Fig. 3-13, a cross-sectional view of a typical engine flywheel (magneto rotor) is shown. The arrows indicate lines of force (flux) of the permanent magnets carried by the flywheel. As indicated by arrows, direction of force of magnetic field is from north pole (N) of left magnet to south pole (S) of right magnet.

Figs. 3-14, 3-15, 3-16 and 3-17 illustrate operational cycle of flywheel type magneto. In Fig. 3-14, flywheel magnets have moved to a position over left and center legs of armature (ignition coil) core. As magnets moved into this position, their magnetic field was attracted by armature core as illustrated in Fig. 3-1 and a potential voltage (emf) was induced in coil windings. However, this emf was not sufficient to cause current to flow across spark plug electrode gap in high tension circuit and points were open in primary circuit.

In Fig. 3-15, flywheel magnets have moved to a new position to where their magnetic field is being attracted by center and right legs of armature core, and is being withdrawn from left and center legs. As indicated by heavy black arrows, lines of force are cutting up through the section of coil windings between left and center legs of armature and are cutting down through coil windings section between center and right legs. If the left hand rule, as explained in a previous paragraph, is applied to the lines of force cutting through coil sections, it is seen resulting emf induced in primary circuit will cause a current to flow through primary coil windings and breaker points which have now been closed by action of the cam.

At instant movement of lines of force cutting through coil winding sections is at maximum rate, maximum flow of current is obtained in primary circuit. At this time, cam opens breaker points interrupting primary circuit and, for an instant, flow of current is absorbed by condenser as illustrated in Fig. 3-16. An emf is also induced in secondary coil windings, but voltage is not sufficient to cause current to flow across spark plug gap.

Flow of current in primary windings created a strong electromagnetic field surrounding coil windings and up through center leg of armature core as shown in Fig. 3-17. As breaker points were opened by cam, interrupting primary circuit, magnetic field starts to collapse cutting coil windings as indicated by heavy black arrows. The emf induced in primary circuit would be sufficient to cause a flow of current across opening breaker points were it not for the condenser absorbing flow of current and bringing it to a controlled stop. This allows electromagnetic field to collapse at such a rapid rate to induce a very high voltage in coil high tension or secondary

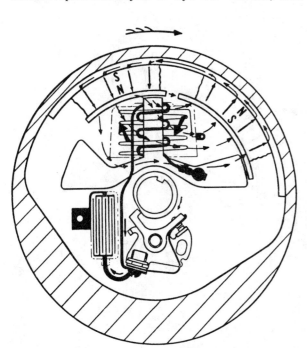

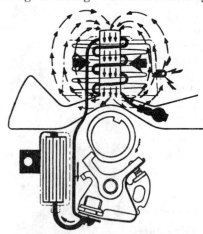

Fig. 3-16 — The flywheel magnets have now turned slightly past the position shown in Fig. 3-15 and the rate of movement of lines of magnetic force cutting through the coil windings is at the maximum. At this instant, the breaker points are opened by the cam and flow of current in the primary circuit is being absorbed by the condenser, bringing the flow of current to a quick, controlled stop. Refer now to Fig. 3-17.

Fig. 3-17 — View showing magneto ignition coil, condenser and breaker points at same instant as illustrated in Fig. 3-16; however, arrows shown above illustrate lines of force of the electromagnetic field established by current in primary coil windings rather than the lines of force of permanent magnets. As the current in the primary circuit ceases to flow, the electromagnetic field collapses rapidly, cutting the coil windings as indicated by heavy arrows and inducing a very high voltage in the secondary coil winding resulting in the ignition spark.

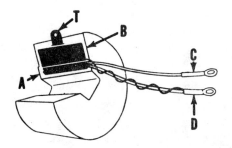

Fig. 3-18—Drawing showing construction of a typical flywheel magneto ignition coil. Primary windings (A) consists of about 200 turns of wire. Secondary winding (B) consists of several thousand turns of fine wire. Coil primary and secondary ground connection is (D); primary connection to breaker point and condenser terminal is (C); and coil secondary (high tension) terminal is (T).

windings. This voltage, in the order of 15,000 to 25,000 volts, is sufficient to break down the resistance air gap between spark plug electrodes and a current will flow across gap. This creates ignition spark which ignites compressed fuel-air mixture in engine cylinder.

Self-Contained Unit Type Magnetos

Some four-stroke cycle engines are equipped with a magneto which is a self-contained unit as shown in Fig. 3–20. This type magneto is driven from engine timing gears via a gear or coupling. All components of the magneto are enclosed in one housing and magneto can be removed from engine as a unit.

UNIT TYPE MAGNETO OPERATING PRINCIPLES. In Fig. 3–21, a schematic diagram of a unit type magneto is shown. Magneto rotor is driven through an impulse coupling (shown at right side of illustration). Function of impulse coupling is to increase rotating speed of rotor, thereby increasing magneto efficiency, at engine cranking speeds.

A typical impulse coupling for a single cylinder engine magneto is shown in Fig. 3–22. When engine is turned at cranking speed, coupling hub pawl engages a stop pin in magneto housing as engine piston is coming up on compression stroke. This stops rotation of coupling hub assembly and magneto rotor. A spring within coupling shell (see Fig. 3–23) connects shell and coupling hub; as engine continues to turn, spring winds up until pawl kickoff contacts pawl and disengages it from stop pin. This occurs at the time an ignition spark is required to ignite compressed fuel-air mixture in engine cylinder. As pawl is released, spring connecting coupling shell and hub unwinds and rapidly spins magneto rotor.

Magneto rotor (see Fig. 3–21) carries

Fig. 3-19—Exploded view of a typical flywheel type magneto used on single cylinder engines in which the breaker points (14) are actuated by a cam on engine camshaft. Push rod (9) rides against cam to open and close points. In this type unit, an ignition spark is produced only on alternate revolutions of the flywheel as the camshaft turns at one-half speed.

1. Flywheel
2. Ignition coil
3. Coil clamps
4. Coil ground lead
5. Breaker point lead
6. Armature core (laminations)
7. Crankshaft bearing retainer
8. High tension lead
9. Push rod
10. Bushing
11. Breaker box cover
12. Point lead strap
13. Breaker point spring
14. Breaker point assy.
15. Condenser
16. Breaker box
17. Terminal bolt
18. Insulators
19. Grounding (stop) spring

permanent magnets. As rotor turns, alternating position of magnets, lines of force of magnets are attracted, then withdrawn from laminations. In Fig. 3–21, arrows show magnetic field concentrated within laminations, or armature core. Slightly further rotation of magnetic rotor will place magnets to where laminations will have greater attraction for opposite poles of magnets. At this instant, lines of force as indicated by arrows will suddenly be withdrawn and an opposing field of force will be established in laminations. Due to this rapid movement of lines of force, a current will be induced in primary magneto circuit as coil windings are cut by lines of force. At instant maximum current is induced in primary windings, breaker points are opened by a cam on magnetic rotor shaft interrupting primary circuit. The lines of magnetic force established by primary current (refer to Fig. 3–5) will cut through secondary windings at

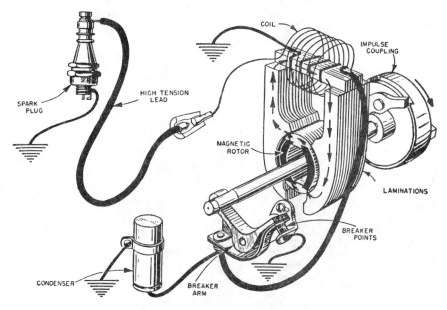

Fig. 3-20—Some engines are equipped with a unit type magneto having all components enclosed in a single housing (H). Magneto is removable as a unit after removing retaining nuts (N). Stop button (B) grounds out primary magneto circuit to stop engine. Timing window is (W).

such a rapid rate to induce a very high voltage in secondary (or high tension) circuit. This voltage will break down

Fig. 3-21 — Schematic diagram of typical unit type magneto for single cylinder engine. Refer to Figs. 3-22, 3-23, and 3-24 for views showing construction of impulse couplings.

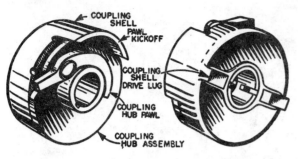

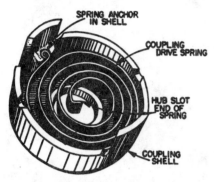

Fig. 3-22 — Views of typical impulse coupling for magneto driven by engine shaft with slotted drive connection. Coupling drive spring is shown in Fig. 3-23. Refer to Fig. 3-24 for view of combination magneto drive gear and impulse coupling used on some magnetos.

Fig. 3-23 — View showing impulse coupling shell and drive spring removed from coupling hub assembly. Refer to Fig. 3-22 for views of assembled unit.

resistance of spark plug electrode gap and a spark across electrodes will result.

At engine operating speeds, centrifugal force will hold impulse coupling hub pawl (See Fig. 3-22) in a position so it cannot engage stop pin in magneto housing and magnetic rotor will be driven through spring (Fig. 3-23) connecting coupling shell to coupling hub. The impulse coupling retards ignition spark, at cranking speeds, as engine piston travels closer to top dead center while magnetic rotor is held stationary by pawl and stop pin. The difference in degrees of impulse coupling shell rotation between position of retarded spark and normal running spark is known as impulse coupling lag angle.

SOLID STATE IGNITION SYSTEM

BREAKERLESS MAGNETO SYSTEM. Solid state (breakerless) magneto ignition system operates somewhat on the same basic principles as conventional type flywheel magneto previously described. The main difference is breaker contact points are replaced by a solid state electronic Gate Controlled Switch (GCS) which has no moving parts. Since, in a conventional system breaker points are closed over a longer period of crankshaft rotation than is the "GCS", a diode has been added to the circuit to provide the same characteristics as closed breaker points.

BREAKERLESS MAGNETO OPERATING PRINCIPLES. The same basic principles for electro-magnetic induction of electricity and formation of magnetic fields by electrical current as outlined for conventional flywheel type magneto also apply to solid state magneto. Therefore principles of different components (diode and GCS) will complete operating principles of solid state magneto.

The diode is represented in wiring diagrams by the symbol shown in Fig. 3-25. The diode is an electronic device that will permit passage of electrical current in one direction only. In electrical schematic diagrams, current flow is opposite direction the arrow part of symbol is pointing.

The symbol shown in Fig. 3-26 is used to represent gate controlled switch (GCS) in wiring diagrams. The GCS acts as a switch to permit passage of current from cathode (C) terminal to anode (A) terminal when in "ON" state and will not permit electric current to flow when in "OFF" state. The GCS can be turned "ON" by a positive surge of electricity at gate (G) terminal and will remain "ON" as long as current remains positive at gate terminal or as long as current is flowing through GCS from cathode (C) terminal to anode (A) terminal.

The basic components and wiring diagram for solid state breakerless magneto are shown schematically in Fig. 3-27. In Fig. 3-28, magneto rotor (flywheel) is turning and ignition coil magnets have just moved into position so their lines of force are cutting ignition coil windings and producing a negative surge of current in primary windings. The diode allows current to flow opposite to the direction of diode symbol arrow and action is same as conventional magneto with breaker points closed. As rotor (flywheel) continues to turn as shown in Fig. 3-29, direction of magnetic flux lines will reverse in armature center leg. Direction of current will change in primary coil circuit and previously conducting diode will be shut off. At this point, neither diode is conducting. As voltage begins to build up as rotor continues to turn, condenser acts

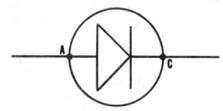

Fig. 3-25 — In a diagram of an electrical circuit, the diode is represented by the symbol shown above. The diode will allow current to flow in one direction only, from cathode (C) to anode (A).

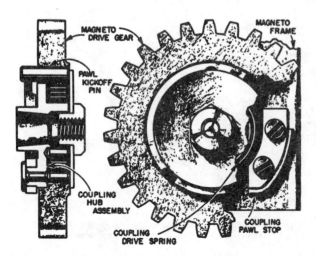

Fig. 3-24 — Views of combination magneto drive gear and impulse coupling used on some magnetos.

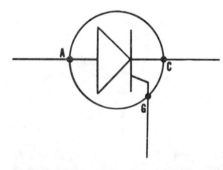

Fig. 3-26 — The symbol used for a Gate controlled Switch (GCS) in an electrical diagram is shown above. The GCS will permit current to flow from cathode (C) to anode (A) when "turned on" by a positive electrical charge at gate (G) terminal.

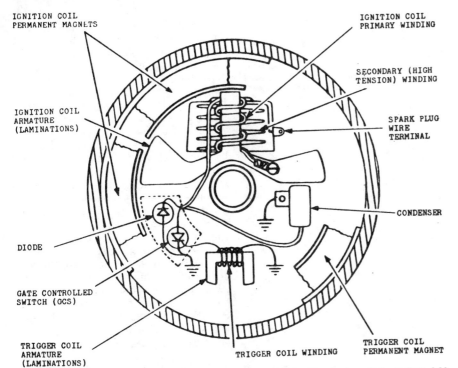

Fig. 3-27—Schematic diagram of typical breakerless magneto ignition system. Refer to Figs. 3-28, 3-29 and 3-30 for schematic views of operating cycle.

CAPACITOR DISCHARGE SYSTEM. Capacitor discharge (CD) ignition system uses a permanent magnet rotor (flywheel) to induce a current in a coil, but unlike conventional flywheel magneto and solid state breakerless magneto described previously, current is stored in a capacitor (condenser). Then, stored current is discharged through a transformer coil to create ignition spark. Refer to Fig. 3–31 for a schematic of a typical capacitor discharge ignition system.

CAPACITOR DISCHARGE OPERATING PRINCIPLES. As permanent flywheel magnets pass by input generating coil (1–Fig. 3–31), current produced charges capacitor (6). Only half of the generated current passes through diode (3) to charge capacitor. Reverse current is blocked by diode (3) but passes through Zener diode (2) to complete reverse circuit. Zener diode (2) also limits maximum voltage of forward current. As flywheel continues to turn and magnets pass trigger coil (4), a small amount of electrical current is generated. This current opens gate controlled switch (5) allowing capacitor to discharge through pulse transformer (7). The rapid voltage rise in transformer primary coil induces a high voltage secondary current which forms ignition spark when it jumps spark plug gap.

THE SPARK PLUG

In any spark ignition engine, the spark plug (See Fig. 3–32) provides means for

as a buffer to prevent excessive voltage build up at GCS before it is triggered.

When rotor reaches approximate position shown in Fig. 3–30, maximum flux density has been achieved in center leg of armature. At this time the GCS is triggered. Triggering is accomplished by triggering coil armature moving into field of permanent magnet which induces a positive voltage on the gate of GCS. Primary coil current flow results in the formation of an electromagnetic field around primary coil which induces a voltage of sufficient potential in secondary coil windings to "fire" spark plug.

When rotor (flywheel) has moved magnets past armature, GCS will cease to conduct and revert to "OFF" state until it is triggered. The condenser will discharge during time GCS was conducting.

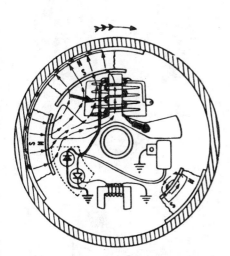

Fig. 3-28—View showing flywheel of breakerless magneto system at instant of rotation where lines of force of ignition coil magnets are being drawn into left and center legs of magneto armature. The diode (see Fig. 3-25) acts as a closed set of breaker points in completing the primary ignition circuit at this time.

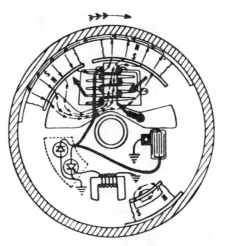

Fig. 3-29—Flywheel is turning to point where magnetic flux lines through armature center leg will reverse direction and current through primary coil circuit will reverse. As current reverses, diode which was previously conducting will shut off and there will be no current. When magnetic flux lines have reversed in armature center leg, voltage potential will again build up, but since GCS is in "OFF" state, no current will flow. To prevent excessive voltage build up, the condenser acts as a buffer.

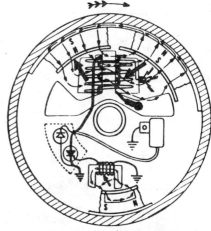

Fig. 3-30—With flywheel in the approximate position shown, maximum voltage potential is present in windings of primary coil. at this time the triggering coil armature has moved into the field of permanent magnet and a positive voltage is induced on the gate of the GCS. The GCS is triggered and primary coil current flows resulting in the formation of an electromagnetic field around the primary coil which induces a voltage of sufficient potential in the secondary windings to "fire" the spark plug.

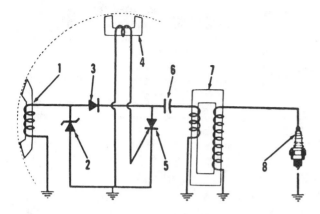

Fig. 3-31—Schematic diagram of typical capacitor discharge ignition system.

1. Generating coil
2. Zener diode
3. Diode
4. Trigger coil
5. Gate controlled switch
6. Capacitor
7. Pulse transformer (coil)
8. Spark plug

igniting compressed fuel-air mixture in cylinder. Before an electric charge can move across an air gap, intervening air must be charged with electricity, or ionized. If spark plug is properly gapped and system is not shorted, not more than 7,000 volts may be required to initiate a spark. Higher voltage is required as spark plug warms up, or if compression pressures or distance of air gap is increased. Compression pressures are highest at full throttle and relatively slow engine speeds, therefore, high voltage requirements or a lack of available secondary voltage most often shows up as a miss during maximum acceleration from a slow engine speed.

There are many different types and sizes of spark plugs which are designed for a number of specific requirements.

THREAD SIZE. The threaded, shell portion of the spark plug and the attaching hole in cylinder are manufactured to meet certain industry established standards. The diameter is referred to as "Thread Size." Those commonly used are: 10mm, 14mm, 18mm, 7/8-inch and 1/2-inch pipe.

REACH. The length of thread, and thread depth in cylinder head or wall are also standardized throughout the industry. This dimension is measured from gasket seat of plug to cylinder end of thread. See Fig. 3-33. Four different reach plugs commonly used are: 3/8-inch, 7/16-inch, 1/2-inch and 3/4-inch.

HEAT RANGE. During engine operation, part of the heat generated during combustion is transferred to the spark plug, and from plug to cylinder through shell threads and gasket. The operating temperature of spark plug plays an important part in engine operation. If too much heat is retained by plug, fuel-air mixture may be ignited by contact with

heated surface before ignition spark occurs. If not enough heat is retained, partially burned combustion products (soot, carbon and oil) may build up on plug tip resulting in "fouling" or shorting out of plug. If this happens, secondary current is dissipated uselessly as it is generated instead of bridging plug gap as a useful spark, and engine will misfire.

Operating temperature of plug tip can be controlled, within limits, by altering length of the path heat must follow to reach threads and gasket of plug. Thus, a plug with a short, stubby insulator around center electrode will run cooler than one with a long, slim insulator. Refer to Fig. 3-34. Most plugs in more popular sizes are available in a number of heat ranges which are interchangeable within the group. The proper heat range is determined by engine design and type of service. Refer to SPARK PLUG SERVICING FUNDAMENTALS for additional information on spark plug selection.

SPECIAL TYPES. Sometimes, engine design features or operating conditions call for special plug types designed for a particular purpose.

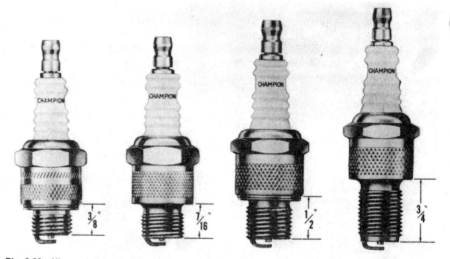

Fig. 3-33—Views showing spark plugs with various "reaches" available. A 3/8-inch reach spark plug measures 3/8-inch from firing end of shell to gasket surface of shell.

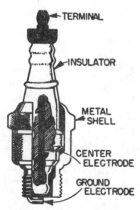

Fig. 3-32—Cross-sectional drawing of spark plug showing construction and nomenclature.

TERMINAL

INSULATOR

METAL SHELL

CENTER ELECTRODE

GROUND ELECTRODE

Fig. 3-34—Spark plug tip temperature is controlled by the length of the path heat must travel to reach cooling surface of the engine cylinder head.

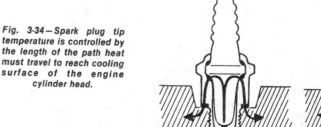
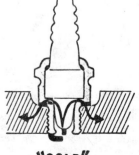

"HOT" "COLD"

ENGINE POWER AND TORQUE RATINGS

The following paragraphs discuss terms used in expressing engine horsepower and torque ratings and explains methods for determining different ratings. Some engine repair shops are now equipped with a dynamometer for measuring engine torque and/or horsepower and the mechanic should be familiar with terms, methods of measurement and how actual power developed by an engine can vary under different conditions.

GLOSSARY OF TERMS

FORCE. Force is an action against an object that tends to move the object from a state of rest, or to accelerate movement of an object. For use in calculating torque or horsepower, force is measured in pounds.

WORK. When a force moves an object from a state of rest, or accelerates movement of an object, work is done. Work is measured by multiplying force applied by distance force moves object, or:

$$\text{work} = \text{force} \times \text{distance}$$

Thus, if a force of 50 pounds moved an object 50 feet, work done would equal 50 pounds times 50 feet, or 2500 pounds-feet (or as it is usually expressed, 2500 foot-pounds).

POWER. Power is the rate at which work is done; thus, if:

$$\text{work} = \text{force} \times \text{distance},$$

then:

$$\text{power} = \frac{\text{force} \times \text{distance}}{\text{time}}$$

From the above formula, it is seen that power must increase if time in which work is done decreases.

HORSEPOWER. Horsepower is a unit of measurement of power. Many years ago, James Watt, a Scotsman noted as inventor of the steam engine, evaluated one horsepower as being equal to doing 33,000 foot-pounds of

work in one minute. This evaluation has been universally accepted since that time. Thus, the formula for determining horsepower is:

$$\text{horsepower} = \frac{\text{pounds} \times \text{feet}}{33,000 \times \text{minutes}}$$

Horsepower (hp.) ratings are sometimes converted to kilowatt (kW) ratings by using the following formula:

$$\text{kW} = \text{hp.} \times 0.745\ 699\ 9$$

When referring to engine horsepower ratings, one usually finds the rating expressed as brake horsepower or rated horsepower, or sometimes as both.

BRAKE HORSEPOWER. Brake horsepower is the maximum horsepower available from an engine as determined by use of a dynamometer, and is usually stated as maximum observed brake horsepower or as corrected brake horsepower. As will be noted in a later paragraph, observed brake horsepower of a specific engine will vary under different conditions of temperature and atmospheric pressure. Corrected brake horsepower is a rating calculated from observed brake horsepower and is a means of comparing engines tested at varying conditions. The method for calculating corrected brake horsepower will be explained in a later paragraph.

RATED HORSEPOWER. An engine being operated under a load equal to the maximum horsepower available (brake horsepower) will not have reserve power for overloads and is subject to damage from overheating and rapid wear. Therefore, when an engine is being selected for a particular load, the engine's brake horsepower rating should be in excess of the expected normal operating load. Usually, it is recommended that the engine not be operated in excess of 80% of the engine maximum brake horsepower rating; thus, the "rated horsepower" of an engine is usually equal to 80% of maximum horsepower the engine will develop.

TORQUE. In many engine specifications, a "torque rating" is given. Engine torque can be defined simply as the turning effort exerted by the engine output shaft when under load.

Torque ratings are sometimes converted to newton-meters (N·m) by using the following formula:

$$\text{N·m} = \text{foot pounds of torque} \times 1.355\ 818$$

It is possible to calculate engine horsepower being developed by measuring torque being developed and engine output speed. Refer to the following paragraphs.

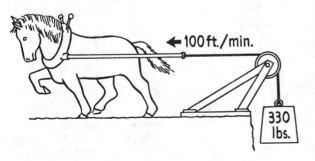

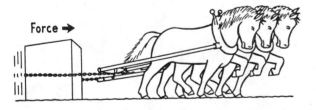

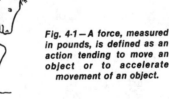

Fig. 4-1 — A force, measured in pounds, is defined as an action tending to move an object or to accelerate movement of an object.

Fig. 4-2 — If a force moves an object from a state of rest or accelerates movement of an object, then work is done.

Fig. 4-3 — This horse is doing 33,000 foot-pounds of work in one minute, or one horsepower.

Force →

Force →

← 100 ft./min.

330 lbs.

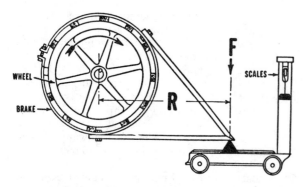

Fig. 4-4 — Diagram showing a prony brake on which the torque being developed by an engine can be measured. By also knowing the rpm of the engine output shaft, engine horsepower can be calculated.

MEASURING ENGINE TORQUE AND HORSEPOWER

PRONY BRAKE. The prony brake is the most simple means of testing engine performance. Refer to diagram in Fig. 4-4. A torque arm is attached to a brake on wheel mounted on engine output shaft. The torque arm, as the brake is applied, exerts a force (F) on scales. Engine torque is computed by multiplying force (F) times length of torque arm radius (R), or:

$$\text{engine torque} = F \times R.$$

If, for example, torque arm radius (R) is 2 feet and force (F) being exerted by torque arm on scales is 6 pounds, engine torque would be 2 feet × 6 pounds, or 12 foot-pounds.

To calculate engine horsepower being developed by use of the prony brake, we must also count revolutions of engine output shaft for a specific length of time. In formula for calculating horsepower:

$$\text{horsepower} = \frac{\text{feet} \times \text{pounds}}{33,000 \times \text{minutes}};$$

Feet in formula will equal circumference transcribed by torque arm radius multiplied by number of engine output shaft revolutions. Thus:

$$\text{feet} = 2 \times 3.14 \times \text{radius} \times \text{revolutions}.$$

Pounds in formula will equal force (F) of torque arm. If, for example, force (F) is 6 pounds, torque arm radius is 2 feet and engine output shaft speed is 3300 revolutions per minute, then:

$$\text{horsepower} = \frac{2 \times 3.14 \times 2 \times 3300 \times 6}{33,000 \times 1}$$

or,

$$\text{horsepower} \times 7.54$$

DYNAMOMETERS. Some commercial dynamometers for testing small engines are now available, although the cost may be prohibitive for all but larger repair shops. Usually, these dynamometers have a hydraulic loading device and scales indicating engine speed and load; horsepower is then calculated by use of a slide rule type instrument. For further information on commercial dynamometers, refer to manufacturers listed in special service tool section of this manual.

HOW ENGINE HORSEPOWER OUTPUT VARIES

Engine efficiency will vary with the amount of air taken into the cylinder on each intake stroke. Thus, air density has a considerable effect on horsepower output of a specific engine. As air density varies with both temperature and atmospheric pressure, any change in air temperature, barometric pressure, or elevation will cause a variance in observed engine horsepower. As a general rule, engine horsepower will:

A. Decrease approximately 3% for each 1000 foot increase above 1000 ft. elevation;

B. Decrease approximately 3% for each 1 inch drop in barometric pressure; or,

C. Decrease approximately 1% for each 10° rise in temperature (Farenheit).

Thus, to fairly compare observed horsepower readings, observed readings should be corrected to standard temperature and atmospheric pressure conditions of 60°F., and 29.92 inches of mercury. The correction formula specified by the Society of Automotive Engineers is somewhat involved; however for practical purposes, the general rules stated above can be used to approximate corrected brake horsepower of an engine when observed maximum brake horsepower is known.

For example, suppose the engine horsepower of 7.54 as found by use of the prony brake was observed at an altitude of 3000 feet and at a temperature of 100°F. At standard atmospheric pressure and temperature conditions, we could expect an increase of 4% due to temperature (100° − 60° × 1% per 10°) and an increase of 6% due to altitude (3000 ft. − 1000 ft. × 3% per 1000 ft.) or a total increase of 10%. Thus, corrected maximum horsepower from this engine would be approximately 7.54 + .75, or approximately 8.25 horsepower.

TROUBLESHOOTING

When servicing an engine to correct a specific complaint, such as engine will not start, is hard to start, etc., a logical step-by-step procedure should be followed to determine cause of trouble before performing any service work. This procedure is "TROUBLESHOOTING."

Of course, if an engine is received in your shop for a normal tune up or specific repair work is requested, troubleshooting procedure is not required and work should be performed as requested. It is wise, however, to fully check the engine before repairs are made and recommend any additional repairs or adjustments necessary to ensure proper engine performance.

The following procedures, as related to a specific complaint or trouble, have proven to be a satisfactory method for quickly determining cause of trouble in a number of engine repair shops.

NOTE: It is not suggested the troubleshooting procedure as outlined in following paragraphs be strictly adhered to at all times. In many instances, customer's comments on when trouble was encountered will indicate cause of trouble. Also, the mechanic will soon develop a diagnostic technique that can only come with experience. In addition to the general troubleshooting procedure, reader should also refer to special notes following this section and to the information included in engine, carburetor and magneto servicing fundamentals sections.

If Engine Will Not Start— Or Is Hard To Start

1. If engine is equipped with a rope or crank starter, turn engine slowly. As engine piston is coming up on compression stroke, a definite resistance to turning should be felt on rope or crank. This resistance should be noted every other crankshaft revolution on a single cylinder engine, on every revolution of a two cylinder engine and on every ½-revolution of a four cylinder engine. If correct cranking resistance is noted, engine compression can be considered as not the cause of trouble at this time.

NOTE: Compression gages for gasoline engines are available and are of value in troubleshooting engine service problems.

Where available from engine manufacturer, specifications will be given for engine compression pressure in engine service sections of this manual. On engines having electric starters, remove spark plug and check engine compression with gage; if gage is not available, hold thumb so spark plug hole is partly covered. An alternating blowing and suction action should be noted as engine is cranked.

If very little or no compression is noted, refer to appropriate engine repair section for repair of engine. If check indicates engine is developing compression, proceed to step 2.

2. Remove spark plug wire and hold wire terminal about 1/8-inch (3.18 mm) away from cylinder (on wires having rubber spark plug boot, insert a small screw or bolt in terminal).

NOTE: A test plug with 1/8-inch (3.18 mm) gap is available or a test plug can be made by adjusting the electrode gap of a new spark plug to 0.125 inch (3.18 mm).

While cranking engine, a bright blue spark should snap across the 1/8-inch (3.18 mm) gap. If spark is weak or yellow, or if no spark occurs while cranking engine, refer to **IGNITION SYSTEM SERVICE FUNDAMENTALS** for information on appropriate type system.

If spark is satisfactory, remove and inspect spark plug. Refer to **SPARK PLUG SERVICE FUNDAMENTALS.** If in doubt about spark plug condition, install a new plug.

NOTE: Before installing plug, make certain electrode gap is set to proper dimension shown in engine repair section of this manual. Refer also to Fig. 5 – 1.

If ignition spark is satisfactory and engine will not start with new plug, proceed with step 3.

3. If engine compression and ignition spark are adequate, trouble within the fuel system should be suspected. Remove and clean or renew air cleaner or cleaner element. Check fuel tank (Fig. 5 – 2) and make certain it is full of fresh fuel as prescribed by engine manufacturer. If equipped with a fuel shut-off valve, make certain valve is open.

If engine is equipped with remote throttle controls that also operate carburetor choke plate, check to be certain that when controls are placed in choke position, carburetor choke plate is fully closed. If not, adjust control linkage so choke will fully close; then, try to start engine. If engine does not start after several turns, remove air cleaner assembly; carburetor throat should be wet with gasoline. If not, check for reason fuel is not getting to carburetor. On models with gravity feed from fuel tank to carburetor (fuel tank above carburetor), disconnect fuel line at carburetor to see that fuel is flowing through the line. If no fuel is flowing, remove and clean fuel tank, fuel line and any fuel filters or shut-off valve.

On models having a fuel pump separate from carburetor, remove fuel line at carburetor and crank engine through several turns; fuel should spurt from open line. If not, disconnect fuel line from tank to fuel pump at pump connection. If fuel will not run from open line, remove and clean fuel tank, line and if so equipped, fuel filter and/or shut-off valve. If fuel runs from open line, remove and overhaul or renew the fuel pump.

After making sure clean, fresh fuel is available at carburetor, again try to start engine. If engine will not start, refer to recommended initial adjustments for carburetor in appropriate engine repair section of this manual and adjust carburetor idle and/or main fuel needles.

If engine will not start when compression and ignition test within specifications and clean, fresh fuel is available to carburetor, remove and clean or overhaul carburetor as outlined in CARBURETOR SERVICING FUNDAMENTALS section of this manual.

4. The preceding troubleshooting techniques are based on the fact that to run, an engine must develop compression, have an ignition spark and receive proper fuel-air mixture. In some instances, there are other factors involved. Refer to special notes following this section for service hints on finding common causes of engine trouble that may not be discovered in normal troubleshooting procedure.

If Engine Starts, Then Stops

This complaint is usually due to fuel starvation, but may be caused by a faulty ignition system. Recommended troubleshooting procedure is as follows:

1. Remove and inspect fuel tank cap; on all engines, fuel tank is vented through breather in fuel tank cap so air can enter tank as fuel is used. If engine stops after running several minutes, a clogged breather should be suspected. On some engines, it is possible to let engine run with fuel tank cap removed and if this permits engine to run without stopping, clean or renew cap.

CAUTION: Be sure to observe safety precautions before attempting to run engine without fuel tank cap in place. If there is any danger of fuel being spilled on engine or spark entering open tank, *do not* attempt to run engine without fuel tank cap in place. If in doubt, try a new cap.

2. If clogged breather in fuel tank cap is eliminated as cause of trouble, a partially clogged fuel filter or fuel line should be suspected. Remove and clean fuel tank and line and if so equipped, clean fuel shut-off valve and/or fuel tank filter. On some engines, a screen or felt type fuel filter is located in carburetor fuel inlet; refer to engine repair section for appropriate engine make and model for carburetor construction.

3. After cleaning fuel tank, line, filter, etc., if trouble is still encountered, a sticking or faulty carburetor inlet needle valve or float may be cause of trouble. Remove, disassemble and clean carburetor using data in engine repair section and in CARBURETOR SERVICE FUNDAMENTALS as a guide.

4. If fuel system is eliminated as cause of trouble by performing procedure outlined in steps 1, 2 and 3, check magneto or battery ignition coil on

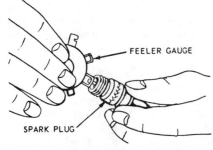

Fig. 5-1 – Be sure to check spark plug electrode gap with proper size feeler gage and adjust gap to specification recommended by manufacturer.

Fig. 5-2 – Condensation can cause water and rust to form in fuel tank even though only clean fuel has been poured into tank.

tester if such equipment is available. If not, check for ignition spark immediately after engine stops. Renew coil, condenser and breaker points if no spark is noted. Also, check for engine compression immediately after engine stops; trouble may be caused by sticking intake or exhaust valve or cam followers (tappets). If no or little compression is noted immediately after engine stops, refer to ENGINE SERVICE FUNDAMENTALS section and to engine repair data in appropriate engine repair section of this manual.

Engine Overheats

When air cooled engines overheat, check for:

1. Air inlet screen in blower housing plugged with grass, leaves, dirt or other debris.

2. Remove blower housing and shields and check for dirt or debris accumulated on or between cooling fins on cylinder.

3. Missing or bent shields or blower housing. (Never attempt to operate an air cooled engine without all shields and blower housing in place.)

4. A too lean main fuel-air adjustment of carburetor.

5. Improper ignition spark timing. Check breaker point gap, and on engine with unit type magneto, check magneto to engine timing. On battery ignition units with timer or distributor, check for breaker points opening at proper time.

6. Engines being operated under loads in excess of rated engine horsepower or at extremely high ambient (surrounding) air temperatures may overheat.

Engine Surges When Running

Trouble with an engine surging is usually caused by improper carburetor adjustment or improper governor adjustment.

1. Refer to CARBURETOR paragraphs in appropriate engine repair section and adjust carburetor as outlined.

2. If adjusting carburetor did not correct surging condition, refer to GOVERNOR paragraph and adjust governor linkage.

3. If any wear is noted in governor linkage and adjusting linkage did not correct problem, renew worn linkage parts.

4. If trouble is still not corrected, remove and clean or overhaul carburetor as necessary. Also check for any possible air leaks between the carburetor to engine gaskets or air inlet elbow gaskets.

Special Notes on Engine Troubleshooting

ENGINES WITH COMPRESSION RELEASE. Several different makes of four-stroke cycle engines now have a compression release that reduces compression pressure at cranking speeds, thus making it easier to crank the engine. Most models having this feature will develop full compression when turned in a reverse direction with throttle in wide open position. Refer to the appropriate engine repair section in this manual for detailed information concerning compression release used on different makes and models.

IGNITION SYSTEM SERVICE

The fundamentals of servicing ignition systems are outlined in the following paragraphs. Refer to appropriate heading for type ignition system being inspected or overhauled.

BATTERY IGNITION SERVICE FUNDAMENTALS

Usually all components are readily accessible and while use of test instruments is sometimes desirable, condition of the system can be determined by simple checks. Refer to following paragraphs.

GENERAL CONDITION CHECK. Remove spark plug wire and if terminal is rubber covered, insert small screw or bolt in terminal. Hold uncovered end of terminal or bolt inserted in terminal about ⅛-inch (3.18 mm) away from engine or connect spark plug wire to test plug. Crank engine while observing gap between spark plug wire terminal and engine; if a bright blue spark snaps across gap, condition of system can be considered satisfactory. However, ignition timing may have to be adjusted. Refer to timing procedure in appropriate engine repair section.

VOLTAGE, WIRING AND SWITCH CHECK. If no spark, or a weak yellow-orange spark occurred when checking system as outlined in preceding paragraph, proceed with following check:

Test battery condition with hydrometer or voltmeter. If check indicates a dead cell, renew battery; recharge battery if a discharged condition is indicated.

NOTE: On models with electric starter or starter-generator unit, battery can be assumed in satisfactory condition if the starter cranks the engine freely.

If battery checks within specifications, but starter unit will not turn engine, a faulty starter unit is indicated and ignition trouble may be caused by excessive current draw of such a unit. If battery and starting unit, if so equipped, are in satisfactory condition, proceed as follows:

Remove battery lead wire from ignition coil and connect a test light of same voltage as battery between disconnected lead wire and engine ground. Light should go on when ignition switch is in "on" position and go off when switch is in "off" position. If not, renew switch and/or wiring and recheck for satisfactory spark. If switch and wiring are funtioning properly, but no spark is obtained, proceed as follows:

BREAKER POINTS AND CONDENSER. Remove breaker box cover and, using small screwdriver, separate and inspect breaker points. If burned or deeply pitted, renew breaker points and condenser. If point contacts are clean to grayish in color and are only slightly pitted, proceed as follows: Disconnect condenser and ignition coil lead wires from breaker point terminal and connect a test light and battery between terminal and engine ground. Light should go on when points are closed and should go out when points are open. If light fails to go out when points are open, breaker arm insulation is defective and breaker points must be renewed. If light does not go on when points are in closed position, clean or renew breaker points. In some instances, new breaker point contact surfaces may have an oily or wax coating or have foreign material between the surfaces so proper contact is prevented. Check ignition timing and breaker point gap as outlined in appropriate engine repair section of this manual.

Connect test light and battery between condenser lead and engine ground; if light goes on, condenser is shorted out and should be renewed. Capacity of condenser can be checked if test instrument is available. It is usually good practice to renew condenser whenever new breaker points are being installed if tester is not available.

IGNITION COIL. If a coil tester is available, condition of coil can be checked. However, if tester is not available, a reasonably satisfactory performance test can be made as follows:

Disconnect high tension wire from spark plug. Turn engine so cam has allowed breaker points to close. With ignition switch on, open and close points with small screwdriver while holding high tension lead about ⅛ to ¼-inch (3.18 to 6.35 mm) away from engine ground. A bright blue spark should snap across gap between spark plug wire and ground each time points are opened. If no spark occurs, or spark is weak and yellow-orange, renewal of ignition coil is indicated.

Sometimes, an ignition coil may perform satisfactorily when cold, but fail after engine has run for some time and coil is hot. Check coil when hot if this condition is indicated.

FLYWHEEL MAGNETO SERVICE FUNDAMENTALS

In servicing a flywheel magneto ignition system, the mechanic is concerned with troubleshooting, service adjustments and testing magneto components. The following paragraphs outline basic steps in servicing a flywheel type magneto. Refer to appropriate engine section for adjustment and test specifications for a particular engine.

Troubleshooting

If engine will not start and malfunction of ignition system is suspected, make the following checks to find cause of trouble.

Check to be sure ignition switch (if so equipped) is in "On" or "Run" position and the insulation on wire leading to ignition switch is in good condition. Switch can be checked with timing and test light as shown in Fig. 5-3. Disconnect lead from switch and attach one clip of test light to switch terminal and remaining clip to engine. Light should go on when switch is in "Off" or "Stop" position, and should go off when switch is in "On" or "Run" position.

Inspect high tension (spark plug) wire for worn spots in insulation or breaks in wire. Frayed or worn insulation can be repaired temporarily with plastic electrician's tape.

If no defects are noted in ignition switch or ignition wires, remove and inspect spark plug as outlined in SPARK PLUG SERVICING section. If spark plug is fouled or is in questionable condition, connect a spark plug of known quality to high tension wire, ground base of spark plug to engine and turn engine rapidly with starter. If spark across electrode gap of spark plug is a bright blue, magneto can be considered in satisfactory condition.

NOTE: Some engine manufacturers specify a certain type spark plug and a specific test gap. Refer to appropriate engine service section; if no specific spark plug type or electrode gap is recommended for test purposes, use spark plug type and electrode gap recommended for engine make and model.

If spark across gap of test plug is weak or orange colored, or no spark occurs as engine is cranked, magneto should be serviced as outlined in the following paragraphs.

Magneto Adjustments

BREAKER CONTACT POINTS. Adjustment of breaker contact points affects both ignition timing and magneto edge gap. Therefore, breaker contact point gap should be carefully adjusted according to engine manufacturer's specifications. Before adjusting breaker contact gap, inspect contact points and renew if condition of contact surfaces is questionable. It is sometimes desirable to check condition of points as follows: Disconnect condenser and primary coil leads from breaker point terminal. Attach one clip of a test light (See Fig. 5-3) to breaker point terminal and remaining clip of test light to magneto ground. Light should be out when contact points are open and should go on when engine is turned to close breaker contact points. If light stays on when points are open, insulation of breaker contact arm is defective. If light does not go on when points are closed, contact surfaces are dirty, oily or are burned.

Adjust breaker point gap as follows unless manufacturer specifies adjusting breaker gap to obtain correct ignition timing. First, turn engine so points are closed to be sure contact surfaces are in alignment and seat squarely. Then, turn engine so breaker point opening is maximum and adjust breaker gap to manufacturer's specification. A wire type feeler gage is recommended for checking and adjusting the breaker contact gap. Be sure to recheck gap after tightening breaker point base retaining screws.

IGNITION TIMING. On some engines, ignition timing is non-adjustable and a certain breaker point gap is specified. On other engines, timing is adjustable by changing position of magneto stator plate with a specified breaker point gap or by simply varying breaker point gap to obtain correct timing. Ignition timing is usually specified either in degrees of engine (crankshaft) rotation or in piston travel before piston reaches top dead center position. In some instances, a specification is given for ignition timing even though timing may be non-adjustable; if a check reveals timing is incorrect on these engines, it is an indication of incorrect breaker point adjustment or excessive wear of breaker cam. Also, on some engines, it may indicate a wrong breaker can be installed or cam has been installed in a reversed position on engine crankshaft.

Some engines may have a timing mark or flywheel locating pin to locate flywheel at proper position for ignition spark to occur (breaker points begin to open). If not, it will be necessary to measure piston travel as illustrated in Fig. 5-4 or install a degree indicating device on engine crankshaft.

A timing light as shown in Fig. 5-3 is a valuable aid in checking or adjusting

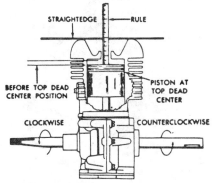

Fig. 5-4 — On some engines, it will be necessary to measure piston travel with rule, dial indicator or special timing gage when adjusting or checking ignition timing.

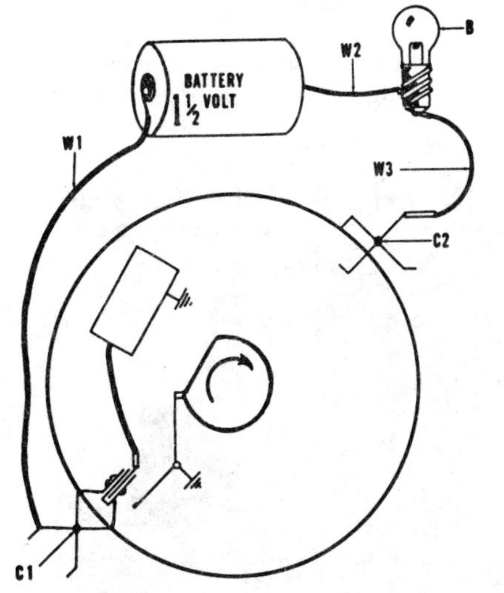

Fig. 5-3 — Drawing showing a simple test lamp for checking ignition timing and/or breaker point opening.

B. 1½ volt bulb	W1. Wire
C1. Spring clamp	W2. Wire
C2. Spring clamp	W3. Wire

engine timing. After disconnecting ignition coil lead from breaker point terminal, connect leads of timing light as shown. If timing is adjustable by moving magneto stator plate, be sure breaker point gap is adjusted as specified. Then, to check timing, slowly turn engine in normal direction of rotation past point at which ignition spark should occur. Timing light should be on, then go out (breaker points open) just as correct timing location is passed. If not, turn engine to proper timing location and adjust timing by relocating magneto stator plate or varying breaker contact gap as specified by engine manufacturer. Loosen screws retaining stator plate or breaker points and adjust position of stator plate or points so points are closed (timing light is on). Then, slowly move adjustment until timing light goes out (points open) and tighten retaining screws. Recheck timing to be sure adjustment is correct.

ARMATURE AIR GAP. To fully concentrate magnetic field of flywheel magnets within armature core, it is necessary that flywheel magnets pass as closely to armature core as possible

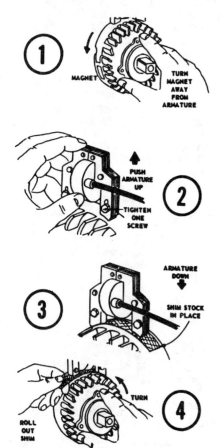

Fig. 5-5—Views showing adjustment of armature air gap when armature is located outside flywheel. Refer to Fig. 5-6 for engines having armature located inside flywheel.

without danger of metal to metal contact. Clearance between flywheel magnets and legs of armature core is called armature air gap.

On magnetos where armature and high tension coil are located outside of the flywheel rim, adjustment of armature air gap is made as follows: Turn engine so flywheel magnets are located directly under legs of armature core and check clearance between armature core and flywheel magnets. If measured clearance is not within manufacturers specifications, loosen armature core mounting screws and place shims of thickness equal to minimum air gap specification between magnets and armature core (Fig. 5-5). The magnets will pull armature core against shim stocks. Tighten armature core mounting screws, remove shim stock and turn engine through several revolutions to be sure flywheel does not contact armature core.

Where armature core is located under or behind flywheel, the following methods may be used to check and adjust armature air gap: On some engines, slots or openings are provided in flywheel through which armature air gap can be checked. Some engine manufacturers provide a cut-away flywheel that can be installed temporarily for checking armature air gap. A test flywheel can be made out of a discarded flywheel (See Fig. 5-6), or out of a new flywheel if service volume on a particular engine warrants such expenditure. Another method of checking armature air gap is to remove flywheel and place a layer of plastic tape equal to minimum specified air gap over legs of armature core. Reinstall flywheel and turn engine through several revolutions and remove flywheel; no evidence of contact between flywheel magnets and plastic tape should be noticed. Then cover legs of armature core with a layer of tape of thickness equal to maximum specified air gap; then, reinstall flywheel and turn engine through several revolutions. Indication of the flywheel magnets contacting plastic tape should be noticed after flywheel is again removed. If magnets contact first thin layer of tape applied to armature core legs, or if they do not contact second thicker layer of tape, armature air gap is not within specifications and should be adjusted.

NOTE: Before loosening armature core mounting screws, scribe a mark on mounting plate against edge of armature core so that adjustment of air gap can be gaged.

In some instances, it may be necessary to slightly enlarge armature core mounting holes before proper air gap adjustment can be made.

MAGNETO EDGE GAP. The point of maximum acceleration of movement of flywheel magnetic field through high tension coil (and therefore, the point of maximum current induced in primary coil windings) occurs when trailing edge of flywheel magnet is slightly past left hand leg of armature core. The exact point of maximum primary current is determined by using electrical measuring devices, distance between trailing edge of flywheel magnet and leg of armature core at this point is measured and becomes a service specification. This distance, which is stated either in thousandths of an inch or in degrees of flywheel rotation, is called the Edge Gap or "E" Gap.

For maximum strength of ignition spark, breaker points should just start to open when flywheel magnets are at specified edge gap position. Usually, edge gap is non-adjustable and will be maintained at proper dimension if contact breaker points are adjusted to recommended gap and correct breaker cam is installed. However magneto edge gap can change (and spark intensity thereby reduced) due to the following:

a. Flywheel drive key sheared
b. Flywheel drive key worn (loose)
c. Keyway in flywheel or crankshaft worn (oversized)
d. Loose flywheel retaining nut which can also cause any above listed difficulty
e. Excessive wear on breaker cam
f. Breaker cam loose on crankshaft
g. Excessive wear on breaker point rubbing block or push rod so points cannot be properly adjusted.

Unit Type Magneto Service Fundamentals

Improper functioning of carburetor, spark plug or other components often

Fig. 5-6—Where armature core is located inside flywheel, check armature gap by using a cut-away flywheel unless other method is provided by manufacturer; refer to appropriate engine repair section. Where possible, an old discarded flywheel should be used to cut-away section for checking armature gap.

causes difficulties that are thought to be an improperly functioning magneto. Since a brief inspection will often locate other causes for engine malfunction, it is recommended one be certain magneto is at fault before opening magneto housing. Magneto malfunction can easily be determined by simple tests as outlined in following paragraph.

Troubleshooting

With a properly adjusted spark plug in good condition, ignition spark should be strong enough to bridge a short gap in addition to actual spark plug gap. With engine running, hold end of spark plug wire not more than 1/16-inch (1.59 mm) away from spark plug terminal. Engine should not misfire.

To test magneto spark if engine will not start, remove ignition wire from magneto end cap socket. Bend a short piece of wire so when it is inserted in end cap socket, other end is about 1/8-inch (3.18 mm) from engine casting. Crank engine slowly and observe gap between wire and engine; a strong blue spark should jump gap the instant impulse coupling trips. If a strong spark is observed, it is recommended magneto be eliminated as source of engine difficulty and spark plug, ignition wire and terminals be thoroughly inspected.

If, when cranking engine, impulse coupling does not trip, magneto must be removed from engine and coupling overhauled or renewed. It should be noted that if impulse coupling will not trip, a weak spark will occur.

Magneto Adjustments and Service

BREAKER POINTS. Breaker points are accessible for service after removing magneto housing end cap. Examine point contact surfaces for pitting or pyramiding (transfer of metal from one surface to the other); a small tungsten file or fine stone may be used to resurface points. Badly worn or badly pitted points should be renewed. After points are resurfaced or renewed, check breaker point gap with rotor turned so points are opened maximum distance. Refer to MAGNETO paragraph in appropriate engine repair section for point gap specifications.

When installing magneto end cap, both end cap and housing mating surfaces should be throughly cleaned and a new gasket be installed.

CONDENSER. Condenser used in unit type magneto is similar to that used in other ignition systems. Refer to MAGNETO paragraph in appropriate engine repair section for condenser test

specifications. Usually, a new condenser should be installed whenever breaker points are being renewed.

COIL. The ignition coil can be tested without removing the coil from housing. Instructions provided with coil tester should have coil test specifications listed.

ROTOR. Usually, service on magneto rotor is limited to renewal of bushings or bearings, if damaged. Check to be sure rotor turns freely and does not drag or have excessive end play.

MAGNETO INSTALLATION. When installing a unit type magneto on an engine, refer to MAGNETO paragraph in appropriate engine repair section for magneto to engine timing information.

SOLID STATE IGNITION SERVICE FUNDAMENTALS

Because of differences in solid state ignition construction, it is impractical to outline a general procedure for solid state ignition service. Refer to specific engine section for testing, overhaul notes and timing of solid state ignition systems.

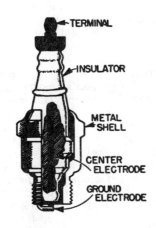

Fig. 5-7 – Cross-sectional drawing of spark plug showing construction and nomenclature.

Fig. 5-8 – Normal plug appearance in four-stroke cycle engine. Insulator is light tan to gray in color and electrodes are not burned. Renew plug at regular intervals as recommended by engine manufacturer.

SPARK PLUG SERVICING

ELECTRODE GAP. Spark plug electrode gap should be adjusted by bending the ground electrode. Refer to Fig. 5-7. Recommended gap is listed in SPARK PLUG paragraph in appropriate engine repair section of this manual.

CLEANING AND ELECTRODE CONDITIONING. Spark plugs are most usually cleaned by abrasive action commonly referred to as "sand blasting." Actually, ordinary sand is not used, but a special abrasive which is non-conductive to electricity even when melted, thus the abrasive cannot short

Fig. 5-9 – Appearance of spark plug indicating cold fouling. Cause of cold fouling may be use of a too-cold plug, excessive idling or light loads, carburetor choke out of adjustment, defective spark plug wire or boot, carburetor adjusted too "rich" or low engine compression.

Fig. 5-10 – Appearance of spark plug indicating wet fouling; a wet, black oily film is over entire firing end of plug. Cause may be oil getting by worn valve guides, worn oil rings or plugged breather or breather valve in tappet chamber.

Fig. 5-11 – Appearance of spark plug indicating overheating. Check for plugged cooling fins, bent or damaged blower housing, engine being operated without all shields in place or other causes for engine overheating. Also can be caused by too lean a fuel-air mixture or spark plug not tightened properly.

out plug current. Extreme care should be used in cleaning plugs after sand blasting, however, as any particles of abrasive left on plug may cause damage to piston rings, piston or cylinder walls. Some engine manufacturers recommend spark plug be renewed rather than cleaned because of possible engine damage from cleaning abrasives.

After plug is cleaned by abrasive, and before gap is set, electrode surfaces between grounded and insulated electrodes should be cleaned and returned as nearly as possible to original shape by filing with a point file. Failure to properly

dress electrodes can result in high secondary voltage requirements, and misfire of the plug.

PLUG APPEARANCE DIAGNOSIS. Appearance of a spark plug will be altered by use, and an examination of plug tip can contribute useful information which may assist in obtaining better spark plug life. Figs. 5-8 and 5-11 are provided by Champion Spark Plug Company to illustrate typical observed conditions. Listed in captions are probable causes and suggested corrective measures.

ENGINE SERVICE
DISASSEMBLY AND ASSEMBLY

Special techniques must be developed in repair of engines of aluminum alloy or magnesium alloy construction. Soft threads in aluminum or magnesium castings are often damaged by carelessness in overtightening fasteners or in attempting to loosen or remove seized fasteners. Manufacturer's recommended torque values for tightening screw fasteners should be followed closely.

NOTE: If damaged threads are encountered, refer to following paragraph, "REPAIRING DAMAGED THREADS."

A given amount of heat applied to aluminum or magnesium will cause it to expand a greater amount than will steel under similar conditions. Because of different expansion characteristics, heat is usually recommended for easy installation of bearings, pins, etc., in aluminum or magnesium castings. Sometimes, heat can be used to free parts that are seized or where an interference fit is used. Heat, therefore, becomes a service tool and the application of heat one of the required service techniques. An open flame is not usually advised because it destroys paint and other protective coatings and because a uniform and controlled temperature with open flame is difficult to obtain. Methods commonly used are heating in oil or water, with a heat lamp, electric hot plate, electric hot air gun, or in an oven or kiln. The use of water or oil gives a fairly accurate temperature control but is somewhat limited as to the size and type of part that can be handled. Thermal crayons are available which can be used to determine temperature of a heated part. These crayons melt when the part reaches a specified temperature, and a number of crayons for different temperatures are available. Temperature indicating crayons are usually available at welding equipment supply houses.

Use only specified gaskets when reassembling, and use an approved gasket cement or sealing compound unless otherwise stated. Seal all exposed threads and repaint or retouch with an approved paint.

CARBURETOR SERVICING FUNDAMENTALS

The bulk of carburetor service consists of cleaning, inspection and adjustment. After considerable service it may become necessary to overhaul the carburetor and renew worn parts to restore original operating efficiency. Although carburetor condition affects engine operating economy and power, ignition and engine compression must also be considered to determine and correct causes of poor performance.

Before dismantling carburetor for cleaning or overhaul, clean all external surfaces and remove accumulated dirt and grease. Refer to appropriate engine repair section for carburetor exploded or cross-sectional views. Dismantle carburetor and note any discrepancies to assure correction during overhaul. Thoroughly clean all parts and inspect for damage or wear. Wash jets and passages and blow clear with clean, dry compressed air. Do not use a drill or wire to clean jets as possible enlargement of calibrated holes will disturb operating balance. Measurement of jets to determine extent of wear is difficult and new parts are usually installed to assure satisfactory results.

Carburetor manufacturers provide for many of their models an assortment of gaskets and other parts usually needed to do a correct job of cleaning and overhaul. These assortments are usually catalogued as Gasket Kits and Overhaul Kits respectively.

On float type carburetors, inspect float pin and needle valve for wear and renew if necessary. Check metal floats for leaks and where a dual type float is installed, check alignment of float sections. Check cork floats for loss of protective coating and absorption of fuel.

NOTE: Do not attempt to recoat cork floats with shellac or varnish or to resolder leaky metal floats. Renew part if defective.

Check fit of throttle and choke valve shafts. Excessive clearance will cause improper valve plate seating and will permit dust or grit to be drawn into engine. Air leaks at throttle shaft bores due to wear will upset carburetor calibration and contribute to uneven engine operation. Rebush valve shaft holes where necessary and renew dust seals. If rebushing is not possible, renew body part supporting shaft. Inspect throttle and choke valve plates for proper installation and condition.

Power or idle adjustment needles must not be worn or grooved. Check condition of needle seal packing or "O" ring and renew packing or "O" ring if necessary.

Reinstall or renew jets, using correct size listed for specific model. Adjust power and idle settings as described for specific carburetors in engine service section of manual.

It is important that carburetor bore at idle discharge ports and in vicinity of throttle valve be free of deposits. A partially restricted idle port will produce a "flat spot" between idle and mid-range rpm. This is because the restriction makes it necessary to open throttle wider than the designed opening to obtain proper idle speed. Opening throttle wider than the design specified amount will uncover more of the port than was intended in calibration of carburetor. As a result an insufficient amount of the port will be available as a reserve to cover transition period (idle to mid-range rpm) when the high speed system begins to function.

When reassembling float type carburetors, be sure float position is properly adjusted. Refer to CARBURETOR paragraph in appropriate engine repair section for float level adjustment specifications.

REPAIRING DAMAGED THREADS

Damaged threads in castings can be renewed by use of thread repair kits which are recommended by a number of equipment and engine manufacturers. Use of thread repair kits is not difficult, but instructions must be carefully

Fig. 5-12—Damaged threads in casting before repair. Refer to Figs. 5-13, 5-14 and 5-15 for steps in installing thread insert. (Series of photos provided by Heli-Coil Corp., Danbury, Conn.)

Fig. 5-13—First step in repairing damaged threads is to drill out old threads using exact size drill recommended in instructions provided with thread repair kit. Drill all the way through an open hole or all the way to bottom of blind hole, making sure hole is straight and that centerline of hole is not moved in drilling process.

Fig. 5-14—Special drill taps are provided in thread repair kit for threading drilled hole to correct size for outside of thread insert. A standard tap cannot be used.

followed. Refer to Figs. 5-12 through 5-15 which illustrate the use of Heli-Coil thread repair kits that are manufactured by the Heli-Coil Corporation, Danbury, Connecticut.

Heli-Coil thread repair kits are

Fig. 5-15—A thread insert and a completed repair are shown above. Special tools are provided in thread repair kit for installation of thread insert.

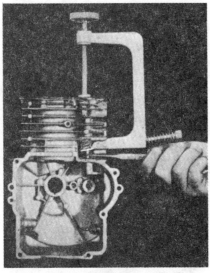

Fig. 5-16—View showing one type of valve spring compressor being used to remove keeper. (Block is cut-away to show valve spring.)

Fig. 5-17—Drawing showing three types of valve spring keepers used.

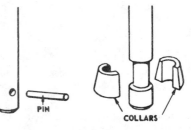

available through parts departments of most engine and equipment manufacturers; thread inserts are available in all National Coarse (USS) sizes from #4 to 1½ inch and National Fine (SAE) sizes from #6 to 1½ inch. Also, sizes for repairing 14mm and 18mm spark plug ports are available.

VALVE SERVICE FUNDAMENTALS

When overhauling engines, obtaining proper valve sealing is of primary importance. The following paragraphs cover fundamentals of servicing intake and exhaust valves, valve seats and valve guides.

REMOVING AND INSTALLING VALVES. A valve spring compressor, one type of which is shown in Fig. 5-16, is a valuable aid in removing and installing intake and exhaust valves. This tool is used to hold spring compressed while removing or installing pin, collars or retainer from valve stem. Refer to Fig. 5-17 for views showing some of the different methods of retaining valve spring to valve stem.

VALVE REFACING. If valve face (See Fig. 5-18) is slightly worn, burned or pitted, valve can usually be refaced providing proper equipment is available. Many shops will usually renew valves, however, rather than invest in somewhat costly valve refacing tools.

Before attempting to reface a valve, refer to specifications in appropriate engine repair section for valve face angle. On some engines, manufacturer recommends grinding the valve face to an angle of ½° to 1° less than that of the valve seat. Refer to Fig. 5-19. Also, nominal valve face angle may be either 30° or 45°.

After valve is refaced, check thickness of valve "margin" (See Fig. 5-18). If margin is less than manufacturer's minimum specification (refer to specifications in appropriate engine repair section), or is less than one-half the margin of a new valve, renew valve. Valves having excessive material removed in refacing operation will not give satisfactory service.

When refacing or renewing a valve, the seat should also be reconditioned, or in engines where valve seat is renewable, a new seat should be installed. Refer to following paragraph "RESEATING OR RENEWING VALVE SEATS." Then, the seating surfaces should be lapped in using a fine valve grinding compound.

RESEATING OR RENEWING VALVE SEATS. On engines having valve seat machined in cylinder block casting, seat can be reconditioned by using a correct angle seat grinding stone or valve seat cutter. When reconditioning valve seat, care should be taken that only enough material is removed to provide a good seating on valve contact surface. The width of seat should then be

measured (See Fig. 5-20) and if width exceeds manufacturer's maximum specifications, seat should be narrowed by using one stone or cutter with an angle 15° greater than seat angle and a second stone or cutter with an angle 15° less than seat angle. When narrowing seat, coat seat lightly with Prussian blue and check where seat contacts valve face by inserting valve in guide and rotating valve lightly against seat. Seat should contact approximately center of valve face. By using only narrow angle seat narrowing stone or cutter, seat contact will be moved towards outer edge of valve face.

On engines having renewable valve seats, refer to appropriate engine repair section in this manual for recommended method of removing old seat and installing new seat. Refer to Fig. 5-21 for one method of installing new valve seats. Seats are retained in cylinder block bore by an interference fit; that is, seat is slightly larger than bore in block. It sometimes occurs that valve seat will become loose in bore, especially on engines with aluminum crankcase. Some manufacturers provide oversize valve seat inserts (insert O.D. larger than standard part) so that if standard size insert fits loosely, bore can be cut oversize and a new insert be tightly installed. After installing valve seat insert in engines of aluminum construction, metal around seat should be peened as shown in Fig. 5-22. Where a loose insert is encountered and an oversize insert is not available, loose insert can usually be tightened by center-punching cylinder block material at three equally spaced points around insert, then peening completely around insert as shown in Fig. 5-22.

For some engines with cast iron cylinder blocks, a service valve seat insert is available for reconditioning valve seat, and is installed by counterboring cylinder block to specified dimensions, then driving insert into place. Refer to appropriate engine repair section in this manual for information on availability and installation of service valve seat inserts for cast iron engines.

INSTALLING OVERSIZE PISTON AND RINGS

Some engine manufacturers have over-size piston and ring sets available for use in repairing engines in which cylinder bore is excessively worn and standard size piston and rings cannot be

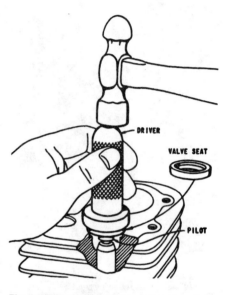

Fig. 5-21 — View showing one method used to install valve seat insert. Refer to appropriate engine repair section for manufacturer's recommended method.

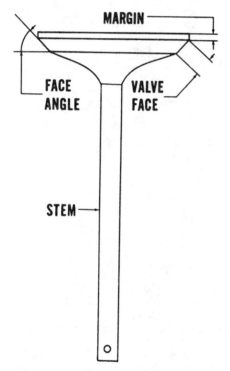

Fig. 5-18 — Drawing showing typical four-stroke cycle engine valve. Face angle is usually 30° or 45°. On some engines, valve face is ground to an angle of ½ or 1 degree less than seat angle.

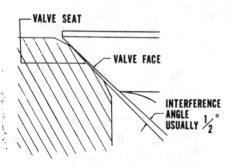

Fig. 5-19 — Drawing showing line contact of valve face with valve seat when valve face is ground at smaller angle than valve seat; this is specified on some engines.

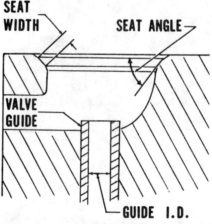

Fig. 5-20 — Cross-sectional drawing of typical valve seat and valve guide as used on some engines. Valve guide may be integral part of cylinder block; on some models so constructed, valve guide I.D. may be reamed out and an oversize valve stem installed. On other models, a service guide may be installed after counterboring cylinder block.

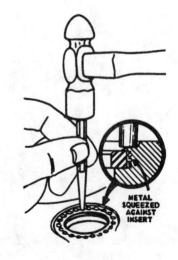

Fig. 5-22 — It is usually recommended that on aluminum block engines, metal be peened around valve seat insert after insert is installed.

used. If care and approved procedure are used in oversizing cylinder bore, installation of an oversize piston and ring set should result in a highly satisfactory overhaul.

Cylinder bore may be oversized by using either a boring bar or a hone; however, if a boring bar is used, it is usually recommended cylinder bore be finished with a hone. Refer to Fig. 5-23.

Where oversize piston and rings are available, it will be noted in appropriate engine repair section of this manual. Also, the standard bore diameter will also be given. Before attempting to rebore or hone the cylinder to oversize, carefully measure the cylinder bore to be sure standard size piston and rings will not fit within tolerance. Also, it may be possible cylinder is excessively worn or damaged and reboring or honing to largest oversize will not clean up worn or scored surface.

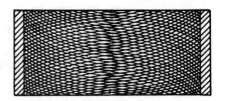

Fig. 5-23 — A cross-hatch pattern as shown should be obtained when honing cylinder. Pattern is obtained by moving hone up and down cylinder bore as it is being turned by slow speed electric drill.

SERVICE SHOP TOOL BUYER'S GUIDE

This listing of Service Shop Tools is solely for the convenience of users of this manual and does not imply endorsement or approval by Intertec Publishing Corporation of the tools and equipment listed. The listing is in response to many requests for information on sources for purchasing special tools and equipment. Every attempt has been made to make the listing as complete as possible at time of publication and each entry is made from the latest material available.

Special engine service tools such as seal drivers, bearing drivers, etc., which are available from the engine manufacturer are not listed in this section of the manual. Where a special service tool is listed in the engine service section of this manual, the tool is available from the central parts or service distributors listed at the end of most engine service sections, or from the manufacturer.

NOTE TO MANUFACTURERS AND NATIONAL SALES DISTRIBUTORS OF ENGINE SERVICE TOOLS AND RELATED SERVICE EQUIPMENT. To obtain either a new listing for your products, or to change or add to an existing listing, write to Intertec Publishing Corporation, Book Division, P.O. Box 12901, Overland Park, Kansas 66212.

Engine Service Tools

Ammco Tools, Inc.
Wacker Park
North Chicago, Illinois 60064
Valve spring compressor, torque wrenches, cylinder hones, ridge reamers, piston ring compressors, piston ring expanders.

Black & Decker Mfg. Co.
701 East Joppa Road
Towson, Maryland 21204
Valve grinding equipment.

Bloom, Inc.
Route Four, Hiway 20 West
Independence, Iowa 50644
Engine repair stand with crankshaft straightening attachment.

Brush Research Mfg. Inc.
4642 East Floral Drive
Los Angeles, California 90022
Cylinder hones.

E-Z Lok
P.O. Box 2069
Gardena, California 90247
Thread repair insert kits for metal, wood and plastic.

Foley-Belsaw Company
Outdoor Power Equipment Parts Division
6301 Equitable Road
P.O. Box 419593
Kansas City, Missouri 64141
Crankshaft straightener and repair stand, valve refacers, valve seat grinders, parts washers, cylinder hones, gages and ridge reamers, piston ring expanders and compressor, flywheel pullers, torque wrenches, rotary mower blade balancers.

Frederick Mfg. Co., Inc.
1400 C., Agnes Avenue
Kansas City, Missouri 64127
Crankshaft straightener.

Heli-Coil Products Division
Heli-Coil Corporation
Shelter Rock Lane
Danbury Connecticut 06810
Thread repair kits, thread inserts, installation tools.

K-D Tools
3575 Hempland Road
Lancaster, Pennsylvania 17604
Thread repair kits, valve spring compressors, reamers, micrometers, dial indicators, calipers.

Keystone Reamer & Tool Co.
Post Office Box 310
Millersburg, Pennsylvania 17061
Valve seat cutter and pilots, adjustable reamers.

Ki-Sol Corporation
100 Larkin Williams Ind. Court
Fenton, Missouri 63026
Cylinder hone, ridge reamer, ring compressor, ring expander, ring groove cleaner, torque wrenches, valve spring compressor, valve refacing equipment.

K-Line Industries, Inc.
315 Garden Avenue
Holland, Michigan 49423
Cylinder hone, ridge reamer, ring compressor, valve guide tools, valve spring compressor, reamers.

K.O. Lee Company
101 South Congress
Aberdeen, South Dakota 57401
Valve refacing, valve seat grinding, valve reseating and valve guide reaming tools.

Kwik-Way Mfg. Co.
500 57th Street
Marion, Iowa 52302
Cylinder boring equipment, valve facing equipment, valve seat grinding equipment.

Lisle Corporation
807 East Main
Clarinda, Iowa 51632
Cylinder hones, ridge reamers, ring compressors, valve spring compressors.

Microdot, Inc.
P.O. Box 3001
800 So. State College Blvd.
Fullerton, California 92631
Thread repair insert kits.

Mighty Midget Mfg. Co.,
Div. of Kansas City Screw Thread Co.
2908 E. Truman Road
Kansas City, Missouri 64127
Crankshaft straightener.

Neway Manufacturing, Inc.
1013 No. Shiawassee
Corunna, Michigan 48817
Valve seat cutters.

Owatonna Tool Company
436 Eisenhower Drive
Owatonna, Minnesota 55060
Valve tools, spark plug tools, piston ring tools, cylinder hones.

Power Lawnmower Parts Inc.
1920 Lyell Avenue
P.O. Box 60860
Rochester, NY 14606-0860
Gasket cutter tool, gasket scraper tool, crankshaft cleaning tool, ridge reamer, valve spring compressor, valve seat cutters, thread repair kits, valve lifter, piston ring expander.

Precision Manufacturing
& Sales Co., Inc.
2140 Range Rd., P.O. Box 149
Clearwater, Florida 33517
Cylinder boring equipment, measuring instruments, valve equipment, hones,

porting, hand tools, test equipment, threading, presses, parts washers, milling machines, lathes, drill presses, glass beading machines, dynos, safety equipment.

Sunnen Product Company
7910 Manchester Avenue
Saint Louis, Missouri 63143
Cylinder hones, rod reconditioning, valve guide reconditioning.

Rexnord Specialty Fastener Division
3000 W. Lomita Blvd.
Torrance, California 90505
Thread repair insert kits (Keenserts) and installation tools.

Vulcan Tools
Div. of TRW, Inc.
2300 Kenmore Avenue
Buffalo, New York 14207
Cylinder hones, reamers, ridge removers, valve spring compressors, valve spring testers, ring compressor, ring groove cleaner.

Waters-Dove Manufacturing
Post Office Box 40
Skiatook, Oklahoma 74070
Crankshaft straightener, oil seal remover.

Test Equipment and Gages

Allen Test Products
2101 North Pitcher Street
Kalamazoo, Michigan 49007
Coil and condenser testers, compression gages.

Applied Power, Inc.,
Auto Division
P.O. Box 27207
Milwaukee, Wisconsin 53227
Compression gage, condenser tester, tachometer, timing light, ignition analyzer.

AW Dynamometer, Inc.
131-1/2 East Main Street
Colfax, Illinois 61728
Engine test dynamometer.

B.C. Ames Company
131 Lexington
Waltham, Massachusetts 02254
Micrometer dial gages and indicators.

The Bendix Corporation
Engine Products Division
Delaware Avenue
Sidney, New York 13838
Condenser tester, magneto test equipment, timing light.

Burco
208 Delaware Avenue
Delmar, New York 12054

Coil and condenser tester, compression gage, carburetor tester, grinders, Pow-R-Arms, chain breakers, rivet spinners.

Dixson, Inc.
Post Office Box 1449
Grand Junction, Colorado 81501
Tachometer, compression gage, timing light.

Foley-Belsaw Company
Outdoor Power Equipment
Parts Division
6301 Equitable Road
P.O. Box 419593
Kansas City, Missouri 64141
Cylinder gage, amp/volt testers, condenser and coil tester, magneto tester, ignition testers for conventional and solid-state systems, tachometers, spark testers, compression gages, timing lights and gages, micrometers and calipers, torque wrenches, carburetor testers, vacuum gages.

Fox Valley Instrument Co.
Route 5, Box 390
Cheboygan, Michigan 49721
Coil and condenser tester, ignition actuated tachometer.

Graham-Lee Electronics, Inc.
4200 Center Avenue NE
Minneapolis, Minnesota 55421
Coil and condenser tester.

K-D Tools
3575 Hempland Road
Lancaster, Pennsylvania 17604
Diode tester and installation tools, compression gage, timing light, timing gages.

Ki-Sol Corporation
100 Larkin Williams Ind. Court
Fenton, Missouri 63026
Micrometers, telescoping gages, compression gages, cylinder gages.

K-Line Industries, Inc.
315 Garden Avenue
Holland, Michigan 49423
Compression gage, tachometers.

Merc-O-Tronic Instruments
Corporation
215 Branch Street
Almont, Michigan 48003
Ignition analyzers for conventional solid-state and magneto systems, electric tachometers, electronic tachometer and dwell meter, power timing lights, ohmmeters, compression gages, mechanical timing devices.

Owatonna Tool Company
436 Eisenhower Drive
Owatonna, Minnesota 55060
Feeler gages, hydraulic test gages.

Power Lawnmower Parts Inc.
1920 Lyell Avenue
P.O. Box 60860
Rochester, NY 14606-0860
Condenser and coil tester, compression gage, flywheel magneto tester, ignition tester.

Prestolite Electronics Div.
An Allied Company
Post Office Box 931
Toledo, Ohio 43694
Magneto test plug.

Simpson Electric Company
853 Dundee Avenue
Elgin, Illinois 60120
Electrical and electronic test equipment.

L.S. Starrett Company
1-165 Crescent Street
Athol, Massachusetts 01331
Micrometers, dial gages, bore gages, feeler gages.

Stevens Instrument Company, Inc.
Post Office Box 193
Waukegan, Illinois 60085
Ignition analyzers, timing lights, tachometers, volt-ohmmeters, spark checkers, CD ignition testers.

Stewart-Warner Corporation
1826 Diversey Parkway
Chicago, Illinois 60614
Compression gage, ignition tachometer, timing light ignition analyzer.

P.A. Sturtevant Co.,
Division Dresser Ind.
3201 North Wolf Road
Franklin Park, Illinois 60131
Torque wrenches, torque multipliers, torque analyzers.

Sun Electric Corporation
Instrument Products Div.
1560 Trimble Road
San Jose, California 95131
Compression gage, hydraulic test gages, coil and condenser tester, ignition tachometer, ignition analyzer.

Westberg Mfg. Inc.
3400 Westach Way
Sonoma, California 95476
Ignition tachometer, magneto test equipment, ignition system analyzer.

Shop Tools and Equipment

A-C Delco Division,
General Motors Corp.
400 Renaissance Center
Detroit, Michigan 48243
Spark plug cleaning and test equipment.

**Applied Power, Inc.,
Auto. Division**
Box 27207
Milwaukee, Wisconsin 53227
Arc and gas welding equipment, welding rods, accessories, battery service equipment.

Black & Decker Mfg. Co.
701 East Joppa Road
Towson, Maryland 21204
Air and electric powered tools.

Bloom, Inc.
Route 4, Hiway 20 West
Independence, Iowa 50644
Lawn mower repair bench with built-in engine cranking mechanism.

**Campbell Chain Division
McGraw-Edison Co.**
Post Office Box 3056
York, Pennsylvania 17402
Chain type engine and utility slings.

Champion Pneumatic Machinery Co.
1301 No. Euclid Avenue
Princeton, Illinois 61356
Air compressors.

Champion Spark Plug Co.
Post Office Box 910
Toledo, Ohio 43661
Spark plug cleaning and testing equipment, gap tools and wrenches.

Chicago Pneumatic Tool Co.
2200 Bleecker Street
Utica, New York 13503
Air impact wrenches, air hammers, air drills and grinders, nut runners, speed ratchets.

Clayton Manufacturing Company
415 North Temple City Boulevard
El Monte, California 91731
Steam cleaning equipment.

E-Z Lok
P.O. Box 2069
240 E. Rosecrans Avenue
Gardena, California 90247
Thread repair insert kits for metal, wood and plastic.

**Foley-Belsaw Company
Outdoor Power Equipment
Parts Division**
6301 Equitable Road
P.O. Box 419593
Kansas City, Missouri 64141
Torque wrenches, parts washers, micrometers and calipers.

G & H Products, Inc.
Post Office Box 770
St. Paris, Ohio 43027
Motorized lawn mower stand.

**General Scientific Equipment
Company**
Limekiln Pike & Williams Avenue
Box 27309
Philadelphia, Pennsylvania 19118-0255
Safety equipment.

Graymills Corporation
3705 North Lincoln Avenue
Chicago, Illinois 60613
Parts washing stand.

**Heli-Coil Products Div.,
Heli-Coil Corp.**
Shelter Rock Lane
Danbury, Connecticut 06810
Thread repair kits, thread inserts and installation tools.

Ingersoll-Rand
253 E. Washington Avenue
Washington, New Jersey 07882
Air and electric impact wrenches, electric drills and screwdrivers.

Ingersoll-Rand Co.
Deerfield Industrial Park
South Deerfield, Massachusetts 01373
Impact wrenches, portable electric tools, battery powered ratchet wrench.

Jaw Manufacturing Co.
39 Mulberry Street, P.O. Box 213
Reading, Pennsylvania 19603
Files for renewal of damaged threads, hex nut size rethreader dies, flexible shaft drivers and extensions, screw extractors, impact drivers.

**Jenny Division
of Homestead Ind., Inc.**
Box 348
Carapolis, Pennsylvania 15108-0348
Steam cleaning equipment, pressure washing equipment.

Keystone Reamer & Tool Co.
Post Office Box 310
Millersburg, Pennsylvania 17061
Adjustable reamers, twist drills, tape, dies, etc.

K-Line Industries, Inc.
315 Garden Avenue
Holland, Michigan 49423
Air and electric impact wrenches.

Microdot, Inc.
P.O. Box 3001
800 So. State College Blvd.
Fullerton, California 92631
Thread repair insert kits.

Owatonna Tool Company
436 Eisenhower Drive
Owatonna, Minnesota 55060
Bearing and gear pullers, hydraulic shop presses.

Power Lawnmower Parts Inc.
1920 Lyell Avenue
P.O. Box 60860
Rochester, NY 14606-0860
Flywheel puller, starter wrench and flywheel holder, blade sharpener, chain breakers, rivet spinners, gear pullers.

**Pronto Tool Division,
Ingersoll-Rand**
2600 East Nutwood Avenue
Fullerton, California 92631
Torque wrenches, gear and bearing pullers.

Shure Manufacturing Corp.
1601 South Hanley Road
Saint Louis, Missouri 63144
Steel shop benches, desks, engine overhaul stand.

Sioux Tools, Inc.
2801-2999 Floyd Blvd.
Sioux City, Iowa 51102
Portable air and electric tools.

**Rexnord Specialty
Fastener Division**
3000 W. Lomita Blvd.
Torrance, California 90505
Thread repair insert kits (Keenserts) and installation tools.

Vulcan Tools
2300 Kenmore Avenue
Buffalo, New York 14207
Air and electric impact wrenches.

Sharpening and Maintenance Equipment for Small Engine Powered Implements

**Bell Industries,
Saw & Machine Division**
Post Office Box 2510
Eugene, Oregon 97402
Saw chain grinder.

Desa Industries, Inc.
25000 South Western Ave.
Park Forest, Illinois 60466
Saw chain breakers and rivet spinners, chain files, file holder and filing guide.

**Foley-Belsaw Company
Outdoor Power Equipment
Parts Division**
6301 Equitable Road
P.O. Box 419593
Kansas City, Missouri 64141
Circular, band and hand saw filers, heavy duty grinders, saw setters, retoothers, circular saw vises, lawn mower sharpener, saw chain sharpening and repair equipment.

Granberg Industries
200 S. Garrard Blvd.

Richmond, California 94804
Saw chain grinder, file guides, chain breakers and rivet spinners, chain saw lumber cutting attachments.

Ki-Sol Corporation
100 Larkin Williams Ind. Court
Fenton, Missouri 63026
Mower blade balancer.

Magna-Matic Div.,
A.J. Karrels Co.
Box 348
Port Washington, Wisconsin 53074
Rotary mower blade balancer, "track" checking tool.

Omark Industries, Inc.
4909 International Way
Portland, Oregon 97222
Saw chain and saw bars, maintenance equipment, to include file holders, filing vises, rivet spinners, chain breakers, filing depth gages, bar groove gages, electric chain saw sharpeners.

Power Lawnmower Parts Inc.
1920 Lyell Avenue
P.O. Box 60860
Rochester, NY 14606-0860
Grinding wheels, parts cleaning brush, tube repair kits, blade balancer.

S.I.P. Grinding Machines
American Marsh Pumps
P.O. Box 23038
722 Porter Street
Lansing, Michigan 48909
Lawn mower sharpeners, saw chain grinders, lapping machine.

Specialty Motors Mfg.
641 California Way,
P.O. Box 157
Longview, Washington 98632
Chain saw bar rebuilding equipment.

Mechanic's Hand Tools

Channellock, Inc.
1306 South Main Street
Meadville, Pennsylvania 16335

John H. Graham & Company, Inc.
617 Oradell Avenue
Oradell, New Jersey 07649

Jaw Manufacturing Company
39 Mulberry Street
Reading, Pennsylvania 19603

K-D Tools
3575 Hempland Road
Lancaster, Pennsylvania 17604

K-Line Industries, Inc.
315 Garden Avenue
Holland, Michigan 49423

Millers Falls Division
Ingersoll-Rand
Deerfield Industrial Park
South Deerfield, Massachusetts 01373

New Britain Tool Company
Division of Litton Industrial
Products
P.O. Box 12198
Research Triangle Park, N.C. 27709

Owatonna Tool Company
436 Eisenhower Drive
Owatonna, Minnesota 55060

Power Lawnmower Parts Inc.
1920 Lyell Avenue
P.O. Box 60860
Rochester, NY 14606-0860
Crankshaft wrench, spark plug wrench, hose clamp pliers.

Proto Tool Division
Ingersoll-Rand
2600 East Nutwood Avenue
Fullerton, California 92631

Snap-On Tools
2801 80th Street
Kenosha, Wisconsin 53140

Triangle Corporation—Tool
Division
Cameron Road
Orangeburg, South Carolina 29115

Vulcan Tools
Division of TRW, Inc.
2300 Kenmore Avenue
Buffalo, New York 14207

J.H. Williams
Division of TRW, Inc.
400 Vulcan Street
Buffalo, New York 14207

Shop Supplies (Chemicals, Metallurgy Products, Seals, Sealers, Common Parts Items, etc.)

ABEX Corp.
Amsco Welding Products
Fulton Industrial Park
3-9610-14
P.O. Box 258
Wauseon, Ohio 43567
Hardfacing alloys.

Atlas Tool & Manufacturing Co.
7100 S. Grand Avenue
Saint Louis, Missouri 63111
Rotary mower blades.

Bendix Automotive Aftermarket
1094 Bendix Drive, Box 1632
Jackson, Tennessee 38301
Cleaning chemicals.

CR Industries
900 North State Street
Elgin, Illinois 60120
Shaft and bearing seals.

Clayton Manufacturing Company
415 North Temple City Blvd.
El Monte, California 91731
Steam cleaning compounds and solvents.

E-Z Lok
P.O. Box 2069,
240 E. Rosecrans Ave.
Gardena, California 90247
Thread repair insert kits for metal, wood and plastic.

Eutectic Welding Alloys Corp.
40-40 172nd Street
Flushing, New York 11358
Specialized repair and maintenance welding alloys.

Foley-Belsaw Company
Outdoor Power Equipment
Parts Division
6301 Equitable Road
P.O. Box 419593
Kansas City, Missouri 64141
Parts washers, cylinder head re-threaders, micrometers, calipers, gasket material, nylon rope, universal lawnmower blades, oil and grease products, bolt, nut, washer and spring assortments.

Frederick Manufacturing Co., Inc.
1400 C Agnes Street
Kansas City, Missouri 64127
Throttle controls and parts.

Heli-Coil Products Division
Heli-Coil Corporation
Shelter Rock Lane
Danbury, Connecticut 06810
Thread repair kits, thread inserts and installation tools.

King Cotton Cordage
617 Oradell Avenue
Oradell, New Jersey 07649
Starter rope, nylon rod.

Loctite Corporation
705 North Mountain Road
Newington, Connecticut 06111
Threading locking compounds, bearing mounting compounds, retaining compounds and sealants, instant and structural adhesives.

McCord Gasket Division
Ex-Cell-O Corporation
2850 West Grand Boulevard
Detroit, Michigan 48202
Gaskets, seals.

Microdot, Inc.
P.O. Box 3001
800 So. State College Blvd.
Fullerton, California 92631
Thread repair insert kits.

Permatex Industrial
705 N. Mountain Road
Newington, Connecticut 06111
Cleaning chemicals, gasket sealers, pipe
sealants, adhesives, lubricants.

Power Lawnmower Parts Inc.
1920 Lyell Avenue
P.O. Box 60860
Rochester, NY 14606-0860
Grinding compound paste, gas tank
sealer stick, shop aprons, rotary mower
blades, oil seals, gaskets, solder, throttle controls and parts, nylon starter
rope.

Radiator Specialty Co.
Box 34689
Charlotte, North Carolina 28234
Cleaning chemicals (Gunk), and solder
seal.

**Rexnord Specialty
Fastener Div.**
3000 W. Lomita Blvd.
Torrance, California 90506
Thread repair insert kits (Keenserts) and
installation tools.

**Union Carbide Corporation
Home & Auto Products Division**
Old Ridgebury Road
Danbury, CT 06817
Cleaning chemicals, engine starting
fluid.

BRIGGS & STRATTON

2-STROKE

BRIGGS & STRATTON CORPORATION
P.O. Box 702
Milwaukee, Wisconsin 53201

Model	Bore	Stroke	Displacement
62030, 62033	2.125 in. (53.975 mm)	1.70 in. (43.18 mm)	6.0 cu. in. (98.3 cc)
95700, 96700	2.362 in. (59.995 mm)	2.054 in. (52.172 mm)	9.0 cu. in. (147.5 cc)

BRIGGS & STRATTON NUMERICAL MODEL NUMBER SYSTEM

CUBIC INCH DISPLACEMENT	FIRST DIGIT AFTER DISPLACEMENT BASIC DESIGN SERIES	SECOND DIGIT AFTER DISPLACEMENT CRANKSHAFT, CARBURETOR GOVERNOR	THIRD DIGIT AFTER DISPLACEMENT BEARINGS, REDUCTION GEARS & AUXILIARY DRIVES	FOURTH DIGIT AFTER DISPLACEMENT TYPE OF STARTER
6	0	0 - Horizontal Diaphragm	0 - Plain Bearing	0 - Without Starter
8	1	1 - Horizontal Vacu-Jet	1 - Flange Mounting Plain Bearing	1 - Rope Starter
9	2	2 - Horizontal Pulsa-Jet	2 - Replaceable Bearing	2 - Rewind Starter
10	3			
11	4	3 - Horizontal Flo-Jet Pneumatic Governor	3 - Flange Mounting Ball Bearing	3 - Electric - 120 Volt, Gear Drive
13	5	4 - Horizontal Flo-Jet Mechanical Governor	4 -	4 - Electric Starter - Generator - 12 Volt, Belt Drive
17	6			
19	7	5 - Vertical Vacu-Jet	5 - Gear Reduction (6 to 1)	5 - Electric Starter Only- 12 Volt, Gear Drive
22	8	6 -	6 - Gear Reduction (6 to 1) Reverse Rotation	6 - Alternator Only
23	9			
24		7 - Vertical Flo-Jet	7 -	7 - Electric Starter - 6 or 12 Volt, Gear Drive, with Alternator
25				
28		8 -	8 - Auxiliary Drive Perpendicular to Crankshaft	8 - Vertical Pull Starter
29				
30				
32		9 - Vertical Pulsa-Jet	9 - Auxiliary Drive Parallel to Crankshaft	*Digit 6 formerly used for "Wind Up" Starter on 60000, 80000 and 92000 Series.
40				
42				

<u>9</u>	<u>5</u>	<u>7</u>	<u>2</u>	<u>2</u>
9 Cubic Inch	Design Series 5	Vertical Flo-Jet	Replaceable Bearing	Rewind Starter

Fig. BS1—Explanation of engine model numerical code used by Briggs & Stratton to identify engine and optional equipment.

ENGINE INFORMATION

All models are two-stroke, single-cylinder, third port loop scavenged design engines. Models can be equipped with chrome plated cylinder bore or a cast iron cylinder bore. Refer to Fig. BS1 for interpretation of Briggs & Stratton engine model numbers.

MAINTENANCE

SPARK PLUG. Recommended spark plug for Models 62030 and 62033 is a Champion RCJ-8 or equivalent. Recommended spark plug for Models 95700 and 96700 is a Champion J19LM or equivalent. Specified spark plug gap for all models is 0.030 inch (0.76 mm).

CARBURETOR. Models 62030 and 62033 are equipped with a diaphragm type carburetor with an integral fuel pump and Models 95700 and 96700 are equipped with a float type carburetor. Refer to the appropriate paragraphs for model being serviced.

Diaphragm Type Carburetor. Refer to Fig. BS3 for an external view of carburetor.

Initial adjustment of carburetor mixture screw (3—Fig. BS4) is 1-1/2 turns counterclockwise from a lightly seated position. Make final adjustment with engine at operating temperature and running. Turn mixture screw (3) clockwise (leaner) until engine runs smoothly, then turn mixture screw counterclockwise (richer) until engine just starts to sputter. This position will provide the best engine performance under load. Mixture screw should never be less than 1-1/4 turns open from a lightly seated position or improper engine lubrication will result.

To disassemble carburetor, remove the five diaphragm cover screws and lift off cover. Be careful not to lose the pump valve springs (4—Fig. BS5) when removing diaphragm. Refer to Fig. BS6 and remove inlet needle assembly. Using a small hook, remove the C-ring holding needle seat to carburetor body. Using a 1/16 inch (1.6 mm) punch or Allen wrench, pry out check valve and seat. With fuel mixture screw (3—Fig. BS4) removed, remove "O" ring seal from shoulder on fuel mixture screw with a small hook. Clean carburetor body and diaphragm cover in carburetor cleaner.

Inspect for worn or damaged parts and renew as needed. Check carburetor body for warpage with a straightedge. Check with straightedge (1—Fig. BS7) placed from corner-to-corner as shown. Renew if more than 0.003 inch (0.076 mm) warpage is evident.

To reassemble, install fuel inlet needle seat (2—Fig. BS8) with grooved side facing toward bottom of cavity. Insert "C" ring (1) to a depth of 5/16 inch (7.92 mm) in cavity. Place spring (7—Fig. BS6) in spring pocket machined in carburetor. Assemble hinge pin (3), inlet needle lever (2) and inlet needle (5). Make certain inlet needle retaining clip (4) holds inlet needle against needle lever. Install assembly in carburetor body making certain spring is centered on lever dimple (6), then install retaining screw (1). Spring lever height should be 1.16 inch (1.57 mm). Measure as shown in Fig. BS10. Carefully bend lever to adjust. Install check valve (2—Fig. BS11) in pocket. Install Welch plug (1) and use a ¼ inch round punch to bend and seat Welch plug. Assemble fuel pump spring and cup, check valve springs, diaphragm, gasket and cover. Tighten diaphragm cover screws firmly follow-

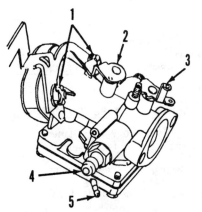

Fig. BS3—View of diaphragm type carburetor used on Models 62030 and 62033.

1. Mounting screws
2. Throttle lever
3. Choke lever
4. Fuel inlet
5. Primer inlet

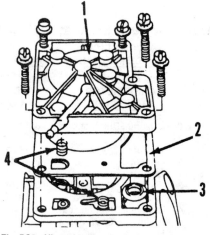

Fig. BS5—View of carburetor fuel pump cover (1), diaphragm (2), cup and spring (3) and valve springs (4).

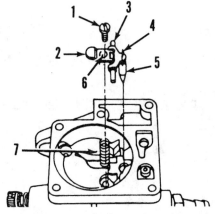

Fig. BS6—Internal view of diaphragm type carburetor.

1. Screw
2. Lever
3. Pin
4. Clip
5. Fuel inlet needle
6. Dimple
7. Spring

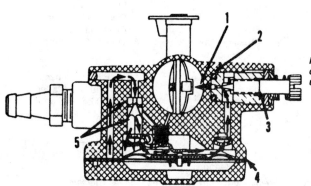

Fig. BS4—Cutaway view diaphragm type carburetor used on Models 62030 and 62033.

1. Metering hole
2. "O" ring
3. Mixture screw
4. Diaphragm
5. Inlet needle & seat

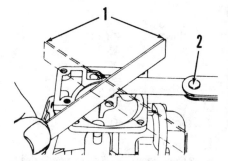

Fig. BS7—Check carburetor body for warpage with straightedge (1) and feeler gage (2). If warpage exceeds 0.003 inch (0.076 mm), renew carburetor.

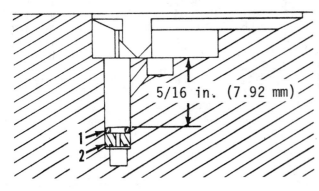

5/16 in. (7.92 mm)

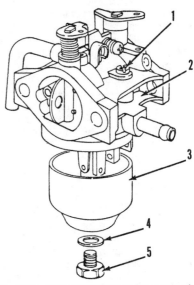

Fig. BS8—Refer to text for proper installation of "C" ring (1) and fuel inlet needle seat (2).

Fig. BS14—View of float type carburetor used on Models 95700 and 96700.

1. Idle jet
2. Carburetor body
3. Float bowl
4. Gasket
5. Screw

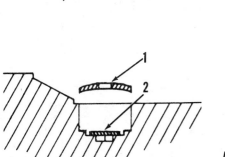

1.16 in. (1.57 mm)

Fig. BS10—Lever height should be 1.16 inch (1.57 mm). Measure as shown.

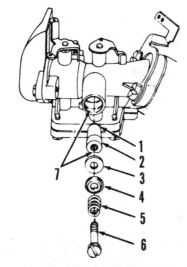

Fig. BS13—View of low speed mixture screw components on diaphragm type carburetor used on Models 62030 and 62033.

1. "O" ring
2. Seat
3. Sealing washer
4. Metal washer
5. Spring
6. Mixture screw
7. Flat area

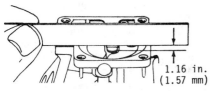

Fig. BS11—Install check valve (2) and Welch plug (1). Use a ¼ inch round punch to bend and seat Welch plug.

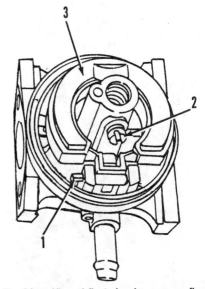

Fig. BS15—View of float chamber area on float type carburetor used on Models 95700 and 96700.

1. Squared side of float pin
2. Main jet
3. Float

ing sequence shown in Fig. BS12. Assemble throttle plate and throttle shaft, choke shaft and choke plate. Install "O" ring (1—Fig. BS13) on shoulder of seat (2). Install spring (5), metal washer (4) and sealing washer (3) on fuel mixture screw (6). Align flat area (7) of seat (2) with flat area of seat bore and use oil fill tube (part 280131) to firmly seat assembly in bore. Screw fuel

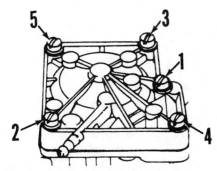

Fig. BS12—Firmly tighten diaphragm cover screws following the sequence shown.

mixture screw (6) in until lightly seated, then back out 1-1/2 turns.

Float Type Carburetor. The float type carburetor used on Models 95700 and 96700 is equipped with a fixed idle and main fuel jet. Standard fixed main jet size is #82.5. Optional #80 and #77.5 fixed main jets are available for high altitude operation. The only adjustment is engine idle rpm which is controlled by the throttle stop screw. Engine idle speed should be 1200 rpm with engine at operating temperature.

To disassemble carburetor, remove float bowl retaining screw (5—Fig. BS14), gasket (4) and float bowl (3). Float pin has one end which is flat on two sides (1—Fig. BS15) and one end that is round. Push pin from round end using a 5/64 inch (1.98 mm) punch to remove pin. Remove float, fuel inlet needle and float bowl gasket. Remove main jet (2) and idle jet (1—Fig. BS14).

Remove throttle plate, throttle shaft, choke plate, choke shaft and all shaft washers and seals.

To reassemble carburetor, reverse disassembly procedure. Float height of plastic float is nonadjustable.

GOVERNOR. Models 62030 and 62033 are equipped with an air-vane type governor and Models 95700 and 96700 are equipped with a centrifugal type governor. Refer to appropriate paragraph for service information.

301

Air-Vane Type Governor. Refer to Fig. BS16 for a view of the air-vane type governor used on Models 62030 and 62033. If adjustment of maximum no-load speed is necessary, tighten spring tension to increase speed and loosen spring tension to decrease speed. Adjust spring tension by bending the tang to which the governor spring is attached using tool 19229 as shown in Fig. BS17.

Centrifugal Type Governor. Refer to Fig. BS18 for a simplified view of the centrifugal type governor linkage used on Models 95700 and 96700. Maximum engine speed is limited by changing spring (4) and varying location of spring in holes located in speed lever (6). Three springs are available. Alternative springs provide maximum speed ranges of 2600-2900, 3000-3300 and 3400-3600 rpm depending upon spring selection and location.

IGNITION SYSTEM. Models 62030 and 62033 can be equipped with either a magneto type or a Magnetron ignition system. Models 95700 and 96700 are equipped with a magnetron ignition system. Refer to the appropriate paragraph for service information.

Magneto Type Ignition. Breaker-point gap on all models with magneto type ignition should be 0.020 inch (0.51 mm). Refer to Fig. BS19. One of the breaker contact points is an integral part of the ignition condenser. The breaker arm pivots on a knife edge retained by a slot in the pivot post. Breaker-point actuating plunger should be renewed if plunger is worn to a length of 0.931 inch (23.6 mm) or less. Specified air gap between armature and flywheel magnet is 0.006-0.010 inch (0.15-0.25 mm). The magneto type ignition uses a zinc flywheel key (1—Fig. BS20). Ignition coil can be tested using an ohmmeter. Resistance of ignition coil on the primary side should be 0.2-0.3 ohm. Resistance of ignition coil on secondary side should be 2500-3500 ohms.

Magnetron Ignition System. Some Models 62030 and 62033 and all Models 95700 and 96700 are equipped with the Magnetron ignition system. Installation of the correct flywheel key is required. Use the aluminum flywheel key (3—Fig. BS20) for Models 62030 and 62033 and aluminum key (2) for Models 95700 and 96700.

To check spark, remove spark plug. Connect spark plug cable to B&S tester 19051, then ground remaining tester lead to engine. Spin engine at 350 rpm or more. If spark jumps the 0.166 inch (4.2 mm) tester gap, system is functioning properly.

To remove armature and Magnetron module, remove flywheel shroud and armature retaining screws. Use a 3/16 inch (4.76 mm) diameter pin punch to release stop switch wire from module. To remove module, unsolder wires, push module retainer away from laminations and remove module.

Resolder wires for reinstallation and use Permatex or equivalent to hold ground wires in position. Ignition armature air gap for Models 62030 and 62033 should be 0.006-0.010 inch (0.15-0.25 mm). Ignition armature air gap for Models 95700 and 96700 should be 0.008-0.016 inch (0.20-0.40 mm).

CYLINDER HEAD AND COMBUSTION CHAMBER. Cylinder head on Models 62030 and 62033 is an integral part of one crankcase half and cylinder head on Models 95700 and 96700 is a separate casting from the crankcase halves. Cylinder head should be removed periodically and carbon and combustion deposits cleaned.

Specified compression for Models 62030 and 62033 without compression release is 90-110 psi (620-758 kPa). Specified compression for Models 62030 and 62033 with compression release mechanism is 80-90 psi (552-620 kPa). Specified compression for Models 95700 and 96700 is 90-120 psi (620-827 kPa).

LUBRICATION. Manufacturer recommends mixing a good quality BIA or NMMA oil certified for TC-W service with regular or low-lead gasoline at a 50:1 ratio.

FLYWHEEL BRAKE. Some 95700 and 96700 models are equipped with a band type flywheel brake which should stop the engine within three seconds, with remote control set at high speed

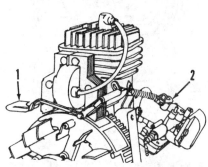

Fig. BS16—View showing air vane (1) and throttle lever (2) used on Models 62030 and 62033.

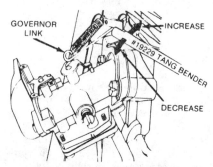

Fig. BS17—Use tool 19229 to adjust maximum engine speed on Models 62030 and 62033. Bend tang as needed to increase or decrease spring tension.

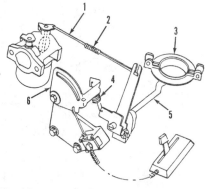

Fig. BS18—Simplified view of governor linkage used on Models 95700 and 96700.

1. Governor lever-to-carburetor rod
2. Spring
3. Flyweight assy.
4. Spring
5. Governor shaft
6. Speed lever

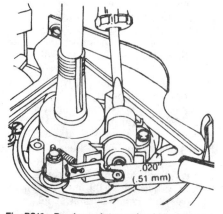

Fig. BS19—Breaker-point gap should be adjusted to 0.020 inch (0.51 mm) as shown.

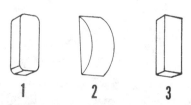

Fig. BS20—View showing the various keys used on different types of ignition systems. Use key style (1) for magneto type ignition, key style (2) for Models 95700 and 96700 and key style (3) for Models 62030 and 62033 with Magnetron ignition system.

position, when operator releases equipment safety control. Stopping time can be checked with tool 19255.

To adjust brake, first loosen both control bracket screws enough to allow movement of bracket. Place bayonet end of gage 19256 in control lever as shown in Fig. BS21. Push on control lever and install the other end of gage in cable clamp hole. Apply pressure to control bracket with a screwdriver and move it in direction shown by arrow until tension on gage is just eliminated. Hold control bracket in this position and tighten both screws to 25-30 in.-lbs. (2.82-3.39 N·m). As gage is removed a slight friction should be felt on gage and control lever should not move.

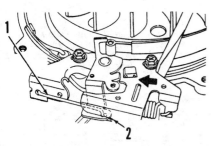

Fig. BS21—View showing flywheel brake adjustment on Models 95700 and 96700. Gage (1), part 19256, is inserted in hole (2). Refer to text.

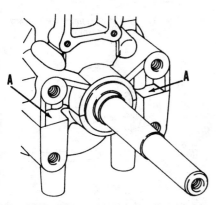

Fig. BS22—When separating crankcase halves, remove components as outlined in text, then tap locations (A) with a soft-faced mallet.

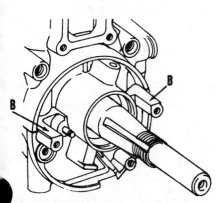

Fig. BS23—When separating crankcase halves, remove components as outlined in text, then tap locations (B) with a soft-faced mallet.

GENERAL MAINTENANCE. Check and tighten all loose bolts, nuts and clamps daily. Check for fuel leakage and repair if necessary. Clean cooling fins and external surfaces at 50 hour intervals.

REPAIRS

TIGHTENING TORQUES. Recommended tightening torque specifications are as follows:

Spark plug:
  All Models . . . . .170 in.-lbs.
    (19.2 N·m)
Flywheel nut:
  Models 95700 &
    96700 . . . . .30 ft.-lbs.
      (40.7 N·m)
Connecting rod:
  Models 62030 &
    62033 . . . . .55 in.-lbs.
      (6.2 N·m)
Muffler:
  Models 62030 &
    62033 . . . . .115 in.-lbs.
      (13 N·m)
  Models 95700 &
    96700 . . . . .85 in.-lbs.
      (9.6 N·m)
Crankcase:
  Models 62030 &
    62033 . . . . .90 in.-lbs.
      (10.2 N·m)
  Models 95700 &
    96700 . . . . .60 in.-lbs.
      (6.8 N·m)
Cylinder:
  Models 95700 &
    96700 . . . . .110 in.-lbs.
      (12.4 N·m)
Carburetor:
  Models 62030 &
    62033 . . . . .100 in.-lbs.
      (11.3 N·m)
  Models 95700 &
    96700 . . . . .50 in.-lbs.
      (5.6 N·m)

CYLINDER AND CRANKCASE. On Models 62030 and 62033, the cylinder is an integral part of one crankcase half. On Models 95700 and 96700 the cylinder is a separate casting and can be separated from the crankcase halves.

To split crankcase halves on Models 62030 and 62033, remove the back plate, flywheel and crankcase cover screws. Using a soft-faced mallet, tap crankcase halves apart at locations (A—Fig. BS22) and (B—Fig. BS23). Lift out crankshaft, piston and rod assembly as shown in Fig. BS24.

Clean carbon from ports and inspect cylinder bore for scoring or other damage. Standard cylinder bore diameter is 2.124-2.125 inches

(53.95-53.97 mm). If cylinder bore diameter is 2.128 inches (54.051 mm) or more, renew cylinder. Maximum allowable cylinder taper is 0.003 inch (0.076 mm) and maximum allowable out-of-round is 0.0025 inch (0.063 mm).

Refer to Fig. BS25 for a view of the compression release valve used on some models. Clean all carbon and foreign material from cavity and valve. When assembling valve on cylinder, seal and tighten the two screws to 30 in.-lbs. (3.4 N·m).

When reassembling crankcase halves, coat surfaces (A—Fig. BS26) with a silicone type sealer. Tighten all screws to specification listed under TIGHTENING TORQUES.

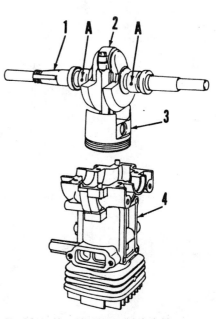

Fig. BS24—View showing crankshaft (1) connecting rod (2), piston (3), crankcase and cylinder (4) and main bearing locating tabs (A) used on Models 62030 and 62033.

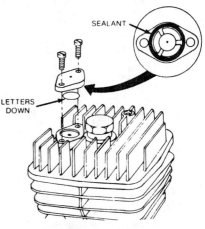

Fig. BS25—View showing compression release device used on cylinder head of some 62030 and 62033 models.

To split crankcase on Models 95700 and 96700, remove the four socket head screws retaining cylinder-to-crankcase. Using a soft-faced mallet, gently tap cylinder (Fig. BS27) to loosen cylinder. Carefully separate cylinder from crankcase. Remove the four cap screws retaining crankcase halves. Place flywheel side on a flat work surface and insert magneto end of crankshaft into flywheel taper. Refer to Fig. BS28 and insert a wooden hammer handle into crankcase to prevent crankshaft rotation. Place puller plate 19316, with "X" toward piston, and puller screw 19318 onto pto end of crankshaft. Thread puller studs 19317 into lower crankcase half. Turn puller screw until crankcase halves are separated, then withdraw puller assembly and lower crankcase. Insert puller studs 19317, with short threads down (Fig. BS29), into upper crankcase half. Protect crankshaft threads with flywheel nut. Place puller plate 19316, with "X" toward piston, onto crankshaft, then install four nuts and washers onto puller studs. Turn puller screw 19318 until upper crankcase is separated from crankshaft main bearing, then withdraw crankcase and puller assembly.

Remove oil seal from each crankcase half. Governor crank, on pto side, must be removed before ball bearing main bearings can be removed. Refer to Fig. BS30 for required tools and removal procedure for main bearing on pto side and to Fig. BS31 for required tools and removal procedure for main bearing on magneto side.

Remove snap rings retaining piston-to-connecting rod, then gently push piston pin from piston using tool shown in Fig. BS32. Separate piston from connecting rod. Remove piston rings from piston.

Connecting rod and crankshaft are considered an assembly and are not serviced separately.

Standard cylinder bore diameter is 2.362 inches (60 mm). Cylinder bore can be cast iron or chrome plated. If cylinder bore is chrome plated and cylinder diameter measures 2.368 inches (60.15 mm) or more, renew cylinder. If cylinder bore is cast iron and cylinder

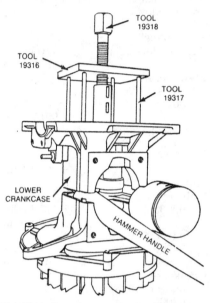

Fig. BS28—To separate crankcase halves, use puller assembly shown and insert a wooden hammer handle into crankcase to prevent crankshaft rotation. Refer to text.

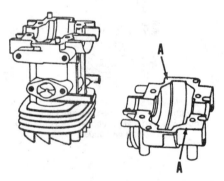

Fig. BS26—Apply sealer on surfaces (A) prior to reassembling crankcase.

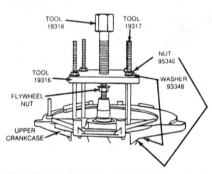

Fig. BS29—To remove crankshaft from crankcase, use puller assembly shown and refer to text.

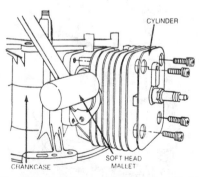

Fig. BS27—To remove cylinder head on Models 95700 and 96700, tap cylinder with a soft-faced mallet, then carefully lift cylinder off piston and connecting rod assembly.

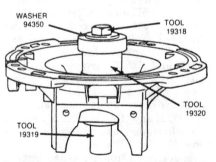

Fig. BS30—Main bearing on pto side must be pulled from crankcase half using tools assembled as shown. If main bearing is removed, it must be renewed.

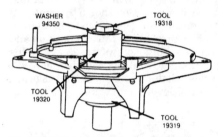

Fig. BS31—Main bearing on magneto side must be pulled from crankcase half using tools assembled as shown. If main bearing is removed, it must be renewed.

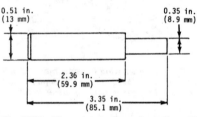

Fig. BS32—View shows dimensions for making piston pin removal tool.

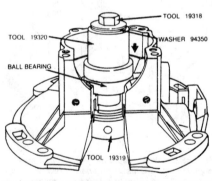

Fig. BS33—Assemble tools as shown to install ball type main bearing into pto side crankcase half on Models 95700 and 96700.

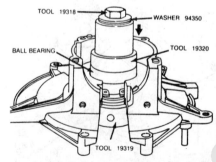

Fig. BS34—Assemble tools as shown to install ball type main bearing into magneto side crankcase half on Models 95700 and 96700.

diameter measures 2.369 inches (60.18 mm) or more, renew cylinder.

To install new main bearing into pto side crankcase half, assemble tools as shown in Fig. BS33. To install new main bearing into magneto side crankcase half, first install snap ring into bearing bore, then assemble tools as shown in Fig. BS34. Main bearing on magneto side seats against snap ring. To install new seals, refer to Fig. BS35 for magneto side seal and to Fig. BS36 for pto side seal. Pull seals into bores until they are seated.

To assemble magneto side crankcase half onto crankshaft, assemble tools as shown in Fig. BS37. Prevent crankshaft from turning, then turn nut to press crankcase half onto crankshaft until bearing seats. Remove tools and place crankshaft into flywheel taper on a suitable work bench. Install new gasket on crankshaft, slide governor assembly down onto governor journal with weights as shown in Fig. BS38. Install new governor oil seal into crankcase until seal is flush with chamfer (Fig. BS39). Slide governor shaft through governor shaft bearing and place washer on shaft. Install "E" ring into groove cut in governor shaft. Position governor shaft as shown in Fig. BS40 and slide pto half of

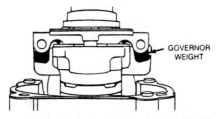

Fig. BS38—Governor weights must be held in position shown during crankcase assembly.

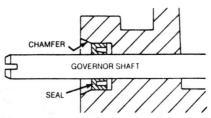

Fig. BS39—Governor shaft oil seal must be pressed into seal bore until outer seal edge is flush with chamfered edge of bore.

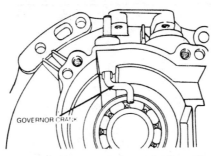

Fig. BS40—Governor crank must be in position shown during crankcase assembly.

crankcase down onto crankshaft while holding governor crank in position. Assemble tools as shown in Fig. BS41. Prevent crankshaft from turning, then turn blade adapter bolt to pull crankcase onto crankshaft. Puller stud 19317 can be installed into crankcase halves to guide them together.

Refer to Fig. BS42 and install new rings on piston. Align ring end gaps with locator pins in ring grooves. Refer to following PISTON AND CONNECTING ROD paragraphs to assemble piston onto connecting rod. Install new gasket on cylinder. Make certain ring end gaps are aligned with locator pins, then slide cylinder and head assembly down over piston. Beveled portion on lower side of cylinder acts as a ring compressor. Install cylinder retaining bolts and tighten to specification listed under TIGHTENING TORQUES.

PISTON AND CONNECTING ROD. Refer to BS43 for a view of piston and

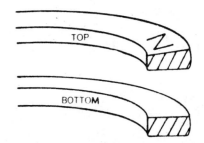

Fig. BS42—When installing piston rings on Models 95700 and 96700, install beveled ring into top ring groove of piston.

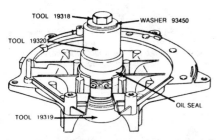

Fig. BS35—Assemble tools as shown to install seal into magneto side crankcase half on Models 95700 and 96700.

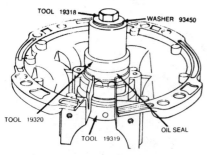

Fig. BS36—Assemble tools as shown to install seal into pto side crankcase half on Models 95700 and 96700.

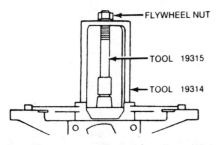

Fig. BS37—Assemble tools as shown to install magneto side crankcase half onto crankshaft. Make certain crankshaft does not turn during assembly procedure.

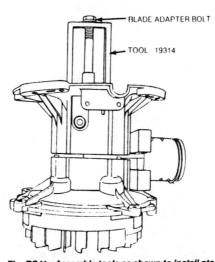

Fig. BS41—Assemble tools as shown to install pto side crankcase half onto crankshaft and magneto side crankcase half assembly.

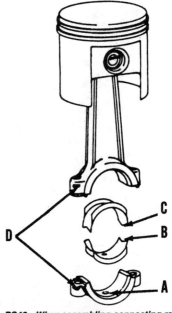

Fig. BS43—When assembling connecting rod on Model 62030 and 62033, align notch (B) and vee (C) in bearing liners. Match marks (D) on connecting rod cap and connecting rod must be aligned during assembly.

rod assembly used on Models 62030 and 62033. Piston pin floats in both piston pin bore and rod small end. Retaining rings are used to hold piston pin in place. Inspect piston for scuffing or excessive wear in ring lands and renew as needed. Connecting rod rides on needle bearings at crankshaft end. New needle bearings are retained with a wax coating on one side. Old needle bearings can be held in position with heavy grease. When assembling connecting rod, align vee (C) and notch (B) in connecting rod liners and match marks (D) on connecting rod and connecting rod cap. Tighten connecting rod bolts to specification listed under TIGHTENING TORQUES.

Refer to Fig. BS44 for a view of piston and rod assembly used on Models 95700 and 96700. Piston pin floats in piston pin bore and needle bearing which is installed in connecting rod small end. Retaining rings are used to hold piston pin in place. Open side of retaining rings should be located at the top or bottom of piston pin bore after installation. During assembly, arrow on top of piston crown must point toward exhaust port side of engine.

CRANKSHAFT. On Models 62030 and 62033, the crankshaft is supported at each end by encased needle bearings which are held in position by bearing locator tabs (Fig. BS45). Each bearing locator tab fits into a notch machined into respective crankcase half. Crankshaft end play should be

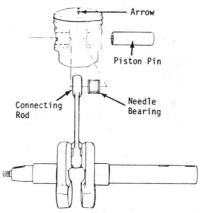

Fig. BS44—View showing connecting rod, piston and crankshaft assembly used on Models 95700 and 96700. Refer to text.

0.002-0.013 in. (0.05-0.33 mm). If crankshaft main bearing journals measure 0.7515 inch (19.088 mm) or less, renew crankshaft. If crankshaft crankpin journal measures 0.7420 inch (18.846 mm) or less, renew crankshaft.

On Models 95700 and 96700, the crankshaft is supported at each end by ball bearing type main bearings. Crankshaft and connecting rod are considered an assembly and are not serviced separately. If crankshaft main bearing journal diameters measure 0.982 inch (24.96 mm) or less, or if journals are out-of-round 0.0005 inch (0.0127 mm) or more, renew crankshaft.

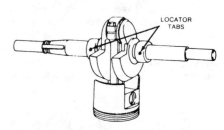

Fig. BS45—Locator tabs on encased needle bearings used on Models 62030 and 62033 must be aligned with tab notches machined into crankcase halves during assembly.

BRIGGS & STRATTON

4-STROKE MODELS

Model	No. Cyls.	Bore	Stroke	Displacement	Power Rating
170000, 171000	1	3 in. (76.2 mm)	2.375 in. (60.3 mm)	16.8 cu. in. 275.1 cc)	7 hp. (5.2 kW)
190000, 191000	1	3 in. (76.2 mm)	2.75 in. (69.85 mm)	19.44 cu. in. (318.5 cc)	8 hp. (6 kW)
220000	1	3.438 in. (87.3 mm)	2.375 in. (60.3 mm)	22.04 cu. in. (361.1 cc)	10 hp. (7.5 kW)
250000, 251000, 252000, 253000	1	3.438 in. (87.3 mm)	2.625 in. (66.68 mm)	24.36 cu. in. (399.1 cc)	11 hp. (8.2 kW)

Engines in this section are four-stroke, one-cylinder engines and both vertical and horizontal crankshaft models are covered. Crankshaft is supported in main bearings which are an integral part of crankcase and cover or ball bearings which are a press fit on crankshaft. Cylinder block and bore are a single aluminum casting.

Connecting rod in all models rides directly on crankpin journal. Horizontal crankshaft models are splash lubricated by an oil dipper attached to connecting rod cap and vertical crankshaft models are lubricated by an oil slinger wheel located on governor gear.

Early models use a magneto type ignition system with points and condenser located underneath flywheel. Late models use a breakerless (Magnetron) ignition system.

A float type carburetor is used on all models and a fuel pump is available as optional equipment for some models.

Refer to page 139 for engine identification and always give engine model and serial number when ordering parts or service material.

MAINTENANCE

SPARK PLUG. Recommended spark plug is Champion J8 or equivalent. To decrease radio interference use Champion XJ8 or equivalent spark plug. Electrode gap for all models is 0.030 inch (0.762 mm).

CARBURETOR. One of two different Flo-Jet carburetors will be used. Carburetor may be identified as Flo-Jet I (Fig. B2) or Flo-Jet II (Fig. B3). Refer to appropriate figure for model being serviced.

FLO-JET I CARBURETOR. Initial adjustment procedure varies according to horsepower rating and crankshaft type.

For initial carburetor adjustment for 7, 8 and 10 horsepower (5.2, 6 and 7.5 kW) horizontal crankshaft models, open idle mixture screw 1½ turns and main fuel mixture screw 2½ turns.

For initial carburetor adjustment for 11 horsepower (8.2 kW) horizontal crankshaft models, open idle mixture screw 1 turn and main fuel mixture screw 1½ turns.

For initial carburetor adjustment for all vertical crankshaft models, open idle mixture screw 1¼ turns and main fuel mixture screw 1½ turns.

Make final adjustment with engine at normal operating temperature and running. Place engine under load and adjust main fuel mixture screw for leanest mixture that will allow satisfactory acceleration and steady governor operation. Set engine at idle speed, no load and adjust idle mixture screw to obtain smoothest idle operation.

As each adjustment affects the other, adjustment procedure may have to be repeated.

FLO-JET II CARBURETOR. For initial adjustment open idle mixture screw 1 turn and main fuel mixture screw 1½ turns. Make final adjustment with engine at normal operating temperature and running. Place engine under load and adjust main fuel mixture screw for leanest mixture that will allow satisfactory acceleration and steady governor operation. Set engine at idle speed, no load and adjust idle mixture screw to obtain smoothest idle operation.

As each adjustment affects the other, adjustment procedure may have to be repeated.

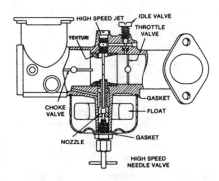

Fig. B2—Cross-sectional view of Flo-Jet I carburetor.

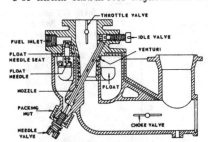

Fig. B3—Cross-sectional view of Flo-Jet II carburetor. Before separating upper and lower body sections, loosen packing nut and power needle valve as a unit and use special screwdriver (tool number 19062) to remove nozzle.

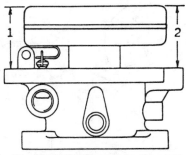

Fig. B4—Dimension (2) must be the same as dimension (1) plus or minus 1/32-inch (0.79 mm).

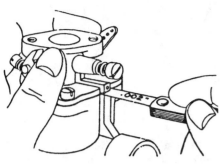

Fig. B5 — Check upper body of Flo-Jet II carburetor for warpage as outlined in text.

To check float level for all models, invert carburetor body and float assembly. Refer to Fig. B4 for proper float level dimensions. Adjust by bending float lever tang that contacts inlet valve.

Check Flo-Jet II carburetor for upper body warpage using an 0.002 inch (0.0508 mm) feeler gage as shown in Fig. B5. If upper body is warped more than 0.002 inch (0.0508 mm) it must be renewed.

CHOKE-A-MATIC CARBURETOR CONTROLS. Engines may be equipped with a control unit with which carburetor choke, throttle and magneto grounding switch are operated from a single lever (Choke-A-Matic carburetors).

To check operation of Choke-A-Matic controls, move control lever to "CHOKE" position; carburetor choke slide or plate must be completely closed. Move control lever to "STOP" position; magneto grounding switch should be making contact. With control lever in "RUN", "FAST" or "SLOW" position, carburetor choke should be completely open. On units with remote controls, synchronize movement of remote lever to carburetor control lever by loosening screw (C – Fig. B6) and moving control wire housing (D) as required. Tighten screw to clamp housing securely. Refer to Fig. B7 to check remote control wire movement.

AUTOMATIC CHOKE (THERMOSTAT TYPE). A thermostat operated choke is used on some models equipped with Flo-Jet II carburetor. To adjust choke linkage, hold choke shaft so thermostat lever is free. At room temperature, stop screw in thermostat collar should be located midway between thermostat stops. If not, loosen stop screw, adjust collar and tighten stop screw. Loosen set screw (S – Fig. B8) on thermostat lever. Slide lever on shaft to insure free movement of choke unit. Turn thermostat shaft clockwise until stop screw contacts thermostat stop. While holding shaft in this position, move shaft lever until choke is open exactly 1/8-inch (3 mm) and tighten lever set screw. Turn thermostat shaft counterclockwise until stop screw contacts thermostat stop as shown in Fig. B9. Manually open choke valve until it stops against top of choke link opening. At this time choke should be open at least 3/32-inch (2.38 mm), but not more than 5/32-inch (4 mm). Hold choke valve in wide open position and check position of counterweight lever. Lever should be in a horizontal position with free end towards right.

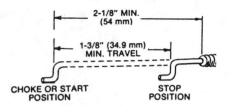

Fig. B7 — For proper operation of Choke-A-Matic controls, remote control wire must extend to dimension shown and have a minimum travel of 1⅜ inches (34.9 mm).

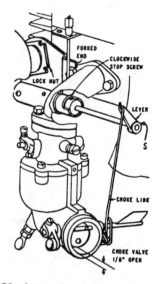

Fig. B8 — Automatic choke used on some models equipped with Flo-Jet II carburetor showing unit in "HOT" position.

FUEL TANK OUTLET. Some models are equipped with a fuel tank outlet as shown in Fig. B10. On other engines, a fuel sediment bowl is incorporated with fuel tank outlet as shown in Fig. B11.

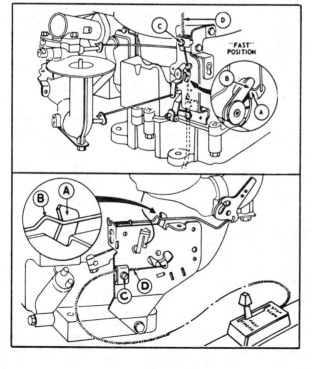

Fig. B6 — On Choke-A-Matic controls shown, choke actuating lever (A) should just contact choke link or shaft (B) when control is at "FAST" position. If not, loosen screw (C) and move control wire housing (D) as required.

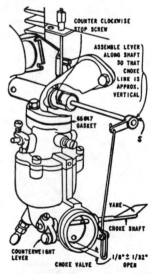

Fig. B9 — Turn thermostat shaft counterclockwise until stop screw contacts thermostat stop as shown.

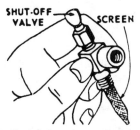

Fig. B10—Fuel tank outlet used on some B&S models.

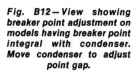

Fig. B12—View showing breaker point adjustment on models having breaker point integral with condenser. Move condenser to adjust point gap.

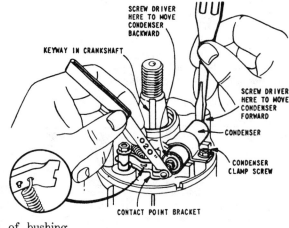

Clean any lint and dirt from tank outlet screens with a brush. Varnish or other gasoline deposits may be removed by use of a suitable solvent. Tighten packing nut or remove nut and shut-off valve, then renew packing if leakage occurs around shut-off valve stem.

FUEL PUMP. A fuel pump is available as optional equipment on some models. Refer to **SERVICING BRIGGS & STRATTON ACCESSORIES** section for service information.

GOVERNOR. Engines are equipped with a gear driven mechanical governor and governor unit is enclosed within engine crankcase and is driven from camshaft gear. Lubrication oil slinger is an integral part of governor unit on all vertical crankshaft engines. All binding or slack due to wear must be removed from governor linkage to prevent "hunting" or unsteady operation. To adjust carburetor to governor linkage, loosen clamp bolt on governor lever. Move link end of governor lever until carburetor throttle shaft is in wide open position. Using a screwdriver, rotate governor lever shaft clockwise as far as possible and tighten clamp bolt.

Governor gear and weight unit can be removed when engine is disassembled. Refer to exploded views of engines in Fig. B21, B22, B23 and B28. Remove governor lever, cotter pin and washer from outer end of governor lever shaft.

Slide governor lever out of bushing towards inside of engine. Governor gear and weight unit can now be removed. Renew governor lever shaft bushing in crankcase, if necessary and ream new bushing after installation to 0.2385-0.239 inch (6.05-6.07 mm).

IGNITION SYSTEM. Early models use a magneto system which incorporates a breaker point set and condenser. Late models use Magnetron breakerless ignition system. Refer to appropriate paragraph for model being serviced.

MAGNETO IGNITION. All models use breaker points and condenser located under flywheel.

One of two different types of ignition points as shown in Fig. B12 and B13 are used. Breaker point gap is 0.020 inch (0.508 mm) for all models with magneto ignition.

On each type, breaker contact arm is actuated by a plunger in a bore in engine crankcase which rides against a cam on engine crankshaft. Plunger can be removed after removing breaker points. Renew plunger if worn to a length of 0.870 inch (22.098 mm) or less. If breaker point plunger bore in crankcase is worn, oil will leak past plunger. Check bore with B&S gage, tool number 19055. If plug gage will enter bore ¼-inch (6.35

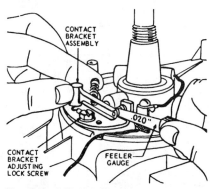

Fig. B13—Adjusting breaker point gap on models having breaker points separate from condenser.

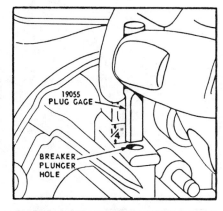

Fig. B14—If B&S gage number 19055 can be inserted in plunger bore ¼-inch (6.35 mm) or more, bore is worn and must be rebushed.

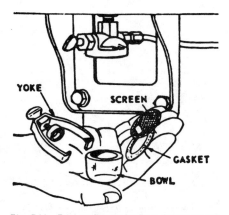

Fig. B11—Fuel sediment bowl and tank outlet used on some models.

Fig. B15—Views showing reaming plunger bore to accept bushing (left view), installing bushing (center) and finish reaming bore of bushing (right).

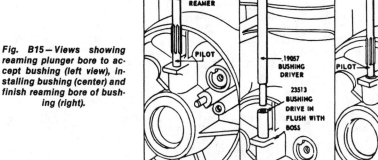

mm) or more, bore should be reamed and a bushing installed. Refer to Fig. B14. To ream bore and install bushing it will be necessary to remove breaker points, armature, ignition coil and crankshaft. Refer to Fig. B15 for steps in reaming bore and installation of bushing.

Plunger must be reinstalled with groove toward top (Fig. B15A) to prevent oil contamination in breaker point box.

For reassembly set armature to flywheel air gap at 0.010-0.014 inch (0.254-0.356 mm) for two-leg armature or 0.016-0.019 inch (0.406-0.483 mm) for three-leg armature. Ignition timing is non-adjustable on these models.

MAGNETRON IGNITION. Magnetron ignition is a self-contained breakerless ignition system and flywheel does not need to be removed except to check or service keyways or crankshaft key.

To check spark, remove spark plug and connect spark plug cable to B&S tester, part number 19051 and ground remaining tester lead to engine cylinder head. Spin engine at 350 rpm or more. If spark jumps the 0.166 inch (4.2 mm) tester gap, system is functioning properly.

To remove armature and Magnetron module, remove flywheel shroud and armature retaining screws. Use a 3/16-inch (4.76 mm) diameter pin punch to release stop switch wire from module. To remove module, unsolder wires, push module retainer away from laminations and push module off. See Fig. B15B.

Resolder wires for reinstallation and use Permatex or equivalent to hold ground wires in position.

Adjust armature air gap to 0.010-0.014 inch (0.25-0.36 mm).

LUBRICATION. Horizontal crankshaft engines have a splash lubrication system provided by an oil dipper attached to connecting rod. Refer to Fig. B16 for view of various types of dippers used.

Vertical crankshaft engines are lubricated by an oil slinger wheel on governor gear which is driven by camshaft gear. See Fig. B16A.

Oils approved by manufacturer must meet requirements of API service classification SC, SD, SE or SF.

Use SAE 10W-40 oil for temperatures above 20°F (-7°C) and SAE 5W-30 oil for temperatures below 20°F (-7°C).

Check oil level at five hour intervals and maintain at bottom edge of filler plug. **DO NOT** overfill.

Recommended oil change interval for all models is every 25 hours of normal operation.

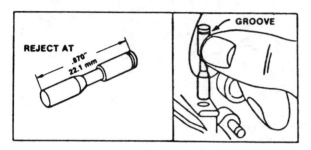

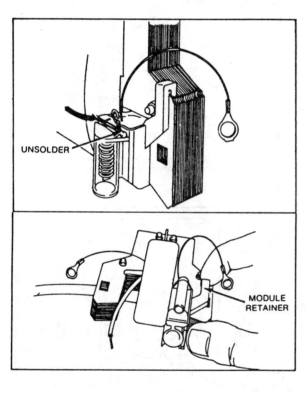

Fig. B15A—Insert plunger into bore with groove toward top.

Fig. B15B—Wires must be unsoldered to remove Magnetron module.

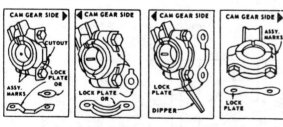

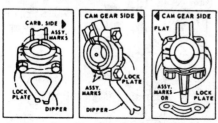

Fig. B16—Install connecting rod in engine as indicated according to type used. Note dipper installation on horizontal crankshaft engine connecting rod.

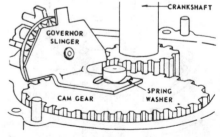

Fig. B16A—Vertical crankshaft engines are lubricated by an oil slinger mounted on governor gear assembly.

Crankcase oil capacity for 16.8 and 19.44 cubic inch (275.1 and 318.5 cc) engines is 2¼ pints (1.1 L) for vertical crankshaft models and 2¾ pints (1.3 L) for horizontal crankshaft models.

Crankcase oil capacity for 22.04 and 24.36 cubic inch (361.1 and 399.1 cc) engines is 3 pints (1.4 L) for both vertical and horizontal crankshaft models.

CRANKCASE BREATHER. A crankcase breather is built into engine valve cover. A partial vacuum must be maintained in crankcase to prevent oil from being forced out past oil seals and gaskets or past breaker point plunger or piston rings. Air can flow out of crankcase through breather, but a one-way valve blocks return flow, maintaining necessary vacuum. Breather mounting holes are offset one way. A vent tube connects breather to carburetor air horn for extra protection against dusty conditions.

REPAIRS

CYLINDER HEAD. When removing cylinder head note location from which different length cap screws are removed as they must be reinstalled in their original positions.

Always use a new gasket when reinstalling cylinder head. Do not use sealer on gasket. Lubricate cylinder head bolt threads with graphite grease, install in correct locations and tighten in several even steps in sequence shown in Fig. B17 until 165 in.-lbs. (19 N·m) torque is obtained. Start and run engine until normal operating temperature is reached, stop engine and retighten head bolts as outlined.

It is recommended carbon and lead deposits be removed at 100 to 300 hour intervals, or whenever cylinder head is removed.

CONNECTING ROD. Connecting rod and piston are removed from cylinder head end of block as an assembly. Aluminum alloy connecting rod rides directly on induction hardened crankshaft crankpin journal. Rod should be rejected if crankpin bore is scored or out-of-round more than 0.0007 inch (0.0178 mm) or if piston pin bore is scored or out-of-round more than 0.0005 inch (0.0127 mm). Renew connecting rod if either crankpin journal or piston pin bore is worn to, or larger than, sizes given in chart.

REJECT SIZES FOR CONNECTING ROD

Model	Crankpin bore	Pin bore*
170000, 171000	1.0949 in. (27.81 mm)	0.674 in. (17.12 mm)
190000, 191000	1.1265 in. (28.61 mm)	0.674 in. (17.12 mm)
All other models	1.252 in. (31.8 mm)	0.802 in. (20.37 mm)

*Piston pins of 0.005 inch (0.127 mm) oversize are available for service. Piston pin bore in rod can be reamed to this size of crankpin bore is within specifications.

Refer to Fig. B16, locate style rod used for model being serviced and install in engine as indicated.

Tighten connecting rod cap screws to 165 in.-lbs. (19 N·m) torque on 170000, 171000, 190000 and 191000 models and to 190 in.-lbs. (22 N·m) torque on all other models.

PISTON, PIN AND RINGS. Chrome plated aluminum piston used in aluminum bore (Kool-Bore) engines should be renewed if top ring side clearance in groove exceeds 0.007 inch (0.18 mm) or if piston is visibly scored or damaged. Piston should also be renewed, or pin bore reamed for 0.005 inch (0.127 mm) oversize pin, if pin bore is 0.673 inch (17.09 mm) or larger for 170000, 171000, 190000 and 191000 models or 0.801 inch (20.32 mm) or larger for all other models.

Renew piston pin if pin is 0.0005 inch (0.0127 mm) or more out-of-round or if pin is worn to a diameter of 0.671 inch (17.04 mm) or smaller for 170000, 171000, 190000 or 191000 models or 0.799 inch (20.29 mm) or smaller for all other models.

Ring end gap should be 0.010-0.025 inch (0.254-0.635 mm) and compression ring should be renewed if end gap is greater than 0.035 inch (0.89 mm) and oil control ring should be renewed if end gap is greater than 0.045 inch (1.14 mm).

Pistons used in 220000 and 250000 engines have a notch and a letter "F" stamped in piston (Fig. B17A) which must face flywheel side of engine after installation.

Pistons and rings are available in a variety of oversizes as well as standard.

A chrome ring set is available for slightly worn standard bore cylinders. Refer to note in **CYLINDER** section.

CYLINDER. Standard cylinder bore diameter is 2.999-3.000 inches (76.175-76.230 mm) for 170000, 171000, 190000 and 191000 models and 3.4365-3.4375 inches (87.287-87.313 mm) for all other models.

If cylinder is worn 0.003 inch (0.076 mm) or more, or out-of-round 0.0025 inch (0.0635 mm) or more, it should be

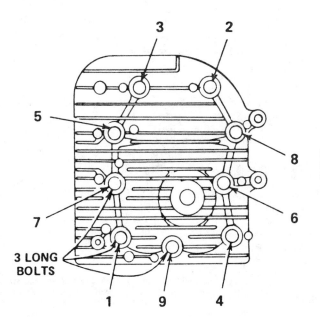

3 LONG BOLTS

Fig. B17—Tighten cylinder head bolts in sequence shown. Note location of three long bolts.

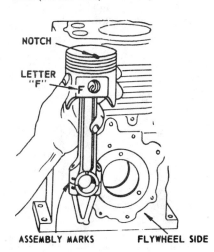

Fig. B17A—Notch and letter "F" stamped in piston of 220000 and 250000 engines must face flywheel side of engine after installation.

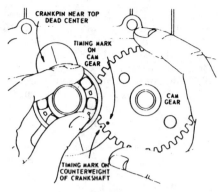

Fig. B18—Align timing mark on cam gear with mark on crankshaft counterweight on ball bearing equipped models.

resized to nearest oversize for which piston and rings are available.

NOTE: A chrome piston ring set is available for slightly worn standard bore cylinders. No honing or cylinder deglazing is required for these rings. Cylinder bore can be a maximum of 0.005 inch (0.127 mm) oversize when using chrome rings.

It is recommended a hone be used for resizing cylinders. Operate hone at 300-700 rpm with an up and down movement which will produce a 45° angle cross-hatch pattern. Clean cylinder after honing with oil or soap and water.

Approved hone for aluminum bore is Ammco No. 3956 for rough and finishing or Sunnen AN200 for rough and Sunnen AN500 for finishing.

CRANKSHAFT AND MAIN BEAR-INGS. Crankshaft may be supported at each end in main bearings which are an integral part of crankcase, cover or sump or ball bearing mains which are a press fit on crankshaft and fit into machined bores in crankcase, cover or sump.

Crankshaft for models with main bearings as an integral part of crank-

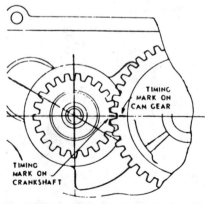

Fig. B19—Align timing marks on cam gear and crankshaft gear on plain bearing models.

case, cover or sump, should be renewed or reground if main bearing journals are worn to, or beyond crankshaft rejection specifications as listed in following chart.

CRANKSHAFT REJECTION SIZES

Model	Magneto end journal	Drive end journal
170000, 190000	0.9975 in. (25.34 mm)	1.1790 in. (29.95 mm)
171000 Synchro-balanced 191000 Synchro-balanced	1.1790 in. (29.95 mm)	1.1790 in. (29.95 mm)
All other models	1.376 in. (34.95 mm)	1.376 in. (34.95 mm)

Crankshaft for models with ball bearing main bearings should be renewed if new bearings are loose on journals. Bearings should be a press fit.

Crankshaft for all models should be renewed or reground if connecting rod crankpin journal diameter is worn to, or beyond crankshaft rejection specifications as listed in following chart.

CRANKSHAFT REJECTION SIZES

Model	Crankpin journal
170000, 171000 Synchro-balanced	1.090 in. (27.69 mm)
190000, 191000 Synchro-balanced	1.122 in. (28.50 mm)
All other models	1.247 in. (31.67 mm)

Connecting rod is available for crankshaft which has had connecting rod crankpin journal reground to 0.020 inch (0.508 mm) undersize.

Service main bearing bushings are available for 170000, 171000, 190000 and 191000 models with main bearings cast as an integral part of crankcase, cover or sump if main bearing bores are worn to, or beyond rejection sizes as listed in following chart. All other models with integral type main bearings must have crankcase, cover or sump renewed if they are worn to, or beyond rejection sizes listed in chart.

MAIN BEARING REJECT SIZES

Model	Magneto end bearing	Drive end bearing
170000, 171000, 190000, 191000	1.0036 in. (25.491 mm)	1.185 in. (30.099 mm)
Synchro-balanced models	1.185 in. (30.099 mm)	1.185 in. (30.099 mm)
All other models	1.383 in. (35.128 mm)	1.383 in. (35.128 mm)

To install service main bushings it is necessary to ream main bearing bores to correct size. Main bearing tool kit, part number 19184, is available and contains all necessary reamers, guides, drivers and supports. Make certain all metal cuttings are removed before pressing bushings into place and align oil notches or holes as bushings are installed.

Ball bearing mains are a press fit on crankshaft and must be removed by pressing crankshaft out of bearing. Renew ball bearing if worn or rough. Expand new bearing by heating in oil and install on crankshaft with shield side towards crankpin journal.

Crankshaft end play should be 0.002-0.008 inch (0.0508-0.2032 mm). At least one 0.015 inch thick cover or sump gasket must be used. Additional cover gaskets in a variety of thicknesses are available if end play is greater than 0.008 inch (0.2032 mm) metal shims are available for use on crankshaft.

Place metal shims between crankshaft gear and cover or sump on models with integral type main bearings or between magneto end of crankshaft and crankcase on ball bearing equipped models.

When reinstalling crankshaft, make certain timing marks are aligned (Fig. B18 or B19) and if equipped with

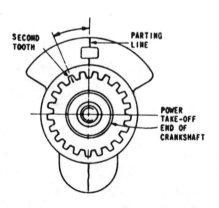

Fig. B20—Location of tooth to align with timing mark on cam gear if mark is not visible on crankshaft gear.

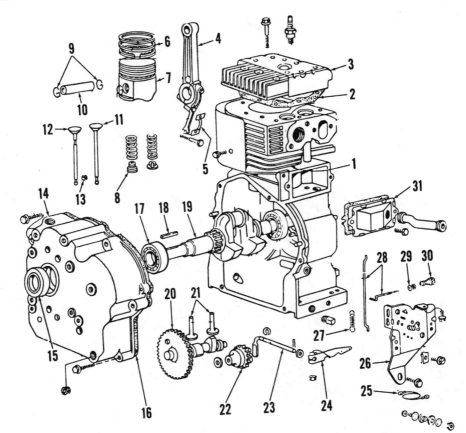

1. Cylinder block
2. Head gasket
3. Cylinder head
4. Connecting rod
5. Rod bolt lock
6. Rings
7. Piston
8. Rotocoil (exhaust valve)
9. Retainer clips
10. Piston pin
11. Intake valve
12. Exhaust valve
13. Retainers
14. Crankcase cover
15. Oil seal
16. Crankcase gasket
17. Main bearings
18. Key
19. Crankshaft
20. Camshaft
21. Tappet
22. Governor gear
23. Governor crank
24. Governor lever
25. Ground wire
26. Governor control plate
27. Spring
28. Governor rod
29. Spring
30. Nut

Fig. B21 — Exploded view of 220000 engine assembly.

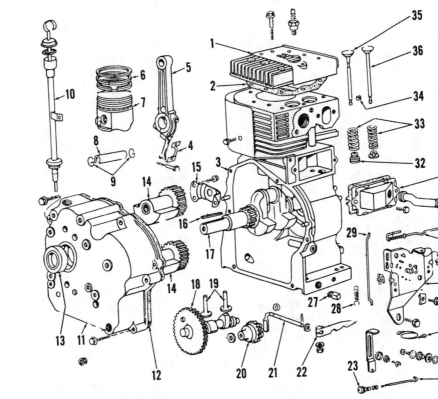

1. Cylinder head
2. Head gasket
3. Cylinder block
4. Rod bolt lock
5. Connecting rod
6. Rings
7. Piston
8. Piston pin
9. Retainer clip
10. Dipstick assy.
11. Crankcase cover
12. Crankcase gasket
13. Oil seal
14. Counterweight & bearing assy.
15. Retainer
16. Key
17. Crankshaft
18. Camshaft assy.
19. Tappet
20. Governor gear
21. Governor crank
22. Governor lever
23. Governor nut & spring
24. Governor control rod
25. Ground wire
26. Governor control plate
27. Drain plug
28. Spring
29. Governor link
30. Choke link
31. Breather assy.
32. Rotocoil (exhaust valve)
33. Valve springs
34. Retainer
35. Exhaust valve
36. Intake valve

Fig. B22 — Exploded view of 170000, 190000, 251000, 252000 or 253000 engine assembly.

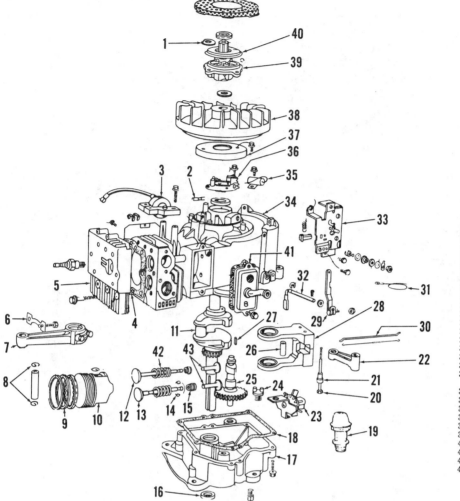

1. Thrust washer
2. Breaker point plunger
3. Armature assy.
4. Head gasket
5. Cylinder head
6. Rod screw lock
7. Connecting rod
8. Piston pin & retaining clips
9. Rings
10. Piston
11. Crankshaft
12. Intake valve
13. Exhaust valve
14. Retainer
15. Rotocoil (exhaust valve)
16. Oil seal
17. Oil sump (engine base)
18. Crankcase gasket
19. Oil minder
20. Cap screw (2 used)
21. Spacer (2 used)
22. Link
23. Governor & oil slinger
24. Plug
25. Cam gear assy.
26. Dowel pin (2 used)
27. Key
28. Counterweight assy.
29. Governor lever
30. Governor link
31. Ground wire
32. Governor crank
33. Choke-A-Matic control
34. Cylinder assy.
35. Condenser
36. Breaker points
37. Cover
38. Flywheel assy.
39. Clutch housing
40. Rewind starter clutch
41. Breather assy.
42. Valve spring
43. Tappets

Fig. B23—Exploded view of 171000, 191000, 251000 or 252000 Synchro-Balanced engine assembly.

counter balance weights they must be positioned properly. See Fig. B27B.

CAMSHAFT. Camshaft gear is an integral part of camshaft which is supported at each end in bearing bores machined into crankcase, cover or sump.

Camshaft should be renewed if either journal is worn to a diameter of 0.498 inch (12.66 mm) or less or if cam lobes are worn or damaged.

Crankcase, cover or sump must be renewed if bearing bores are worn to 0.506 inch (12.852 mm) or larger or if tool number 19164 enters bearing bore ¼-inch (6.35 mm) or more.

Compression release mechanism on camshaft gear holds exhaust valve slightly open at very low engine rpm as a starting aid. Mechanism should work freely and spring holds actuator cam against pin.

When installing camshaft in engines with integral type main bearings, align timing mark on camshaft gear with timing mark on crankshaft gear (Fig. B19).

Fig. B24—View showing operating principle of Synchro-Balancer used on some vertical crankshaft engines. Counterweight oscillates in opposite direction of piston.

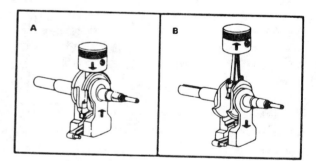

Fig. B25—Exploded view of Synchro-Balancer assembly. Counterweights ride on eccentric journals on crankshaft.

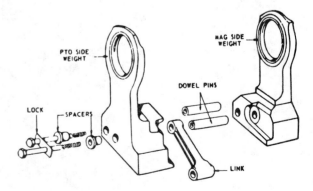

When installing camshaft in engines with ball bearing mains, align timing mark on camshaft gear with timing mark on crankshaft counterweight (Fig. B18).

If timing mark is not visible on crankshaft gear, align camshaft gear

timing mark with second tooth to the left of crankshaft counterweight parting line (Fig. B20).

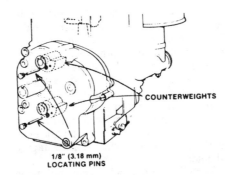

1/8" (3.18 mm)
LOCATING PINS

Fig. B27B—To properly align counterweights, remove two small screws from crankcase cover and insert ⅛-inch (3.18 mm) diameter locating pins.

VALVE SYSTEM. Valve seats are machined directly into cylinder block and are ground at a 45° angle. Seat width should be 3/64 to 1/16-inch (1.19 to 1.58 mm).

Valve face is ground at 45° angle and valve should be renewed if margin is 1/64-inch (0.4 mm) or less after refacing.

Valve guides should be checked for wear using valve guide gage number 19151. If gage enters valve guide 5/16-inch (7.9 mm) or more, guides should be reamed using reamer number 19183 and bushing number 19192 should be installed.

Briggs & Stratton also has a tool kit, part number 19232, available for removing factory or field installed guide bushings so new bushing, part number 231218 may be installed.

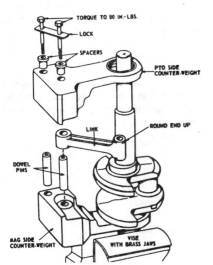

Fig. B26—Assemble balance units on crankshaft as shown. Install link with rounded edge on free end toward pto end of crankshaft.

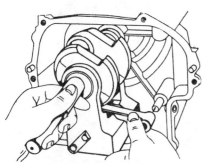

Fig. B27—When installing crankshaft and balancer assembly, place free end of link on anchor pin in crankcase.

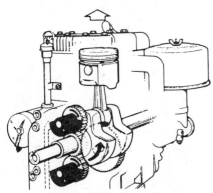

Fig. B27A—View of rotating counterbalance system used on some models. Counterweight gears are driven by crankshaft.

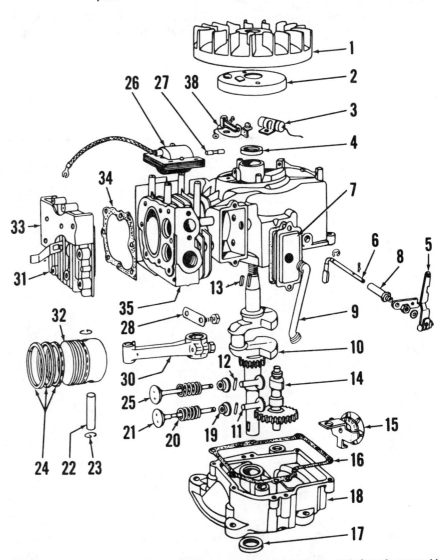

Fig. B28—Exploded view of 170000, 190000 and 220000 series vertical crankshaft engine assembly.

1. Flywheel
2. Breaker point cover
3. Condenser
4. Oil seal
5. Governor lever
6. Governor crank
7. Breather & valve
8. Bushing
9. Breather vent tube
10. Crankshaft
11. Tappets
12. Valve retaining pins
13. Flywheel key
14. Camshaft and gear
15. Governor & oil slinger assy.
16. Gasket
17. Oil seal
18. Oil sump (engine base)
19. Valve spring retainer
20. Valve spring
21. Exhaust valve
22. Piston pin
23. Retaining rings
24. Piston rings
25. Intake valve
26. Armature & coil assy.
27. Breaker plunger
28. Rod bolt lock
30. Connecting rod
31. Cylinder head
32. Piston
33. Air baffle
34. Head gasket
35. Cylinder block
38. Breaker points

Valve tappet gap (cold) for all models is 0.005-0.007 inch (0.13-0.18 mm) for intake valve and 0.009-0.011 inch (0.23-0.28 mm) for exhaust valve.

To adjust valves, piston must be at "top dead center" on compression stroke. Grind end of valve stem off squarely to obtain clearance.

SYNCHRO—BALANCER. All vertical crankshaft engines, except 220000 models, may be equipped with an oscillating Synchro-Balancer. Balance weight assembly rides on eccentric journals on crankshaft and moves in opposite direction of piston (Fig. B24).

To disassemble balancer unit, first remove flywheel, engine base, cam gear, cylinder head and connecting rod and piston assembly. Carefully pry off crankshaft gear and key. Remove the two cap screws holding halves of counterweight together. Separate weights and remove link, dowel pins and spacers. Slide weights from crankshaft (Fig. B25).

To reassemble, install magneto side weight on magneto end of crankshaft. Place crankshaft (pto end up) in a vise (Fig. B26). Install both dowel pins and place link on pin as shown. Note rounded edge on free end of link must be up. Install pto side weight, spacers, lock and cap screws. Tighten cap screws to 80 in.-lbs. (9 N·m) and secure with lock tabs. Install key and crankshaft gear with chamfer on inside of gear facing shoulder on crankshaft.

Install crankshaft and balancer assembly in crankcase, sliding free end of link on anchor pin as shown in Fig. B27. Reassemble engine.

ROTATING COUNTERBALANCE SYSTEM. All horizontal crankshaft engines, except 220000 models, may be equipped with two gear driven counterweights in constant mesh with crankshaft gear. Gears, mounted in crankcase cover, rotate in opposite direction of crankshaft (Fig. B27A).

To properly align counterweights when installing cover, remove two small screws from cover and insert 1/8-inch (3.18 mm) diameter locating pins through holes and into holes in counterweights as shown in Fig. B27B.

With piston at TDC, install cover assembly. Remove locating pins, coat threads of timing hole screws with non-hardening sealer and install screws with fiber sealing washers.

NOTE: If counterweights are removed from crankcase cover, exercise care in handling or cleaning to prevent losing needle bearings.

BRIGGS & STRATTON

4-STROKE MODELS

Model	Bore	Stroke	Displacement	Model	Bore	Stroke	Displacement
6B, Early 60000	2.3125 in. (58.7 mm)	1.500 in. (38.1 mm)	6.3 cu. in. (103 cc)	100000	2.5000 in. (63.1 mm)	2.125 in. (54.0 mm)	10.4 cu. in. (169 cc)
Late 60000	2.3750 in. (60.3 mm)	1.500 in. (38.1 mm)	6.7 cu. in. (109 cc)	110900, 111200, 111900	2.7813 in. (70.6 mm)	1.875 in. (47.6 mm)	11.4 cu. in. (186 cc)
8B, 80000	2.3750 in. (60.3 mm)	1.750 in. (44.5 mm)	7.8 cu. in. (127 cc)	112200	2.7813 in. (70.6 mm)	1.875 in. (47.6 mm)	11.4 cu. in. (186 cc)
81000, 82000	2.3750 in. (60.3 mm)	1.750 in. (44.5 mm)	7.8 cu. in. (127 cc)	113900	2.7813 in. (70.6 mm)	1.875 in. (47.6 mm)	11.4 cu. in. (186 cc)
92000, 93500	2.5625 in. (65.1 mm)	1.750 in. (44.5 mm)	9.0 cu. in. (148 cc)	130000	2.5625 in. (65.1 mm)	2.438 in. (60.9 mm)	12.6 cu .in. (203 cc)
94000, 94500, 94900	2.5625 in. (65.1 mm)	1.750 in. (44.5 mm)	9.0 cu. in. (148 cc)	131400, 131900	2.5625 in. (65.1 mm)	2.438 in. (60.9 mm)	12.6 cu. in. (203 cc)
95500	2.5625 in. (65.1 mm)	1.750 in. (44.5 mm)	9.0 cu. in. (148 cc)	140000	2.7500 in. (69.9 mm)	2.375 in. (60.3 mm)	14.1 cu. in. (231 cc)

FIRST DIGIT AFTER DISPLACEMENT		SECOND DIGIT AFTER DISPLACEMENT	THIRD DIGIT AFTER DISPLACEMENT	FOURTH DIGIT AFTER DISPLACEMENT
CUBIC INCH DISPLACEMENT	BASIC DESIGN SERIES	CRANKSHAFT, CARBURETOR GOVERNOR	BEARINGS, REDUCTION GEARS & AUXILIARY DRIVES	TYPE OF STARTER
6	0	0 -	0 - Plain Bearing	0 - Without Starter
8	1	1 - Horizontal Vacu-Jet	1 - Flange Mounting Plain Bearing	1 - Rope Starter
9	2	2 - Horizontal Pulsa-Jet	2 - Ball Bearing	2 - Rewind Starter
10	3	3 - Horizontal Flo-Jet (Pneumatic Governor)	3 - Flange Mounting Ball Bearing	3 - Electric - 110 Volt, Gear Drive
11	4	4 - Horizontal Flo-Jet (Mechanical Governor)	4 -	4 - Elec. Starter-Generator - 12 Volt, Belt Drive
13	5			
14	6	5 - Vertical Vacu-Jet	5 - Gear Reduction (6 to 1)	5 - Electric Starter Only - 12 Volt, Gear Drive
17	7			
19	8	6 -	6 - Gear Reduction (6 to 1) Reverse Rotation	6 - Alternator Only *
20	9			
22		7 - Vertical Flo-Jet	7 -	7 - Electric Starter, 12 Volt Gear Drive, with Alternator
23				
24		8 -	8 - Auxiliary Drive Perpendicular to Crankshaft	8 - Vertical-pull Starter
25				
30		9 - Vertical Pulsa-Jet	9 - Auxiliary Drive Parallel to Crankshaft	* Digit 6 formerly used for "Wind-Up" Starter on 60000, 80000 and 92000 Series
32				

EXAMPLES

To identify Model 100202:

10	0	2	0	2
10 Cubic Inch	Design Series 0	Horizontal Shaft - Pulsa-Jet Carburetor	Plain Bearing	Rewind Starter

Similarly, a Model 92998 is described as follows:

9	2	9	9	8
9 Cubic Inch	Design Series 2	Vertical Shaft - Pulsa-Jet Carburetor	Auxiliary Drive Parallel to Crankshaft	Vertical Pull Starter

Fig. B59 — Explanation of numerical code used by Briggs & Stratton to identify engine and optional equipment from model number.

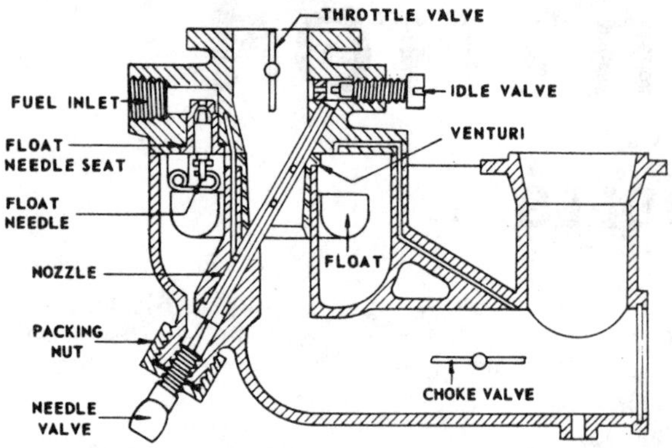

THROTTLE VALVE

IDLE VALVE

FUEL INLET

VENTURI

FLOAT
NEEDLE SEAT

FLOAT
NEEDLE

NOZZLE

FLOAT

PACKING
NUT

NEEDLE
VALVE

CHOKE VALVE

Fig. B60—Cross-sectional view of typical B&S "two-piece" carburetor. Before separating upper and lower body sections, loosen packing nut and unscrew nut and needle valve as a unit. Then, using special screwdriver, remove nozzle. Refer to text.

MAINTENANCE

SPARK PLUG. Recommended spark plug for all models is a Champion 8, or equivalent. If resistor type plugs are necessary to decrease radio interference, use Champion RJ8 or equivalent. Electrode gap for all models is 0.030 inch (0.76 mm).

CAUTION: Briggs & Stratton does not recommend using abrasive blasting method to clean spark plugs as this may introduce some abrasive material into the engine which could cause extensive damage.

ENGINE IDENTIFICATION

Engines covered in this section have aluminum cylinder blocks with either plain aluminum cylinder bore or with a

cast iron sleeve integrally cast into the block.

Early production of the 60000 model engine were of the same bore and stroke as the 6B model engine. The bore on 60000 engine was changed from 2.3125 inches (58.7 mm) to 2.3750 inches (60.3 mm) at serial number 5810060 on engines with plain aluminum bore, and at serial number 5810030 on engines with a cast iron sleeve.

Refer to Fig. B59 for chart explaining engine numerical code to identify engine model. Always furnish correct engine model and model number when ordering parts or service information.

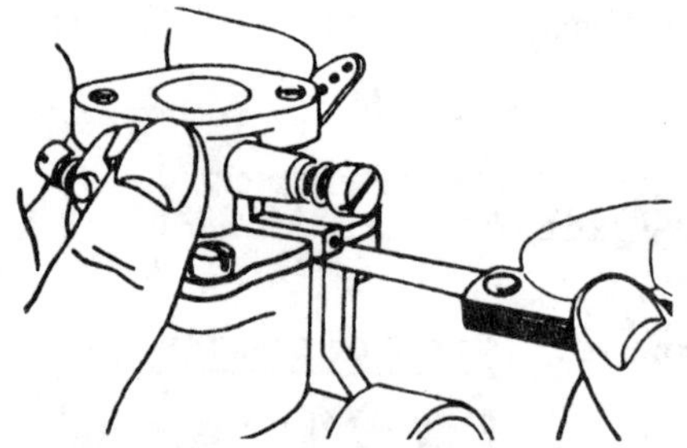

FLOAT TYPE (FLO-JET) CAR-BURETORS. Three different float type carburetors are used. They are called a "two-piece" (Fig. B60), a small "one-piece" (Fig. B64) or a large "one-piece" (Fig. B66) carburetor depending upon the type of construction.

Float type carburetors are equipped with adjusting needles for both idle and power fuel mixtures. Counterclockwise rotation of the adjusting needles richens the mixture. For initial starting adjustment, open the main needle valve (power fuel mixture) 1½ turns on the two-piece carburetor and 2½ turns on the small one-piece carburetor. Open the idle needle ½ to ¾ turn on the two-piece carburetor and 1½ turns on the small one-piece carburetor. On the large one-piece carburetor, open both needle valves 1-1/8 turns.

Make final adjustments with engine at operating temperature and running. Set the speed control for desired operating speed, turn main needle clockwise until engine misses, and then turn it counterclockwise just past the smooth operating point until the engine begins to run unevenly. Return the speed control to idle position and adjust the idle speed

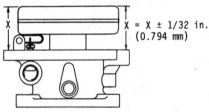

Fig. B61—Checking upper body of "two-piece" carburetor for warpage. Refer to text.

$X = X \pm 1/32$ in.
(0.794 mm)

Fig. B62—Carburetor float setting should be within specifications shown. To adjust float setting, bend tang with needlenose pliers as shown in Fig. B63.

NEEDLE VALVE

IDLE VALVE

VENTURI

THROTTLE
BUTTERFLY

CHOKE
BUTTERFLY

GASKET

FLOAT

NOZZLE

GASKET

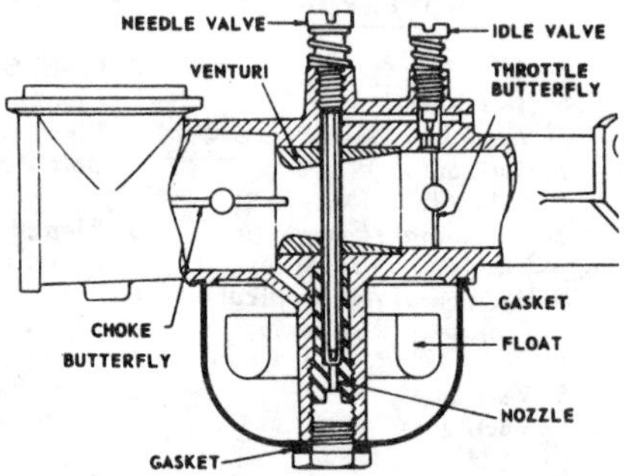

Fig. B64—Cross-sectional view of typical B&S small "one-piece" float type carburetor. Refer to Fig. B65 for disassembly views.

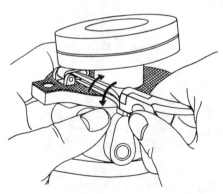

Fig. B63—Bending tang with needlenose pliers to adjust float setting. Refer to Fig. B62 for method of checking float setting.

Fig. B65—Disassembling the small "one-piece" float type carburetor. Pry out welch plug, remove choke butterfly (disc), remove choke shaft and needle valve; venturi can then be removed as shown in left view.

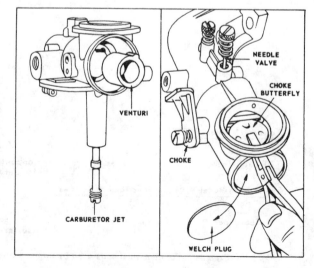

NEEDLE
VALVE

CHOKE
BUTTERFLY

VENTURI

CHOKE

CARBURETOR JET

WELCH PLUG

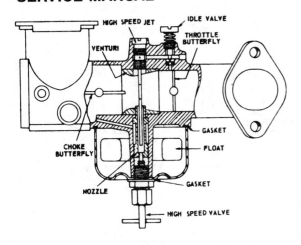

Fig. B66—Cross-sectional view of B&S large "one-piece" float type carburetor.

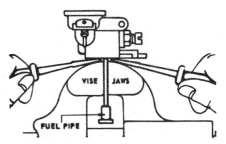

Fig. B68—Removing brass fuel feed pipe from suction type carburetor. Press new brass pipe into carburetor until it projects 2-9/32 to 2-5/16 inches (57.9-58.7 mm) from carburetor face. Nylon fuel feed pipe is threaded into carburetor.

stop screw until the engine idles at 1750 rpm. Adjust the idle needle valve until the engine runs smoothly. Reset the idle speed stop screw if necessary. The engine should then accelerate without hesitation. If engine does not accelerate properly, turn the main needle valve counterclockwise slightly to provide a richer fuel mixture.

The float setting on all float type carburetors should be within dimensions shown in Fig. B62. If not, bend the tang on float as shown in Fig. B63 to adjust float setting. If any wear is visible on the inlet valve or the inlet valve seat, install a new valve and seat assembly. On large one-piece carburetors, the renewable inlet valve seat is pressed into the carburetor body until flush with the body.

NOTE: The upper and lower bodies of the two-piece float type carburetor are locked together by the main nozzle. Refer to cross-sectional view of carburetor in Fig. B60. Before attempting to separate the upper body from the lower body, loosen packing nut and unscrew nut and needle valve. Then, using special screwdriver (B&S tool 19061 or 19062), remove nozzle.

If a 0.002 inch (0.05 mm) feeler gage can be inserted between upper and lower bodies of the two-piece carburetor as shown in Fig. B61, the upper body is warped and should be renewed.

Check the throttle shaft for wear on all float type carburetors. If 0.010 inch (0.25 mm) or more free play (shaft-to-bushing clearance) is noted, install new throttle shaft and/or throttle shaft bushings. To remove worn bushings, turn a ¼ inch × 20 tap into bushing and pull bushing from body casting with the tap. Press new bushings into casting by using a vise and if necessary, ream bushings with a 7/32 inch drill bit.

SUCTION TYPE (VACU-JET) CARBURETORS. A typical suction type (Vacu-Jet) carburetor is shown in Fig. B67. This type carburetor has only one fuel mixture adjusting needle. Turning the needle clockwise leans the air:fuel mixture. Adjust suction type carburetors with fuel tank approximately one-half full and with the engine at operating temperature and running at approximately 3000 rpm, no-load. Turn needle valve clockwise until engine begins to lose speed; then, turn needle slowly counterclockwise until engine begins to run unevenly from a too rich air:fuel mixture. This should result in a correct adjustment for full load operation. Adjust idle speed to 1750 rpm.

To remove the suction type carburetor, first remove carburetor and fuel tank as an assembly, then remove carburetor from fuel tank. When reinstalling carburetor on fuel tank, use a new gasket and tighten retaining screws evenly.

The suction type carburetor has a fuel feed pipe extending into fuel tank. The pipe has a check valve to allow fuel to feed up into the carburetor but prevents fuel from flowing back into the tank. If check valve is inoperative and cleaning in alcohol or acetone will not free the check valve, renew the fuel feed pipe. If feed pipe is made of brass, remove as shown in Fig. B68. Using a vise, press new pipe into carburetor so it extends from 2-9/32 to 2-5/16 inches (57.9-58.7 mm) from carburetor body. If pipe is made of nylon (plastic), screw pipe out of carburetor body with wrench. When installing new nylon feed pipe, be careful not to overtighten.

NOTE: If soaking carburetor in cleaner for more than one-half hour, be sure to remove all nylon parts and "O" ring, if used, before placing the carburetor in cleaning solvent.

PUMP TYPE (PULSA-JET) CARBURETORS. The pump type (Pulsa-Jet) carburetor is basically a suction type carburetor incorporating a fuel

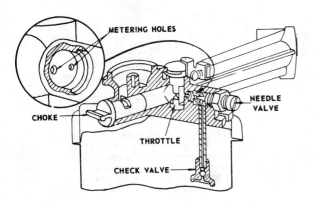

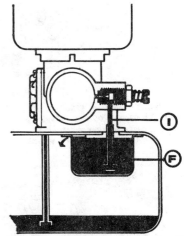

Fig. B67—Cutaway view of typical suction type (Vacu-Jet) carburetor. Inset shows fuel metering holes which are accessible for cleaning after removing needle valve. Be careful not to enlarge the holes when cleaning them.

Fig. B69—Fuel flow in Pulsa-Jet carburetor. Fuel pump incorporated in carburetor fills constant level sump (F) below carburetor and excess fuel flows back into tank. Fuel is drawn from sump through inlet (I) past fuel mixture adjusting needle by vacuum in carburetor.

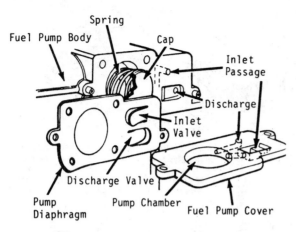

Fig. B70—Exploded view of fuel pump that is incorporated in Pulsa-Jet carburetor except those used on 82900, 92900, 94900, 110900, 111900, 112200 and 113900 models; refer to Fig. B71.

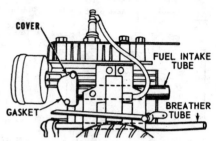

Fig. B74—On 82000 models, intake tube is threaded into intake port of engine; a gasket is placed between intake port cover and intake port.

pump to fill a constant level fuel sump in top of fuel tank. Refer to schematic view in Fig. B69. This makes a constant air:fuel mixture available to engine regardless of fuel level in tank. Adjustment of the pump type carburetor fuel mixture needle valve is the same as outlined for suction type carburetors in previous paragraph, except that fuel level in tank is not important.

To remove the pump type carburetor, first remove the carburetor and fuel tank as an assembly; then, remove carburetor from fuel tank. When reinstalling carburetor on fuel tank, use a new gasket or pump diaphragm as required and tighten retaining screws evenly.

Fig. B70 shows an exploded view of the .pump unit used on all carburetors except those for Models 82900, 92900, 94900, 110900, 111900, 112200 and 113900 the pump diaphragm is placed between the carburetor and fuel tank as shown in Fig. B71.

The pump type carburetor has two fuel feed pipes. The long pipe feeds fuel into the pump portion of the carburetor from which fuel then flows to the constant level fuel sump. The short pipe extends into the constant level fuel sump and feeds fuel into the carburetor venturi via fuel mixture needle valve.

As check valves are incorporated in the pump diaphragm, fuel feed pipes on pump type carburetors do not have a check valve. However, if the fuel screen in lower end of pipe is broken or clogged and cannot be cleaned, the pipe or screen housing can be renewed. If pipe is made of nylon, pipe snaps into place and considerable force is required to remove or install pipe. Be careful to not damage new pipe. If pipe is made of brass, clamp pipe lightly in a vise and drive old screen housing from pipe with a screwdriver or small chisel as shown in Fig. B72. Drive a new screen housing onto pipe with a soft faced hammer.

NOTE: If soaking carburetor in cleaner for more than one-half hour, be sure to remove all nylon parts and "O" ring, if used, before placing carburetor in cleaning solvent.

NOTE: On engine Models 82900, 92900, 94900, 95500, 110900, 111900 and 113900, be sure air cleaner retaining screw is in place if engine is being operated (during tests) without air cleaner installed. If screw is not in place, fuel will lift up

through the screw hole and enter carburetor throat as the screw hole leads directly into the constant level fuel sump.

INTAKE TUBE. Models 82000, 92000, 93500, 94500, 94900, 95500, 100900, 110900, 111900, 113900, 130900 and 131900 have an intake tube between carburetor and engine intake port. Carburetor is sealed to intake tube with an "O" ring as shown in Fig. B73.

On Model 82000 engines, the intake tube is threaded into the engine intake port. A gasket is used between the engine intake port cover and engine casting. Refer to Fig. B74.

On Models 92000, 93500, 94500, 94900, 95500, 110900, 111900 and 113900 engines, the intake tube is bolted to the engine intake port and gasket is used between the intake tube and engine casting. Refer to Fig. B75. On Models 100900, 130900 and 131900 intake tubes are attached to engine in similar manner.

CHOKE-A-MATIC CARBURETOR CONTROLS. Engines equipped with float, suction or pump type carburetors may be equipped with a control unit with which the carburetor choke, throttle and magneto grounding switch are operated from a single lever (Choke-A-Matic carburetors). Refer to Figs. B76 through B82 for views showing the different

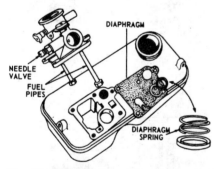

Fig. B71—On Models 82900, 92900, 94900, 110900, 111900, 112200 and 113900, pump type (Pulsa-Jet) carburetor diaphragm is installed between carburetor and fuel tank.

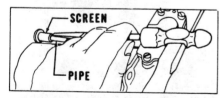

Fig. B72—To renew screen housing on pump type carburetor with brass feed pipes, drive old screen housing from pipe as shown. To hold pipe, clamp lightly in a vise.

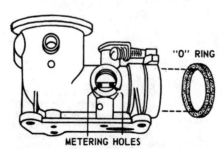

Fig. B73—Metering holes in pump type carburetors are accessible for cleaning after removing fuel mixture needle valve. On models with intake pipe, carburetor is sealed to pipe with "O" ring.

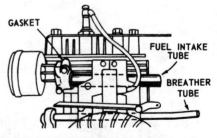

Fig. B75—On 92000, 93500, 94500, 94900, 95500, 110900, 111900 and 113900 models, fuel intake tube is bolted to engine intake port and a gasket is placed between tube and engine. On 100900, 130900 and 131900 vertical crankshaft models, intake tube and gasket are similar.

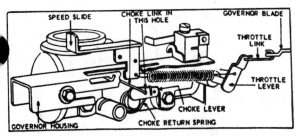

Fig. B76—Choke-A-Matic control on float type carburetor. Remote control can be attached to speed slide.

Fig. B77—Typical Choke-A-Matic control on suction type carburetor. Remote control can be attached to speed lever.

Fig. B78—Choke-A-Matic control in choke and stop positions on float carburetor.

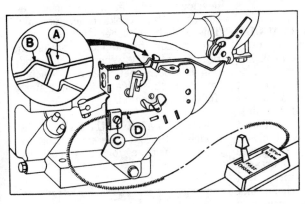

Fig. B79—Choke-A-Matic controls in choke and stop positions on suction carburetor. Bend choke link if necessary to adjust control.

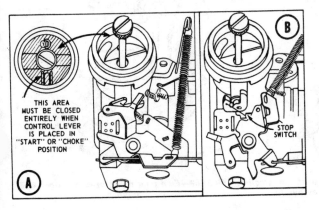

Fig. B80—On Choke-A-Matic control shown, choke actuating lever (A) should just contact choke link or shaft (B) when control is at "FAST" position. If not, loosen screw (C) and move control wire housing (D) as required.

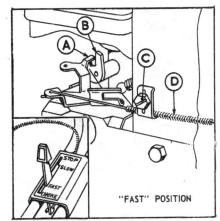

"FAST" POSITION

Fig. B82—On Choke-A-Matic controls shown, lever (A) should just contact choke shaft arm (B) when control is in "FAST" position. If not, loosen screw (C) and move control wire housing (D) as required, then tighten screw.

types of Choke-A-Matic carburetor controls.

To check operation of Choke-A-Matic carburetor controls, move control lever to "CHOKE" position. Carburetor choke slide or plate must be completely closed. Then, move control lever to "STOP" position. Magneto grounding switch should be making contact. With the control lever in "RUN", "FAST" or "SLOW" position, carburetor choke should be completely open. On units with remote controls, synchronize movement of remote lever to carburetor control lever by loosening screw (C – Fig. B80 or Fig. B82) and moving control wire housing (D) as required; then, tighten screw to clamp the housing securely. Refer to Fig. B83 to check remote control wire movement.

AUTOMATIC CHOKE (THERMOSTAT TYPE). A thermostat operated choke is used on some models equipped with the two-piece carburetor. To adjust choke linkage, hold choke shaft so thermostat lever is free. At room temperature, stop screw in thermostat collar should be located midway between thermostat stops. If not, loosen stop screw, adjust the collar and tighten stop screw. Loosen set screw (S – Fig.

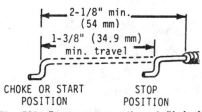

Fig. B81—When Choke-A-Matic control is in "START" or "CHOKE" position, choke must be completely closed as shown in view A. When control is in "STOP" position, arm should contact stop switch (view B).

Fig. B83—For proper operation of Choke-A-Matic controls, remote control wire must extend to dimension shown and have a minimum travel of 1-3/8 inches (34.9 mm).

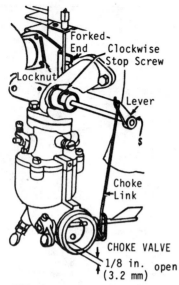

Fig. B84—Automatic choke used on some models equipped with "two-piece" Flo-Jet carburetor showing unit in "HOT" position.

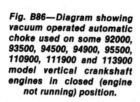

Fig. B86—Diagram showing vacuum operated automatic choke used on some 92000, 93500, 94500, 94900, 95500, 110900, 111900 and 113900 model vertical crankshaft engines in closed (engine not running) position.

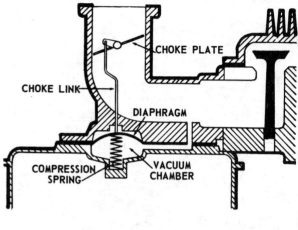

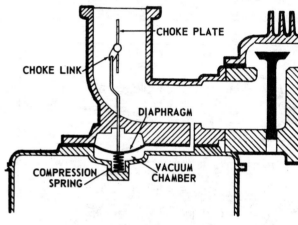

Fig. B87—Diagram showing vacuum operated automatic choke in open (engine running) position.

B84) on thermostat lever. Then, slide lever on shaft to ensure free movement of choke unit. Turn thermostat shaft clockwise until stop screw contacts thermostat stop. While holding shaft in this position, move shaft lever until choke is open exactly 1/8 inch (3.17 mm) and tighten lever set screw. Turn thermostat shaft counterclockwise until stop screw contacts thermostat stop as shown in Fig. B85. Manually open choke valve until it stops against top of choke link opening. At this time, choke valve should be open at least 3/32 inch (2.38 mm), but not more than 5/32 inch (3.97 mm). Hold choke valve in wide open position and check position of counter-

weight lever. Lever should be in a horizontal position with free end towards right.

AUTOMATIC CHOKE (VACUUM TYPE). A spring and vacuum operated automatic choke is used on some 92000, 93500, 94500, 94900, 95500, 110900, 111900 and 113900 vertical crankshaft engines. A diaphragm under carburetor is connected to the choke shaft by a link. The compression spring works against the diaphragm, holding choke in closed position when engine is not running. See Fig. B86. As engine starts, increased vacuum works against the spring and pulls the diaphragm and choke link down, holding choke in open (running) position shown in Fig. B87.

During operation, if a sudden load is applied to engine or a lugging condition develops, a drop in intake vacuum occurs, permitting choke to close partially. This provides a richer fuel mixture to meet the condition and keeps the engine running smoothly. When the load condition has been met, increased vacuum returns choke valve to normal running (fully open) position.

FUEL TANK OUTLET. Small models with float type carburetors are equipped with a fuel tank outlet as

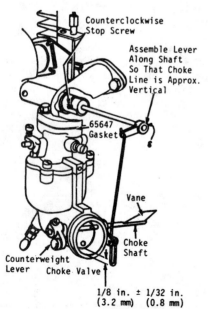

Fig. B85—Automatic choke on "two-piece" Flo-Jet carburetor in "COLD" position.

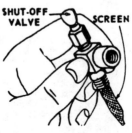

Fig. B88—Fuel tank outlet used on smaller engines with float type carburetor.

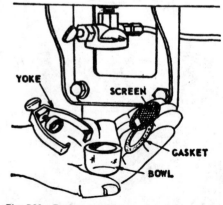

Fig. B89—Fuel tank outlet used on larger B&S engines.

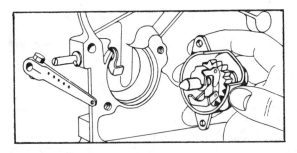

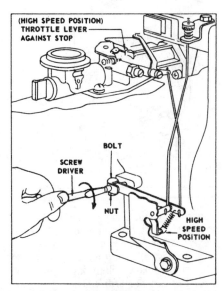

Fig. B90—Removing governor unit (except on 100000, 112200, 130000, 131000 and late 140000 models) from inside crankcase cover on horizontal crankshaft models. Refer to Fig. B91 for exploded view of governor.

shown in Fig. B88. On larger engines, a fuel sediment bowl is incorporated with the fuel tank outlet as shown in Fig. B89. Clean any lint and dirt from tank outlet screens with a brush. Varnish or other gasoline deposits may be removed by using a suitable solvent. Tighten packing nut or remove nut and shut-off valve, then renew packing if leakage occurs around shut-off valve stem.

FUEL PUMP. A fuel pump is available as optional equipment on some models. Refer to SERVICING BRIGGS & STRATTON ACCESSORIES section in this manual for fuel pump service information.

GOVERNOR. All models are equipped with either a mechanical (flyweight type) or an air vane (pneumatic) governor. Refer to the appropriate paragraph for model being serviced.

Mechanical Governor. Three different designs of mechanical governors are used.

On all engines except 100000, 112200, 130000, 131000 and all late 140000 models, a governor unit as shown in Fig. B90 is used. An exploded view of this governor unit is shown in Fig. B91. The governor housing is attached to inner

Fig. B95—Linkage adjustment on 100000, 112200, 130000, 131000 and late 140000 horizontal crankshaft mechanical governor models; refer to text for procedure.

side of crankcase cover and the governor gear is driven from the engine camshaft gear. Use Figs. B90 and B91 as a

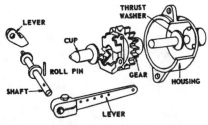

Fig. B91—Exploded view of governor unit used on all horizontal crankshaft models (except 100000, 112200, 130000, 131000 and late 140000 models) with mechanical governor. Refer also to Figs. B90 and B92.

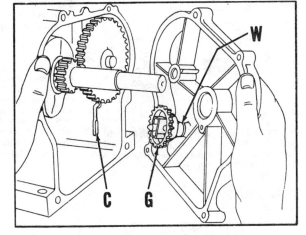

Fig. B93—Installing crankcase cover on 100000, 112200, 130000, 131000 and late 140000 models with mechanical governor. Governor crank (C) must be in position shown. A thrust washer (W) is placed between governor (G) and crankcase cover.

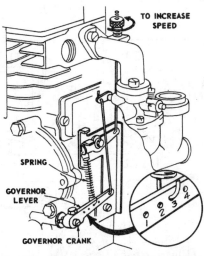

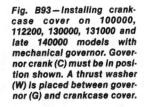

Fig. B92—View of governor linkage used on horizontal crankshaft mechanical governor models except 100000, 112200, 130000, 131000 and late 140000 models. Governor spring should be hooked in governor lever as shown in inset.

Fig. B94—Cutaway drawing of governor and linkage used on 100000, 112200, 130000, 131000 and late 140000 horizontal crankshaft models.

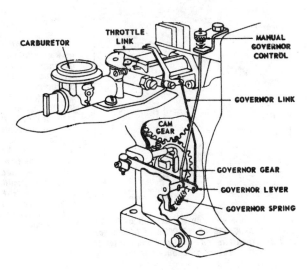

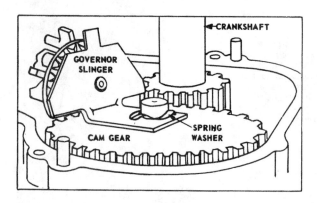

Fig. B96—View showing 95500, 100000, 113900, 130000, 131000 and 140000 vertical crankshaft model mechanical governor unit. Drawing is of lower side of engine with oil sump (engine base) removed; note spring washer location used on 100000, 130000 and 131000 models only.

disassembly and assembly guide. Renew any parts that bind or show excessive wear. After governor is assembled, refer to Fig. B92 and adjust linkage by loosening screw clamping governor lever to governor crank. Turn governor lever counterclockwise so carburetor throttle is in wide open position. Hold lever and turn governor crank as far counterclockwise as possible, then tighten screw clamping lever to crank. Governor crank can be turned with screwdriver. Check linkage to be sure it is free and that the carburetor throttle will move from idle to wide open position.

On 100000, 112200, 130000, 131000 and late 140000 horizontal crankshaft models, the governor gear and weight unit (G – Fig. B93) is supported on a pin in engine crankcase cover and the governor crank is installed in a bore in the engine crankcase. A thrust washer (W) is placed between governor gear and crankcase cover. When assembling crankcase cover to crankcase, be sure governor crank (C) is in position shown in Fig. B93. After governor unit and crankcase cover is installed, refer to Fig. B94 for installation of linkage. Before attempting to start engine, refer to Fig. B95 and adjust linkage by loosening bolt clamping governor lever to governor crank. Set control lever in high speed position. Using a screwdriver, turn governor crank as far clockwise as possible and tighten governor lever clamp bolt. Check to be sure carburetor throttle can be moved from idle to wide open position and that linkage is free.

On vertical crankshaft 95500, 100000, 113900, 130000, 131000 and 140000 with mechanical governor, the governor weight unit is integral with the lubricating oil slinger and is mounted on lower end of camshaft gear as shown in Fig. B96. With engine upside down, place governor and slinger unit on camshaft gear as shown, place spring washer (100000, 130900 and 131900 models only) on camshaft gear and install engine base on crankcase. Assemble linkage as shown in Fig. B97; then, refer to Fig. B98 and adjust linkage by loosening governor lever to governor crank clamp bolt, place control lever in high speed position, turn governor crank with screwdriver as far as possible and tighten governor lever clamping bolt.

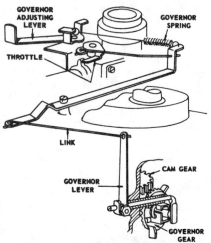

Fig. B97—Schematic drawing of 95500, 100000, 113900, 130000, 131000 and 140000 mechanical governor and linkage used on vertical crankshaft mechanical governed models.

Fig. B99—Views showing operating principle of air vane (penumatic) governor. Air from flywheel fan acts against air vane to overcome tension of governor spring; speed is adjusted by changing spring tension.

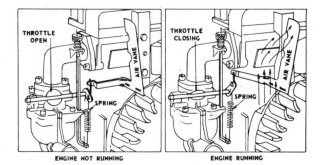

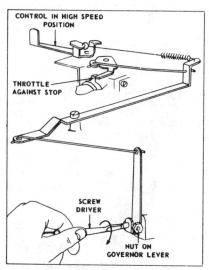

Fig. B98—View showing adjustment of 95500, 100000, 113900, 130000, 131000 and 140000 vertical crankshaft mechanical governor; refer to text for procedure.

Fig. B100—Air vane governors and linkage. ILL. 1; the governor vane should be checked for clearance in all positions. ILL. 2; the vane should top 1/8 to 1/4 inch (3.175-6.35 mm) from the magneto coil. ILL. 3; with wide open throttle, the link connecting vane arm to throttle lever should be in a vertical position on vertical cylinder engines and in a horizontal position (ILL. 4) on horizontal cylinder engines. Bend link slightly (ILL. 5) to remove any binding condition in linkage.

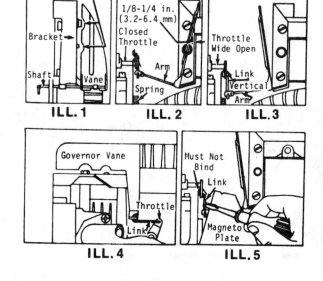

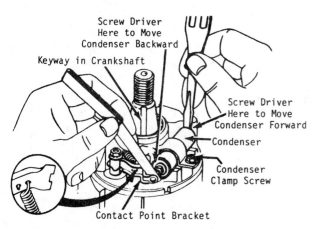

Fig. B101—View showing breaker point adjustment on models having breaker point integral with condenser. Move condenser to adjust point gap.

Air Vane (Pneumatic) Governor. Refer to Fig. B99 for views showing typical air vane governor operation.

The vane should stop 1/8 to 1/4 inch (3.17-6.35 mm) from the magneto coil (Fig. B100, illustration 2) when the linkage is assembled and attached to carburetor throttle shaft. If necessary to adjust, spring the vane while holding the shaft. With wide open throttle, the link from the air vane arm to the carburetor throttle should be in a vertical position on horizontal crankshaft models and in a horizontal position on vertical crankshaft models. (Fig. B100, illustration 3 and 4.) Check linkage for binding. If binding condition exists, bend links slightly to correct. Refer to Fig. B100, illustration 5.

NOTE: Some engines are equipped with a nylon governor vane which does not require adjustment.

MAGNETO. Breaker contact gap on all models with magneto type ignition is 0.020 inch (0.51 mm). On all models except "Sonoduct" (vertical crankshaft with flywheel below engine), breaker points and condenser are accessible after removing engine flywheel and breaker cover. On "Sonoduct" models, breaker points are located below breaker cover on top side of engine.

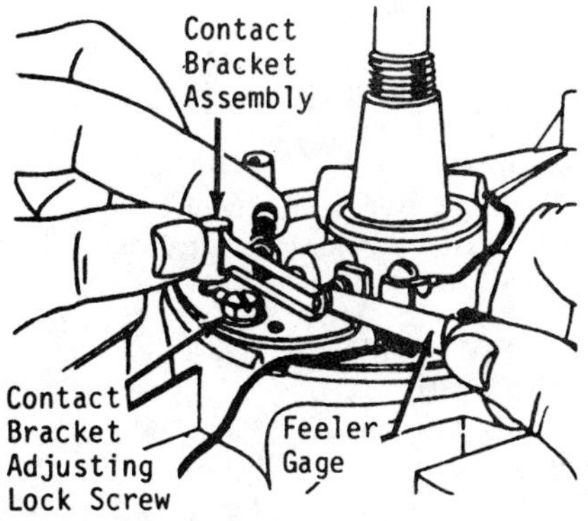

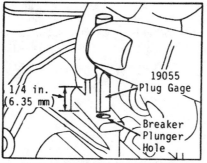

Fig. B103—If Briggs & Stratton plug gage #19055 can be inserted in breaker plunger bore a distance of 1/4 inch (6.35 mm) or more, bore is worn and must be rebushed.

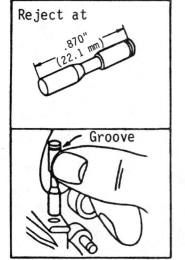

Fig. B102—Adjustment of breaker point gap on models having breaker point separate from condenser.

Fig. B102A—Insert plunger into bore with groove toward top. Refer to text.

On some models, one breaker contact point is an integral part of the ignition condenser and the breaker arm pivots on a knife edge retained by a slot in pivot post. On these models, breaker contact gap is adjusted by moving the condenser as shown in Fig. B101. On other models, breaker contact gap is adjusted by relocating position of breaker contact bracket. Refer to Fig. B102.

On all models, breaker contact arm is actuated by a plunger held in a bore in engine crankcase and rides against a cam on engine crankshaft. Plunger can be removed after removing breaker points. Renew plunger if worn to a length of 0.870 inch (22.1 mm) or less. Plunger must be installed with grooved end to the top to prevent oil seepage (Fig. B102A). Check breaker point plunger bore wear in crankcase with B&S plug gage 19055. If plug gage will enter bore 1/4 inch (6.35 mm) or more, bore should be reamed and a bushing installed. Refer to Fig. B103 for method of checking bore and to Fig. B104 for steps in reaming bore and installing bushing if bore is worn. To ream bore and install bushing, it is necessary that the breaker points, armature and ignition coil and the crankshaft be removed.

On "Sonoduct" models, armature and ignition coil are inside flywheel on bottom side of engine. Armature air gap is correct if armature is installed flush with mounting boss as shown in Fig. B105. On all other models, armature and ignition coil are located outside flywheel. Armature-to-flywheel air gap should be as follows:

Models 6B, 8B, 6000—

Two Leg Armature . . . 0.006-0.010 in. (0.15-0.25 mm)

Three Leg Armature . . 0.012-0.016 in. (0.30-0.41 mm)

Models 80000, 82000, 92000, 93000, 93500, 94000, 94500, 94900, 95000, 95500, 110000, 110900—

Two Leg Armature . . . 0.006-0.010 in. (0.15-0.25 mm)

Three Leg Armature . . 0.012-0.016 in. (0.30-0.41 mm)

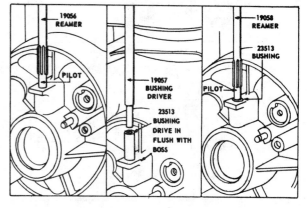

Fig. B104—Views showing reaming plunger bore to accept bushing (left view), installing bushing (center) and finish reaming bore (right) of bushing.

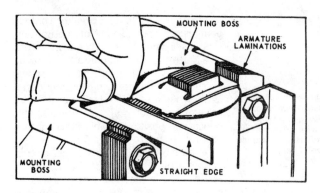

Fig. B105—On "Sonoduct" models, align armature core with mounting boss for proper magneto air gap.

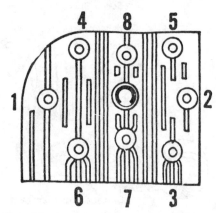

Fig. B106—Cylinder head screw tightening sequence. Long screws are used in positions 2, 3 and 7.

Models 111200, 111900,
112200, 113900 . . . . .0.010-0.016 in.
　　　　　　　　　　　　(0.25-0.41 mm)
Models 100000,
130000, 131000—
　Two Leg Armature . . .0.010-0.014 in.
　　　　　　　　　　　　(0.25-0.36 mm)
　Three Leg Armature . .0.012-0.016 in.
　　　　　　　　　　　　((0.30-0.41 mm)
Model 140000—
　Two Leg Armature . . .0.010-0.014 in.
　　　　　　　　　　　　(0.25-0.36 mm)
　Three Leg Armature . .0.016-0.019 in.
　　　　　　　　　　　　(0.41-0.48 mm)

MAGNETRON IGNITION. Magnetron ignition is a self-contained breakerless ignition system. Flywheel does not need to be removed except to check or service keyways or crankshaft key.

To check spark, remove spark plug. Connect spark plug cable to B&S tester, part 19051, and ground remaining tester lead to cylinder head. Spin engine at 350 rpm or more. If spark jumps the 0.166 inch (4.2 mm) tester gap, system is functioning properly.

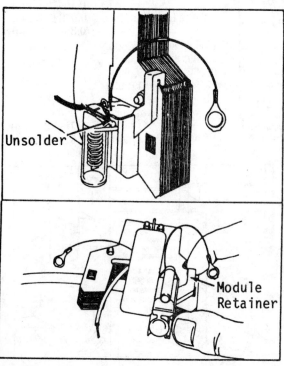

Fig. B105A—Wires must be unsoldered to remove Magnetron ignition module. Refer to text.

To remove armature and Magnetron module, remove flywheel shroud and armature retaining screws. Use a 3/16 inch (4.76 mm) diameter pin punch to release stop switch wire from module. To remove module, unsolder wires, push module retainer away from laminations and remove module. See Fig. B105A.

Resolder wires for reinstallation and use Permatex or equivalent to hold ground wires in position.

Adjust armature air gap to 0.010-0.014 inch (0.25-0.36 mm).

LUBRICATION. Vertical crankshaft engines are lubricated by an oil slinger wheel driven by the cam gear. On early 6B and 8B models, the oil slinger wheel was mounted on a bracket attached to the crankcase. Renew the bracket if pin on which gear rotates is worn to 0.490 inch (12.45 mm). Renew steel bushing in hub or gear if worn. On later model vertical crankshaft engines, the oil slinger wheel, pin and bracket are an integral unit with the bracket being retained by the lower end of the engine camshaft.

On 100000, 130000 and 131000 models, a spring washer is placed on lower end of camshaft between bracket and oil sump boss. Renew the oil slinger assembly if teeth are worn on slinger gear or gear is loose on bracket. On horizontal crankshaft engines, a splash system (oil dipper on connecting rod) is used for engine lubrication.

Check oil level at five hour intervals and maintain at bottom edge of filler plug or to FULL mark on dipstick.

Recommended oil change interval for all models is every 25 hours of normal operation.

Manufacturer recommends using oil with an API service classification of SC, SD, SE, or SF. Use SAE 10W-40 oil for temperatures above 20°F (−7°C) and SAE 5W-30 oil for temperatures below 20°F (−7°C).

Crankcase capacity for all aluminum cylinder engines with displacement of 9 cubic inches (147 cc) or below, is 1¼ pint (0.6 L).

Crankcase capacity of vertical crankshaft aluminum cylinder engines with displacement of 11 cubic inches (186 cc) is 1¼ pint (0.6 L).

Crankcase capacity for all 10 and 13 cubic inch (169 and 203 cc) vertical crankshaft, aluminum cylinder engines is 1¾ pint (0.8 L).

Crankcase capacity for all 10 and 13 cubic inch (169 and 203 cc) horizontal crankshaft, aluminum cylinder engines is 1¼ pint (0.6 L).

Crankcase capacity for 14 cubic inch (231 cc) vertical crankshaft, aluminum cylinder engines is 2¼ pint (1.1 L).

Crankcase capacity for 14 cubic inch (231 cc) horizontal crankshaft, aluminum cylinder engines is 2¾ pint (1.3 L).

Crankcase capacity for engines with cast iron cylinder liner is 3 pints (1.4 L).

CRANKCASE BREATHER. The crankcase breather is built into the

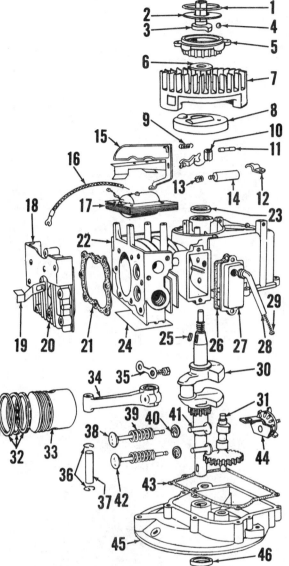

Fig. B107 — Exploded view of typical vertical crankshaft model with air vane (pneumatic) governor. To remove flywheel, remove blower housing and starter unit; then, unscrew starter clutch housing (5). Flywheel can then be pulled from crankshaft.

1. Snap ring
2. Washer
3. Ratchet
4. Steel balls
5. Starter clutch
6. Washer
7. Flywheel
8. Breaker cover
9. Breaker point spring
10. Breaker arm & pivot
11. Breaker plunger
12. Condenser clamp
13. Coil spring (primary wire retainer)
14. Condenser
15. Governor air vane & bracket assy.
16. Spark plug wire
17. Armature & coil assy.
18. Air baffle
19. Spark plug grounding switch
20. Cylinder head
21. Cylinder head gasket
22. Cylinder block
23. Crankshaft oil seal
24. Cylinder shield
25. Flywheel key
26. Gasket
27. Breather & tappet chamber cover
28. Breather tube assy.
29. Coil spring
30. Crankshaft
31. Cam gear & shaft
32. Piston rings
33. Piston
34. Connecting rod
35. Rod bolt lock
36. Piston pin retaining rings
37. Piston pin
38. Intake valve
39. Valve springs
40. Valve spring keepers
41. Tappets
42. Exhaust valve
43. Gasket
44. Oil slinger assy.
45. Oil sump (engine base)
46. Crankshaft oil seal

engine valve cover. The mounting holes are offset so the breather can only be installed one way. Rinse breather in solvent and allow to drain. A vent tube connects the breather to the carburetor air horn on certain model engines for extra protection against dusty conditions.

REPAIRS

CYLINDER HEAD. When removing cylinder head, be sure to note the position from which each of the different length screws was removed. If screws are not reinstalled in the same holes when installing the head, it will result in screws bottoming in some holes and not enough thread contact in others. Lubricate the cylinder head screws with graphite grease before installation. Do not use sealer on head gasket. When installing cylinder head, tighten all screws lightly and then retighten them in sequence shown in Fig. B106 to 165 in.-lbs.

(19 N·m) on 140000 models and to 140 in.-lbs. (16 N·m) on all other models. Run the engine for 2 to 5 minutes to allow it to reach operating temperature and retighten the head screws again following the sequence and torque value specified.

NOTE: When checking compression on models with "Easy-Spin" starting, turn engine opposite the direction of normal rotation. See CAMSHAFT paragraph.

OIL SUMP REMOVAL, AUXILARY PTO MODELS. On 92000, 93500, 94900, 95500, 110900 or 113900 models with auxiliary pto, one of the oil sump (engine base) to cylinder retaining screws is installed in the recess in sump for the pto auxiliary drive gear. To remove the oil sump, refer to Fig. B108 and remove the cover plate (upper view) and then remove the shaft stop (lower left view). The gear and shaft can then be moved as shown in lower right view

to allow removal of the retaining screws. Reverse procedure to reassemble.

OIL BAFFLE PLATE. 140000 MODEL ENGINES. Model 140000 engines with mechanical governor have a baffle located in the cylinder block (crankcase). When servicing these engines, it is important that the baffle be correctly installed. The baffle must fit tightly against the valve tappet boss in the crankcase. Check for this before installing oil sump (vertical crankshaft models) or crankcase cover (horizontal crankshaft models).

CONNECTING ROD. The connecting rod and piston are removed from cylinder head end of block as an assembly. The aluminum alloy connecting rod rides directly on the induction hardened crankpin. The rod should be rejected if the crankpin hole is scored or out-of-round more than 0.0007 inch (0.018 mm) or if the piston pin hole is scored or out-of-round more than 0.0005 inch (0.013 mm). Wear limit sizes are given in the following chart. Reject the connecting rod if either the crankpin or piston pin hole is worn to, or larger than the sizes given in the following table.

REJECT SIZES FOR CONNECTING ROD

Model	Bearing Bore	Pin Bore
6B, 60000 . . .	0.876 in. (22.25 mm)	0.492 in. (12.50 mm)
8B, 80000 . . .	1.001 in. (25.43 mm)	0.492 in. (12.50 mm)
81000, 82000, 110000 . .	1.001 in. (25.43 mm)	0.492 in. (12.50 mm)
92000, 93500 . . .	1.001 in. (25.43 mm)	0.492 in. (12.50 mm)
94500, 94900, 95500 . . .	1.001 in. (25.43 mm)	0.492 in. (12.50 mm)
100000 . .	1.001 in. (25.43 mm)	0.555 in. (14.10 mm)
110900, 111200, 111900, 112200, 113900, 130000, 131000 . .	1.001 in. (25.43 mm)	0.492 in. (12.50 mm)
140000 . .	1.095 in. (27.81 mm)	0.674 in. (17.12 mm)

NOTE: Piston pins of 0.005 inch (0.13 mm) oversize are available for service. Piston pin hole in rod can be reamed to this size if crankpin hole in rod is within specifications.

Connecting rod must be reassembled so match marks on rod and cap are aligned. Tighten connecting rod bolts to 165 in.-lbs. (19 N·m) for 140000 models and to 100 in.-lbs. (11 N·m) for all other models.

PISTON, PIN AND RINGS. Pistons for use in engines having aluminum bore ("Kool Bore") are not interchangeable with those for use in cylinders having cast-iron sleeve. Pistons may be identified as follows: Those for use in cast-iron sleeve cylinders have a plain, dull aluminum finish, have an "L" stamped on top and use an oil ring expander. Those for use in aluminum bore cylinders are chrome plated (shiny finish), do not have an identifying letter and do not use an oil ring expander.

Reject pistons showing visible signs of wear, scoring and scuffing. If, after cleaning carbon from top ring groove, a new top ring has a side clearance of 0.007 inch (0.18 mm) or more, reject the piston. Reject piston or hone piston pin hole to 0.005 inch (0.13 mm) oversize if pin hole is 0.0005 inch (0.013 mm) or more out-of-round, or is worn to a diameter of 0.554 inch (14.07 mm) or more on 100000 engines, 0.673 inch (17.09 mm) or more on 140000 engines, or 0.491 inch (12.47 mm) or more on all other models.

If the piston pin is 0.0005 inch (0.013 mm) or more out-of-round, or is worn to a diameter of 0.552 inch (14.02 mm) or smaller on 100000 engines, 0.671 inch (17.04 mm) or smaller on 140000 engines, or 0.489 inch (12.42 mm) or smaller on all other models, reject pin.

The piston ring gap for new rings should be 0.010-0.025 inch (0.25-0.64 mm) for models with aluminum cylinder bore and 0.010-0.018 inch (0.25-0.46 mm) for models with cast-iron cylinder bore. On aluminum bore engines, reject compression rings having an end gap of 0.035 inch (0.80 mm) or more and reject oil rings having an end gap of 0.045 inch (1.14 mm) or more. On cast-iron bore engines, reject compression rings having an end gap of 0.030 inch (0.75 mm) or more and reject oil rings having an end gap of 0.035 inch (0.90 mm) or more.

Pistons and rings are available in several oversizes as well as standard.

A chrome ring set is available for slightly worn standard bore cylinders. Refer to note in CYLINDER section.

CYLINDER. If cylinder bore wear is 0.003 inch (0.76 mm) or more or is 0.0025 inch (0.06 mm) or more out-of-round, cylinder must be rebored to next larger oversize.

The standard cylinder bore sizes for each model are given in the following table.

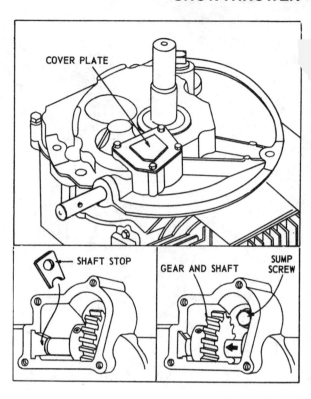

Fig. B108 — To remove oil sump (engine base) on 92000, 93500, 94900, 95500, 110900 or 113900 models with auxiliary pto, remove cover plate, shaft stop and retaining screw as shown.

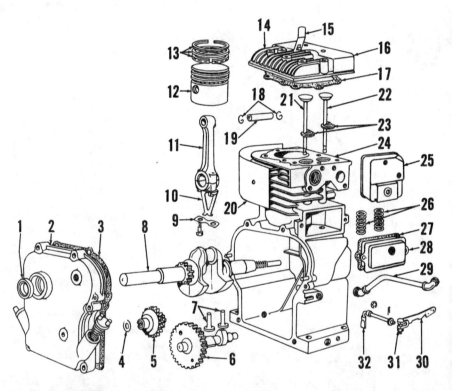

Fig. B109 — Exploded view of 100000 horizontal crankshaft engine assembly. Except for 112200, 130000, and late 140000 models, other horizontal crankshaft models with mechanical governor will have governor unit as shown in Fig. B90; otherwise, construction of all other horizontal crankshaft models is similar.

1. Crankshaft oil seal
2. Crankcase cover
3. Gasket
4. Thrust washer
5. Governor assy.
6. Cam gear & shaft
7. Tappets
8. Crankshaft
9. Rod bolt lock
10. Oil dipper
11. Connecting rod
12. Piston
13. Piston rings
14. Cylinder head
15. Spark plug ground switch
16. Air baffle
17. Cylinder head gasket
18. Piston pin retaining rings
19. Piston pin
20. Air baffle
21. Exhaust valve
22. Intake valve
23. Valve spring retainers
24. Cylinder block
25. Muffler
26. Valve springs
27. Gaskets
28. Breather & tappet chamber cover
29. Breather pipe
30. Governor lever
31. Clamping bolt
32. Governor crank

STANDARD CYLINDER BORE SIZES

Model	Cylinder diameter
6B, Early 60000	2.3115-2.3125 in. (58.71-58.74 mm)
Late 60000	2.3740-2.3750 in. (60.30-60.33 mm)
8B, 80000, 81000, 82000	2.3740-2.3750 in. (60.30-60.33 mm)
92000, 93500	2.5615-2.5625 in. (65.06-65.09 mm)
94500, 94900, 95500	2.5615-2.5625 in. (65.06-65.09 mm)
100000	2.4990-2.5000 in. (63.47-63.50 mm)
110900, 111200, 111900, 112200, 113900	2.7802-2.7812 in. (70.62-70.64 mm)
130000, 131400, 131900	2.5615-2.5625 in. (65.06-65.09 mm)
140000	2.7490-2.7500 in. (69.82-69.85 mm)

A hone is recommended for resizing cylinders. Operate hone at 300-700 rpm and with an up and down movement that will produce a 45° crosshatch pattern. Clean cylinder after honing with oil or soap suds. Always check availability of oversize piston and ring sets before honing cylinder.

Approved hones are as follows: For aluminum bore, use Ammco 3956 for rough and finishing or Sunnen AN200 for rough and Sunnen AN500 for finishing. For sleeved bores, use Ammco 4324 for rough and finishing, or Sunnen AN100 for rough and Sunnen AN300 for finishing.

NOTE: A chrome piston ring set is available for slightly worn standard bore cylinders. No honing or cylinder deglazing is required for these rings. The cylinder bore can be a maximum of 0.005 inch (0.01 mm) oversize when using chrome rings.

CRANKSHAFT AND MAIN BEARINGS. Except where equipped with ball bearings, the main bearings are an integral part of the crankcase and cover or sump. The bearings are renewable by reaming out the crankcase and cover or sump bearing bores and installing service bushings. The tools for reaming the crankcase and cover or sump, and for installing the service bushings are available from Briggs & Stratton. If the bearings are scored, out-of-round 0.0007 inch (0.018 mm) or more, or are worn to or larger than the reject sizes in the following table, ream the bearings and install service bushings.

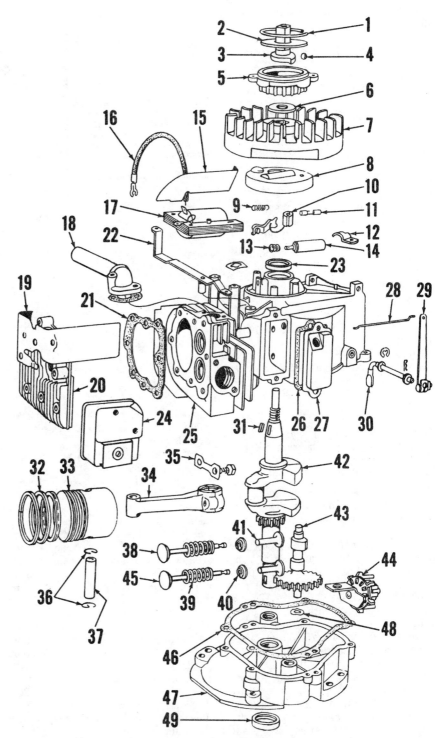

Fig. B110 — Exploded view of 100000 vertical crankshaft engine with mechanical governor. Models 130000, 131000 and 140000 vertical crankshaft models with mechanical governor are similar. Refer to Fig. B107 for typical vertical crankshaft model with air vane governor.

1. Snap ring
2. Washer
3. Starter ratchet
4. Steel balls
5. Starter clutch
6. Washer
7. Flywheel
8. Breaker cover
9. Breaker arm spring
10. Breaker arm & pivot
11. Breaker plunger
12. Condenser clamp
13. Primary wire retainer spring
14. Condenser
15. Air baffle
16. Spark plug wire
17. Armature & coil assy.
18. Intake pipe
19. Air baffle
20. Cylinder head
21. Cylinder head gasket
22. Linkage lever
23. Crankshaft oil seal
24. Muffler
25. Cylinder block
26. Gasket
27. Breather & tappet chamber cover
28. Governor link
29. Governor crank
30. Governor crank
31. Flywheel key
32. Piston rings
33. Piston
34. Connecting rod
35. Rod bolt lock
36. Piston pin retaining rings
37. Piston pin
38. Intake valve
39. Valve springs
40. Valve spring retainers
41. Tappets
42. Crankshaft
43. Cam gear
44. Governor & oil slinger assy.
45. Exhaust valve
46. Gasket
47. Oil sump (engine base)
48. Thrust washer
49. Crankshaft oil seal

MAIN BEARING REJECT SIZES

Model	Magneto Bearing	PTO Bearing
6B, 60000	0.878 in. (22.30 mm)	0.878 in.* (22.30 mm)*
8B, 80000	0.878 in. (22.30 mm)	0.878 in.* (22.30 mm)*
81000, 82000	0.878 in. (22.30 mm)	0.878 in.* (22.30 mm)*
92000, 93500, 94000, 94500, 94900, 95500	0.878 in. (22.30 mm)	0.878 in.* (22.30 mm)*
100000	0.878 in. (22.30 mm)	1.003 in. (25.48 mm)
110900, 111200, 111900, 112200, 113900	0.878 in. (22.30 mm)	0.878 in.* (22.30 mm)*
130000, 131000	0.878 in. (22.30 mm)	1.003 in. (25.48 mm)
140000	1.004 in. (25.50 mm)	1.185 in. (30.10 mm)

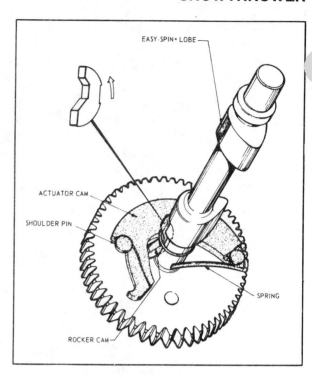

Fig. B111—Compression release camshaft used on 111200, 111900 and 112200 engines. At cranking speed, spring holds actuator cam inward against the rocker cam and rocker cam is forced above exhaust cam surface.

*All models equipped with auxiliary drive unit have a main bearing rejection size for main bearing at pto side of 1.003 inch (25.48 mm)

Bushings are not available for all models; therefore, on some models the sump must be renewed if necessary to renew the bearing.

Main bearing journal diameter for flywheel and pto side main bearing journals on all standard models with plain bearings, except 130000, 131000 and 140000 models, is 0.873 inch (22.17 mm). On models equipped with auxiliary drive unit, pto side main bearing journal diameter is 0.998 inch (25.35 mm). Main bearing journal diameter at flywheel side of 130000 and 131000 models is 0.873 inch (22.17 mm) and journal diameter at pto side is 0.998 inch (25.35 mm). Main bearing journal diameter at flywheel side of 140000 model is 0.997 inch (25.32 mm) and journal diameter at pto side is 1.179 inch (29.95 mm).

Crankpin journal rejection size for 6B and 60000 models is 0.870 inch (22.10 mm), crankpin journal rejection size for 140000 model is 1.090 inch (27.69 mm) and crankpin journal rejection size for all other models if 0.996 inch (25.30 mm).

Ball bearing mains are a press fit on the crankshaft and must be removed by pressing the crankshaft out of the bearing. Reject ball bearing if worn or rough.

Expand new bearing by heating it in oil and install it on crankshaft with seal side towards crankpin journal.

Crankshaft end play on all models is 0.002-0.008 inch (0.05-0.20 mm). At least one 0.015 inch cover or sump gasket must be in place. Additional gaskets in several sizes are available to aid in end play adjustment. If end play is over 0.008 inch (0.20 mm), metal shims are available for use on crankshaft between crankshaft gear and cylinder.

Refer to VALVE TIMING section for proper timing procedure when installing crankshaft.

CAMSHAFT. The camshaft and camshaft gear are an integral casting on all models. The camshaft and gear should be inspected for wear on the journals, cam lobes and gear teeth. On all standard models except 110900, 111200, 111900, 112200 and 113900 models, if camshaft journal diameter is 0.498 inch (12.65 mm) or less, reject camshaft.

On 110900, 111200, 111900, 112200 and 113900 models, camshaft journal rejection size for journal on flywheel side is 0.436 inch (11.07 mm) and for pto side journal rejection size is 0.498 inch (12.65 mm).

On models with "Easy-Spin" starting, the intake cam lobe is designed to hold the intake valve slightly open on part of the compression stroke. Therefore, to check compression, the engine must be turned backwards.

"Easy-Spin" camshafts (cam gears) can be identified by two holes drilled in the web of the gear. Where part number of an older cam gear and an "Easy-Spin" cam gear are the same (except for an "E" following the "Easy-Spin" part number), the gears are interchangeable.

The camshaft used on 11200, 11900 and 112200 models is equipped with the "Easy-Spin" intake lobe and a mechanically operated compression release on the exhaust lobe. With engine stopped or at cranking speed, the spring holds the actuator cam weight inward against the rocker cam. See Fig. B111. The rocker cam is held slightly above the exhaust cam surface which in turn holds the exhaust valve open slightly during compression stroke. This compression release greatly reduces the power needed for cranking.

When engine starts and rpm is increased, the actuator cam weight moves outward overcoming the spring pressure. See Fig. B111A. The rocker cam is rotated below the cam surface to provide normal exhaust valve operation.

Refer to VALVE TIMING section for proper timing procedure when installing camshaft.

VALVE SYSTEM. Valve tappet clearance (cold) for 94500, 94900 and 95500 models is 0.004-0.006 inch (0.10-0.15 mm) for intake valve and 0.007-0.009 inch (0.18-0.23 mm) for exhaust valve.

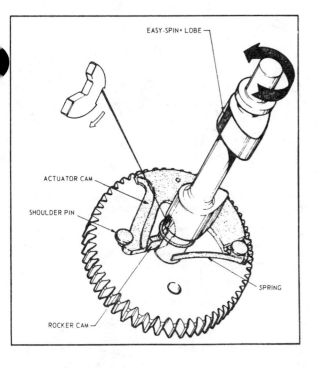

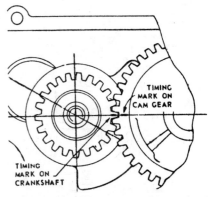

Fig. B111A—When engine starts and rpm is increased, actuator cam weight moves outward allowing rocker cam to rotate below exhaust cam surface.

Fig. B113—Align timing marks on cam gear and crankshaft gear on plain bearing models.

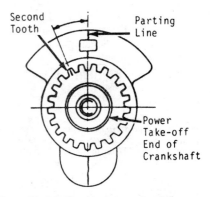

Fig. B114—location of tooth to align with timing mark on cam gear if mark is not visible on crankshaft gear.

Valve tappet clearance (cold) for all other models is 0.005-0.007 inch (0.13-0.18 mm) for intake valve and 0.009-0.011 inch (0.23-0.28 mm) for exhaust valve.

Valve tappet clearance is adjusted on all models by carefully grinding off end of valve stem to increase clearance or by grinding valve seats deeper and/or renewing valve or lifter to decrease clearance.

Valve face and seat angle should be ground at 45°. Renew valve if margin is 1/16 inch (1.588 mm) or less. Seat width should be 3/64 to 1/16 inch (1.119-1.588 mm).

The valve guides on all engines with aluminum blocks are an integral part of the cylinder block. To renew the valve guides, they must be reamed out and a bushing installed.

On all models except 140000 model, ream guide out first with reamer (B&S part 19064) only about 1/16 inch (1.588 mm) deeper than length of bushing (B&S part 63709). Press in bushing with driver (B&S part 19065) until it is flush with top of guide bore. Finish ream the

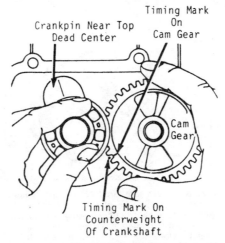

Fig. B112—Align timing marks on cam gear with mark on crankshaft counterweight on ball bearing equipped models.

bushing with a finish reamer (B&S part 19066).

On 140000 model, ream guide out with reamer (B&S part 19183) only about 1/16 inch (1.588 mm) deeper than top end of flutes on reamer. Press in

bushings with soft driver as bushing is finish reamed at the factory.

VALVE TIMING. On engines equipped with ball bearing mains, align the timing mark on the cam gear with the timing mark on the crankshaft counterweight as shown in Fig. B112. On engines with plain bearings, align the timing marks on the camshaft gear with the timing mark on the crankshaft gear as shown in Fig. B113. If the timing mark is not visible on the crankshaft gear, align the timing mark on the camshaft gear with the second tooth to the left of the crankshaft counterweight parting line as shown in Fig. B114.

SERVICING BRIGGS & STRATTON ACCESSORIES

WINDUP STARTER

STARTER OPERATION (EARLY UNITS). Refer to Figs. B115 and B116. Turning the knob (17) to "CRANK" position holds the engine flywheel stationary. Turning the crank (1) winds up the power spring within the spring and

cup assembly (8). Spring tension is held by the clutch ratchet (15) and the crank ratchet pawl (2). On some starters, an additional ratchet pawl (2A – Fig. B116) allows the crank to be reversed when there is not enough room for a full turn of crank. Approximately 5½ turns of crank are required to fully wind the

power spring. Turning the holding knob (17) to "START" position releases the engine flywheel allowing the power spring to turn the engine.

NOTE: Do not turn knob to "CRANK" position when engine is running.

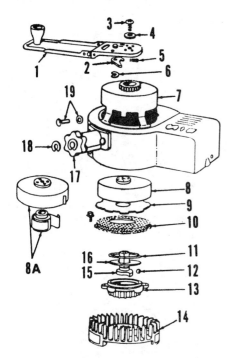

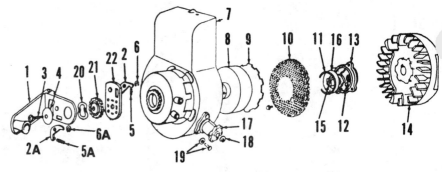

Fig. B116—Exploded view of typical Briggs & Stratton windup starter equipped with ratchet type windup crank. Starter is used where space does not permit full turn of crank.

1. Crank	6. Snap ring
2. Pawl	6A. Snap ring
2A. Pawl	7. Blower housing
3. Screw	8. Power spring &
4. Washer	cup assy.
5. Spring	9. Retainer plate
5A. Spring	10. Screen

11. Snap ring	17. Flywheel holder
12. Clutch balls	18. Snap ring
13. Clutch housing	19. Rivet & burr
14. Flywheel	20. Wave washer
15. Clutch ratchet	21. Ratchet
16. Ratchet cover	22. Pawl housing

Fig. B115—Exploded view of typical early production windup starter. Refer to Fig. B116 for exploded view of starter using a ratchet type crank and to Fig. B117 for late production starter.

1. Crank	9. Retainer plate
2. Pawl	10. Screen
3. Screw	11. Snap ring
4. Washer	12. Clutch balls
5. Spring	13. Clutch housing
6. Snap ring	14. Flywheel
7. Blower housing	15. Starter ratchet
8. Power spring &	16. Ratchet cover
cup assy.	17. Flywheel holder
8A. Power spring &	18. Snap ring
cup assy.	19. Rivet & burr

If engine does not start readily, check to be sure choke is in fully closed position and that ignition system is delivering a spark.

STARTER OPERATION (LATE UNITS). Late production windup starter assemblies have a control lever (Fig. B117) instead of a flywheel holder (17—Fig. B115 and B116). Moving the control lever to "CRANK" position locks the hub of the starter spring assembly (see lower view, Fig. B117) allowing starter spring to be wound up by turning crank. Moving the control lever to "START" position releases the spring assembly hub. The spring unwinds, engaging the starter ratchet, and cranks up engine.

CAUTION: Never work on engine or driven machinery if starter is wound up. If attached machinery such as mower blade is jammed up so starter will not turn engine, remove blower housing and starter assembly from engine to allow starter to unwind.

OVERHAUL (EARLY UNIT). Remove starter and blower housing from engine. Check condition of

Fig. B117—On late production units, control lever (top view) locks hub of spring assembly (bottom) to allow spring to be wound up by turning crank. To remove spring assembly, first remove screw (3—Fig. B118), washer (4) and crank (1). Bend tangs of blower housing out as shown in bottom view and lift cup, spring and control lever assembly from housing. Note spring washer placed between cup and washer.

flywheel and flywheel holder assembly. Check action of clutch ratchet (15—Fig. B115 or B116). Remove snap ring (11) to disassemble clutch unit. Bend tangs of blower housing back to remove retainer (9). Remove screw (3) retaining crank and ratchet assembly to power spring cup (8).

Flywheel holder assembly may be renewed as a unit using new rivets and washers (19) after cutting old assembly from blower housing. Knob (17) and snap ring (18) are available separately.

On early models, power spring and cup (8A) are available separately. On later units, spring and cup (8) are available only as an assembly.

Renew parts as necessary and reassemble as follows: Place ratchet (15)

Fig. B118—Exploded view of late production windup starter assembly. Control lever (L) locks ratchet drive hub to allow starter to be wound up and releases hub to crank engine.

L. Control lever	9. Cup, spring &
1. Crank	release assy.
2. Pawl	10. Sems screws
3. Screw	11. Rotating screen
4. Washer	12. Seal
5. Spring	13. Starter ratchet
6. Snap ring	14. Ratchet cover
7. Blower housing	15. Steel balls
8. Spring washer	16. Clutch housing

in cup (13), drop the balls (12) in beside ratchet and install washer (16) and snap ring (11). Reassemble blower housing and starter unit in reverse of disassembly procedure.

OVERHAUL (LATE UNIT). Refer to Fig. B118 and move control lever (L)

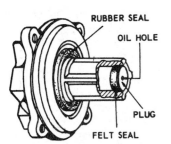

Fig. B119—Cutaway view of late type starter ratchet showing sealing felt and plug in outer end. Oil ratchet through hole in plug. A rubber seal is also used at ratchet cover.

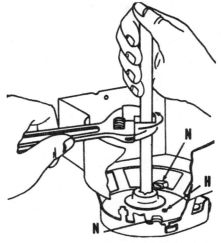

Fig. B121—Using square shaft and wrench to windup the rewind starter spring. Refer to text.

H. Hole in starter pulley N. Nylon bumpers

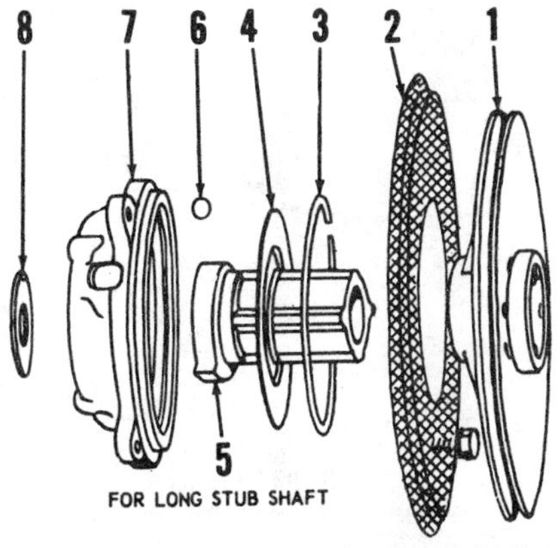

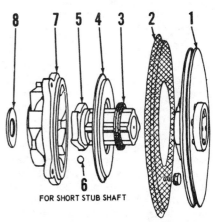

Fig. B123—Exploded view of early production starter clutch unit. Refer to Fig. B126 for view of "long stub shaft". A late type unit (Fig. B125) should be installed when renewing "long" crankshaft with "short" (late production) shaft.

1. Starter rope pulley
2. Rotating screen
3. Snap ring
4. Ratchet cover
5. Starter ratchet
6. Steel balls
7. Clutch housing (flywheel nut)
8. Spring washer

to "START" position. If starter is wound up and will not turn engine, hold crank (1) securely and loosen screw (3). Remove blower housing and starter unit from engine and, if not already done, remove screw (3) and crank (1). Turn blower housing over and bend tangs outward as shown in bottom view of Fig. B117; then lift cup, spring and control lever assembly from the blower housing.

CAUTION: Do not attempt to disassemble the cup and spring assembly; unit is serviced as a complete assembly only.

When reassembling unit, renew the spring washer (8–Fig. B118) if damaged in any way. Grease the mating surfaces of spring cup and blower housing and stick spring washer in place with grease. Renew pawl (2) or pawl spring (5) on crank assembly (1) if necessary. Check condition of ratchet teeth on blower housing. Renew housing if teeth are broken or worn off. Reassemble by reversing disassembly procedure.

Starter ratchet cover (14) and ratchet (13) can be removed after removing the Sems screws (10) and rotating screen (11). Be careful not to lose the steel balls (15). Clutch housing (16) also is the flywheel retaining nut. To remove housing, hold flywheel and turn housing counterclockwise. A special wrench (B&S 19114) is available for removing and installing housing.

To reassemble, first be sure spring washer is in place on crankshaft with

cup (hollow) side towards the flywheel. Install starter clutch housing and tighten securely. Place starter ratchet on crankshaft and drop the steel balls in place in housing. Reinstall ratchet cover with new seal (12) if required. Reinstall the rotating screen and install blower housing and starter assembly.

REWIND STARTERS

OVERHAUL. To renew broken rewind spring, grasp free outer end of spring (S–Fig. B120) and pull broken end from starter housing. With blower housing removed, bend nylon bumpers (N) up and out of the way and remove starter pulley from housing. Untie knot in rope (R) and remove rope and inner end of broken spring from pulley. Apply a small amount of grease on inner face of pulley, thread inner end of new spring in pulley hub (on older models, install retainer in hub with split side of retainer away from spring hook) and place pulley in housing. Renew nylon bumpers if necessary and bend bumpers down to within 1/8 inch (3.175 mm) of the pulley. Insert a 3/4-inch (19 mm) square bar in pulley hub and turn pulley approximately 13 1/2 turns in a counterclockwise

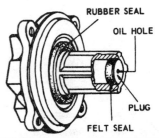

Fig. B124—View of late production sealed starter clutch unit. Late unit can be used with "short stub shaft" only. Refer to Fig. B126. Refer to Fig. B125 for cutaway view of ratchet (5).

1. Starter rope pulley
2. Rotating screen
3. Rubber seal
4. Ratchet cover
5. Starter ratchet
6. Steel balls
7. Clutch housing (flywheel nut)
8. Spring washer

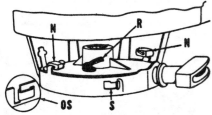

Fig. B120—View of Briggs & Stratton rewind starter assembly.

N. Nylon bumpers R. Starter rope
OS. Old style spring S. Rewind spring

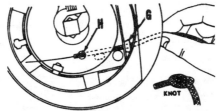

Fig. B122—Threading starter rope through guide (G) in blower housing and hole (H) in starter pulley with wire hooked in end of rope. Tie knot in end of rope as shown.

Fig. B125—Cutaway view showing felt seal and plug in end of late production starter ratchet (5–Fig. B124).

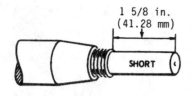

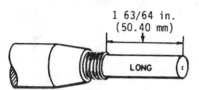

Fig. B126 — Crankshaft with short stub (top view) must be used with late production starter clutch assembly. Early crankshaft (bottom view) can be modified by cutting off stub end to the dimension shown in top view and beveling end of shaft to allow installation of late type clutch unit.

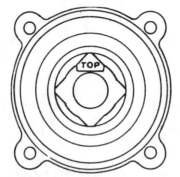

Fig. B128 — When installing blower housing and starter assembly, turn starter ratchet so word "TOP" stamped on outer end of ratchet is towards engine cylinder head.

direction as shown in Fig. B121. Tie wrench to blower housing with wire to hold pulley so hole (H) in pulley is aligned with rope guide (G—Fig. B122) in housing. Hook a wire in inner end of rope and thread rope through guide and hole in pulley. Tie a knot in rope and release the pulley allowing the spring to wind the rope into the pulley groove.

To renew starter rope only, it is not generally necessary to remove starter pulley and spring. Wind up the spring and install new rope as outlined in preceding paragraph.

Two different types of starter clutches have been used. Refer to exploded view of early production unit in Fig. B123 and exploded view of late production unit in Fig. B124. The outer end of the late production ratchet (refer to cutaway view in Fig. B125) is sealed with a felt and a retaining plug and a rubber ring is used to seal ratchet to ratchet cover.

To disassemble early type starter clutch unit, refer to Fig. B123. Remove snap ring (3). Lift ratchet (5) and cover (4) from starter housing (7) and crankshaft. Be careful not to lose the steel balls (6). Starter housing (7) is also flywheel retaining nut. To remove housing, first remove screen (2) and using B&S flywheel wrench 19114, unscrew

housing from crankshaft in counterclockwise direction. When reinstalling housing, be sure spring washer (8) is placed on crankshaft with cup (hollow) side towards flywheel. Install starter housing and tighten securely. Reinstall rotating screen. Place ratchet on crankshaft and into housing, then insert the steel balls. Reinstall cover and retaining snap ring.

To disassemble late starter clutch unit, refer to Fig. B124. Remove rotating screen (2) and starter ratchet

cover (4). Lift ratchet (5) from housing and crankshaft and extract the steel balls (6). If necessary to remove housing (7), hold flywheel and unscrew housing in counterclockwise direction using B&S flywheel wrench 19114. When installing housing, be sure spring washer (8) is in place on crankshaft with cup (hollow) side towards flywheel. Tighten housing securely. Inspect felt seal and plug in outer end of ratchet. Renew ratchet if seal or plug is damaged as these parts are not serviced separately. Lubricate the felt with oil and place ratchet on crankshaft. Insert the steel balls and install ratchet cover, rubber seal and rotating screen.

NOTE: Crankshafts used with early and late starter clutches differ. Refer to Fig. B126. If renewing early (long) crankshaft with late (short) shaft, also install late type starter clutch unit. If renewing early starter clutch with late type unit, the crankshaft must be shortened to the dimension shown for short shaft in Fig. B126. Also, hub of starter rope pulley must be shortened to ½-inch (12.7 mm) dimension shown in Fig. B127. Bevel end of crankshaft after removing the approximate ⅜ inch (9.525 mm) from shaft.

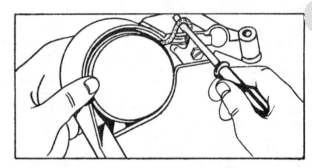

Fig. B129 — Use a screwdriver to pull rope up to provide about one foot (305 mm) of slack.

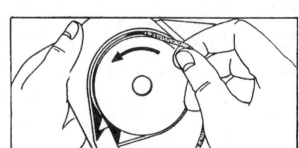

Fib. B130 — Rotate pulley and rope 3 turns counterclockwise to remove spring tension from rope.

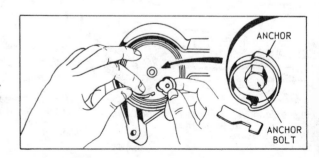

Fig. B131 — Remove anchor bolt and spring anchor to release inner end of spring.

Fig. B127 — When installing a late type starter clutch unit as replacement for early type, either install new starter rope pulley or cut hub of old pulley to dimension shown.

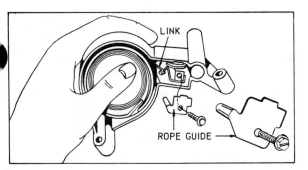

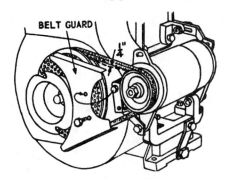

Fig. B132 — Rope guide removed from Vertical Pull starter. Note position of friction link.

Fig. B137 — View showing starter-generator belt adjustment on models so equipped. Refer to text.

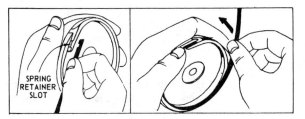

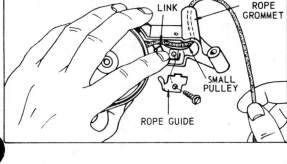

Fig. B133 — Hook outer end of spring in retainer slot, then coil the spring counter-clockwise in the housing.

Fig. B134 — When installing pulley assembly in housing, make sure friction link is positioned as shown, then install rope guide.

ly pry off the plastic spring cover. Refer to Fig. B131 and remove anchor bolt and spring anchor. Carefully remove spring from housing. Unbolt and remove rope guide. Note position of the friction link in Fig. B132. Remove old rope from pulley.

It is not necessary to remove the gear retainer unless pulley or gear is damaged and renewal is necessary. Clean and inspect all parts. The friction link should move the gear to both extremes of its travel. If not, renew the linkage assembly.

Install new spring by hooking end of retainer slot and winding until spring is coiled in housing. See Fig. B133. Insert new rope through the housing and into the pulley. Tie a small knot, heat seal the knot and pull it tight into the recess in pulley. Install pulley assembly in the housing with friction link in pocket of casting as shown in Fig. B134. Install rope guide. Rotate pulley counterclockwise until rope is fully wound. Hook free end of spring to anchor, install screw and tighten to 75-90 in.-lbs. (8-11 N·m). Lubricate spring with a small amount of engine oil. Snap the plastic spring cover in place. Preload spring by pulling rope up about one foot (305 mm), then winding rope and pulley 2 or 3 turns clockwise.

When installing blower housing and starter assembly, turn starter ratchet so word "TOP" on ratchet is towards cylinder head.

VERTICAL PULL STARTER

OVERHAUL. To renew rope or spring, first remove all spring tension from rope. Using a screwdriver as shown in Fig. B129, pull rope up about one foot (305 mm). Wind rope and pulley counterclockwise 3 turns as shown in Fig. B130 to remove all tension. Careful-

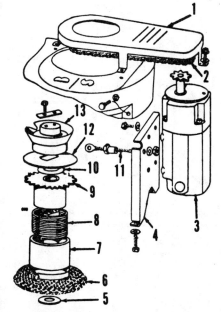

Fig. B135 — Exploded view of early production 110 volt electric starter used on vertical crankshaft models. Starter used on horizontal crankshaft models was similar.

1. Chain shield
2. Drive chain
3. Electric motor
4. Mounting bracket
5. Washer
6. Rotating screen
7. Shaft adapter
8. Spring
9. Sprocket & hub
10. Thrust washer
11. Adjusting screw
12. Guard washer
13. Rope starter pulley

Fig. B136 — Exploded view of late production 110 volt electric starter unit. Starters for horizontal crankshaft models are similar.

1. Belt shield
2. Belt guide
3. Clutch assy.
4. Cover
5. Commutator brush assy.
6. Connector plug
7. Electric motor
8. Drive belt
9. Drive pulley
10. Blower housing

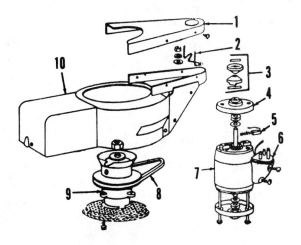

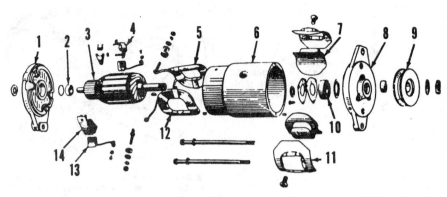

Fig. B138 – Exploded view of Delco-Remy starter-generator unit used on some B&S engines.

1. Commutator end frame
2. Bearing
3. Armature
4. Ground brush holder
5. Field coil
6. Frame
7. Pole shoe
8. Drive end frame
9. Pulley
10. Bearing
11. Field coil insulator
12. Field coil
13. Brush
14. Insulated brush holder

110-VOLT ELECTRIC STARTERS (CHAIN AND BELT DRIVE)

Early production 110-volt electric starter with chain drive is shown in Fig. B135. Later type with belt drive is shown in Fig. B136. Starters for horizontal crankshaft engines are similar to units shown for vertical shaft engines.

Chain (2 – Fig. B135) on early models is adjusted by changing position of nuts on adjusting stud (11) so chain deflection is approximately ¼ inch (6.35 mm) between sprockets. There should be about 1/32-inch (0.8 mm) clearance between spring (8) and sprocket (9) when unit is assembled. Clutch unit (7, 8 and 9) should disengage when engine starts.

Belt (8 – Fig. B135) tension is adjusted by shifting starter motor in slotted mounting holes so clutch unit (3) on starter motor will engage belt and turn engine, but so belt will not turn starter when clutch is disengaged.

CAUTION: Always connect starter cord to starter before plugging cord into 110-volt outlet and disconnect cord from outlet before removing cord from starter connector. Do not run the electric starter for more than 60 seconds at a time.

12-VOLT STARTER-GENERATOR UNIT

The combination starter-generator functions as a cranking motor when the starting switch is closed. When engine is operating and with starting switch open, the unit operates as a generator. Generator output and circuit voltage for the battery and various operating requirements are controlled by a current-voltage regulator. On units where voltage regulator is mounted separately from generator unit, do not mount regulator with cover down as regulator

will not function in this position. To adjust belt tension, apply approximately 30 pounds (14 kg) pull on generator adjusting flange and tighten mounting bolts. Belt tension is correct when a pressure of 10 pounds (5 kg) applied midway between pulleys will deflect belt ¼ inch (6.35 mm). See Fig. B137. On units equipped with two drive belts, always renew belts in pairs. A 50-ampere hour capacity battery is recommended. Starter-generator units are intended for use in temperatures above 0° F (–18° C). Refer to Fig. B138 for exploded view of starter-generator. Parts and service on the

starter-generator are available at authorized Delco-Remy service centers.

12-VOLT ELECTRIC STARTER (BELT DRIVE)

Refer to Fig. B139. Adjust position of starter (10) so clutch will engage belt and turn engine when starter is operated, but so belt will not turn starter when engine is running. The 12-volt electric starter is intended for use in temperatures above 15° F (–9° C). Driven equipment should be disengaged before using starter. Parts and service on starter motor are available at Autolite service centers.

CAUTION: Do not use jumper (booster) battery as this may result in damage to starter motor.

GEAR DRIVE STARTERS

Two types of gear drive starters may be used, a 110 volt AC starter or a 12 volt DC starter. Refer to Fig. B140 for an exploded view of starter motor. A properly grounded receptacle should be used with power cord connected to 110 volt AC starter motor. A 32 ampere hour capacity battery is recommended for use with 12 volt DC starter motor.

To renew a worn or damaged flywheel ring gear, drill out retaining rivets using a 3/16-inch drill. Attach new ring gear

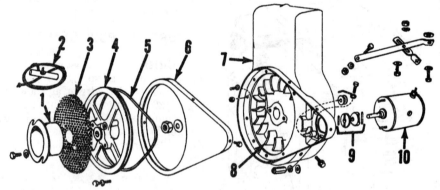

Fig. B139 – Exploded view of Briggs & Stratton 12 volt starter used on some engines. Parts and/or service on starter motor are available through an authorized Autolite service center.

1. Rope pulley
2. Starter rope
3. Rotating screen
4. Starter driven pulley
5. Drive belt
6. Belt guard
7. Blower housing
8. Flywheel
9. Starter clutch
10. Starter motor.

Fig. B140 – View of 110 volt AC starter motor. 12 volt DC starter motor is similar. Rectifier and switch unit (8) is used on 110 volt motor only.

1. Pinion gear
2. Helix
3. Armature shaft
4. Drive cap
5. Thrust washer
6. Housing
7. End cap
8. Rectifier & switch
9. Bolt
10. Nut

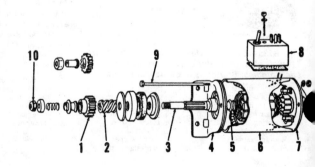

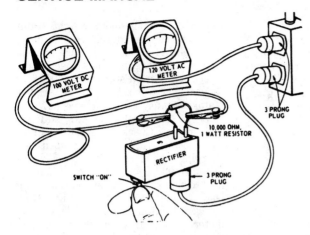

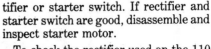

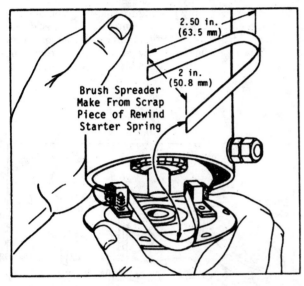

Fig. B141—Test connections for 110 volt rectifier. Refer to text for procedure.

Brush Spreader Make From Scrap Piece of Rewind Starter Spring

2.50 in. (63.5 mm)

2 in. (50.8 mm)

Fig. B142—Tool shown may be fabricated to hold brushes when installing motor and cap.

tifier or starter switch. If rectifier and starter switch are good, disassemble and inspect starter motor.

To check the rectifier used on the 110 volt AC starter motor, remove rectifier unit from starter motor. Solder a 10,000 ohm, 1 watt resistor to the DC internal terminals of the rectifier as shown in Fig. B141. Connect a 0-100 range DC voltmeter to resistor leads. Measure the voltage of the AC outlet to be used. With starter switch in "OFF" position, a zero reading should be shown on DC voltmeter. With starter switch in "ON" position, the DC voltmeter should show a reading that is 0-14 volts lower than AC line voltage measured previously. If voltage drop exceeds 14 volts, renew rectifier unit.

Disassembly of starter motor is evident after inspection of unit and referral to Fig. B140. Note position of bolts (9) during disassembly so they can be installed in their original positions during reassembly. When reassembling motor, lubricate end cap bearings with SAE 20 oil. Be sure to match the drive cap keyway to the stamped key in the housing when sliding the armature into the motor housing. Brushes may be held in their holders during installation by making a brush spreader tool from a piece of metal as shown in Fig. B142. Splined end of helix (2–Fig. B140) must be towards end of armature shaft as shown in Fig. B143. Tighten armature shaft nut to 170 in-lbs. (19 N·m).

using screws provided with new ring gear.

To check for correct operation of starter motor, remove starter motor from engine and place motor in a vise or other holding fixture. Install a 0-5 amp ammeter in power cord to 110 volt AC starter motor. On 12 volt DC motor, connect a 6 volt battery to motor with a 0-50 amp ammeter in series with positive (+) line from battery starter motor. Connect a tachometer to drive end of starter. With starter activated on 110

volt motor, starter motor should turn at 5200 rpm minimum with a maximum current draw of 3½ amps. The 12 volt motor should turn at 5000 rpm minimum with a current draw of 25 amps maximum. If starter motor does not operate satisfactorily, check operation of rec-

FLYWHEEL ALTERNATOR

Early Type External

Refer to Fig. B144 for assembled view and to Fig. B145 for exploded view of

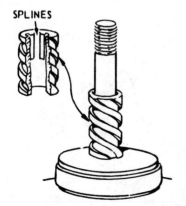

SPLINES

Fig. B143—Install helix on armature so splines of helix are to the outer end as shown.

Fig. B144—Assembled and installed view of early type 12 volt flywheel alternator. Refer to Fig. B145 for exploded view of unit.

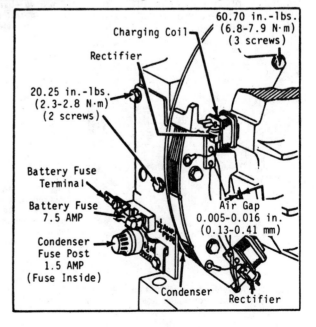

Charging Coil

Rectifier

60.70 in.-lbs. (6.8-7.9 N·m) (3 screws)

20.25 in.-lbs. (2.3-2.8 N·m) (2 screws)

Battery Fuse Terminal

Battery Fuse 7.5 AMP

Condenser Fuse Post 1.5 AMP (Fuse Inside)

Air Gap 0.005-0.016 in. (0.13-0.41 mm)

Condenser

Rectifier

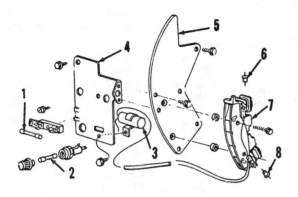

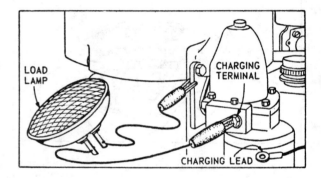

Fig. B145 — Exploded view of early type 12 volt flywheel alternator designed for use with the 12 volt starter motor unit shown in Fig. B139.

1. Fuse (3AG 7½ amp)
2. Fuse (AG 1½ amp)
3. Condenser
4. Support plate
5. Alternator plate
6. Rectifier
7. Armature & coil assy.
8. Rectifier

alternator. The 12 volt flywheel alternator is designed for use with the 12 volt electric starter and a 20 to 24 ampere hour capacity battery.

Armature air gap should be 0.005 inch (0.13 mm). Rectifiers (diodes) are available separately from the armature and coil unit which is available as an assembly only. When renewing the condenser or fuses, be sure to use the correct Briggs & Stratton parts.

No cut-out is required with this alternator as the rectifiers (6) prevent reverse flow of electricity through the alternator.

1½ Amp Nonregulated

Some engines are equipped with 1½ amp nonregulated flywheel alternator. This alternator (Fig. B146) with a solid state rectifier, is designed for use with a compact battery. A 12 ampere hour battery is suggested for warm temperature operation and a 24 ampere hour battery should be used in cold service. The alternator is rated at 3600 rpm. At lower speeds, available output is reduced.

To test for alternator output, disconnect charging lead from charging terminal. Connect a 12 volt lamp between charging terminal and ground as shown in Fig. B147. Start engine. If lamp lights, alternator is functioning. If lamp

does not light, alternator is defective.

To check for faulty stator, disconnect charging lead from battery and rectifier. Remove rectifier box mounting screw. Turn box to expose eyelets to which red and black stator leads are soldered. Start engine and with engine operating, touch load lamp leads to eyelets as in

Fig. B148. If lamp does not light, stator or flywheel is defective. Remove flywheel and check to be sure magnet ring is in place and has magnetism. Check wires from stator to rectifier. If flywheel and wiring is good, renew stator assembly which includes new rectifier. If the load lamp lights, stator is satisfactory. Check for faulty rectifier as follows: With engine stopped and charging lead disconnected from charging terminal, connect one ohmmeter lead to charging terminal and remaining lead to engine block. See Fig. B149. Check for continuity, then reverse ohmmeter leads and again check for continuity. Ohmmeter should show a continuity reading for one direction only. If tests show no continuity in either direction or continuity in both directions, rectifier is faulty and must be renewed. Rectifier assembly is serviced separately.

Fig. B147 — Disconnect charging lead and connect load lamp to test alternator output.

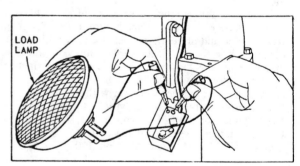

Fig. B148 — Use 12 volt load lamp to test for defective stator. Refer to text.

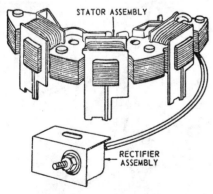

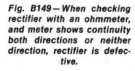

Fig. B146 — Stator and rectifier assembly used on the 1½ amp nonregulated flywheel alternator.

Fig. B149 — When checking rectifier with an ohmmeter, and meter shows continuity both directions or neither direction, rectifier is defective.

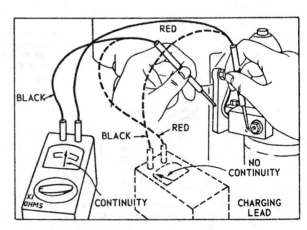

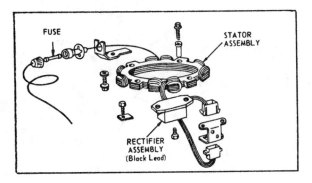

Fig. B150—Stator and rectifier assemblies used on the 4 amp nonregulated flywheel alterator. Fuse is 7½ amp AGC or 3AG.

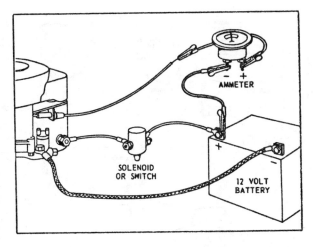

Fig. B151—install ammeter as shown for output test.

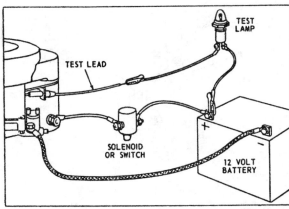

Fig. B152—Connect a test lamp as shown to test for shorted stator or defective rectifier. Refer to text.

4 Amp Nonregulated

Some engines are equipped with the 4 amp nonregulated flywheel alternator shown in Fig. B150. A solid-state rectifier and 7½ amp fuse is used with this alternator.

If battery is run down and no output from alternator is suspected, first check the 7½ amp fuse. If fuse is good, clean and tighten all connections. Disconnect charging lead and connect an ammeter as shown in Fig. B151. Start engine and check for alternator output. If ammeter shows no charge, stop engine, remove ammeter and install a test lamp as shown in Fig. B152. Test lamp should not light. If it does light, stator or rectifier is defective. Unplug rectifier plug under blower housing. If test lamp goes out, rectifier is defective. If test lamp does not go out, stator is shorted.

If shorted stator is indicated, use an ohmmeter and check continuity as follows: Touch one test lead to lead inside of fuse holder as shown in Fig. B153. Touch the remaining test lead to each of the four pins in rectifier connector. Unless the ohmmeter shows continuity at each of the four pins, stator winding is open and stator must be renewed.

If defective rectifier is indicated, unbolt and remove the flywheel blower housing with rectifier. Connect one ohmmeter test lead to blower housing and remaining lead to the single pin connector in rectifier connector. See Fig. B154. Check for continuity, then reverse leads and again test for continuity. If tests show no continuity in either direction or continuity in both directions, rectifier is faulty and must be renewed.

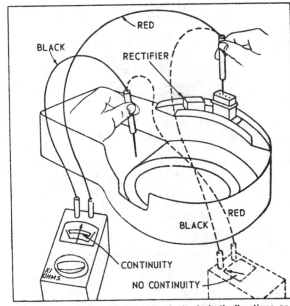

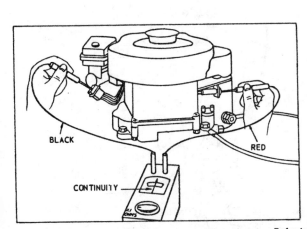

Fig. B153—Use an ohmmeter to check condition of stator. Refer to text.

Fig. B154—If ohmmeter shows continuity in both directions or in neither direction, rectifier is defective.

7 Amp Regulated

A 7 amp regulated flywheel alternator is used with the 12 volt gear drive starter motor. The alternator is equipped with a solid state rectifier and regulator. An isolation diode is also used on most models.

If engine will not start, using electric start system, and trouble is not in starting motor, install an ammeter in circuit as shown in Fig. B156. Start engine manually. Ammeter should indicate charge. If ammeter does not show battery charging taking place, check for defective wiring and if necessary proceed with troubleshooting.

If battery charging occurs with engine running, but battery does not retain charge, then isolation diode may be defective. The isolation diode is used to prevent battery drain if alternator circuit malfunctions. After troubleshooting diode, remainder of circuit should be inspected to find reason for excessive battery drain. To check operation of diode, disconnect white lead of diode from fuse holder and connect a test lamp from the diode white lead to the negative terminal of battery. Test lamp should not light. If test lamp lights, diode is defective. Disconnect test lamp and disconnect red lead of diode. Test continuity of diode with ohmmeter by connecting leads of ohmmeter to leads of diode, then reverse lead connections. The ohmmeter should show continuity in one direction and an open circuit in the other direction. If readings are incorrect, then diode is defective and must be renewed.

To troubleshoot alternator assembly, disconnect white lead of isolation diode from fuse holder and connect a test lamp between positive (+) terminal of battery and fuse holder on engine. Engine must not be started. With connections made, test lamp should not light. If isolation diode operates correctly and test lamp does light, stator, regulator or rectifier is defective. Unplug rectifier-regulator plug under blower housing. If lamp remains lighted, stator is grounded. If lamp goes out, regulator or rectifier is shorted.

If previous test indicated stator is grounded, check stator leads for defects and repair if necessary. If shorted leads are not found, renew stator. Check stator for an open circuit by using an ohmmeter. Connect positive (+) ohmmeter lead to fuse holder as shown in Fig. B157 and remaining lead to one of the pins in the rectifier and regulator connector. Check each of the four pins in the connector. The ohmmeter should show continuity at each pin; if not, there is an open in stator and stator must be renewed.

To test rectifier, unplug rectifier and regulator connector plug and remove

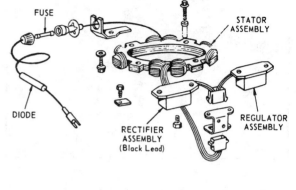

Fig. B155 — Stator, rectifier and regulator assemblies used on the 7 amp regulated flywheel alternator.

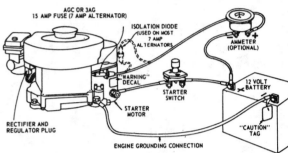

Fig. B156 — Typical wiring used on engines equipped with 7 amp flywheel alternator.

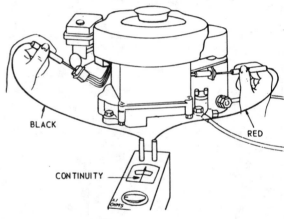

Fig. B157 — Use an ohmmeter to check condition of stator. Refer to text.

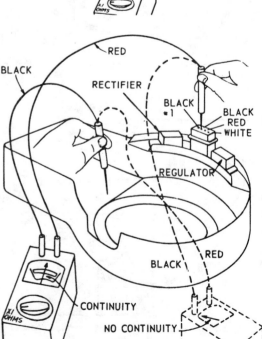

Fig. B158 — Be sure good contact is made between ohmmeter test lead and metal cover when checking rectifier and regulator.

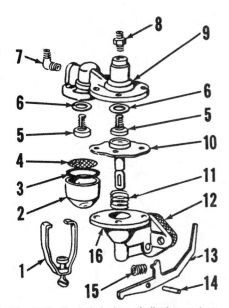

Fig. B159—Exploded view of diaphragm type fuel pump used on some Briggs & Stratton engines. Refer to Fig. B160 for assembly and disassembly views.

1. Yoke assy.	9. Fuel pump head
2. Filter bowl	10. Pump diaphragm
3. Gasket	11. Diaphragm spring
4. Filter screen	12. Gasket
5. Pump valves	13. Pump lever
6. Gaskets	14. Lever pin
7. Elbow fitting	15. Lever spring
8. Connector	16. Fuel pump body

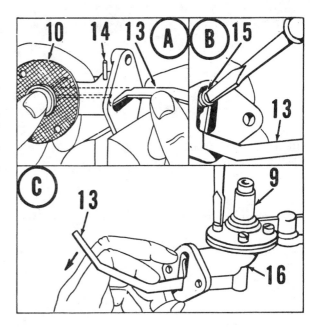

Fig. B160—Views showing disassembly and reassembly of diaphragm type fuel pump. Refer to text for procedure and to Fig. B159 for exploded view of pump and for legend.

blower housing from engine. Using an ohmmeter, check for continuity between connector pins connected to black wires and blower housing as shown in Fig. B158. Be sure good contact is made with metal of blower housing. Reverse ohmmeter leads and check continuity again. The ohmmeter should show a continuity reading for one direction only on each plug. If either pin shows a continuity reading for both directions, or if either pin shows no continuity for either direction, rectifier must be renewed.

To test regulator unit, repeat procedure used to test rectifier unit, except connect ohmmeter lead to pins connected to red wire and white wire. If ohmmeter shows continuity in either direction for red lead pin, regulator is defective and must be renewed. White lead pin should read as an open on the ohmmeter in one direction and a weak reading in the other direction. Otherwise, the regulator is defective and must be renewed.

FUEL PUMP

A diaphragm type fuel pump is available on many models as optional equipment. Refer to Fig. B159 for exploded view of pump.

To disassemble pump, refer to Figs. B159 and B160. Remove clamp (1), fuel bowl (2), gasket (3) and screen (4). Remove screws retaining upper body (9) to lower body (16). Pump valves (5) and gaskets (6) can now be removed. Drive lever retaining pin (14) out to either side of body (16). Press diaphragm (10) against spring (11) as shown in view A, Fig. B160, and remove lever (13). Diaphragm and spring (11–Fig. B159) can now be removed.

To reassemble, place diaphragm spring in lower body and place diaphragm on spring, being sure spring enters cup on bottom side of diaphragm and that slot in shaft is at right angle to pump level. Compress diaphragm against spring as in view A, Fig. B160, and insert hooked end of lever into slot in shaft. Align hole in lever with hole in lower body and drive pin into place. Insert lever spring (15) into body and push outer end of spring into place over hook on arm of lever as shown in view B. Hold lever downward as shown in view C while tightening screws holding upper body to lower body. When installing pump on engine, apply a liberal amount of grease on lever (13) at point where it contacts groove in crankshaft.

BRIGGS AND STRATTON
DISTRIBUTORS

(Arranged Alphabetically by States)

These franchised firms carry extensive stocks of repair parts. Contact them for name of the nearest service distributor who may have the parts you need.

BEBCO, Incorporated
2221 Second Avenue, South
Birmingham, Alabama 35233

Power Equipment Company
3141 North 35th Avenue, Unit 101
Phoenix, Arizona 85107

Pacific Power Equipment Company
1565 Adrain Road
Burlingame, California 94010

Power Equipment Company
1045 Cindy Lane
Carpinteria, California 93013

Pacific Power Equipment Company
5000 Oakland Street
Denver, Colorado 80239

Spencer Engine, Incorporated
1114 West Cass Street
Tampa, Florida 33606

Sedco, Incorporated
1414 Red Plum Road
Norcross, Georgia 30093

Small Engine Clinic, Incorporated
98019 Kam highway
Aiea, Hawaii 96701

Midwest Engine Warehouse
515 Romans Road
Elmhurst, Illinois 60126

Commonwealth Engine, Incorporated
1361 South 15th Street
Louisville, Kentucky 40210

Suhren Engine Company, Inc.
8330 Earhart Boulevard
New Orleans, Louisiana 70118

Grayson Company of Louisiana, Inc.
100 Fannin Street
Shreveport, Louisiana 71101

W. J. Connell Company
65 Green Street
Foxboro, Massachusetts 02035

Carl A. Anderson, Inc. of Minnesota
2737 West Service Road
Eagan, Minnesota 55121

Medart Engines & Parts
100 Larkin Williams Ind. Ct.
Fenton, Missouri 63026

Automotive Equipment Service
3117 Holmes
Kansas City, Missouri 64109

Original Equipment, Incorporated
905 Second Avenue, North
Billings, Montana 59101

Carl A. Anderson Inc. of Nebraska
7410 "L" Street
Omaha, Nebraska 68127

W. J. Connel Co., Bound Brook Div.
Chimney Rock Road, Route 22
Bound Brook, New Jersey 08805

Power Equipment Company
7209 Washington Street, North East
Alburquerque, New Mexico 87109

AEA, Incorporated
700 West 28th Street
Charlotte, North Carolina 28206

Gardner, Incorporated
1150 Chesapeake Avenue
Columbus, Ohio 43212

American Electric Ignition Company
124 North West Eighth
Oklahoma City, Oklahoma 73101

Brown & Wiser, Incorporated
9991 South West Avery Street
Tualatin, Oregon 97062

Pitt Auto Electric Company
2900 Stayton Street
Pittsburgh, Pennsylvania 15212

Automotive Electric Corporation
3250 Millbranch Road
Memphis, Tennessee 38116

American Electric Ignition Co., Inc.
618 Jackson Street
Amarillo, Texas 79101

Grayson Company, Incorporated
1234 Motor Street
Dallas, Texas 75207

Engine Warehouse, Incorporated
7415 Empire Central Drive
Houston, Texas 77040

Frank Edwards Company
110 South 300 West
Salt Lake City, Utah 84101

RBI Corporation
101 Cedar Ridge Drive
Ashland, Virginia 23005

Bitco Western, Incorporated
4030 1st Avenue, South
Seattle, Washington 98134

Air Cooled Engine Supply
South 127 Walnut Street
Spokane, Washington 99204

Wisconsin Magnetic, Incorporated
4727 North Teutonia Avenue
Milwaukee, Wisconsin 53209

CANADIAN DISTRIBUTORS

Canadian Curtiss-Wright, Ltd.
2-3571 Viking Way
**Richmond, British Columbia
V6V 1W1**

Canadian Curtiss-Wright, Ltd.
89 Paramount Road
Winnipeg, Manitoba R2X 2W6

Canadian Curtiss-Wright, Ltd.
1815 Sismet Road
Mississauga, Ontario L4W 1P9

Canadian Curtiss-Wright, Ltd.
8100-N Trans Canada Hwy.
St. Laurent, Quebec H4S 1M5

HONDA

AMERICAN HONDA MOTOR CO., INC.
100 W. Alondra Blvd.
Gardena, CA 90247

Model	Bore	Stroke	Displacement	Rated Power
G150	64 mm (2.5 in.)	45 mm (1.8 in.)	144 cc (8.8 cu. in.)	2.6 kW (3.5 hp)
G200	67 mm (2.6 in.)	56 mm (2.2 in.)	197 cc (12.0 cu. in.)	3.7 kW (5.0 hp)
G300	72 mm (3.0 in.)	60 mm (2.4 in.)	272 cc (16.6 cu. in.)	5.2 kW (7.0 hp)

ENGINE IDENTIFICATION

Honda G series engines are horizontal crankshaft four-stroke, air-cooled, single-cylinder engines. Valves are located in cylinder block. Power rating is at 3600 rpm. Specifications for engine models listed are for snowthrower applications.

Always furnish complete engine model and serial number information when ordering parts. Refer to Fig. HN1 for number location.

MAINTENANCE

SPARK PLUG. Recommended spark plug is NGK BR4HS for all models. Spark plug should be renewed after every 100 hours of operation. Set electrode gap to 0.6-0.7 mm (0.024-0.028 in.).

CARBURETOR. All models are equipped with a float-type side-draft carburetor (Fig. HN2). Carburetor is equipped with an idle fuel mixture screw. High speed fuel mixture is controlled by a fixed jet.

On Model G150, engine no-load speed should be 2000 rpm and is adjusted by repositioning spring in holes on muffler retainer. Adjust idle fuel mixture screw for highest engine rpm; nominal setting is 1-3/8 turns open. Then adjust operating speed to 2000 rpm by changing spring to different hole in muffler retainer if speed is not correct.

On Model G200, engine idle speed is 1000-1100 rpm and is adjusted by repositioning stop plate for engine throttle lever. With engine idling, adjust idle fuel mixture screw to obtain highest rpm; nominal setting is 1-1/4 turns open. Then adjust idle speed to 1000-1100 rpm by loosening cap screw retaining stop plate and repositioning plate so throttle lever will maintain proper idle speed.

On Model G300, operating engine idle speed is 1950-2250 rpm set by adjusting throttle lever stop. To adjust carburetor, loosen throttle lever stop so engine is idling with throttle arm of carburetor against carburetor throttle stop screw. Then, adjust idle fuel mixture screw to obtain highest idle rpm; nominal setting is 1-1/2 turns open. After adjusting idle fuel mixture, set carburetor throttle stop screw so engine is idling at 1550-1850 rpm, then set throttle lever stop to obtain operating idle speed.

Main jet (5—Fig. HN2) controls fuel mixture for high speed operation. Standard main jet is #60 for G150, #72 for G200 and #88 for G300.

Float level should be 6.7-9.7 mm (0.26-0.38 in.) for all models. To check float level, invert carburetor throttle body and measure distance from top side of float to float bowl. Renew float if distance is not within limits; float level is not adjustable.

FUEL FILTER. On Model G300, a fuel sediment bowl is located below fuel shut-off valve. To remove sediment bowl, shut off fuel, unscrew threaded ring and remove ring, sediment bowl and gasket. Clean sediment cup and reinstall using new gasket.

On Model G150, fuel filter is located in fuel line between fuel shut-off valve and carburetor. Close fuel valve, then remove filter. On Models G200 and G300, fuel filter is located in fuel tank outlet. Drain fuel, disconnect fuel line, and unscrew fuel line fitting (Model G200) or fuel filter assembly (Model G300). Check filter and renew if clogged.

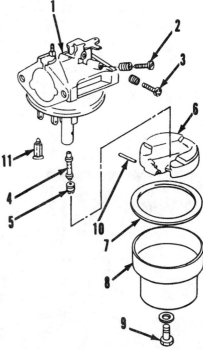

Fig. HN2—Exploded view of carburetor typical of the type used on all models.

1. Throttle body
2. Idle mixture screw
3. Throttle stop screw
4. Nozzle
5. Main jet
6. Float
7. Gasket
8. Float bowl
9. Bowl bolt
10. Float pin
11. Fuel inlet valve

Fig. HN1—Engine serial number is located at (A) on all models.

AIR FILTER. All models are equipped with a dry type air filter attached to carburetor inlet. Remove the filter and clean with low pressure air directed from inside filter. Renew filter if condition is questionable.

GOVERNOR. The internal centrifugal flyweight governor assembly is mounted on engine camshaft gear.

To adjust governor, stop engine and make certain all linkage is in good condition and tension spring (2—Fig. HN3) is not stretched or damaged. Spring (4) (not fitted on Model G300) must pull governor lever (3) toward throttle pivot (6). Loosen clamp bolt (8) and turn governor shaft (1) as far clockwise as it will go and move governor lever to right to hold throttle in wide open position, then tighten clamp bolt.

Start engine and allow to fully warm up. Set governed speed on Model G150 by moving throttle spring to different hole in muffler retainer to obtain governed no-load speed of 2000 rpm. On Model G200, adjust stop screw on throttle lever to obtain engine PTO shaft (camshaft) no-load speed of 1850-1950 rpm. On Model G300, adjust throttle lever stop screw to obtain engine no-load speed of 3450-3750 rpm.

IGNITION SYSTEM. All models are equipped with a breaker point ignition system with breaker points and ignition coil located under flywheel. Breaker points should be checked after every 300 hours of operation. Apply oil to wiper felt if dry. Initial breaker point gap should be adjusted to 0.3-0.4 mm (0.012-0.016 in.) and may be varied to obtain specified 20 degrees BTDC ignition timing setting. Timing tool, part number 07974-8830001, is available from Honda to allow timing adjustment with flywheel removed.

To check ignition timing with flywheel installed, connect positive ohmmeter lead to engine stop switch wire and remaining lead to ground. Turn flywheel until ohmmeter needle deflects. "F" mark on flywheel should align with index mark on crankcase as shown in Fig. HN4. If using Honda tool with flywheel removed, tip of tool should align with index mark "F."

Ignition coil resistance should be 6.6 ohms and capacity of ignition condenser (capacitor) should be 24 microfarad.

LUBRICATION. Recommended engine oil for snowthrower operation is SAE 5W-30 specification SE or SF motor oil for general all-temperature use. Oil capacity is 0.7 L (3/4 qt.) for Models G150 and G200, and 1.2 L (1-1/4 qt.) for Model G300.

REPAIRS

TIGHTENING TORQUES. Recommended tightening torque specifications are as follows:

Flywheel nut:
G150, G200 . . . . 70-80 N·m
(51-58 ft.-lbs.)
G300 . . . . 110-120 N·m
(80-87 ft.-lbs.)
Crankcase cover:
G150 . . . . 8-12 N·m
(6-9 ft.-lbs.)
G200 . . . . 12-16 N·m
(9-12 ft.-lbs.)
G300 . . . . 20-24 N·m
(15-17 ft.-lbs.)
Cylinder head bolts . . . . 24-26 N·m
(17-19 ft.-lbs.)
Connecting rod bolts:
G150, G200 . . . . 11-13 N·m
(8-9 ft.-lbs.)
G300 . . . . 12.6-15.4 N·m
(9-11 ft.-lbs.)

CYLINDER HEAD. To remove cylinder head, first remove cooling shrouds. Clean engine to prevent entrance of foreign material. Remove spark plug. Loosen cylinder head bolts in 1/4 turn increments following sequence shown in Fig. HN5 until they are loose enough to be removed by hand. Remove cylinder head and clean carbon deposits.

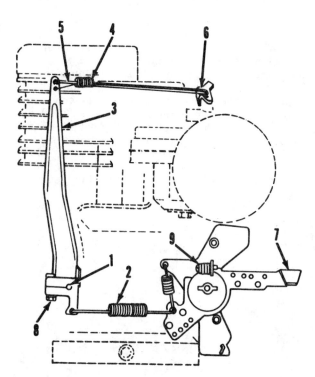

Fig. HN3—Drawing showing typical external governor linkage. Model G300 does not use throttle rod spring (4). Refer to text for adjustment procedure.

1. Governor shaft
2. Tension spring
3. Governor lever
4. Spring
5. Throttle rod
6. Carburetor throttle shaft
7. Throttle lever
8. Clamp bolt
9. Throttle lever stop

"F" Mark on Flywheel

Timing (Index) Mark

Fig. HN4—View showing flywheel timing mark (F) and index mark on cylinder block.

Reinstall cylinder head with new head gasket. Tighten bolts in several steps following sequence shown in Fig. HN5 until specified torque is obtained.

CONNECTING ROD. Connecting rod rides directly on crankpin journal. Piston and connecting rod are removable after removing cylinder head and crankcase side cover. Unbolt and remove rod oil dipper and cap, then push piston and rod assembly out top of block. Remove snap rings and piston pin and separate piston from rod.

Standard connecting rod piston pin bore is 15.0 mm (0.590 in.) on Models G150 and G200, and 16.005-16.020 mm (0.630-0.631 in.) on Model G300. Wear limit is 15.07 mm (0.593 in.) on Models G150 and G200, and 16.08 mm (0.633 in.) on Model G300.

Standard oil clearance between connecting rod and crankpin journal is 0.048-0.066 mm (0.002-0.003 in.) for Models G150 and G200, and 0.048-0.065 mm (0.002-0.0026 in.) for Model G300. Maximum allowable clearance is 0.12 mm (0.0047 in.) for Models G150 and G200, and 0.25 mm (0.0098 in.) on Model G300.

Standard connecting rod side play on crankpin journal is 0.1-0.8 mm (0.004-0.031 in.) for G150 and G200, and 0.6-1.0 mm (0.024-0.039 in.) for Model G300. Maximum allowable wear limit for side play is 1.2 mm (0.047 in.) for Models G150 and G200, Model G300 should not be above standard maximum clearance.

Piston may be installed on connecting rod facing either way. Install piston and rod assembly in engine so mark on top of piston is toward valve side of cylinder block. Align matching marks when installing rod cap. Install oil dipper and tighten connecting rod nuts to specified torque.

PISTON, PIN AND RINGS. Refer to connecting rod paragraph for piston and connecting rod removal procedure. After separating piston from connecting rod, carefully remove rings and clean deposits from piston top and ring grooves. Be careful not to damage piston when cleaning ring grooves. Measure piston diameter at thrust surfaces at bottom of piston skirt and 90 degrees from piston pin. Standard piston dimensions and wear limits are as follows:

Model	Standard Diameter	Wear Limit
G150	64 mm (2.52 in.)	63.88 mm (2.515 in.)
G200	67 mm (2.638 in.)	66.8 mm (2.630 in.)
G300	75.97-76 mm (2.991-2.992 in.)	75.85 mm (2.986 in.)

Standard pin bore in piston for Models G150 and G200 is 15 mm (0.59 in.) with wear limit of 15.046 mm (0.592 in.). For Model G300, standard pin bore is 16 mm (0.630 in.) with wear limit of 16.047 mm (0.6317 in.).

Piston pin diameter wear limit is 14.954 mm (0.589 in.) for Models G150 and G200 and 15.97 mm (0.6287 in.) for Model G300.

Maximum allowable piston ring side clearance in piston ring groove is 0.15 mm (0.0059 in.) for all rings on all models. The maximum allowable ring end gap is 1.0 mm (0.039 in.).

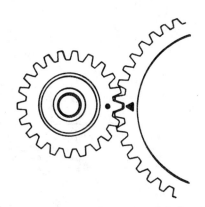

Fig. HN6—View of camshaft and crankshaft gear timing marks for Models G150 and G300. Refer to Fig. HN7 for Model G200.

Before deciding on whether to renew piston and/or rings, check cylinder bore as outlined under CYLINDER AND CRANKCASE and compare cylinder bore diameter measurement with piston skirt diameter. Piston skirt-to-cylinder wall clearance and wear limits are as follows:

Model	Standard Clearance	Wear Limit
G150	...	...
G200	0.045 mm (0.0018 in.)	0.285 mm (0.0112 in.)
G300	0.045 mm (0.0018 in.)	0.25 mm (0.01 in.)

CYLINDER AND CRANKCASE. Cylinder and crankcase are integral casting. Standard cylinder bore diameters and wear limit diameters are as follows:

Model	Standard Diameter	Wear Limit Diameter
G150	64 mm (2.52 in.)	64.165 mm (2.526 in.)
G200	67 mm (2.64 in.)	67.165 mm (2.644 in.)
G300	76.017 mm (2.9928 in.)	76.1 mm (2.9961 in.)

CRANKSHAFT, MAIN BEARINGS AND SEALS. Crankshaft is supported by ball bearing type main bearings at each end. To remove crankshaft, remove all cooling shrouds, flywheel, cylinder head, valve assemblies, crankcase cover and the piston and connecting rod assembly. Turn engine upside down to let valve lifters fall away from cam and carefully remove crankshaft and camshaft. On Model G200, the camshaft is chain driven and chain is removed with crankshaft and camshaft. Remove crankshaft bearings and crankshaft oil seals.

Standard crankpin diameter is 26 mm (1.0236 in.) on Models G150 and G200 with wear limit of 25.917 mm (1.020 in.). Model G300 standard crankpin diameter is 29.95 mm (1.179 in.) with wear limit of 29.85 mm (1.175 in.).

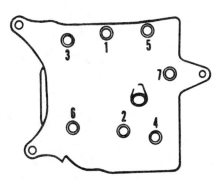

Fig. HN5—Loosening and tightening sequence for cylinder head bolts.

Fig. HN7—View of crankshaft and camshaft sprocket timing marks for Model G200; timing marks for G150 and G300 are shown in Fig. HN6.

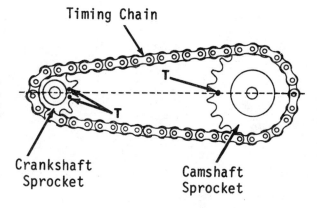

Main bearings are a light press fit on crankshaft and in crankcase and cover bearing bores. It may be necessary to heat crankcase and cover slightly when reinstalling bearings.

Inspect main bearings for roughness or looseness. Also check bearings for loose fit on crankshaft or in crankcase and cover bore.

Install new crankshaft oil seals and, on Model G200, new camshaft oil seal. Be sure to align timing gear timing marks (see Fig. HN6) or timing sprocket timing marks (Fig. HN7) when reassembling engine.

CAMSHAFT, BEARINGS AND SEALS. Camshaft is supported at each end by bearings which are integral part of crankcase and cover. On Model G200, the camshaft is extended through the crankcase cover as a PTO shaft. Refer to CRANKSHAFT, MAIN BEARINGS AND SEALS for camshaft removal procedure.

Renew camshaft if lobe wear is evident. Extensive wear on camshaft bearing journals and in bearing bores of crankcase and cover may require replacement of all of these parts.

GOVERNOR. The internal centrifugal flyweight governor is mounted on the camshaft gear or sprocket. Refer to GOVERNOR paragraph in MAINTENANCE section for external governor adjustments.

To remove governor assembly, refer to CAMSHAFT, BEARINGS AND SEAL. When reassembling, expand the governor weights and place slider on camshaft with notch in slider engaged with lug on camshaft gear or sprocket.

VALVE SYSTEM. Intake valve (tappet) gap should be 0.08-0.16 mm (0.003-0.006 in.) on all models. Exhaust valve gap should be 0.16-0.24 mm (0.006-0.009 in.) on Models G150 and G200. On Model G300, exhaust valve tappet gap should be 0.11-0.19 mm (0.0043-0.0075 in.). Valve tappet gap is increased by grinding off end of valve stem. If gap is excessive, it may be reduced by installing a new valve or by refacing the valve and/or by grinding the valve seat.

Recommended valve seat width is 0.7 mm (0.028 in.) on Models G150 and G200; G300 valve seat width should be 1.06 mm (0.042 in.). Recommended valve face and seat angle is 45 degrees.

Standard valve guide diameter for all models is 7 mm (0.276 in.). Check valve stems and guides for wear and determine which should be renewed to reduce excessive clearance. If valve stem to guide clearance is excessive with new valve, renew the guide as shown in Fig. HN8.

Valve spring free length should be 36.7 mm (1.445 in.) on Models G150 and G200, and 42.7 mm (1.681 in.) on Model G300.

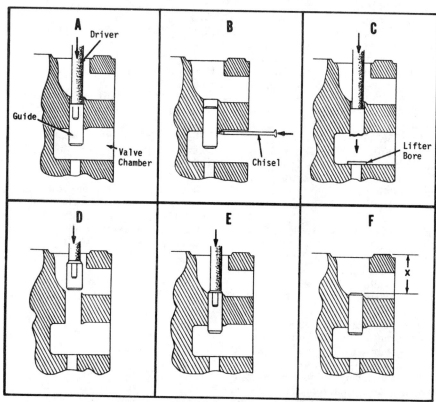

Fig. HN8—To remove valve guide, use 7 mm driver with shank smaller than guide O.D. to drive guide down (A). Break off lower end of guide with chisel (B). Drive remainder of guide down into valve chamber (C). Drive new guide into place (D and E). Check to see that guide is properly located (F) and ream guide using 7 mm valve guide reamer. Guide may also be removed using suitable guide puller.

HONDA

Model	Bore	Stroke	Displacement	Rated Power
GX140	64 mm	45 mm	144 cc	3.8 kW
	(2.5 in.)	(1.8 in.)	(8.8 cu. in.)	(5.0 hp)
GX240	73 mm	58 mm	242 cc	6.0 kW
	(2.87 in.)	(2.3 in.)	(14.9 cu. in.)	(8.0 hp)

ENGINE IDENTIFICATION

Honda GX series engines are horizontal crankshaft 4-stroke, air-cooled, single-cylinder engines. Valves are located in cylinder head. Power rating is at 4200 rpm on Model GX140 and at 3600 rpm on Model GX240. Specifications for engines listed are for snowthrower application.

Always furnish complete engine model and serial number information when ordering parts. Refer to Fig. HN10 for model number location and to Fig. HN11 for serial number location.

MAINTENANCE

SPARK PLUG. Recommended spark plug is NGK BPR5ES for both models. Spark plug should be renewed after every 100 hours of operation. Set electrode gap to 0.7-0.8 mm (0.028-0.031 in.).

CARBURETOR. All models are equipped with a Keihin float-type side-draft carburetor (Fig. HN12). Carburetor is equipped with low speed mixture screw (LS) and a throttle stop screw (TS). High speed fuel mixture is controlled by a fixed jet.

On both models, operating engine idle speed is 1950-2250 rpm set by adjusting throttle lever stop. To adjust carburetor, start engine and bring to operating temperature, then loosen throttle lever stop so that engine is idling with throttle arm of carburetor against carburetor throttle stop screw. Adjust low speed mixture screw (LS) to obtain highest idle rpm; nominal setting is 2 turns open on Model GX140 and 1-1/2 turns open on Model GX240. After adjusting low speed mixture, set carburetor throttle stop screw (TS) so engine is idling at 1250-1550 rpm, then set throttle lever stop to obtain operating idle speed.

A fixed main jet controls fuel mixture for high speed operation. Standard main jet size is #70 for GX140 and #88 for GX240.

Float level should be 12.2-15.2 mm (0.48-0.60 in.) on Model GX140 and 11.9-14.5 mm (0.47-0.57 in.) on Model GX240. To check float level, invert carburetor throttle body and measure distance from top side of float to float bowl. Renew float if distance is not within limits; float level is not adjustable.

FUEL FILTER. On both models, a fuel sediment cup is located below fuel shut-off valve in carburetor body. To remove sediment cup, shut off fuel and unscrew threaded cup and remove gasket. Clean sediment cup and reinstall using new gasket.

Fig. HN10—Engine model number (MN) is cast into side of engine crankcase.

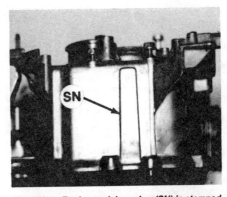

Fig. HN11—Engine serial number (SN) is stamped on engine crankcase boss.

AIR FILTER. Both models are equipped with a dry-type air filter attached to carburetor inlet elbow. Remove the filter and clean with low pressure air from inside filter. Renew filter if condition is questionable.

GOVERNOR. The internal centrifugal flyweight governor assembly is mounted on a stud shaft in engine crankcase and is driven by a gear on engine crankshaft. To adjust governor, stop engine and make certain all linkage is in good condition and tension spring (5—Fig. HN13) is not stretched or damaged. Spring (2) must pull governor lever (3) and carburetor throttle arm toward each other. Loosen clamp bolt (7) and move governor lever (3) to hold carburetor throttle in wide open position. Hold governor lever in this position and rotate governor shaft (6) in same direction until it stops, then tighten clamp screw.

Start engine and allow to fully warm up. Set governed speed adjust stop screw on carburetor throttle lever to obtain engine no-load speed of 1950-2250 rpm. Then, adjust throttle lever stop to obtain engine maximum no-load speed

Fig. HN12—View showing Keihin float-type carburetor typical of the type used on Models GX140 and GX240. View identifies location of low speed mixture screw (LS) and throttle stop screw (TS).

of 3650-3950 rpm on Model GX140 and 3450-3750 rpm on Model GX240.

IGNITION SYSTEM. Both models are equipped with a breakerless ignition system with ignition coil located outside flywheel. Air gap between ignition coil core and flywheel should be 0.2-0.6 mm (0.008-0.024 in.).

To check ignition coil primary side, connect one ohmmeter lead to black coil lead and other ohmmeter lead to coil laminations. Ohmmeter should register 0.7-0.9 ohms. To check secondary coil, attach one ohmmeter lead to spark plug wire and other lead to coil laminations. Ohmmeter should register 6.3-7.7 ohms. Renew coil if readings are not as specified.

VALVE ADJUSTMENT. Refer to Fig. HN14 and remove rocker arm cover. Turn engine crankshaft so piston is at top of compression stroke (both valves fully closed). Using a feeler gage positioned between rocker arm and end of valve stem, check clearance on each valve. Intake valve clearance should be 0.13-0.17 mm (0.005-0.007 in.) and exhaust valve clearance should be 0.18-0.22 mm (0.007-0.008). If necessary, loosen rocker arm jam nut (1—Fig. HN14) and turn adjusting nut (2) to obtain desired clearance. Tighten jam nut and recheck adjustment.

LUBRICATION. Recommended engine oil for snowthrower operation is SAE 5W-30 specification SE or SF motor oil for general all-temperature use. Oil capacity is 0.6 L (0.63 qt.) for Model GX140 and 1.1 L (1.16 qt.) for Model GX240. Refer to Fig. HN15.

REPAIRS

TIGHTENING TORQUES. Recommended tightening torque specifications are as follows:

Flywheel nut:
GX140 . . . . .70-80 N·m
(51-58 ft.-lbs.)
GX240 . . . . .110-120 N·m
(80-87 ft.-lbs.)
Crankcase cover . . . . .22-26 N·m
(16-19 ft.-lbs.)

Cylinder head bolts:
GX140 . . . . .22-26 N·m
(16-19 ft.-lbs.)
GX240 . . . . .32-38 N·m
(23-27 ft.-lbs.)
Connecting rod bolts:
GX140 . . . . .11-13 N·m
(8-9 ft.-lbs.)
GX240 . . . . .22-26 N·m
(16-19 ft.-lbs.)
Rocker arm stud:
GX140 . . . . .8-12 N·m
(6-9 ft.-lbs.)
GX240 . . . . .22-26 N·m
(16-19 ft.-lbs.)
Rocker arm jam nut . . . . . . .8-12 N·m
(6-9 ft.-lbs.)

CYLINDER HEAD AND VALVES. To remove cylinder head, first remove cooling shroud and clean engine to prevent entrance of foreign material. Disconnect and remove carburetor linkage and carburetor. Remove spark plug and rocker arm cover. Remove rocker arm jam nuts, adjusting nuts, rocker arms and push rods. Loosen cylinder head bolts in 1/4-turn increments following sequence shown in Fig. HN16 until they are loose enough to be removed by hand. Remove cylinder head and remove valve spring retainers, springs and valves. Remove push rod guide plate if necessary.

Recommended valve seat width is 0.8 mm (0.032 in.) on Model GX140; Model GX240 valve seat width should be 1.1 mm (0.043 in.). Valve face and seat angle is 45 degrees.

Standard valve guide diameter is 5.5 mm (0.216 in.) on Model GX140 and 6.6 mm (0.260 in.) on Model GX240. Check valve stems and guides for wear and determine which should be renewed to reduce excessive clearance. If valve stem-to-guide clearance is excessive with a new valve, renew the guide. Remove old

Fig. HN13—View of governor linkage.
1. Governor-to-carburetor rod
2. Spring
3. Governor lever
4. Choke rod
5. Tension spring (behind plate & lever)
6. Governor shaft
7. Clamp bolt

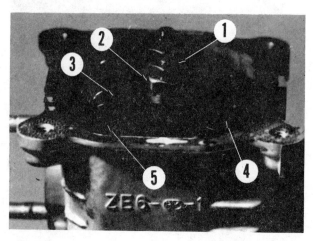

Fig. HN14—View with rocker arm cover removed showing rocker arm adjustment points.
1. Jam nut
2. Adjustment nut
3. Rocker arm
4. Valve stem clearance
5. Push rod

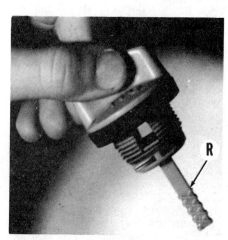

Fig. HN15—Do not screw oil filler cap with gage into crankcase when checking oil level. Level should be at top of reference marks (R) on dipstick.

guide and install new guide using suitable drivers. Valve seat end of valve guide should be 25.5 mm (1 in.) from bottom of head on Model GX140 and 24 mm (0.945 in.) from combustion chamber surface on Model GX240. Ream new guide with a suitable reamer after installation.

Valve spring free length should be 34 mm (1.34 in.) on Model GX140 and 39 mm (1.54 in.) on Model GX240.

Reassemble valves, springs and retainers in cylinder head. Reinstall cylinder head with a new head gasket. Tighten bolts in several steps following sequence shown in Fig. HN16 until specified torque is obtained.

CONNECTING ROD. Connecting rod rides directly on crankpin journal. Piston and connecting rod are removable after removing cylinder head and crankcase side cover. Unbolt and remove rod cap, then push piston and rod assembly out top of block. Remove snap rings and piston pin and separate piston from rod.

Standard connecting rod piston pin bore is 18.005 mm (0.7089 in.). Wear limit is 18.07 mm (0.711 in.).

Standard oil clearance between connecting rod and crankpin journal is 0.040-0.063 mm (0.0016-0.0025 in.). Maximum allowable clearance is 0.12 mm (0.0047 in.).

Standard connecting rod side play on crankpin journal is 0.1-0.7 mm (0.004-0.028 in.). Maximum allowable wear limit for side play is 1.1 mm (0.043 in.).

Fig. HN16—Loosen and tighten cylinder head bolts in sequence shown.

Piston must be installed on connecting rod with mark on top of piston facing toward long (bottom) end of rod as shown in Fig. HN17. Install piston and rod assembly in engine so mark on top of piston is facing down toward push side of cylinder block. Install cap on rod with oil dipper down (toward long end of rod) and tighten connecting rod nuts to specified torque.

PISTON, PIN AND RINGS. Refer to connecting rod paragraph for piston and connecting rod removal procedure. After separating piston from connecting rod, carefully remove rings and clean deposits from piston top and ring grooves. Be careful not to damage piston when cleaning ring grooves.

Measure piston diameter at thrust surfaces at bottom of piston skirt and 90 degrees from piston pin. Standard piston skirt dimension is 63.965-63.985 mm (2.518-2.519 in.) for Model GX140 and 72.74 mm (2.864 in.) for Model GX240. Minimum skirt diameter wear limit is 63.55 mm (2.502 in.) for Model GX140 and 72.62 mm (2.859 in.) for Model GX240.

Standard pin bore in piston for Model GX140 is 18.002-18.008 mm (0.7087-0.7089 in.) with wear limit of 18.048 mm (0.7105 in.). For Model GX240, standard pin bore is 18.002 mm (0.7087 in.) with wear limit of 18.042 mm (0.7103 in.). Piston pin diameter wear limit is 17.954 mm (0.7068 in.).

Maximum allowable piston ring side clearance in piston ring groove is 0.15 mm (0.006 in.) for all rings on piston. The maximum allowable ring end gap is 1.0 mm (0.039 in.).

Before deciding on whether to renew piston and/or rings, check cylinder bore as outlined under CYLINDER AND CRANKCASE and compare cylinder bore diameter measurement with piston skirt diameter. Standard piston skirt clearance is 0.015-0.050 mm (0.0006-0.0020 in.) and maximum allowable clearance is 0.12 mm (0.005 in.).

Install piston rings with marked side toward top of piston and stagger ring end gaps equally around circumference of piston.

CYLINDER AND CRANKCASE. Cylinder and crankcase are integral casting. Standard cylinder bore diameter is 64 mm (2.520 In.) for Model GX140 and 73 mm (2.874 in.) for Model GX240. Maximum allowable wear limit over standard bore diameter is 0.15 mm (0.0060 in.) for Model GX140 and 0.17 mm (0.0067 in.) for Model GX240.

CRANKSHAFT, MAIN BEARINGS AND SEALS. Crankshaft is supported

by ball bearing type main bearings at each end. To remove crankshaft, remove all cooling shrouds, flywheel, cylinder head, valve assemblies, crankcase cover and the piston and connecting rod assembly. Turn engine upside down to let valve lifters fall away from cam and carefully remove crankshaft and camshaft. Remove crankshaft bearings and crankshaft oil seals.

Standard crankpin diameter is 30 mm (1.181 in.) on Model GX140 with wear limit of 29.92 mm (1.178 in.). Model GX240 standard crankpin diameter is 32.985 mm (1.2986 in.) with wear limit of 32.92 mm (1.296 in.).

Main bearings are a light press fit on crankshaft and in crankcase and cover bearing bores. It may be necessary to heat crankcase and cover slightly when reinstalling bearings.

Inspect main bearings for roughness or looseness. Also check bearings for loose fit on crankshaft or in crankcase and cover bore.

Install new crankshaft oil seals. Be sure to align timing gear timing marks (Fig. HN18) when reassembling engine.

CAMSHAFT, BEARINGS AND SEALS. Camshaft is supported at each end by bearings which are an integral part of crankcase and cover. Refer to CRANKSHAFT, MAIN BEARINGS AND SEALS for camshaft removal procedure.

Standard camshaft bearing journal diameter is 14 mm (0.551 in.) on Model

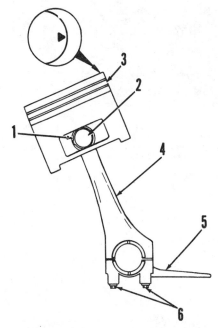

Fig. HN17—Drawing showing proper assembly of piston and connecting rod. Arrow on piston and oil dipper side of connecting rod cap must be installed to the same side.

1. Retaining rings
2. Piston pin
3. Piston
4. Connecting rod
5. Rod cap & dipper
6. Rod bolts

GX140 and 15.984 mm (0.6293 in.) on Model GX240. Renew camshaft if lobe wear is evident. Extensive wear on cam-shaft bearing journals and in bearing bores of crankcase and cover may require replacement of all of these parts.

Standard camshaft lobe height (see Fig. HN20) for Model GX140 is 27.7 mm (1.091 in.) on intake lobe and 27.75 mm (1.093 in.) on exhaust lobe. Model GX240 cam lobe height is 31.2 mm (1.228 in.) on intake lobe and 31.1 mm (1.224 in.) on exhaust lobe. Renew cam if lobe height measures more than 0.25 mm (0.010 in.) below the specified lobe height.

GOVERNOR. The internal centrifugal flyweight governor is mounted on a stub shaft inside the engine crankcase on Model GX140 and on a stub shaft inside the engine crankcase cover on Model GX240. The governor gear is driven by a gear on the crankshaft. Refer to GOVERNOR in MAINTENANCE section for external governor adjustments.

To remove Model GX140 governor assembly, refer to CRANKSHAFT, MAIN BEARINGS AND SEALS. Model GX240 governor can be removed after removing crankcase cover. Remove slider and washer from stub shaft, then remove snap ring, governor assembly and second washer. When reinstalling governor on stub shaft, expand the governor weights and place slider on stub shaft.

Fig. HN18—Align timing marks (F) on camshaft gear and crankshaft gear.

Fig. HN19—Compression release mechanism mounted on camshaft gear reduces compression for cranking engine. Weight (2) overcomes spring (1) at engine operating speed.

Fig. HN20—Measure camshaft lobe height as shown.

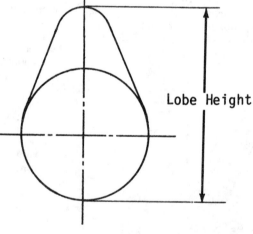

Lobe Height

JACOBSEN

JACOBSEN DIV. OF TEXTRON, INC.
1721 Packard Avenue
Racine, Wisconsin 53403

Model	Bore	Stroke	Displ.
J501	2⅛ in. (54 mm)	1¾ in. (44.5 mm)	6.21 cu. in. (102 cc)

MAINTENANCE

SPARK PLUG. Recommended spark plug is a Champion UJ-12. Electrode gap should be 0.030 inch (0.76 mm).

CARBURETOR. A Walbro Model SDC diaphragm type carburetor is used.

Initial setting of idle and high speed mixture screws is 1 turn open. Final adjustment of mixture screws should be done with engine warm and running. Adjust idle mixture so engine will accelerate without stumbling. Adjust high speed mixture to obtain optimum performance under load.

Carburetor may be disassembled after inspection of unit and referral to exploded view in Fig. J30. Care should be taken not to lose ball and spring which will be released when choke shaft is withdrawn.

Clean and inspect all components. Inspect diaphragms for defects which may affect operation. Examine fuel inlet needle and seat. Inlet needle is renewable, but carburetor body must be renewed if needle seat is excessively worn or damaged. Sharp objects should not be used to clean orifices or passages as fuel flow may be altered. Compressed air should not be used to clean main nozzle as check valve may be damaged. A check valve repair kit is available to renew a damaged valve. Fuel mixture needles must be renewed if grooved or broken. Inspect mixture needle seats in carburetor body and renew body if seats are damaged or excessively worn. Screens should be clean.

To reassemble carburetor, reverse disassembly procedure. Fuel metering lever should be flush with bosses (B – Fig. J31) on chamber floor. Be sure lever spring correctly contacts locating dimple on lever before measuring lever height. Bend lever to obtain correct lever height.

GOVERNOR. Speed control is maintained by a pneumatic (air vane) type governor. Make certain that governor linkage does not bind in any position when moved through full range of travel. Maximum governed top no-load speed should be 4100-4500 rpm. Governor adjustment screw is adjacent to carburetor.

MAGNETO AND TIMING. All models are equipped with a breaker point type flywheel magneto ignition system. Magneto components are accessible after removing flywheel. Breaker point gap is 0.020 inch (0.508 mm). Ignition timing (25°BTDC) is correct with stator plate rotated to full clockwise position.

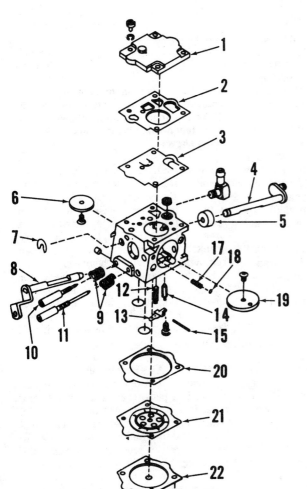

Fig. J30 – Exploded view of Walbro SDC carburetor.

1. Fuel pump cover
2. Gasket
3. Fuel pump diaphragm
4. Throttle shaft
5. Seal
6. Throttle plate
7. "E" ring
8. Choke shaft
9. Springs
10. Idle mixture screw
11. High speed mixture screw
12. Metering diaphragm lever spring
13. Metering diaphragm lever
14. Inlet fuel valve
15. Pin
16. Screw
17. Detent spring
18. Choke detent ball
19. Choke plate
20. Gasket
21. Metering diaphragm
22. Cover

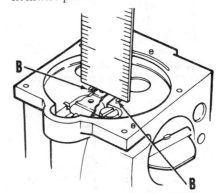

Fig. J31 – Diaphragm lever on Model SDC carburetor should just straightedge placed on bosses (B) adjacent to lever.

LUBRICATION. The engine is lubricated by oil mixed with the fuel. Recommended fuel:oil ratio is 50:1 using Jacobsen two-stroke engine oil mixed with unleaded or regular gasoline.

CLEANING CARBON. Power loss can often be corrected by cleaning carbon from exhaust ports and muffler. To clean, remove spark plug and muffler. Turn engine so that piston is below bottom of exhaust ports and remove carbon from ports with dull knife or similar tool. Clean out the muffler openings, then reinstall muffler and spark plug.

REPAIRS

TIGHTENING TORQUES. Recommended tightening torques are as follows:

Back plate screws ....80-90 in.-lbs.
(8.9-10.2 N·m)
Carburetor adapter
screws ....60-80 in.-lbs.
(6.8-8.9 N·m)
Connecting rod screws ....45-50 in.-lbs.
(5.1-5.6 N·m)
Bearing plate bolts ......80-100 in.-lbs.
(8.9-11.2 N·m)
Cylinder head bolts
or nuts ....150-180 in.-lbs.
(16.9-20.3 N·m)
Engine mounting bolts ..235-260 in.-lbs.
(26.6-29.3 N·m)
Fan housing screws ......60-80 in.-lbs.
(6.9-8.9 N·m)
Flywheel nut ....300-360 in.-lbs.
(33.9-40.7 N·m)
Spark plug...........180-200 in.-lbs.
(20.3-22.6 N·m)
Stator plate screws ......60-80 in.-lbs.
(6.8-8.9 N·m)

PISTON, PIN AND RINGS. To remove piston, remove any shrouding preventing access to cylinder head or

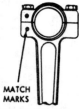

Fig. J32 – Be sure that match marks are aligned when reassembling cap to connecting rod.

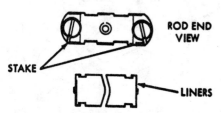

Fig. J33 – Bearing liners should mate together as shown. Stake cap retaining screws as shown.

Fig. J34 – Exploded view of engine.

1. Seal
2. Bearing retainer
3. Needle bearing
4. Gasket
5. Key
6. Crankshaft
7. Snap ring
8. Cylinder & crankcase
9. Ball bearing
10. Gasket
11. Snap ring
12. Crankcase head
13. Seal
14. Bearing liners
15. Bearing rollers (28)
16. Bearing roller guides
17. Rod cap
18. Connecting rod
19. Pin retainers
20. Piston pin
21. Piston
22. Piston rings
23. Gasket
24. Cylinder head
25. Air baffle

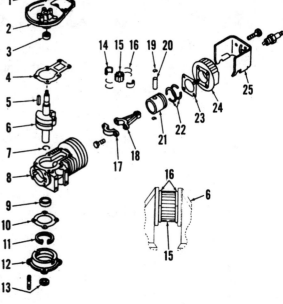

carburetor. Disconnect governor link to carburetor and remove carburetor and reed valve plate. Remove cylinder head. Unscrew connecting rod cap screws and remove connecting rod and piston. Be careful not to lose loose bearing rollers on models so equipped. Refer to CONNECTING ROD section to service connecting rod. The cam ground aluminum piston is fitted with two compression rings. Renew piston if badly worn or scored, or if side clearance of new ring in top ring groove is 0.010 inch (0.25 mm) or more. Install top ring with chamfer down and second ring with chamfer up.

Piston skirt diameter measured at right angle to piston pin is 2.1220-2.1227 inches (53.45-53.92 mm). Piston skirt clearance should be 0.0033-0.0045 inch (0.084-0.114 mm). Piston ring width is 0.0925-0.0935 inch (2.35-2.37 mm) and side clearance in piston ring groove should be 0.0015-0.0035 inch (0.038-0.089 mm). Piston ring end gap should be 0.005-0.013 inch (0.13-0.33 mm).

Piston pin diameter is 0.4999-0.5001 inch (12.697-12.702 mm) and clearance in piston is 0.0003 inch (0.0076 mm) tight to 0.0002 inch (0.0051 mm) loose. Piston pin clearance in rod should be 0.0007-0.0016 inch (0.018-0.041 mm).

CONNECTING ROD. Connecting rod and piston unit can be removed from above after removing cylinder head and carburetor adapter.

The aluminum connecting rod has renewable steel bearing inserts for needle roller outer race. Twenty-eight bearing rollers are used. Bearing guides (16 – Fig. J34) are used to center bearing rollers on crankpin. Bearing rollers and guides may be held on crankpin with grease before installation of rod. Refer to following specifications:

Piston pin
diameter........0.4999-0.5001 inch
(12.697-12.702 mm)
Pin to rod
clearance ....0.0007-0.0016 inch
(0.018-0.041 mm)
Crankpin
diameter........0.7500-0.7505 inch
(19.05-19.06 mm)
Rod side play on
crankpin..........0.005-0.018 inch
(0.13-0.46 mm)

NOTE: Be sure that match marks on rod and cap are aligned as shown in Fig. J32. Stake the screw heads as shown in Fig. J33; steel bearing inserts must be installed as shown.

CRANKSHAFT AND SEALS. The crankshaft is supported by a ball bearing and a needle bearing. The ball bearing is a press fit in crankcase head and on crankshaft. The needle bearing is a press fit in back plate. Be sure crankshaft is properly supported when removing or installing bearing.

Inspect crankshaft and check for straightness. Install crankcase head so that flat spot on inner portion of crankcase head will be towards cylinder. Install seals with lip towards inside of engine.

REED VALVE. On all models the carburetor is mounted on an adapter plate. The plate carries a small spring steel leaf or reed, on the engine side, which acts as an inlet valve. The reed has a bend in it and must be installed so as to force the reed firmly against the mounting plate. Blow-back through the carburetor may be caused by dirt or foreign matter holding reed valve open or by an improperly installed reed.

REWIND STARTER. To disassemble dog type starter shown in Fig. J35, remove starter from engine and pull starter rope out of starter approximately 12 inches (305 mm). Prevent pulley (3) from rewinding and pull slack rope back through rope outlet. Hold rope away from pulley and allow rope pulley to unwind. Unscrew retaining screw (11) and remove dog assembly and rope pulley being careful not to disturb rewind spring (2) in cover. Note direction of spring winding and carefully remove

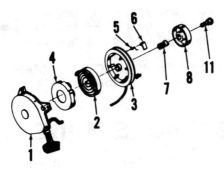

Fig. J35 — Exploded view of rewind starter.

1. Cover	
2. Rewind spring	6. Dog
3. Rope pulley	7. Brake spring
4. Spring holder	8. Retainer
5. Spring	11. Screw

spring from cover. To reassemble starter, reverse disassembly procedure. Rewind spring must be preloaded as follows: Install rope through top of housing and hole in rope pulley. Tie a knot at each end of rope. Turn rope pulley approximately four turns in direction that places load on rewind spring. Hold pulley and pull remainder of rope through rope pulley. Attach handle to rope end and allow rope to rewind into starter. If spring is properly preloaded, rope will fully rewind.

TECUMSEH

TECUMSEH PRODUCTS COMPANY
Grafton, Wisconsin 53024

2-STROKE MODELS

Model	Bore	Stroke	Displacement
AH520 (1500 & 1600)	2.09 in.	1.50 in.	5.16 cu. in.
	(50.8 mm)	(38.1 mm)	(77 cc)
AH600 (1620)	2.09 in.	1.75 in.	6.00 cu. in.
	(50.8 mm)	(44.5 mm)	(90 cc)

ENGINE IDENTIFICATION

All models are two-stroke, single cylinder air-cooled engines with a unit block type construction.

All available engine numbers are required when ordering parts or other service materials.

MAINTENANCE

SPARK PLUG. Recommended spark plug is Champion J17LM or equivalent. Electrode gap should be 0.030 inch (0.76 mm).

CARBURETOR. Tecumseh diaphragm and Tecumseh float type carburetors have been used. Refer to the appropriate following paragraphs for information on specific carburetors.

Tecumseh Diaphragm Type Carburetor. Refer to Fig. TP1 for identification and exploded view of Tecumseh diaphragm type carburetor. Carburetor indentification numbers are

also stamped on carburetor as shown in Fig. TP2.

Initial adjustment of idle mixture and main fuel mixture screws from a lightly seated position for models with solid adjustment (L and H – Fig. TP1) is 1 turn open for each mixture screw. There is no initial adjustment for idle and main fuel mixture screws which have a hole in inner end as hole size determines initial idle fuel mixture. When renewing adjustment screws, make certain hole diameter of replacement screw matches hole diameter of new screw.

Final adjustments are made with engine at operating temperature and running. Operate engine at idle speed and adjust idle mixture screw to obtain smoothest engine idle. Operate engine at rated speed, under load and adjust main fuel mixture screw for smooth engine operation. If engine cannot be operated under load, adjust main fuel mixture needle so engine runs smoothly at rated rpm and accelerates properly when throttle is opened quickly.

When overhauling, observe the following: The fuel inlet fitting is pressed into

bore of body of some models. On these models, the fuel strainer behind inlet fitting can be cleaned by reverse flushing with compressed air after inlet needle and seat are removed. The inlet needle seat fitting is metal with a neoprene seat, so fitting (and enclosed seat) should be removed before carburetor is cleaned with a commercial solvent. The throttle plate (3) should be installed with short line stamped on plate to top of carburetor and facing out. The choke plate (9) should be installed with flat toward fuel inlet side of carburetor as shown.

When installing diaphragm (18), head of rivet should be against fuel inlet valve (16) regardless of size or placement of washers around rivet. On carburetor with "F" embossed on carburetor body, gasket (17) must be installed between diaphragm (18) and cover (19). Other models assembled with gasket, diaphragm and cover positioned as shown in Fig. TP1.

Tecumseh Float Type Carburetor. Refer to Fig. TP3 for identification and exploded view of Tecumseh float type carburetor.

Initial adjustment of idle and main fuel mixture screws from a lightly seated position is 1 turn open for each screw.

Final adjustments are made with engine at operating temperature and running. Operate engine at rated speed and adjust main fuel mixture screw (H) for smoothest engine operation. Operate engine at idle speed and adjust idle mixture screw for smoothest engine idle. Set idle speed at idle stop screw (6). If engine fails to accelerate smoothly, slight adjustment of main fuel mixture screw may be necessary.

When overhauling, check adjustment screws for excessive wear or other damage. The inlet valve fuel needle seats against the Viton rubber seat (16) which is pressed into carburetor body. Remove the rubber seat before cleaning carburetor in commercial cleaning solvent. The seat is removed using a 10-32

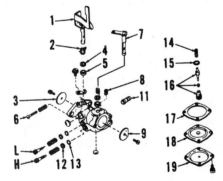

Fig. TP1—Exploded view of typical Tecumseh diaphragm type carburetor.

H. High speed
   mixture screw
L. Idle mixture screw
1. Throttle shaft
2. Spring
3. Throttle plate
4. Felt washer
5. Flat washer
6. Idle stop screw
7. Choke shaft
8. Choke retainer
9. Choke plate
11. Inlet fitting
12. Washer
13. "O" ring
14. Spring
15. Gasket
16. Inlet valve
17. Gasket
18. Diaphragm
19. Cover

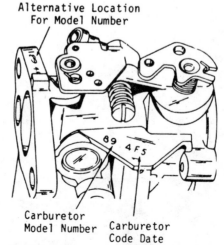

Alternative Location
For Model Number

Carburetor
Model Number

Carburetor
Code Date

Fig. TP2—View showing location of carburetor identification number on Tecumseh diaphragm type carburetor.

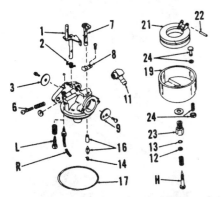

Fig. TP3—Exploded view of typical Tecumseh float type carburetor used on some models.

H. High speed mixture screw	12. Brass washer
L. Idle mixture screw	13. "O" ring
1. Throttle shaft	14. Spring
2. Spring	16. Fuel inlet needle & seat
3. Throttle plate	17. Seal
6. Idle stop screw	19. Fuel bowl
7. Choke shaft	21. Float
8. Choke retainer	22. Pivot pin
9. Choke plate	23. Bowl retaining nut
11. Inlet fitting	24. Bowl drain

or 10-24 tap and must be renewed after removing. Install new seat using a punch that will fit into bore of seat and is large enough to catch the shoulder inside the seat. Drive the fuel inlet needle seat into bore until Viton rubber seat is against bottom of bore. Install throttle plate (3) with the two stamped lines facing out and at 12 and 3 o'clock positions. Install choke plate (9) with flat side toward bottom of carburetor.

Float setting should be 7/32 inch (5.56 mm), measured with body and float assembly in inverted position, between free end of float and rim on carburetor body.

On some models, the fuel inlet fitting (11) is pressed into body. When installing and fitting, start fitting into bore, then apply a light coat of Loctite to the shank and press the fitting into position.

When installing float bowl (19), make certain that correct "O" ring (17) is used. Some "O" rings are round section, others are square.

Fuel hole and the annular groove in retaining nut (23) must be clean. The flat stepped section of fuel bowl (19) should be below the fuel inlet fitting (11). Tighten retaining nut (23) to 50-60 in.-lbs. (6-7 N·m). The high speed mixture screw (H) must not be installed when tightening nut (23).

GOVERNOR. An air vane type governor located on the carburetor throttle shaft is used on some models. On models with variable speed governor, the high speed (maximum) stop screw is located on the speed control lever. On models with fixed engine speed, rpm is adjusted by moving the governor spring bracket (B–Fig. TP5). To increase engine speed, bracket must be moved to increase

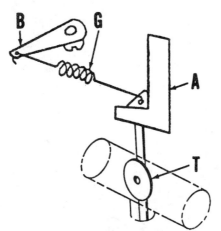

Fig. TP5—On some models, the governor air vane (A) is attached to the carburetor throttle shaft and tension of governor spring (G) hold throttle (T) open.

governor spring (G) tension holding throttle (T) open.

IGNITION SYSTEM. Engines may be equipped with either a magneto type (breaker point type) or a solid state ignition system. Refer to the appropriate paragraph for model being serviced.

Breaker point gap at maximum opening should be set at 0.020 inch (0.51 mm) before adjusting the ignition timing. On some models, ignition timing is not adjustable.

Air gap between external ignition coil laminations and flywheel magnet should be 0.0125 inch (0.32 mm) for all models. The flywheel retaining nut should be tightened to 22-27 ft.-lbs. (30-36 N·m) torque. On breaker point ignition models, points and condenser are located under the flywheel and flywheel must be removed for service.

Use Tecumseh gage 670297 or equivalent thickness plastic sheeting to set air gap as shown in Fig. TP9.

Be sure to use the correct flywheel key or sleeve recommended by the manufacturer. Gray key or sleeve is used on models with breaker point ignition and gold key or sleeve is used for solid state ignition.

If solid state ignition system fails to produce a spark at the spark plug, first check the low tension stop wire and switch for short circuit. Spark plug high tension lead and ignition module are available only as an assembly.

LUBRICATION. Engines are lubricated by mixing a good quality two-stroke air-cooled engine oil with regular grade gasoline.

Follow oil mix ratio suggested by the snowthrower manufacturer. Disregard all conflicting oil mixing instructions listed on oil containers. If correct mixture is not known, ½ pint (24 mL) can be mixed with each gallon of gasoline.

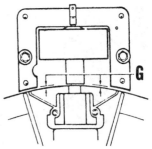

Fig. TP9—Set air gap (G) between coil laminations and flywheel to 0.0125 inch (0.32 mm). Refer to text.

REPAIRS

DISASSEMBLY. Unbolt and remove the cylinder shroud, cylinder head and crankcase covers. Remove connecting rod cap and push the rod and piston unit out through top of cylinder. It may be necessary to remove the ridge from top of cylinder bore before removing piston and connecting rod assembly.

To remove the crankshaft, remove starter housing, flywheel and the bolts securing shroud base (bearing housing) to crankcase. Strike drive end of crankshaft with a leather mallet to dislodge crankshaft and bearing housing.

CONNECTING ROD. A steel connecting rod equipped with needle roller bearings at the crankpin and at the piston pin ends is used on some models.

An aluminum connecting rod equipped with a steel insert and needle bearing rollers at the crankpin end. The piston end rod is equipped with a press in cartridge needle bearing.

Standard crankpin journal diameter is 0.6922-0.6927 inch (17.582-17.595 mm) for AH520 models, 0.8113-0.8118 inch (20.607-20.620 mm) for AH600 models. New roller bearings and steel insert assembly should be installed whenever crankpin bearing is disassembled. New rollers are serviced in a strip and can be wrapped around crankpin during assembly. Ends of steel inserts must correctly engage and match marks on rod and cap must be aligned when assembling. Tighten cap retaining screws to 40-50 in.-lbs. (4.5-5.6 N·m) torque.

PISTON, PINS AND RINGS. Refer to CONNECTING ROD section for piston removal procedure. Piston and cylinder bore specifications are as follows:

Cylinder bore diameter . . 2.093-2.094 in.
(53.162-53.188 mm)

Piston-to-cylinder
clearance . . . . 0.0045-0.0065 in.
(0.12-0.16 mm)

Ring end gap . . . . 0.006-0.016 in.
0.15-0.41 mm

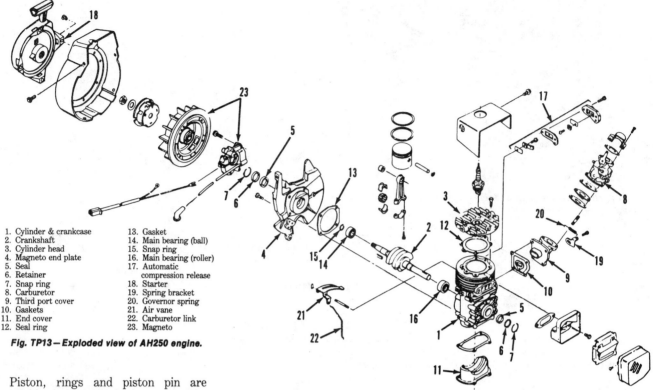

1. Cylinder & crankcase
2. Crankshaft
3. Cylinder head
4. Magneto end plate
5. Seal
6. Retainer
7. Snap ring
8. Carburetor
9. Third port cover
10. Gaskets
11. End cover
12. Seal ring
13. Gasket
14. Main bearing (ball)
15. Snap ring
16. Main bearing (roller)
17. Automatic
    compression release
18. Starter
19. Spring bracket
20. Governor spring
21. Air vane
22. Carburetor link
23. Magneto

Fig. TP13 — Exploded view of AH250 engine.

Piston, rings and piston pin are available in standard size only. The piston pin should be a press fit in heated piston.

Use a cylinder head sealing ring and a ring compressor to compress piston rings, when sliding piston into cylinder.

NOTE: Make certain rings do not catch in recess at top of cylinder.

Always renew the cylinder head metal sealing ring. The cylinder head retaining screws should be tightened to 80-100 in.-lbs. (9-11 N·m) torque. Refer to CONNECTING ROD section for installation of the connecting rod.

CRANKSHAFT AND CRANK-CASE. The crankshaft can be removed after the piston, connecting rod, flywheel and the magneto end bearing plate are removed.

A ball type main bearing is used at the flywheel end and a cartridge needle roller bearing is used at the power take off end of engine crankshaft. It should be necessary to bump or press ball type bearings from crankshaft and outer races of all bearings should fit tightly in housing bores. Crankcase can be heated around bearings to facilitate removal. Bearing should be installed with printed face on race toward center of engine.

If the crankshaft is equipped with thrust washers at ends, make certain they are installed when assembling. Crankshaft end play should be ZERO for all models.

It is important to exercise extreme care when renewing crankshaft seals to prevent their being damaged during installation. If a protector sleeve is not available, use tape to cover any splines, keyways, shoulders or threads over which the seal must pass during installation. Seals should be installed with channel groove of seal toward inside (center) of engine.

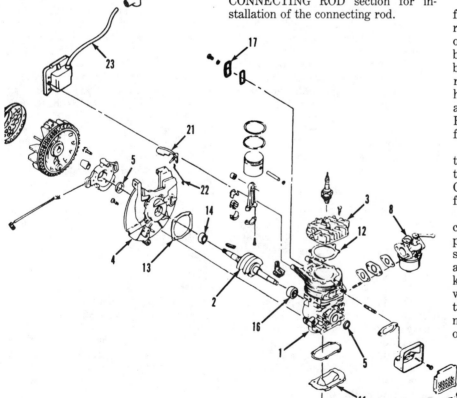

Fig. TP14 — Exploded view of typical AH600 model engine. Refer to Fig. TP13 for legend.

TECUMSEH
4-STROKE MODELS

Model	Bore	Stroke	Displacement
LV35, LAV35, TVS90, H35-prior 1983, ECH90	2.500 in. (63.50 mm)	1.844 in. (46.84 mm)	9.05 cu. in. (184 cc)
H35 after 1983	2.500 in. (63.50 mm)	1.983 in. (49.23 mm)	9.51 cu. in. (156 cc)
LAV40, TVS105, HS40	2.625 in. (66.68 mm)	1.938 in. (49.23 mm)	10.49 cu. in. (172 cc)
TNT100	2.625 in. (66.68 mm)	1.844 in. (46.84 mm)	9.98 cu. in. (164 mm)
V40, VH40, H40, HH40	2.500 in. (63.50 mm)	2.250 in. (57.15 mm)	11.04 cu. in. (181 cc)
LAV50, TNT120, TVS120, HS50	2.812 in. (71.43 mm)	1.938 in. (49.23 mm)	12.04 cu. in. (197 cc)
H50, HH50, VH50, TVM125	2.625 in. (66.68 mm)	2.250 in. (57.15 mm)	12.18 cu. in. (229 cc)
H60	2.625 in. (66.68 mm)	2.500 in. (63.50 mm)	13.53 cu. in. (222 cc)

ENGINE IDENTIFICATION

Engines must be identified by the complete model number, including the specification number in order to obtain correct repair parts. These numbers are located on the name plate and/or tags that are positioned as shown in Fig. T1 or T1A. It is important to transfer identification tags from the original engine to replacement short block assemblies so unit can be identified when servicing.

If selecting a replacement engine and the model or type number or the old engine is not known, refer to chart in Fig. T1B and proceed as follows:

1. List the corresponding number which indicates the crankshaft position.

2. Determine the horsepower needed. (Two-stroke engines are indicated by 00 in the second and third digit positions).

3. Determine the primary features needed. (Refer to the Tecumseh Engines Specification Book No. 692531 for specific engine variations.

4. Refer to Fig. T1C for Tecumseh engine model number interpretation.

The number following the letter code is the horsepower or cubic inch displacement. The number following the model number is the specification number. The last three digits of the specification number indicate a variation to the basic engine specification.

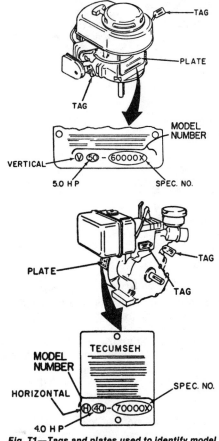

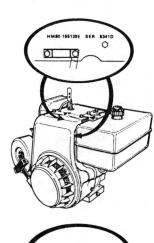

Fig. T1—Tags and plates used to identify model will most often be located in one of the positions shown.

Fig. T1A—Locations of tags and plates used to identify later model engines.

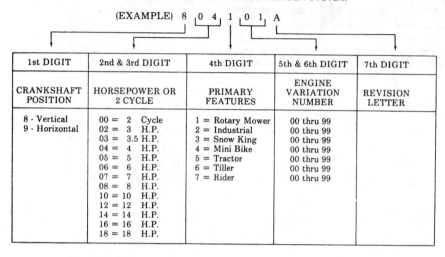

TECUMSEH NUMERICAL SERVICE NUMBER SYSTEM

(EXAMPLE) 8 04 1 01 A

1st DIGIT	2nd & 3rd DIGIT	4th DIGIT	5th & 6th DIGIT	7th DIGIT
CRANKSHAFT POSITION	HORSEPOWER OR 2 CYCLE	PRIMARY FEATURES	ENGINE VARIATION NUMBER	REVISION LETTER
8 - Vertical 9 - Horizontal	00 = 2 Cycle 02 = 3 H.P. 03 = 3.5 H.P. 04 = 4 H.P. 05 = 5 H.P. 06 = 6 H.P. 07 = 7 H.P. 08 = 8 H.P. 10 = 10 H.P. 12 = 12 H.P. 14 = 14 H.P. 16 = 16 H.P. 18 = 18 H.P.	1 = Rotary Mower 2 = Industrial 3 = Snow King 4 = Mini Bike 5 = Tractor 6 = Tiller 7 = Rider	00 thru 99 00 thru 99 00 thru 99 00 thru 99 00 thru 99 00 thru 99 00 thru 99	

Fig. T1B—Reference chart used to select or identify replacement engines.

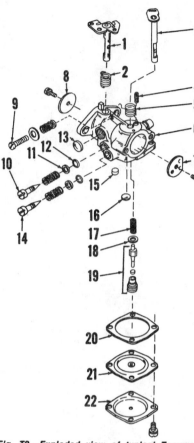

V - Vertical Shaft
LAV - Lightweight Aluminum Vertical
VM - Vertical Medium Frame
VH - Vertical Heavy Duty (Cast Iron)
TVS - Tecumseh Vertical Styled
TNT - Toro N' Tecumseh
ECV - Exclusive Craftsman Vertical
H - Horizontal Shaft
HS - Horizontal Small Frame
HM - Horizontal Medium Frame
HHM - Horizontal Heavy Duty (Cast Iron) Medium Frame
HH - Horizontal Heavy Duty (Cast Iron)
ECH - Exclusive Craftsman Horizontal

Fig. T1C—Chart showing model number interpretation.

(EXAMPLE)

TVS90-43056A - is the model and specification number

TVS - Tecumseh Vertical Styled
90 - Indicates a 9 cubic inch displacement
43056A - is the specification number used for properly identifying the parts of the engine
8310C - is the serial number
8 - first digit is the year of manufacture (1978)
310 - indicates calendar day of that year (310th day or November 6, 1978)
C - represents the line and shift on which the engine was built at the factory.

Fig. T2—Exploded view of typical Tecumseh diaphragm carburetor.

1. Throttle shaft	13. Welch plug
2. Return spring	14. Main fuel mixture screw
3. Choke shaft	
4. Choke stop spring	15. Cup plug
5. Return spring	16. Welch plug
6. Carburetor body	17. Inlet needle spring
7. Choke plate	18. Gasket
8. Throttle plate	19. Inlet needle & seat assy.
9. Idle speed screw	20. Gasket
10. Idle mixture screw	21. Diaphragm
11. Washers	22. Cover
12. "O" rings	

MAINTENANCE

SPARK PLUG. Spark plug recommendations are shown in the following chart.

14mm – 3/8 inch reach
Gasoline . . . . Champion J8
LP-Gas . . . . Champion J8
Kerosene . . . . Champion UJ12
18mm – 1/2 inch reach
Gasoline . . . . Champion D16 or MD16
Kerosene . . . . Champion D21
All 7/8 inch reach . . . . . , . Champion W18

CARBURETOR. Several different carburetors are used on these engines. Refer to the appropriate paragraph for model being serviced.

Tecumseh Diaphragm Carburetor. Refer to model number stamped on the carburetor mounting flange and to Fig. T2 for identification and exploded view of Tecumseh diaphragm type carburetor.

Initial adjustment of idle mixture and main fuel mixture screws from a lightly seated position, is 1 turn open. Clockwise rotation leans mixture and counterclockwise rotation richens mixture.

Final adjustments are made with engine at operating temperature and running. Operate engine at rated speed and adjust main fuel mixture screw (14) for smoothest engine operation. Operate engine at idle speed and adjust idle mixture screw (10) for smoothest engine idle. If engine does not accelerate smoothly, slight adjustment of main fuel mixture screw may be required. Engine idle speed should be approximately 1800 rpm.

The fuel strainer in the fuel inlet fitting can be cleaned by reverse flushing with compressed air after the inlet needle and seat (19 – Fig. T2) are removed. The inlet needle seat fitting is metal with a neoprene seat, so the fitting (and enclosed seat) should be removed before carburetor is cleaned with a commercial solvent. The stamped line on carburetor throttle plate should be toward top of carburetor, parallel with throttle shaft

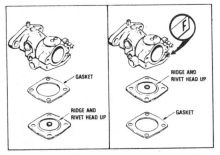

Fig. T2A—Illustration showing correct position of diaphragm on carburetors unmarked and marked with a "F".

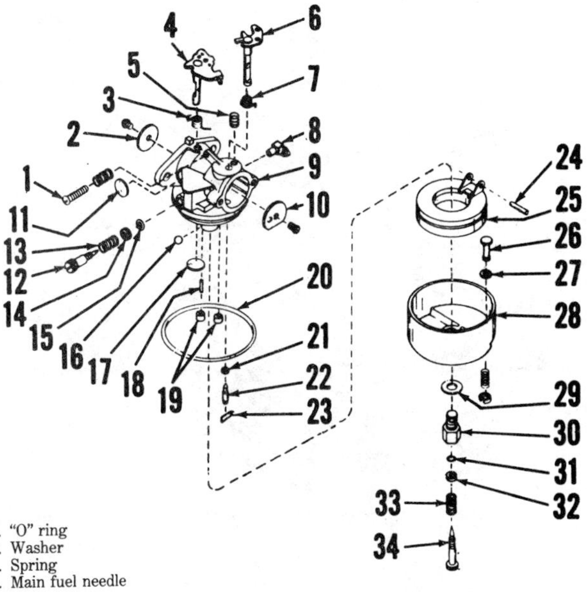

and facing OUTWARD as shown in Fig. T3. Flat side of choke plate should be toward the fuel inlet fitting side of carburetor. Mark on choke plate should be parallel to shaft and should face INWARD when choke is closed. Diaphragm (21–Fig. T2) should be installed with rounded head of center rivet up toward the inlet needle (19) regardless of size or placement of washers around the rivet.

On carburetor Models 0234-252, 265, 266, 269, 270, 271, 282, 293, 303, 322, 327, 333, 334, 344, 345, 348, 349, 350, 351, 352, 356, 368, 371, 374, 378, 379, 380, 404 and 405, or carburetors marked with a "F" as shown in Fig. T2A, gasket (20–Fig. T2) must be installed between diaphragm (21) and cover (22). All other models are assemblies as shown, with gasket between diaphragm and carburetor body.

Tecumseh Standard Float Carburetor. Refer to Fig. T4 for identification and exploded view of Tecumseh standard float type carburetor.

Initial adjustment of idle mixture and main fuel mixture screws from a lightly seated position, is 1 turn open. Clockwise rotation leans mixture and counterclockwise rotation richens mixture.

Final adjustments are made with engine at operating temperature and running. Operate engine at rated speed and adjust main fuel mixture screw (34) for smoothest engine operation. Operate

Fig. T4—Exploded view of standard Tecumseh float type carburetor.

1. Idle speed screw
2. Throttle plate
3. Return spring
4. Throttle shaft
5. Choke stop spring
6. Choke shaft
7. Return spring
8. Fuel inlet fitting
9. Carburetor body
10. Choke plate
11. Welch plug
12. Idle mixture needle
13. Spring
14. Washer
15. "O" ring
16. Ball plug
17. Welch plug
18. Pin
19. Cup plugs
20. Bowl gasket
21. Inlet needle seat
22. Inlet needle
23. Clip
24. Float shaft
25. Float
26. Drain stem
27. Gasket
28. Bowl
29. Gasket
30. Bowl retainer
31. "O" ring
32. Washer
33. Spring
34. Main fuel needle

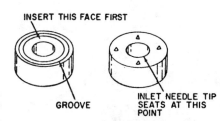

engine at idle speed and adjust idle mixture screw for smoothest engine idle. If engine does not accelerate smoothly, slight adjustment of main fuel mixture screw may be required.

Carburetor must be disassembled and all neoprene or Viton rubber parts removed before carburetor is immersed in cleaning solvent. Do not attempt to reuse any expansion plugs. Install new plugs if any are removed for cleaning.

INSERT THIS FACE FIRST

GROOVE

INLET NEEDLE TIP SEATS AT THIS POINT

Fig. T7—The Viton seat used on some Tecumseh carburetors must be installed correctly to operate properly. All-metal needle is used with seat shown.

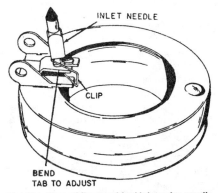

INLET NEEDLE

CLIP

BEND TAB TO ADJUST

Fig. T5—View of float and fuel inlet valve needle. The valve needle shown is equipped with a resilient tip and a clip. Bend tab shown to adjust float height.

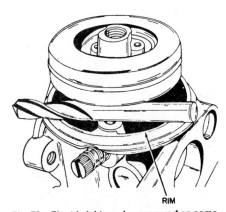

RIM

Fig. T8—Float height can be measured on some models by using a drill as shown. Refer to text for correct specifications.

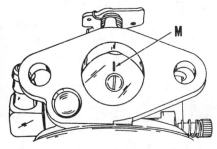

M

Fig. T3—The mark (M) on throttle plate should be parallel to the throttle shaft and outward as shown. Some models may also have mark at 3 o'clock position.

Fig. T6—A 10-24 or 10-32 tap is used to pull the brass seat fitting and fuel inlet valve seat from some carburetors. Use a close fitting flat punch to install new seat and fitting.

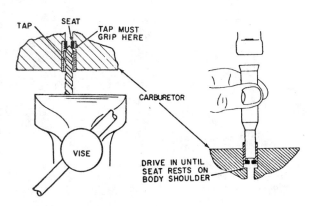

TAP SEAT TAP MUST GRIP HERE

CARBURETOR

VISE

DRIVE IN UNTIL SEAT RESTS ON BODY SHOULDER

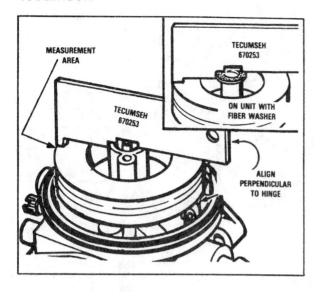

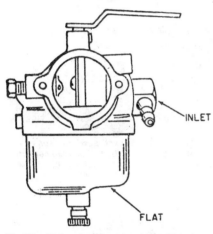

Fig. T8A—Float height can be set using Tecumseh float tool 670253 as shown.

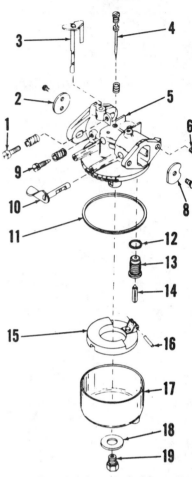

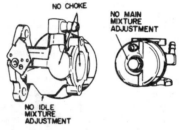

Fig. T9—Flat part of float bowl should be located under the fuel inlet fitting.

Fig. T11—Tecumseh "Automagic" carburetor is similar to standard float models, but can be identified by absence of choke and adjusting needles.

Fig. T13 — Exploded view of typical Carter N carburetor.

1. Idle speed screw
2. Throttle plate
3. Throttle shaft
4. Main adjusting screw
5. Carburetor body
6. Choke shaft spring
7. Ball
8. Choke plate
9. Idle mixture screw
10. Choke shaft
11. Bowl gasket
12. Gasket
13. Inlet valve seat
14. Inlet valve
15. Float
16. Float shaft
17. Fuel bowl
18. Gasket
19. Bowl retainer

The fuel inlet needle valve closes against a neoprene or Viton seat which must be removed before cleaning.

A resilient tip on fuel inlet needle is used on some carburetors (Fig. T5). The soft tip contacts the seating surface machined into the carburetor body to shut off the fuel. Do not attempt to remove the inlet valve seat.

A Viton seat (21 – Fig. T4) is used on some models. The rubber seat can be removed by blowing compressed air in

from the fuel inlet fitting or by using a hooked wire. The grooved face of valve seat should be IN toward bottom of bore and the valve needle should seat on smooth side of the Viton seat. Refer to Fig. T7.

A Viton seat contained in a brass seat fitting is used on some models. Use a 10-24 or 10-32 tap to pull the seat and fitting from carburetor bore as shown in Fig. T6. Use a flat, close fitting punch to install new seat and fitting.

Install the throttle plate (2 – Fig. T4) with the two stamped marks out and at

12 and 3 o'clock positions. The 12 o'clock line should be parallel with the throttle shaft and toward top of carburetor. Install choke plate (10) with flat side down toward bottom of carburetor. Float setting should be 0.200-0.220 inch (5.08-5.6

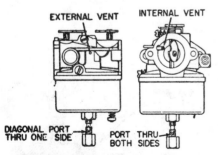

Fig. T10—The bowl retainer contains a drilled fuel passage which is different for carburetors with external and internal fuel bowl vent.

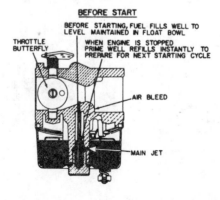

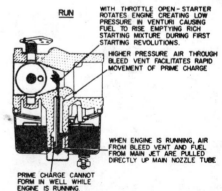

Fig. T12—The "Automagic" carburetor provides a rich starting mixture without using a choke plate. Mixture is changed by operating with a dirty air filter and by incorrect float setting.

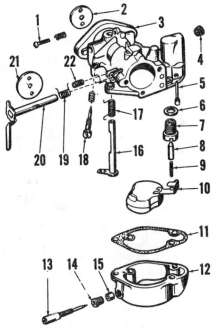

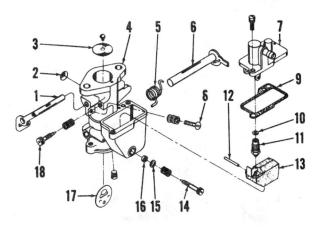

Fig. T16—Exploded view of typical Tillotson E carburetor.

1. Throttle shaft
2. Welch plug
3. Throttle plate
4. Carburetor body
5. Return spring
6. Choke shaft
7. Bowl cover
8. Idle speed stop screw
9. Bowl gasket
10. Gasket
11. Inlet needle & seat assy.
12. Float shaft
13. Float
14. Main fuel needle
15. Washer
16. Packing
17. Choke plate
18. Idle mixture screw

Fig. T14—Exploded view of Marvel-Schebler carburetor.

1. Idle speed screw	12. Fuel bowl
2. Throttle plate	13. Main adjusting screw
3. Carburetor body	14. Retainer
4. Fuel screen	15. Packing
5. Float	16. Throttle shaft
6. Gasket	17. Throttle spring
7. Inlet valve seat	18. Idle mixture screw
8. Inlet valve	19. Choke spring
9. Spring	20. Choke shaft
10. Float	21. Choke ratchet spring
11. Gasket	

mm) and can be measured with a number 4 drill as shown in Fig. T8 or using Tecumseh float gage 670253 as shown in Fig. T8A. Remove float and bend tab at float hinge to change float setting. The fuel inlet fitting (8 – Fig. T4) is pressed into body on some models. Start fitting, then apply a light coat of Loctite to shank and press fitting into

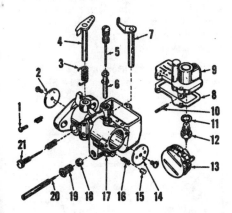

Fig. T15—Exploded view of typical Tillotson MT carburetor.

1. Idle speed screw	12. Inlet needle & seat
2. Throttle plate	assy.
3. Throttle spring	13. Float
4. Throttle shaft	14. Choke plate
5. Idle tube	15. Friction pin
6. Main nozzle	16. Choke shaft spring
7. Choke shaft	17. Carburetor body
8. Gasket	18. Packing
9. Bowl cover	19. Retainer
10. Float shaft	20. Main adjusting screw
11. Gasket	21. Idle mixture screw

position. The flat on fuel bowl should be under the fuel inlet fitting. Refer to Fig. T9.

Be sure to use correct parts when servicing the carburetor. Some gaskets used as (20 – Fig. T4) are square section, while others are round. The bowl retainer (30) contains a drilled passage for fuel to the high speed metering needle (34). A diagonal port through one side of the bowl retainer is used on carburetors with external vent. The port is through both sides on models with internal vent. Refer to Fig. T10.

Tecumseh "Automagic" Float Carburetor. Refer to Fig. T11 and note carburetor can be identified by the absence of the idle mixture screw, choke and main mixture screw.

Float setting is 0.210 inch (5.33 mm) or float may be set with a number 4 drill as shown in Fig. T8 or Tecumseh float gage 670253 as shown in Fig. T8A. Air filter maintenance is important in order to obtain the correct fuel mixture. Refer to notes for servicing standard Tecumseh float carburetors. Refer to Fig. T12 for operating principles.

Carter Carburetor. Refer to Fig. T13 for identification and exploded view of Carter carburetor.

Initial adjustment of idle mixture screw and main fuel mixture screw from a lightly seated position is 1½ turns open for idle mixture screw and 1¾ turns open for main fuel mixture screw. Clockwise rotation leans mixture and counterclockwise rotation richens mixture.

Final adjustments are made with engine at operating temperature and running. Operate engine at rated speed and adjust main fuel mixture screw (4) for smoothest engine operation. Operate engine at idle speed and adjust idle mixture screw (9) for smoothest engine idle. If engine acceleration is not smooth, main fuel mixture screw may have to be adjusted slightly.

To check float level, invert carburetor body and float assembly. There should be 11/64 inch (4.37 mm) clearance between free side of float and machined

surface of body casting. Bend float lever tang to provide correct measurement.

Marvel-Schebler. Refer to Fig. T14 for identification and exploded view of Marvel-Schebler Series AH carburetor.

Initial adjustment of idle mixture screw and main fuel mixture screw from a lightly seated position, is 1 turn open for idle mixture screw and 1¼ turn open for main fuel mixture screw. Clockwise rotation leans mixture and counterclockwise rotation richens mixture.

Final adjustments are made with engine at operating temperature and running. Operate engine at rated speed and adjust main fuel mixture screw (13) for smoothest engine operation. Operate engine at idle speed and adjust idle fuel mixture needle (18) for smoothest engine idle. Adjust idle speed screw (1) to obtain 1800 rpm.

To adjust float level, invert carburetor throttle body and float assembly. Float clearance should be 3/32 inch (2.38 mm) between free end of float and machined surface of carburetor body. Bend tang on float which contacts fuel inlet needle as necessary to obtain correct float level.

Tillotson Type "MT" Carburetor. Refer to Fig. T15 for identification and exploded view of Tillotson type MT carburetor.

Initial adjustment of idle mixture screw and main fuel mixture screw from

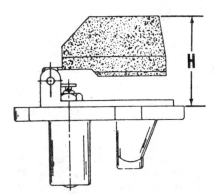

Fig. T17—Float height (H) should be measured as shown for Tillotson type "E" carburetors. Gasket should not be installed when measuring.

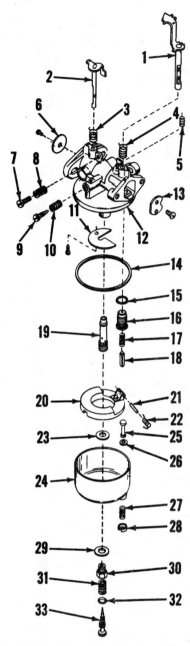

Fig. T18—Exploded view of Walbro LMG carburetor.

1. Choke shaft	18. Inlet valve
2. Throttle shaft	19. Main nozzle
3. Throttle return spring	20. Float
4. Choke return spring	21. Float shaft
5. Choke stop spring	22. Spring
6. Throttle plate	23. Gasket
7. Idle speed stop screw	24. Bowl
8. Spring	25. Drain stem
9. Idle mixture screw	26. Gasket
10. Spring	27. Spring
11. Baffle	28. Retainer
12. Carburetor body	29. Gasket
13. Choke plate	30. Bowl retainer
14. Bowl gasket	31. Spring
15. Gasket	32. "O" ring
16. Inlet valve seat	33. Main fuel screw
17. Spring	

a lightly seated position, is ¾ turn open for idle mixture screw (21) and 1 turn open for main fuel mixture screw (20). Clockwise rotation leans fuel mixture and counterclockwise rotation richens fuel mixture. Final adjustments are made with engine at operating

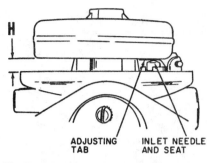

Fig. T19—Float height (H) should be measured as shown on Walbro float carburetors. Bend the adjusting tab to adjust height.

temperature and running. Operate engine at rated speed and adjust main fuel mixture screw for smoothest engine operation. Operate engine at idle speed and adjust idle mixture screw for smoothest engine idle. Adjust throttle stop screw (1) to obtain 1800 rpm. If engine acceleration is not smooth, main fuel mixture screw may have to be adjusted slightly.

To check float setting, invert carburetor throttle body and measure distance from top of float to carburetor body float bowl mating surface. Distance should be 1-13/32 inch (35.72 mm). Carefully bend float tang which contacts fuel inlet needle as necessary to obtain correct float level.

Tillotson Type "E" Carburetor. Refer to Fig. T16 for identification and exploded view of Tillotson type E carburetor.

Initial adjustment of idle mixture screw and main fuel mixture screw from a lightly seated position, is ¾ turn open for idle mixture screw (18) and 1 turn open for main fuel mixture screw (14). Clockwise rotation leans mixture and counterclockwise rotation richens mixture.

Final adjustments are made with engine at operating temperature and running. Operate engine at rated speed and adjust main fuel mixture screw for smoothest engine operation. Operate engine at idle speed and adjust idle mixture screw for smoothest engine idle. Adjust throttle stop screw (8) to obtain desired idle speed. If engine does not accelerate smoothly, it may be necessary to adjust main fuel mixture screw slightly.

To check float setting, invert carburetor throttle body and measure distance between top of float at free end to float bowl mating surface on carburetor. Distance should be 1-5/64 inch (27.38 mm). Refer to Fig. T17. Gasket should not be installed when measuring float height. Carefully bend float tang which contacts fuel inlet needle as necessary to obtain correct float level.

Walbro. Refer to Fig. T18 for iden-

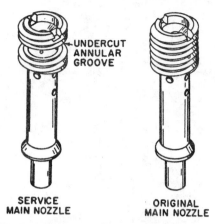

Fig. T20—The main nozzle originally installed is drilled after installation through hole in body. Service main nozzles are grooved so alignment is not necesasry.

tification and exploded view of Walbro carburetor.

Initial adjustment of idle mixture and main fuel mixture screws from a lightly seated position, is 1 turn open. Clockwise rotation leans fuel mixture and counterclockwise rotation richens fuel mixture.

Final adjustments are made with engine at operating temperature and running. Operate engine at rated speed and adjust main fuel mixture screw (33) for smoothest engine operation. Operate engine at idle speed and adjust idle mixture screw for smoothest engine idle. Adjust throttle stop screw (7) so engine idles at 1800 rpm.

To check float level, invert carburetor throttle body and measure clearance between free end of float and machined surface of carburetor. Clearance should be ⅛ inch (3.18 mm). Refer to Fig. T19. Carefully bend float tang which contacts fuel inlet needle as necessary to obtain correct float setting.

NOTE: If carburetor has been disassembled and main nozzle (19—Fig. T18) removed, do not reinstall the original

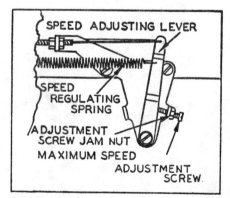

Fig. T21—Speed adjusting screw and lever. The no-load speed should not exceed 3600 rpm.

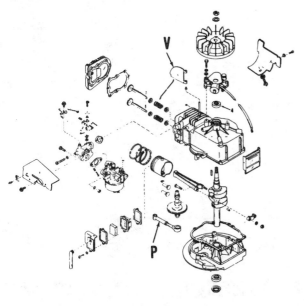

Fig. T21A—Exploded view of Light Frame vertical crankshaft engine. Model shown has plunger type oil pump (P) and air vane (V) governor.

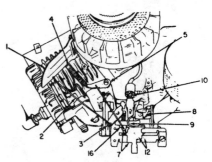

Fig. T24—View of one type of speed control with pneumatic governor. Refer also to Figs. T25, T26 and T27.

1. Air vane
2. Throttle control link
3. Carburetor throttle lever
4. Governor spring
5. Governor spring linkage
7. Speed control lever
8. Carburetor choke lever
9. Choke control
10. Stop switch
12. Alignment holes
16. Idle speed stop screw
18. High speed stop screw (Fig. T27)

equipment nozzle. Install a new service nozzle. Refer to Fig. T20 for differences between original and service nozzles.

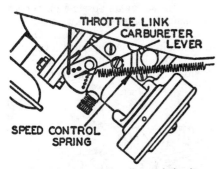

Fig. T22—Outer spring anchorage holes in carburetor throttle lever may be used to reduce speed fluctuations or surging.

PNEUMATIC GOVERNOR. Some engines are equipped with a pneumatic (air vane) type governor. On fixed linkage hookups, the recommended idle speed is 1800 rpm. Standard no-load speed is 3300 rpm. Operating speed range is 2600 to 3600 rpm. On engines not equipped with slide control (Fig. T23), obtain desired speed by varying the tension on governor speed regulating spring by moving speed adjusting lever (Fig. T21) in or out as required. If engine speed fluctuates due to governor hunting or surging, move carburetor end of governor spring into next outer hole on carburetor throttle arm shown in Fig. T22.

A too-lean mixture or friction in governor linkage will also cause hunting or unsteady operation.

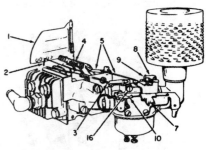

Fig. T25—Pneumatic governor linkage.

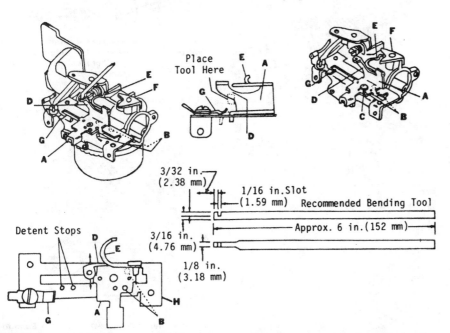

Fig. T23—When carburetor is equipped with slide control, governed speed is adjusted by bending slide arm at point "D". Bend arm outward from engine to increase speed.

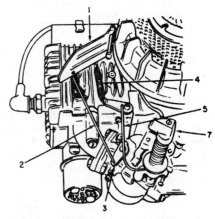

Fig. T26—Pneumatic governor linkage.

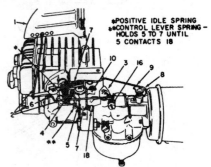

Fig. T27—Pneumatic governor linkage.

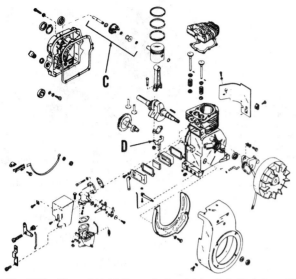

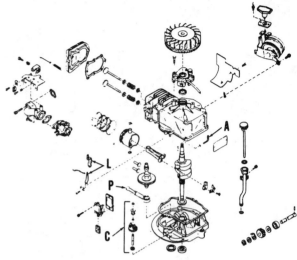

Fig. T27A — View of Light Frame horizontal crankshaft engine with mechanical governor and splash lubrication. Notice that connecting rod cap on model shown is away from camshaft side of engine.

Fig. T28A—View of Light Frame vertical crankshaft engine. Mechanical governor (A, C and L) and plunger type oil pump (P) are shown.

On engines with slide control carburetors, speed adjustment is made with engine running. Remove slide control cover (Fig. T23) marked "choke, fast, slow, etc.". Lock the carburetor in high speed "run" position by matching the hole in slide control member (A) with hole (B) closest to choke end of bracket. Temporarily hold in this position by inserting a tapered punch or pin of suitable size into the hole.

At this time the choke should be wide open and the choke activating arm (E) should be clear of choke lever (F) by 1/64 inch (0.40 mm). Obtain clearance gap by bending arm (E). To increase engine speed, bend slide arm at point (D) outward from engine. To decrease speed, bend arm inward toward engine.

Make certain on models equipped with wiper grounding switch that wiper (G) touches slide (A) when control is moved to "STOP" position.

Refer to Figs. T24, T25, T26 and T27 for assembled views of pneumatic governor systems.

MECHANICAL GOVERNOR. Some engines are equipped with a mechanical (flyweight) type governor. To adjust the governor linkage, refer to Fig. T28 and loosen governor lever screw. Twist pro-truding end of governor shaft counterclockwise as far as possible on vertical crankshaft engines; clockwise on horizontal crankshaft engines. On all models, move the governor lever until carburetor throttle shaft is in wide open position, then tighten governor lever clamp screw.

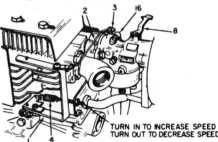

Fig. T29 — View of mechanical governor with one type of constant speed control. Refer also to Fig. T30 through T43 for other mechanical governor installations.

1. Governor lever
2. Throttle control
3. Carburetor throttle lever
4. Governor spring
8. Choke lever
16. Idle speed stop screw

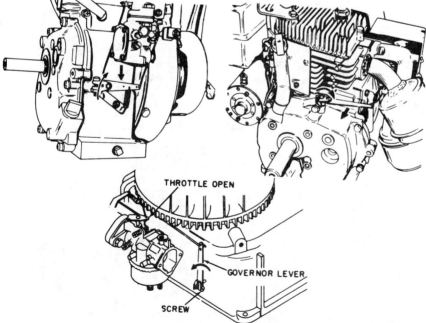

Fig. T28—Views showing location of mechanical governor lever and direction to turn when adjusting position on governor shaft.

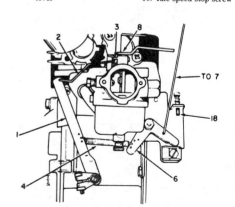

Fig. T30—Mechanical governor linkage. Refer to Fig. T29 for legend except for the following.

6. Bellcrank
7. Speed control lever
18. High speed stop screw

Binding or worn governor linkage will result in hunting or unsteady engine operation. An improperly adjusted car-buretor will also cause a surging or hunting condition.

Refer to Figs. T29 through T42 for views of typical mechanical governor speed control linkage installations. The governor gear shaft must be pressed into bore in cover until the correct amount of the shaft protrudes. Refer to illustration and chart in Fig. T44.

IGNITION SYSTEM. A magneto ignition system with breaker points or

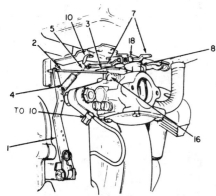

Fig. T31—Mechanical governor linkage.

1. Governor lever	7. Speed control lever
2. Throttle control link	8. Choke lever
3. Carburetor throttle	10. Stop switch
4. Governor spring	16. Idle speed stop screw
5. Governor spring	18. High speed stop screw
linkage	

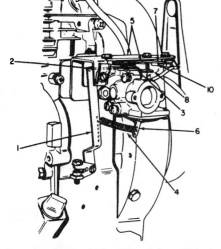

Fig. T32—Mechanical governor linkage. Refer to Fig. T31 for legend. Bellcrank is shown at (6).

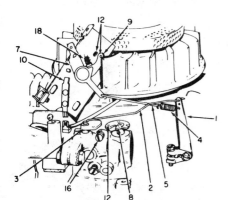

Fig. T33—Mechanical governor linkage. Control cover is raised to view underside.

1. Governor lever	7. Speed control lever
2. Throttle control link	8. Choke lever
3. Carburetor throttle	9. Choke control
lever	10. Stop switch
4. Governor spring	12. Alignment holes
5. Gover spring	16. Idle speed stop screw
linkage	18. High speed stop screw

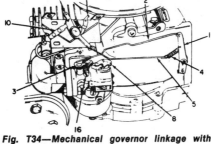

Fig. T34—Mechanical governor linkage with control cover raised. Refer to Fig. T33 for legend.

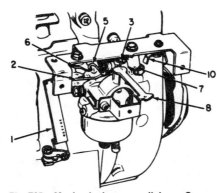

Fig. T35—Mechanical governor linkage. Governed speed of engine is increased by closing loop in linkage (5); decrease speed by spreading loop. Refer to Fig. T33 for legend.

Fig. T36—Mechanical governor linkage.

1. Governor lever	7. Speed control lever
2. Throttle control link	8. Choke lever
3. Throttle lever	9. Choke control link
4. Governor spring	10. Stop switch
5. Governor spring linkage	16. Idle speed stop screw
6. Bellcrank	18. High speed stop screw

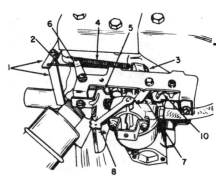

Fig. T37—View of control linkage used on some engines. To increase governed engine speed, close loop (5); to decrease speed, spread loop (5). Refer to Fig. T36 for legend.

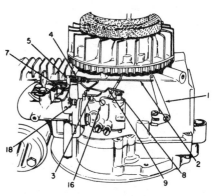

Fig. T38—View of mechanical governor control linkage. Governor spring (4) is hooked onto loop in link (2).

1. Governor lever	
2. Throttle control link	7. Speed control lever
3. Throttle lever	8. Choke lever
4. Governor spring	9. Choke control link
5. Governor spring	16. Idle speed stop screw
linkage	18. High speed stop screw

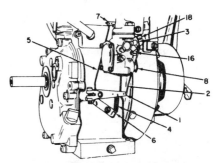

Fig. T39—Linkage for mechanical governor. Refer to have Fig. T38 for legend. Bellcrank is shown at (6).

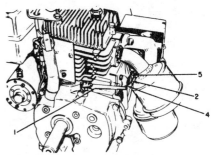

Fig. T40—Mechanical governor linkage. Refer to Fig. T38 for legend.

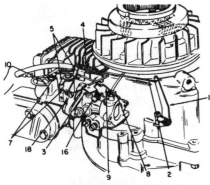

Fig. T41—Mechanical governor linkage. Refer to Fig. T38 for legend.

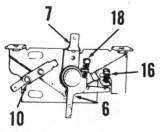

Fig. T42—Models with "Automatic" carburetor use control shown.

6. Governor spring bellcrank	10. Idle speed stop screw
7. Control lever	16. Idle speed stop
	18. High speed stop screw

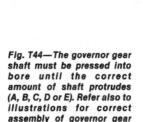

Fig. T44—The governor gear shaft must be pressed into bore until the correct amount of shaft protrudes (A, B, C, D or E). Refer also to illustrations for correct assembly of governor gear and associated parts.

B. 1-5/16 inches (33.338 mm)
C. 1-5/16 inches (33.338 mm)
D. 1-3/8 inches (34.925 mm)
E. 1-19/32 inches (40. 481 mm)

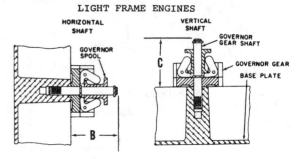

LIGHT FRAME ENGINES

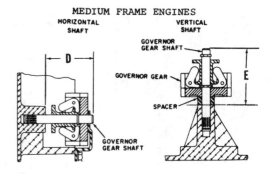

MEDIUM FRAME ENGINES

capacitor-discharge ignition (CDI) may have been used according to model and application. Refer to appropriate paragraph for model being serviced.

Breaker Point Ignition System. Breaker point gap at maximum opening should be 0.020 inch (0.51 mm) for all models. Marks are usually located on stator and mounting post to facilitate timing.

Ignition timing can be checked and adjusted to occur when piston is at specific location (BTDC) if marks are missing. Refer to the following specifications for recommended timing.

Models	Piston Position BTDC
HS40, LAV40, TVS105, TNT100, TNT120	0.035 in. (0.89 mm)
V40, VH40, LAV50, TVS120, H40, HH40, HS50	0.050 in. (1.27 mm)
LV35, LAV35, TVS90, H35, ECH90	0.065 in. (1.65 mm)

Models	Piston Position BTDC
V50, VH50, V60, VH60, H50, HH50, H60, HH60, TVM125, TVM140	0.080 in. (2.03 mm)

Some models may be equipped with the coil and laminations mounted outside the flywheel. Engines equipped

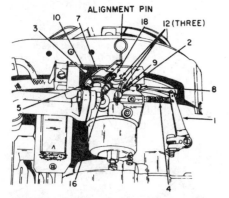

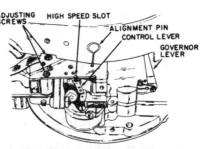

Fig. T43—On model shown, adjust location of cover so control lever is aligned with high speed slot and alignment holes are aligned.

1. Governor lever		8. Choke lever	12. Alignment holes
2. Throttle control link	5. Governor spring linkage	9. Choke linkage	16. Idle speed stop screw
3. Throttle lever	7. Control lever	10. Stop switch	18. High speed stop screw
4. Governor spring			

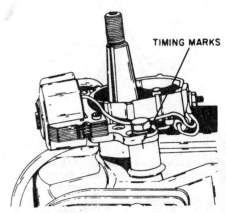

Fig. T45—Align timing marks as shown on magneto ignition system.

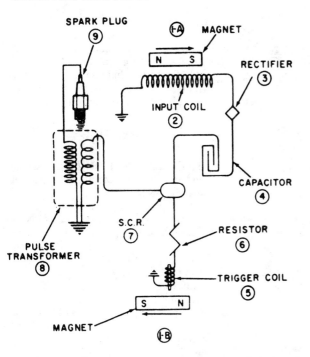

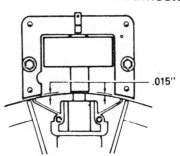

Fig. T46—Wiring diagram of solid state ignition system used on some models.

Fig. T47A—Set coil lamination air gap at 0.015 inch (0.38 mm) at points shown.

with breaker point ignition have the ignition points and condenser mounted under the flywheel with the coil and laminations mounted outside the flywheel. This system is identified by the round shape of the coil and a stamping "Grey Key" in the coil to identify the correct flywheel key.

The correct air gap setting between the flywheel magnets and the coil laminations is 0.015 inch (0.38 mm). Use Tecumseh gage 670259 or equivalent thickness sheet plastic to set gap as shown in Fig. T47A.

Solid State Ignition System. The Tecumseh solid state ignition system does not use ignition breaker points. The only moving part of the system is the rotating flywheel with the charging magnets. As the flywheel magnet passes position (1A – Fig. T46), a low voltage AC current is induced into input coil (2). Current passes through rectifier (3) converting this current to DC current. It then travels to capacitor (4) where it is stored. The flywheel rotates approximately 180° to position (1B). As it passes trigger coil (5), it induces a very small electric charge into the coil. This charge passes through resistor (6) and turns on the SCR (silicon controlled rectifier) switch (7). With the SCR switch closed, low voltage current stored in capacitor (4) travels to pulse transformer (8). Voltage is stepped up instantaneously and current is discharged across the electrodes of spark plug (9), producing a spark before top dead center.

Some units are equipped with a second trigger coil and resistor set to turn the SCR switch on at a lower rpm. This sec-

ond trigger pin is closer to the flywheel and produces a spark at TDC for earlier starting. As engine rpm increases, the first (shorter) trigger pin picks up the small electric charge and turns the SCR switch on, firing the spark plug at TDC.

If system fails to produce a spark to the spark plug, first check high tension lead (Fig. T47). If condition of high tension lead is questionable, renew pulse transformer and high tension lead assembly. Check low tension lead and renew if insulation is faulty. The magneto charging coil, electronic triggering system and mounting plate are available only as an assembly. If necessary to renew this assembly, place unit in position on engine. Start retaining screws, turn mounting plate counterclockwise as far as possible, then tighten retaining screws to 5-7 ft.-lbs. (7-10 N·m).

Engines with solid state (CDI) ignition have all the ignition components sealed in a module and located outside the flywheel. There are no components

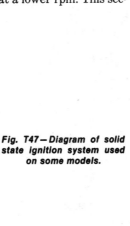

Fig. T47—Diagram of solid state ignition system used on some models.

Fig. T48—Various types of barrel and plunger oil pumps have been used. Chamfered face of collar should be toward camshaft if drive collar has only one chamfered side. If drive collar has a float boss, the boss should be next to the engine lower cover, away from camshaft gear.

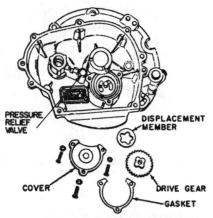

Fig. T49—Disassembled view of typical gear-driven rotor type oil pump.

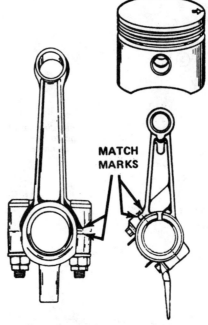

Fig. T50—Match marks on connecting rod and cap should be aligned and should be toward pto end of crankshaft.

under the flywheel except a spring clip to hold the flywheel key in position. This system is identified by the square shape module and a stamping "Gold Key" to identify the correct flywheel key.

The correct air gap setting between the flywheel magnets and the laminations on ignition module is 0.015 inch (0.38 mm). Use Tecumseh gage 670259 or equivalent thickness sheet plastic to set gap as shown in Fig. T47A.

LUBRICATION. Vertical crankshaft engines may be equipped with a barrel and plunger type oil pump or a gear driven rotor type oil pump. Horizontal crankshaft engines may be equipped with a gear-driven rotor type pump or with a dipper type oil slinger attached to the connecting rod.

Oil level should be checked after every five hours of operation: Maintain oil level at lower edge of filler plug or at "FULL" mark on dipstick.

Manufacturer recommends oil with an API service classification SC, SD, SE or SF. Use SAE 30 motor oil for temperatures above 32° F (0° C), SAE 5W-30 motor oil for temperatures between 32°F (0°C) and 0°F (−18°C) and a 90% SAE 10W oil-10% kerosene mixture or SAE 5W-30 for temperatures below 0° F (−18° C). Manufacturer explicity states: DO NOT USE SAE 10W-40 motor oil.

Oil should be changed after the first two hours of engine operation and after every 25 hours of operation thereafter.

REPAIRS

TIGHTENING TORQUES. Recommended tightening torque specifications are as follows:
Spark plug . . . . .250-360 in.-lbs.
(28-41 N·m)
Cylinder head . . . . .160-200 in.-lbs.
(18-23 N·m)

Connecting rod nuts
(except Durlok nuts):
5 hp. (3.7 kW)
medium frame,
6 hp, (4.5 kW) . . . . 86-110 in.-lbs.
(10-12 N·m)

Connecting rod bolts
(Durlok):
5 hp. (3.7 kW) light
frame . . . . .110-130 in.-lbs.
(12-15 N·m)

5 hp. (3.7 kW)
medium frame, 6 hp.
(4.5 kW) . . . . 130-150 in.-lbs.
(15-17 N·m)

All other models . . . . . .95-110 in.-lbs.
(11-12 N·m)

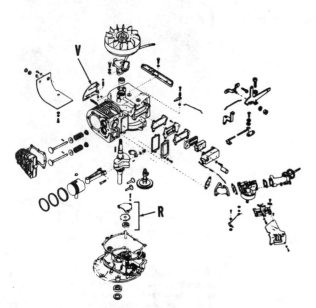

Fig. T49A—View of vertical crankshaft engine. Rotor type oil pump is shown at (R); governor air vane at (V).

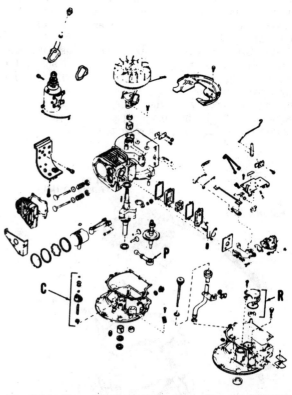

Fig. T50A—View of Medium Frame engine with vertical crankshaft. Some models use plunger type oil pump (P); others are equipped with rotor type oil pump (R).

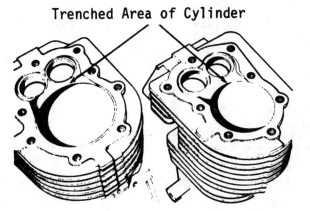

Trenched Area of Cylinder

Fig. T51—Cylinder block on Models H50, HH50, VH50, TVM125 and H60 has been "trenched" to improve fuel flow and power.

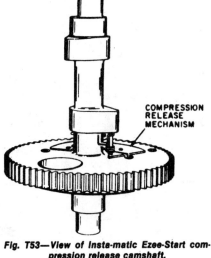

COMPRESSION RELEASE MECHANISM

Fig. T53—View of Insta-matic Ezee-Start compression release camshaft.

Flywheel:

Light frame ....360-400 in.-lbs.
(41-45 N·m)
Medium frame ....430-480 in.-lbs.
(49-54 N·m)
External ignition .....400-440 in.-lbs.
(45-50 N·m)
Magneto stator ....40-90 in.-lbs.
(5-10 N·m)
Mounting flange ....75-110 in.-lbs.
(9-12 N·m)
Carburetor to intake pipe..48-72 in.-lbs.
(5-8 N·m)
Intake pipe to cylinder ....72-96 in.-lbs.
(8-11 N·m)
Gear reduction
housing.............100-144 in.-lbs.
(11-16 N·m)
Gear reduction cover ....75-110 in.-lbs.
(9-12 N·m)

Oil Seal Removed

Snap Ring

Fig. T52—On Models H30 through H60 it is necessary to remove snap ring under oil seal before removing crankcase cover.

H35-after 1983, LAV40,
TVS105, HS40, TNT120,
LAV50, HS50,
TVS120............1.0005-1.0010 in.
(25.413-25.425 mm)
V40, VH40, H40,
HH40, H50, HH50,
VH50, H60.........1.0630-1.0635 in.
(27.000-27.013 mm)

CONNECTING ROD. Piston and connecting rod assembly is removed from cylinder head end of engine. The aluminum alloy connecting rod rides directly on crankshaft crankpin.

Refer to the following table for standard crankpin journal diameter.

Standard inside diameter for connecting rod crankpin bearing journal is shown in the following table.

Model	Diameter
LV35, LAV35, TVS90, H35 prior 1983, ECH90, TNT100............	0.8620-0.8625 in. (21.895-21.908 mm)

Connecting rod bearing-to-crankpin journal clearance should be 0.0005-0.0015 inch (0.013-0.38 mm) for all models.

When installing connecting rod and piston assembly, align the match marks on connecting rod and cap as shown in

Model	Diameter
LV35, LAV35, TVS90, H35-prior 1983, ECH90, TNT100.....	0.8610-0.8615 in. (21.869-21.882 mm)
H35 after 1983, LAV40, TVS105, HS40, TNT120, LAV50, HS50, TVS120.......	0.9995-1.0000 in. (25.390-25.400 mm)
V40, VH40, H40, HH40, H50, HH50, VH50, TVM140	1.0615-1.0620 in. (26.962-26.975 mm)

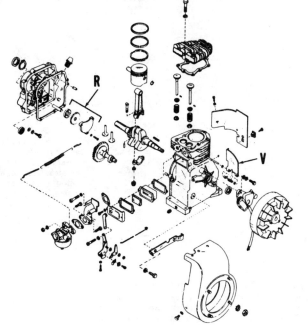

Fig. T52A—View of Light Frame engine with horizontal crankshaft. Air vane (V) type governor and rotor type oil pump (R) are used on model shown.

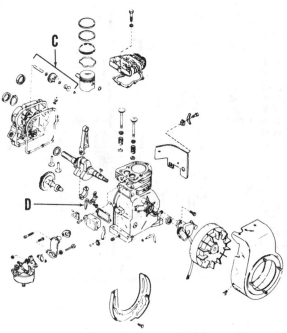

Fig. T53A—View of Light Frame horizontal crankshaft engine with mechanical governor and splash lubrication. Governor centrifugal weights are shown at (C) and lubrication dipper at (D).

Fig. T50. On some models, the piston pin hole is offset in piston and arrow on top of piston should be toward valves. On all engines, the match marks on connecting rod and cap must be toward power takeoff (pto) end of crankshaft. Lock plates, if so equipped, for connecting rod cap retaining screws should be renewed each time cap is removed.

PISTON, PIN AND RINGS. Aluminum alloy pistons are equipped with two compression rings and one oil control ring. Ring end gap for all models is 0.007-0.017 inch (0.18-0.43 mm).

Piston skirt-to-cylinder clearances are listed in the following table.

Model	Clearance
H50, HH50, VH50, H60	0.0035-0.005 in. (0.089-0.127 mm)
LV35, LAV35, TVS90, H35 (all), ECH90, LAV40, TVS105, HS40, TNT100, LAV50, TVS120, TNT120, HS50	0.0045-0.0060 in. (0.114-0.0152 mm)
V40, VH40, H40, HH40	0.0055-0.0070 in. (0.140-0.178 mm)

Standard piston diameters measured at piston skirt 90° from piston pin bore are listed in the following table.

Model	Diameter
LV35, LAV35, H35 (all), TVS90, ECH90	2.4950-2.4955 in. (63.373-63.386 mm)
LAV40, TVS105, HS40, TNT100	2.6200-2.6205 in. (66.548-66.560 mm)
H50, HH50, VH50, H60	2.6210-2.6215 in. (66.575-66.586 mm)
V40, VH40, H40, HH40	2.4945-2.4950 in. (63.363-63.373 mm)
HS50, LAV50, TVS120, TNT120	2.8070-2.8075 in. (71.298-71.311 mm)

Standard ring side clearance in ring grooves are shown in the following table.

Model	Clearance
LV35, LAV35, TVS90, H35 (all), V40, VH40, H40, HH40	0.002-0.003 in. (0.05-0.08 mm)

LAV40, TVS105, HS40:

Compression rings	0.002-0.004 in. (0.05-0.10 mm)
Oil control ring	0.001-0.004 in. (0.03-0.10 mm)

LAV50, HS50, TVS120:

Compression rings	0.003-0.004 in. (0.08-0.10 mm)
Oil control ring	0.002-0.003 in. (0.05-0.08 mm)

TNT120:

Compression rings	0.003-0.004 in. (0.08-0.10 mm)
Oil control ring	0.002-0.004 in. (0.05-0.10 mm)

Model	Clearance
H50, HH50, VH50, H60	0.002-0.004 in. (0.05-0.10 mm)

Refer to CONNECTING ROD section for correct piston to connecting rod

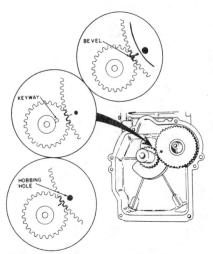

Fig. T54—The camshaft and crankshaft must be correctly timed to ensure valves open at correct time. Different types of marks have been used, but marks should be aligned when assembling.

Fig. T55—View of Medium Frame horizontal crankshaft engine with mechanical governor (C). An oil dipper for splash lubrication is cast onto the connecting rod cap instead of using the rotor type oil pump (R).

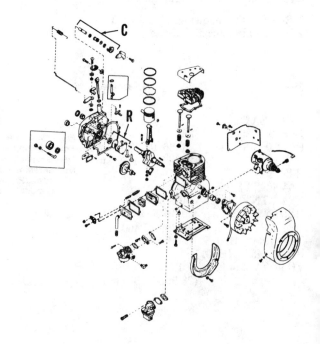

assembly and correct piston installation procedure. Piston pin should be a tight push fit in piston pin bore and connecting rod pin bore and is retained by snap rings at each end of piston pin bore. Install marked side of piston rings up and stagger ring end gaps equally around circumference of piston during installation.

CYLINDER AND CRANKCASE.
Cylinder and crankcase are an integral casting on all models. Cylinder should be honed and fitted to nearest oversize for which piston and ring set are available if cylinder is scored, tapered or out-of-round more than 0.005 inch (0.13 mm).

Standard cylinder bore diameters are shown in the following table.

Model	Diameter
LV35, LAV35, H35 (all), TVS90, ECH90, V40, VH40, H40, HH40	2.5000-2.5010 in. (63.500-63.525 mm)
LAV40, TVS105, HS40, TNT100, H50, HH50, VH50, TVM125, H60	2.6250-2.6260 in. (66.675-66.700 mm)
LAV50, TVS120, HS50, TNT120	2.8120-2.8130 in. (71.425-71.450 mm)

Refer to PISTON, PIN AND RINGS section for correct piston-to-cylinder block clearance.

Note also that cylinder block used on Models H50, HH50, VH50, TVM125 and H60 has been trenched to improve fuel flow and power (Fig. T51).

Standard main bearing bore diameters for both main bearings, or as indicated otherwise, are shown in the following chart.

Model	Diameter
LV35, LAV35, TVS90, H35 (prior to 1983), ECH90, TNT100	0.8755-0.8760 (22.238-22.250 mm)
H35 (after 1983), LAV40, TVS105, HS40, V40, VH40, H40, HH40, LAV50, TVS120, HS50, H50, H60	1.0005-1.0010 in. (25.413-25.424 mm)

A special tool kit is available from Tecumseh to ream the cylinder crankcase and cover (mounting flange) main bearing bores to install service bushing for some models.

CRANKSHAFT, MAIN BEARINGS AND SEALS.
Crankshaft main bearing journals on some models ride directly in the aluminum alloy bores in the cylinder block and the crankcase cover (mounting flange). Other engines are originally equipped with renewable steel backed bronze bushings and some are originally equipped with a ball type main bearing at the pto end of crankshaft.

Refer to CONNECTING ROD section for standard crankshaft crankpin journal diameters. Standard diameters for crankshaft main bearing journals are shown in the following chart.

Model	Diameter
LV35, TVS90, H35 (after 1983), ECH90, TNT100	0.8735-0.8740 in. (22.187-22.200 mm)
H35 (after 1983), LAV40, TVS105, HS40, TNT100, V40, VH40, H40, HH40, LAV50, TVS120, HS50, TNT120, H50, HH50, VH50, H60	0.9985-0.9990 in. (25.362-25.375 mm)

Clearance between crankshaft main bearing journals and main bearing bushings should be 0.0010-0.0025 inch (0.025-0.064 mm) for 0.8735-0.8740 inch (22.187-22.200 mm) diameter journal of 0.0015-0.0025 inch (0.038-0.064 mm) for 0.9985-0.9990 inch (25.362-25.375 mm) journal.

Crankshaft end play for all models is 0.005-0.027 inch (0.13-0.69 mm).

On Models H35 through HS50, it is necessary to remove a snap ring (Fig. T52) which is located under oil seal. Note oil seal depth before removal of seal, remove seal and snap ring. Crankcase cover may not be removed. When installing cover, press new seal in to same depth as old seal before removal.

Ball bearing should be inspected and renewed if rough, loose or damaged. Bearing must be pressed on or off of crankshaft journal using a suitable press or puller.

Always note oil seal depth and direction before removing old seal from crankcase or cover. New seals must be pressed into seal bores to the same depth as old seal before removal on all models.

When installing crankshaft, align crankshaft and camshaft gear timing marks as shown in Fig. T54.

CAMSHAFT.
The camshaft and camshaft gear are an integral part which rides on camshaft journals at each end of camshaft. Camshaft on some models also has a compression release mechanism mounted on camshaft gear which lifts exhaust valve at low cranking rpm to reduce compression and aid starting (Fig. T53).

Renew camshaft if lobes or journals are worn or scored. Spring on compression release mechanism should snap weight against camshaft. Compression release mechanism and camshaft are serviced as an assembly only.

Standard camshaft journal diameter is 0.6230-0.6235 inch (15.824-15.837 mm) for Models V40, VH40, H40, HH40, H50, HH50, VH50, TVM125 and H60 and 0.4975-0.4980 inch (12.637-12.649 mm) for all other models.

Note that exhaust cam follower (lifter) is longer on some models and that on models equipped with barrel and plunger type oil pump, the pump is operated by an eccentric on camshaft.

When installing camshaft, align crankshaft and camshaft timing marks as shown in Fig. T54.

OIL PUMP.
Vertical crankshaft engines may be equipped with a barrel and plunger type oil pump or a gear-driven rotor type oil pump. Horizontal crankshaft engines may be equipped with a gear-driven rotor type pump or with a dipper type oil slinger attached to the connecting rod.

The barrel and plunger type oil pump is driven by an eccentric on the camshaft. Chamfered side of drive collar (Fig. T48) should be toward engine lower cover. Oil pumps may be equipped with two chamfered sides, one chamfered side or with flat boss as shown. Be sure installation is correct.

On engines equipped with gear-driven rotor oil pump, check drive gear and rotor for excessive wear or other damage. End clearance of rotor in pump body should be within limits of 0.006-0.007 inch (0.15-0.18 mm) and is controlled by cover gasket. Gaskets are available in a variety of thicknesses.

On all models with oil pump, be sure to prime pump during assembly to ensure immediate lubrication of engine.

VALVE SYSTEM.
Valve tappet clearance (cold) is 0.010 inch (0.25 mm) for intake and exhaust valves for Models V50, VH50, H50, HH50, TVM125 and H60. Valve tappet clearance for all other models is 0.008 inch (0.03 mm) for intake and exhaust valves.

Valve face angle is 45° and valve seat angle is 46°.

Valve seat width should be 0.035-0.045 inch (0.89-1.14 mm) for all models.

Valve stem guides are cast into cylinder block and are nonrenewable. If excessive clearance exists between valve stem and valve guide, guide should be reamed and new valve with oversize stem installed.

TECUMSEH

4-STROKE MODELS

Model	No. Cyls.	Bore	Stroke	Displacement	Power Rating
VM70*	1	3.062 in. (77.8 mm)	2.531 in. (64.3 mm)	18.65 cu. in. (305.7 cc)	7 hp. (5.2 kW)
VM70**	1	2.94 in. (74.61 mm)	2.531 in. (64.3 mm)	17.16 cu. in. (281 cc)	7 hp. (5.2 kW)
HH70	1	2.75 in. (69.9 mm)	2.531 in. (64.3 mm)	15 cu. in. (246.8 cc)	7 hp. (5.2 kW)
VH70	1	2.75 in. (69.9 mm)	2.531 in. (64.3 mm)	15 cu. in. (246.8 cc)	7 hp. (5.2 kW)
HM70*	1	3.062 in. (77.8 mm)	2.531 in. (64.3 mm)	18.65 cu. in. (305.7 cc)	7 hp. (5.2 kW)
HM70**	1	2.94 in. (74.61 mm)	2.531 in. (64.3 mm)	17.16 cu. in. (281 cc)	7 hp. (5.2 kW)
TVM170	1	2.94 in. (74.61 mm)	2.531 in. (64.3 mm)	17.16 cu. in. (281 cc)	7 hp. (5.2 kW)
V80	1	3.062 in. (77.8 mm)	2.531 in. (64.3 mm)	18.65 cu. in. (305.7 cc)	8 hp. (6 kW)
VM80***	1	3.062 in. (77.8 mm)	2.531 in. (64.3 mm)	18.65 cu. in. (305.7 cc)	8 hp. (6 kW)
VM80****	1	3.125 in. (79.38 mm)	2.531 in. (64.3 mm)	19.4 cu. in. (318 cc)	8 hp. (6 kW)
HM80***	1	3.062 in. (77.8 mm)	2.531 in. (64.3 mm)	18.65 cu. in. (305.7 cc)	8 hp. (6 kW)
HM80****	1	3.125 in. (79.38 mm)	2.531 in. (64.3 mm)	19.4 cu. in. (318 cc)	8 hp. (6 kW)
HH80†	1	3.00 in. (76.2 mm)	2.75 in. (69.9 mm)	19.4 cu. in. (318.7 cc)	8 hp. (6 kW)
HH80††	1	3.313 in. (84.15 mm)	2.75 in. (69.9 mm)	23.75 cu. in. (388.6 cc)	8 hp. (6 kW)
VH80	1	3.313 in. (84.15 mm)	2.75 in. (69.9 mm)	23.75 cu. in. (388.6 cc)	8 hp. (6 kW)
TVM195	1	3.125 in. (79.38 mm)	2.531 in. (64.3 mm)	19.4 cu. in. (318 cc)	8 hp. (6 kW)
HM100†††	1	3.187 in. (80.95 mm)	2.531 in. (64.3 mm)	20.2 cu. in. (330.9 cc)	10 hp. (7.5 kW)
HM100††††	1	3.313 in. (84.15 mm)	2.531 in. (64.3 mm)	21.82 cu. in. (357.6 cc)	10 hp. (7.5 kW)
VM100	1	3.187 in. (80.95 mm)	2.531 in. (64.3 mm)	20.2 cu. in. (330.9 cc)	10 hp. (7.5 kW)
VH100	1	3.313 in. (84.15 mm)	2.75 in. (69.9 mm)	23.75 cu. in. (388.6 cc)	10 hp. (7.5 kW)
HH100	1	3.313 in. (84.15 mm)	2.75 in. (69.9 mm)	23.75 cu. in. (388.6 cc)	10 hp. (7.5 kW)
TVM220	1	3.313 in. (84.15 mm)	2.531 in. (64.3 mm)	21.82 cu. in. (357.6 cc)	10 hp. (7.5 kW)
HH120	1	3.5 in. (88.9 mm)	2.875 in. (72.9 mm)	27.66 cu. in. (452.5 cc)	12 hp. (9 kW)

   * VM70 or HM70 models prior to type letter A
  ** VM70 or HM70 models, type letter A and after
 *** VM80 or HM80 models prior to type letter E
**** VM80 or HM80 models, type letter E and after
    † Prior to type letter B
   †† Type letter B and after
  ††† Early HM100 models
 †††† Late HM100 models

Medium and heavy frame, horizontal or vertical crankshaft, four cycle, one cylinder engines are covered in this section.

Care must be taken to correctly identify engine model as outlined in **TECUMSEH ENGINE IDENTIFICATION INFORMATION** section.

Connecting rod for all models rides directly on crankshaft crankpin journal. Lubrication is provided by splash lubrication for some models or a barrel and plunger type oil pump is used on some models.

A variety of ignition systems, magneto, battery or solid state, have been used. Refer to appropriate paragraph for model being serviced.

Tecumseh or Walbro float type carburetor is used according to model and application.

Refer to **TECUMSEH ENGINE IDENTIFICATION INFORMATION** section for correct engine identification and always give complete engine model, specification and serial numbers when ordering parts or service material.

MAINTENANCE

SPARK PLUG. Recommended spark plug is Champion J8 or equivalent. Electrode gap is 0.030 inch (0.762 mm) for all models.

CARBURETOR. Either a Tecumseh or Walbro float type carburetor is used. Refer to appropriate paragraph for model being serviced.

TECUMSEH CARBURETOR. Refer to Fig. T1 for exploded view of Tecumseh float type carburetor and

location of fuel mixture adjustment screws.

For initial carburetor adjustment for VM70, VM80, VM100, HM70, HM80 and HM100 models, open idle mixture and main fuel mixture screws 1½ turns each.

For initial carburetor adjustment for VH70 model, open idle mixture screw 1 turn and main fuel mixture screw 1¼ turns.

For initial carburetor adjustment for HH80, HH100, HH120 and VH100 models, open idle mixture screw 1¼ turns and main fuel mixture screw 1¾ turns.

Make final adjustment on all models with engine at normal operating temperature and running. Place engine under load and adjust main fuel mixture screw for leanest mixture that will allow satisfactory acceleration and steady governor operation. Set engine at idle speed, no load and adjust idle mixture screw to obtain smoothest idle operation.

As each adjustment affects the other, adjustment procedure may have to be repeated.

To check float level, invert carburetor throttle body and float assembly. Float setting should be 7/32-inch (5.556 mm) measured between free end of float and rim on carburetor body.

Adjust float by carefully bending float lever tang that contacts inlet valve.

Refer to Fig. T1 for exploded view of carburetor during disassembly. When reinstalling Viton inlet valve seat, grooved side of seat must be installed in bore first so inlet valve will seat against smooth side (Fig. T2). Some later models have a Viton tipped inlet needle

(Fig. T3) and a brass seat.

Install throttle plate (2 – Fig. T1) with the two stamped lines facing out and at 12 and 3 o'clock position. The 12 o'clock line should be parallel to throttle shaft and to top of carburetor.

Install choke plate (10) with flat side towards bottom of carburetor.

Fuel fitting (8) is pressed into body. When installing fuel inlet fitting, start fitting into bore; then, apply a light coat of "Loctite" (grade A) to shank and press fitting into place.

WALBRO CARBURETOR. Refer to Fig. T4 for exploded view of Walbro float type carburetor and location of fuel mixture adjustment screws.

For initial carburetor adjustment for VM80 and HM80 models, open idle mixture screw 1¾ turns and main fuel mixture screw 2 turns.

For initial carburetor adjustment for all remaining models, open idle mixture and main fuel mixture adjustment screws 1 turn each.

Make final adjustment on all models with engine at normal operating temperature and running. Place engine under load and adjust main fuel mixture screw for leanest mixture that will allow satisfactory acceleration and steady governor operation. Set engine at idle speed, no load and adjust idle mixture screw to obtain smoothest idle operation.

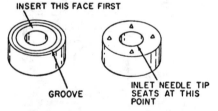

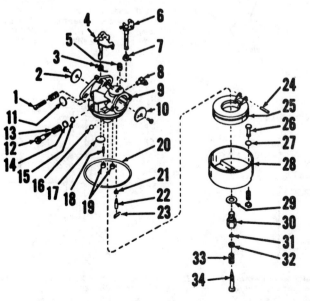

Fig. T1 – Exploded view of Tecumseh carburetor.

1. Idle speed screw
2. Throttle plate
3. Return spring
4. Throttle shaft
5. Choke stop spring
6. Choke shaft
7. Return spring
8. Fuel inlet fitting
9. Carburetor body
10. Choke plate
11. Welch plug
12. Idle mixture needle
13. Spring
14. Washer
15. "O" ring
16. Ball plug
17. Welch plug
18. Pin
19. Cup plugs
20. Bowl gasket
21. Inlet needle seat
22. Inlet needle
23. Clip
24. Float shaft
25. Float
26. Drain stem
27. Gasket
28. Bowl
29. Gasket
30. Bowl retainer
31. "O" ring
32. Washer
33. Spring
34. Main fuel needle

Fig. T2 – Viton seat used on some Tecumseh carburetors must be installed correctly to operate properly. All metal needle is used with seat shown.

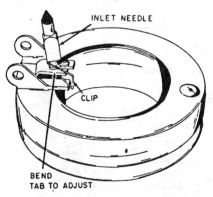

Fig. T3 – View of float and fuel inlet valve needle. Valve needle shown is equipped with resilient tip and a clip. Bend tab shown to adjust float level.

As each adjustment affects the other, adjustment procedure may have to be repeated.

To check float level, invert carburetor throttle body and float assembly (Fig.

T5). Float setting, dimension (H) should be 1/8-inch (3.175 mm) for horizontal crankshaft engines and 3/32-inch (2.381 mm) for vertical crankshaft engines.

Adjust float by carefully bending float lever tang that contacts inlet valve.

Float drop (travel) for all models should be 9/16-inch (14.288 mm) and is adjusted by carefully bending limiting tab on float.

GOVERNOR. A mechanical flyweight type governor is used on all models. Governor weight and gear assembly is driven by camshaft gear and rides on a renewable shaft which is pressed into engine crankcase or crankcase cover.

If renewal of governor shaft is necessary, press governor shaft in until shaft end is located as shown in Fig. T7, T8, T9 or T10.

To adjust governor lever position on vertical crankshaft models, refer to Fig. T11. Loosen clamp screw on governor lever. Rotate governor lever shaft counter-clockwise as far as possible. Move governor lever to left until throttle is fully open, then tighten clamp screw.

On horizontal crankshaft models, loosen clamp screw on lever, rotate governor lever shaft clockwise as far as possible. See Fig. T12. Move governor lever clockwise until throttle is wide open, then tighten clamp screw.

For external linkage adjustments, refer to Figs. T13 and T14. Loosen screw (A), turn plate (B) counter-clockwise as far as possible and move lever (C) to left until throttle is fully open. Tighten screw (A). Governor spring must be hooked in hole (D) as shown. Adjusting screws on bracket shown in Figs. T13 and T14 are used to adjust fixed or variable speed settings.

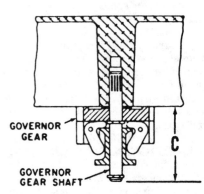

Fig. T8 — Governor gear and shaft installation on VH80 and VH100 models. Dimension (C) is 1 inch (25.4 mm).

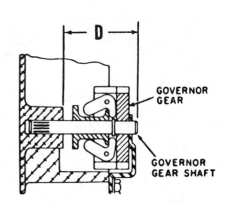

Fig. T9 — Correct installation of governor shaft, gear and weight assembly on HH70, HM70, HM80 and HM100 models. Dimension (D) is 1-3/8 inch (34.925 mm) on HM70, HM80 and HM100 models or 1-17/64 inch (32.147 mm) on HH70 model.

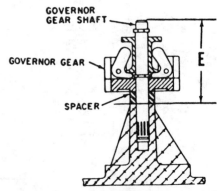

Fig. T10 — Governor gear and shaft installation on VH70, VM70, VM80, VM100, TVM170, TVM195 and TVM220 models. Dimension (E) is 1-19/32 inch (40.481 mm).

Fig. T4 — Exploded view of Walbro carburetor.

1. Choke shaft	18. Inlet valve
2. Throttle shaft	19. Main nozzle
3. Throttle return spring	20. Float
4. Choke return spring	21. Float shaft
5. Choke stop spring	22. Spring
6. Throttle plate	23. Gasket
7. Idle speed stop screw	24. Bowl
8. Spring	25. Drain stem
9. Idle mixture needle	26. Gasket
10. Spring	27. Spring
11. Baffle	28. Retainer
12. Carburetor body	29. Gasket
13. Choke plate	30. Bowl retainer
14. Bowl gasket	31. Spring
15. Gasket	32. "O" ring
16. Inlet valve seat	33. Main fuel adjusting
17. Spring	needle

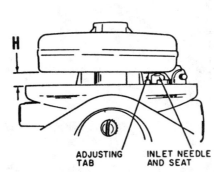

Fig. T5 — Float height (H) should be measured as shown on Walbro float carburetors. Bend adjusting tab to adjust height.

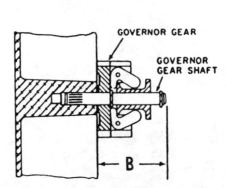

Fig. T7 — View showing installation of governor shaft and governor gear and weight assembly on HH80, HH100 and HH120 models. Dimension (B) is 1 inch (25.4 mm).

IGNITION SYSTEM. A variety of ignition systems, magneto, battery or solid state, have been used. Refer to appropriate paragraph for model being serviced.

MAGNETO IGNITION. Tecumseh flywheel type magnetos are used on some models. On VM70, HM70, VM80, HM80, VM100, HM100, HH70 and VH70 models, breaker points are enclosed by the flywheel. Breaker point gap must be adjusted to 0.020 inch (0.508 mm). Timing is correct when timing mark on stator plate is in line with mark on bearing plate as shown in Fig. T15. If timing marks are defaced, points should start to open when piston is 0.085-0.095 inch (2.159-2.413 mm) BTDC.

Breaker points on HH80, VH80, HH100, VH100 and HH120 models are located in crankcase cover as shown in Fig. T16. Timing should be correct when points are adjusted to 0.020 inch (0.508 mm). To check timing with a continuity light, refer to Fig. T17. Remove "pop" rivets securing identification plates to blower housing. Remove plate to expose a 1¼ inch hole. Connect continuity light to terminal screw (78–Fig. T16) and suitable engine ground. Rotate engine clockwise until piston is on compression stroke and timing mark is just below stator laminations as shown in Fig. T17. At this time, points should be ready to open and continuity light should be on.

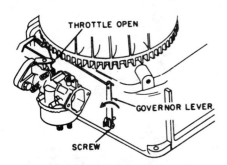

Fig. T11—When adjusting governor linkage on VH70, VM70, VM80, VM100, TVM170, TVM195 or TVM220 models, loosen clamp screw and rotate governor lever shaft and lever counter-clockwise as far as possible.

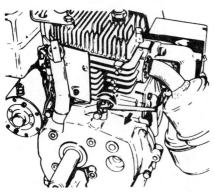

Fig. T12—On HH70, HM70, HM80 and HM100 models, rotate governor lever shaft and lever clockwise when adjusting linkage.

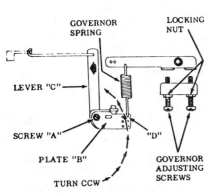

Fig. T13—External governor linkage on VH80 and VH100 models. Refer to text for adjustment procedure.

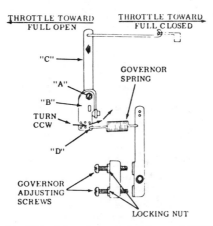

Fig. T14—External governor linkage on HH80, HH100 and HH120 models. Refer to text for adjustment procedure.

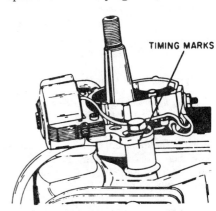

Fig. T15—On VM70, VH70, HM70, HH70, VM80, HM80, VM100 and HM100 equipped with magneto ignition, adjust breaker point gap to 0.020 inch (0.508 mm) and align timing marks as shown.

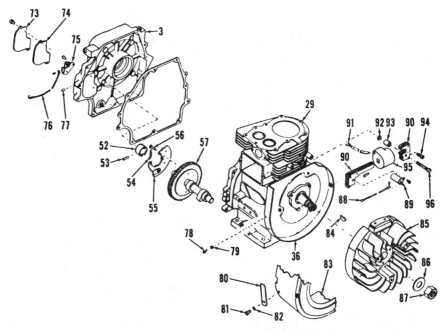

Fig. T16—Exploded view of magneto ignition components used on HH80, HH100 and HH120 models. Timing advance and breaker points used on engines equipped with battery ignition are identical.

3. Crankcase cover	57. Camshaft assy.	88. Condenser wire
29. Cylinder block	73. Breaker box cover	89. Condenser
36. Blower air baffle	74. Gasket	90. Armature core
52. Breaker cam	75. Breaker points	91. High tension lead
53. Push rod	76. Ignition wire	92. Washer
54. Spring	77. Pin	93. Spacer
55. Timing advance weight	78. Screw	94. Screw
56. Rivet	79. Clip	95. Coil
	80. Ground switch	96. Screw
	81. Screw	
	82. Washer	
	83. Blower housing	
	84. Flywheel key	
	85. Flywheel	
	86. Washer	
	87. Nut	

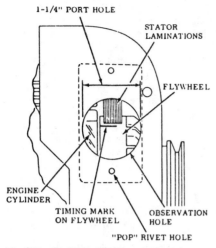

Fig. T17—On HH80, HH100 and HH120 models, remove identification plate to observe timing mark on flywheel through the 1¼-inch hole in blower housing.

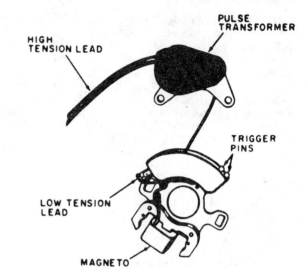

Fig. T19—View of solid state ignition system used on some models not equipped with flywheel alternator.

Rotate flywheel until mark just passes under edge of laminations. Points should open and light should be out. If not, adjust points slightly until light goes out. Points are actuated by push rod (53–Fig. T16) which rides against breaker cam (52). Breaker cam is driven by a tang on advance weight (55). When cranking, spring (54) holds advance weight in retarded position (TDC). At operating speeds, centrifugal force overcomes spring pressure and weight moves cam to advance ignition so spark occurs when piston is at 0.095 inch (2.413 mm) BTDC.

An air gap of 0.006-0.010 inch (0.1524-0.254 mm) should be between flywheel and stator laminations. To adjust gap, turn flywheel magnet into position under coil core. Loosen holding screws and place shim stock or feeler gage between coil and magnet. Press coil against gage and tighten screws.

BATTERY IGNITION. Models HH80, HH100 and HH120 may be equipped with a battery ignition system. Coil and condenser are externally mounted while points are located in crankcase cover. See Fig. T18. Points should be adjusted to 0.020 inch (0.508 mm). To check timing, disconnect primary wire between coil and points and follow the same procedure as described in **MAGNETO IGNITION** section.

SOLID STATE IGNITION (WITHOUT ALTERNATOR). Tecumseh solid state ignition system shown in Fig. T19 may be used on some models not equipped with flywheel alternator. This system does not use ignition breaker points. The only moving part of system is rotating flywheel with charging magnets. As flywheel magnet passes position (1A–Fig. T20), a low voltage AC current is induced into input coil (2). Current passes through rectifier (3) converting this current to DC. It then travels to capacitor (4) where it is stored. Flywheel rotates approximately 180° to position (1B). As it passes trigger coil (5), it induces a very small electric charge into coil. This charge passes through resistor (6) and turns on SCR (silicon controlled rectifier) switch (7). With SCR switch closed, low voltage current stored in capacitor (4) travels to pulse transformer (8). Voltage is step-

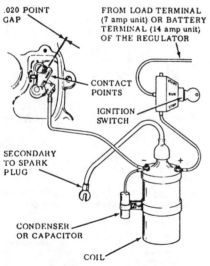

Fig. T18—Typical battery ignition wiring diagram used on some HH80, HH100 and HH120 engines.

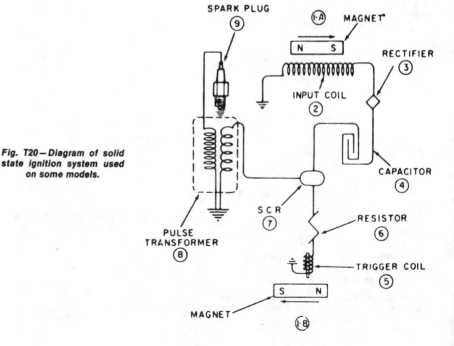

Fig. T20—Diagram of solid state ignition system used on some models.

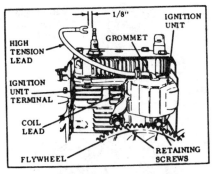

Fig. T21 — View of solid state ignition unit used on some models equipped with flywheel alternator. System should produce a good blue spark 1/8-inch (3.175 mm) long at cranking speed.

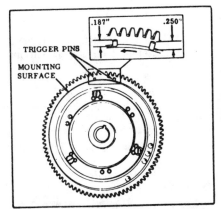

Fig. T23 — Remove flywheel and drive trigger pins in or out as necessary until long pin is extended 0.250 inch (6.35 mm) and short pin is extended 0.187 inch (4.763 mm) above mounting surface.

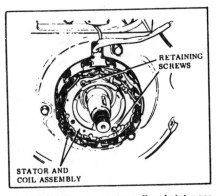

Fig. T25 — Ignition generator coil and stator serviced only as an assembly.

ped up instantaneously and current is discharged across electrodes of spark plug (9), producing a spark before TDC.

Some units are equipped with a second trigger coil and resistor set to turn SCR switch on at a lower rpm. This second trigger pin is closer to flywheel and produces a spark at TDC for easier starting. As engine rpm increases, first (shorter) trigger pin picks up small electric charge and turns SCR switch on, firing spark plug BTDC.

If system fails to produce a spark to spark plug, first check high tension lead (Fig. T19). If condition of high tension lead is questionable, renew pulse transformer and high tension lead assembly. Check low tension lead and renew if insulation is faulty. Magneto charging coil, electronic triggering system and mounting plate are available only as an assembly. If necessary to renew this assembly, place unit in position on engine. Start retaining screws, turn mounting plate counter-clockwise as far as possible, then tighten retaining screw to a torque of 5-7 ft.-lbs. (7-10 N·m).

SOLID STATE IGNITION (WITH ALTERNATOR). Tecumseh solid state ignition system used on some models equipped with flywheel alternator does not use ignition breaker points. The only

moving part of system is rotating flywheel with charging magnets and trigger pins. Other components of system are ignition generator coil and stator assembly, spark plug and ignition unit.

The long trigger pin induces a small charge of current to close SCR (silicon controlled rectifier) switch at engine cranking speed and produces a spark at TDC for starting. As engine rpm increases, first (shorter) trigger pin induces current which produces a spark when piston is 0.095 inch (2.413 mm) BTDC.

Test ignition system by holding high tension lead 1/8-inch (3.175 mm) from spark plug (Fig. T21), crank engine and check for a good blue spark. If no spark is present, check high tension lead and coil lead for loose connections or faulty insulation. Check air gap between trigger pin and ignition unit as shown in Fig. T22. Air gap should be 0.006-0.010 inch (0.1524-0.254 mm). To adjust air gap, loosen the two retaining screws and move ignition unit as necessary, then tighten retaining screws.

NOTE: Long trigger pin should extend 0.250 inch (6.35 mm) and short trigger pin should extend 0.187 inch (4.763 mm), measured as shown in Fig. T23. If not, remove flywheel and drive pins in or out as required.

Remove coil lead from ignition terminal and connect an ohmmeter as shown in Fig. T24. If series resistance test of ignition generator coil is below 400 ohms, renew stator and coil assembly (Fig. T25). If resistance is above 400 ohms, renew ignition unit.

LUBRICATION. On VH70, VM70, VM100, TVM170, TVM195 and TVM220 models, a barrel and plunger type oil pump (Fig. T26 or T27) driven by an eccentric on camshaft, pressure lubricates upper main bearing and connecting rod journal. When installing early type pump (Fig. T26), chamfered side of drive collar must be against thrust bearing surface on camshaft gear. When installing late type pump, place side of drive collar with large flat surface shown in Fig. T27 away from camshaft gear.

An oil slinger (59 – Fig. T28), installed on crankshaft between gear and lower bearing is used to direct oil upward for

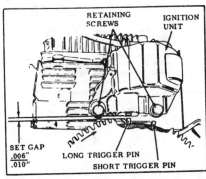

Fig. T22 — Adjust air gap between long trigger pin and ignition unit to 0.006-0.010 inch (0.1524-0.254 mm).

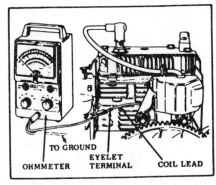

Fig. T24 — View showing an ohmmeter connected for resistance test of ignition generator coil.

Fig. T26 — View of early type oil pump used on VH70, VM70 and VM80 models. Chamfered face of drive collar should be toward camshaft gear.

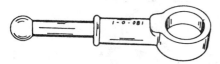

Fig. T27—Install late type oil pump so large flat surface on drive collar is away from camshaft gear.

complete engine lubrication on VH80 and VH100 models. A tang n slinger hub, when inserted in slot in crankshaft gear, correctly positions slinger on crankshaft as shown in Fig. T28.

Splash lubrication system on all other models is provided by use of an oil dipper on connecting rod. See Figs. T30 and T31.

Oils approved by manufacturer must meet requirements of API service classification SC, SD, SE or SF.

Use an SAE 30 oil for temperatures above 32° F (0° C) and SAE 5W20 or 5W30 oil for temperatures below 32° F (0° C).

Check oil level at 5 hour intervals or before initial start-up of engine. Maintain oil level at edge of filler hole or at "FULL" mark on dipstick, as equipped.

Recommended oil change interval for all models is every 25 hours of normal operation.

Fig. T28—Oil slinger (59) on VH80 and VH100 models must be installed on crankshaft as shown.

REPAIRS

TIGHTENING TORQUE. Recommended tightening torque specifications are as follows:

Cylinder head—
VM70, HM70, VM80, HM80, VM100, HM100, HH70, VH70, TVM170, TVM195, TVM220 . . . . 170 in.-lbs. (19 N·m)
HH80, VH80, HH100, VH100, HH120 . . . . 200 in.-lbs. (23 N·m)

Connecting rod—
VM70, HM70, VM80, HM80, VM100, HM100, HH70, VH70, TVM170, TVM195, TVM220 . . . . 120 in.-lbs. (13 N·m)
HH80, VH80, HH100, VH100, HH120 . . . . 110 in.-lbs. (12 N·m)

Crankcase cover—
All models . . . . 65-110 in.-lbs. (5-12 N·m)

Bearing retainer—
HH80, VH80, HH100, VH100, HH120 . . . . 65-110 in.-lbs. (5-12 N·m)

Ball bearing retainer nut—
All models so equipped . . . . 15-22 in.-lbs. (1-3 N·m)

Flywheel nut—
Light frame . . . . 400-440 in.-lbs. (45-50 N·m)
Medium frame . . . . 430-500 in.-lbs. (49-57 N·m)
Heavy frame . . . . 600-660 in.-lbs. (68-75 N·m)

CYLINDER HEAD. When removing cylinder head, note location of different length cap screws for aid in correct reassembly. Always install new head gasket and tighten cap screws evenly in sequence shown in Figs. T32, T33, T34

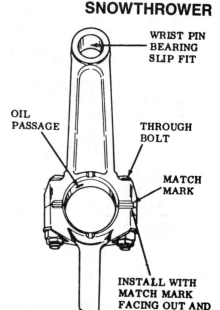

Fig. T31—Connecting rod assembly used on HH80, HH100 and HH120 models.

or T35. Refer to **TIGHTENING TORQUE** section for correct torque specifications for model being serviced.

CONNECTING ROD. Piston and connecting rod assembly is removed from cylinder head end of engine. Connecting rod rides directly on crankpin journal of crankshaft.

Standard crankpin journal diameter is 1.1865-1.1870 inches (30.14-30.15 mm) for VM70, HM70, VM80, HM80, VM100, HM100, HH70, VH70, TVM170, TVM195 and TVM220 models and 1.3750-1.3755 inches (34.93-34.94 mm) for all remaining models.

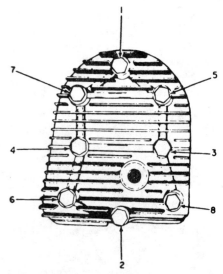

Fig. T32—On VM70, HM70, VH70 and HH70 models, tighten cylinder head cap screws evenly to a torque of 170 in.-lbs. (19 N·m) using tightening sequence shown.

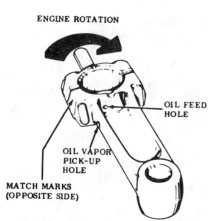

Fig. T29—Connecting rods used on VH80 and VH100 have two oil holes.

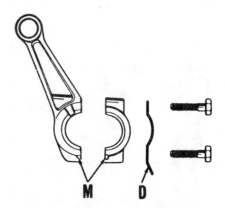

Fig. T30—Connecting rod assemble used on VH70, VM70, VM80, VM100, HH70, HM70, HM80, HM100, TVM170, TVM195 and TVM220 models. Note position oil dipper (D) and match marks (M).

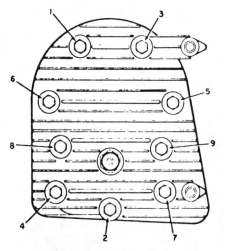

Fig. T33 — Tighten cylinder head cap screws on HM80, VM80, HM100, VM100, TVM170, TVM195 and TVM220 models in sequence shown to 170 in.-lbs. (19 N·m) torque.

Connecting rod to crankpin journal running clearance should be 0.002 inch (0.0508 mm) for all models.

Connecting rods are equipped with match marks which must be aligned and face pto end of crankshaft after installation. See Figs. T29, T30 or T31.

Connecting rod is available in a variety of sizes for undersize crankshafts, as well as standard.

PISTON, PIN AND RINGS. Aluminum alloy piston is fitted with two compression rings and one oil control ring.

Piston skirt diameter and piston skirt to cylinder wall clearance is measured at bottom edge of skirt at a right angle to piston pin for all models.

Piston skirt diameter for HH70 and VH70 models is 2.7450-2.7455 inches (69.72-69.74 mm) and skirt to cylinder bore clearance is 0.0045-0.0060 inch (0.1143-0.1524 mm).

Piston skirt diameter for HH80 model prior to type letter B, VM70 and HM70 models with type letter A and after and TVM170 models is 2.9325-2.9335 inches (74.49-74.51 mm) and skirt to cylinder bore clearance is 0.004-0.006 inch (0.1016-0.1524 mm).

Piston skirt diameter for VM70 and HM70 models prior to type letter A, V80 model, VM80 and HM80 models prior to type letter E is 3.0575-3.0585 inches (77.66-77.69 mm) and skirt to cylinder bore clearance is 0.0035-0.0055 inch (0.0889-0.1397 mm).

Piston skirt diameter for VM80 and HM80 models with type letter E or after and TVM195 models is 3.1195-3.1205 inches (79.24-79.26 mm) and skirt to cylinder bore clearance is 0.004-0.006 inch (0.1016-0.1524 mm).

Piston skirt diameter for HM100 (early production) and VM100 models is 3.1817-3.1842 inches (80.815-80.823 mm) and skirt to cylinder bore clearance is 0.0028-0.0063 inch (0.0711-0.1600 mm).

Piston skirt diameter for HH80 models with type letter B and after, VH80, VH100, HM100 (late production), HH100 and TVM220 models is 3.308-3.310 inches (84.02-84.07 mm) and skirt to cylinder bore clearance is 0.002-0.005 inch (0.0508-0.1270 mm).

Piston skirt diameter for HH120 model is 3.4950-3.4970 inches (88.77-88.82 mm) and skirt to cylinder bore clearance is 0.003-0.006 inch (0.0762-0.1524 mm).

Side clearance of top ring in piston groove for VM70 (prior to type letter A), V80 model and VM80 model (prior to

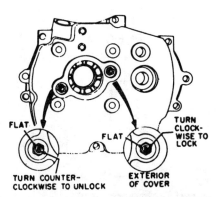

Fig. T36 — View showing bearing locks on HM70, HH70, HM80 and HM100 models equipped with ball bearing main bearings. Locks must be released before removing crankcase cover. Refer to Fig. T37 for interior view of cover and locks.

type letter E) is 0.003-0.004 inch (0.0762-0.1016 mm), for HM and TVM170 models is 0.0028-0.0051 inch (0.0711-0.1296 mm), for VM80 (type letter E and after), HM80, TVM195, HM100, VM100, TVM220, H80, VH100, HH100 and VH80 models is 0.002-0.005 inch (0.0508-0.1270 mm).

Standard piston pin diameter is 0.625-0.6254 inch (15.88-15.89 mm) for VM70, HM70, VM80, HM80, HH70, VH70, TVM170 and TVM195 models and 0.6873-0.6875 inch (17.457-17.462 mm) for all other models.

Piston pin clearance should be 0.0001-0.0008 inch (0.0025-0.0203 mm) in connecting rod and 0.0002-0.0005 inch (0.0051-0.0127 mm) in piston. If excessive clearance exists, both piston and pin must be renewed as pin is not available separately.

Pistons should be assembled on connecting rod so arrow on top of piston is pointing toward carburetor side of engine after installation.

Pistons and rings are available in a variety of oversizes as well as standard.

CYLINDER. If cylinder is scored or excessively worn, or if taper or out-of-round exceeds 0.004 inch (0.1016 mm), cylinder should be rebored to nearest

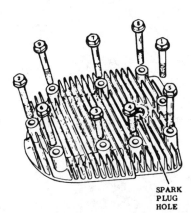

Fig. T34 — View showing cylinder head cap screw tightening sequence used on early HH80, HH100 and HH120 engines. Tighten cap screw to 200 in.-lbs. (23 N·m) torque. Note type and length of cap screws.

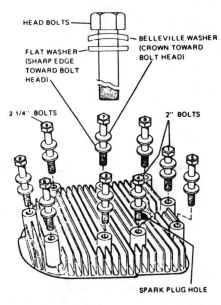

Fig. T35 — Flat washers and Belleville Washers are used on cylinder head cap screws on late HH80, HH100 and HH120 and all VH80 and VH100 engines. Tighten cap screws in sequence shown to a torque of 200 in.-lbs. (23 N·m).

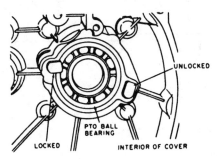

Fig. T37 — Interior view of crankcase cover and ball bearing locks used on HM70, HH70, HM80 and HM100 models.

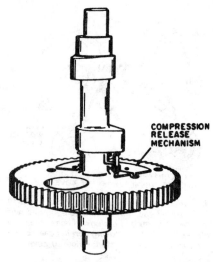

Fig. T38 — View of Insta-matic Ezee-Start compression release camshaft assembly used on all models except HH80, HH100 and HH120.

oversize for which piston and rings are available.

Standard cylinder bore for HH70 and VH70 models is 2.750-2.751 inches (69.850-69.854 mm).

Standard cylinder bore for HH80 (prior to type letter B), VM70 and HH70 models with type letter A and after and TVM170 models is 2.9375-2.9385 inches (74.61-74.64 mm).

Standard cylinder bore for VM70 and HM70 models prior to type letter A, VM80 and HM80 models prior to type letter E and V80 models is 3.062-3.063 inches (77.78-77.80 mm).

Standard cylinder bore for VM80 and HM80 models with type letter E or after and TVM195 models is 3.125-3.126 inches (79.38-79.40 mm).

Standard cylinder bore for HM100 (early production) and VM100 models is 3.187-3.188 inches (80.95-80.98 mm).

Standard cylinder bore for HH80 (type letter B and after), VH80, HM100 (late production), VH100, HH100 and TVM220 models is 3.312-3.313 inches (84.13-84.15 mm).

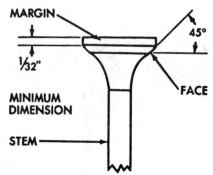

Fig. T39 — Valve face angle should be 45°. Minimum valve head margin is 1/32-inch (0.794 mm).

Standard cylinder bore for HH120 model is 3.500-3.501 inches (88.90-88.93 mm).

CRANKSHAFT AND MAIN BEARINGS. Crankshaft main journals ride directly in aluminum alloy bearings in crankcase and mounting flange (engine base) on vertical crankshaft engines or in two renewable steel backed bronze bushings. On some horizontal crankshaft engines, crankshaft rides in a renewable sleeve bushing at flywheel end and a ball bearing or bushing at pto end. Models HH80, VH80, HH100, VH100 and HH120 are equipped with taper roller bearings at both ends of crankshafts.

Standard main bearing bore diameter for bushing type main bearings for VM70, VH70 and HH70 models should be 1.0005-1.0010 inches (25.41-25.43 mm) for either bearing.

Standard main bearing bore diameter for bushing type main bearings for all remaining models should be 1.0005-1.0010 inches (25.41-25.43 mm) for main bearing in cylinder block section and 1.1890-1.1895 inches (30.20-30.21 mm) for main bearing in cover section.

Normal running clearance of crankshaft journals in aluminum bearings or bronze bushings is 0.0015-0.0025 inch (0.038-0.064 mm). Renew crankshaft if main journals are more than 0.001 inch (0.025 mm) out-of-round or renew or regrind crankshaft if connecting rod journal is more than 0.0005 inch (0.0127 mm) out-of-round.

Check crankshaft gear for wear, broken teeth or loose fit on crankshaft. If gear is damaged, remove from crankshaft with an arbor press. Renew gear pin and press new gear on shaft making certain timing mark is facing pto end of shaft.

On models equipped with ball bearing at pto end of shaft, refer to Figs. T36 and T37 before attempting to remove crankcase cover. Loosen locknuts and rotate protruding ends of lock pins counter-clockwise to release bearing and remove cover. Ball bearing will remain on crankshaft. When reassembling, turn lock pins clockwise until flats on pins face each other, then tighten locknuts to 15-22 in-lbs. (2-3 N·m) torque.

Crankshaft end play for all models except those with taper roller bearing main bearings is 0.005-0.027 inch (0.1270-0.6858 mm) and is controlled by varying thickness of thrust washers between crankshaft and cylinder block or crankcase cover.

Crankshaft main bearing preload for all models with taper roller bearing main bearings is 0.001-0.007 inch (0.0254-0.1778 mm) and is controlled by varying thickness of shim gasket (32 – Fig. T42) on all models.

Taper roller bearings are a press fit on crankshaft and must be renewed if removed. Heat bearings in hot oil to aid installation.

Fig. T40 — Exploded view of typical vertical crankshaft engine. Renewable bushings (13 and 36) are not used on VM70, VM80 and VM100 models.

1. Cylinder head
2. Head gasket
3. Exhaust valve
4. Intake valve
5. Pin
6. Spring cap
7. Valve spring
8. Spring cap
9. Cylinder block
10. Magneto
11. Flywheel
12. Oil seal
13. Crankshaft bushing
14. Breather assy.
15. Carburetor
16. Intake pipe
17. Top compression ring
18. Second compression ring
19. Oil ring expander
20. Oil control ring
21. Piston pin
22. Piston
23. Retaining ring
24. Connecting rod
25. Thrust washer
26. Crankshaft
27. Thrust washer
28. Rod cap
29. Rod bolt lock
30. Camshaft assy.
31. Valve lifters
32. Oil pump
33. Gasket
34. Mounting flange (engine base)
35. Oil screen
36. Crankshaft bushing
37. Oil seal
38. Spacer
39. Governor shaft
40. Governor gear assy.
41. Spool
42. Retaining rings

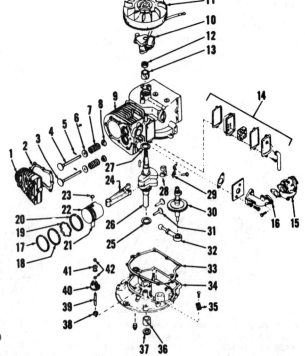

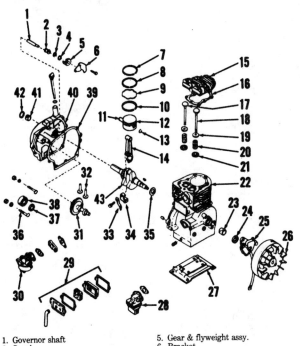

Fig. T41—Exploded view of typical horizontal crankshaft engine. Engines may be equipped with crankshaft bushing (41) or ball bearing (38) at pto end of shaft.

9. Oil ring expander
10. Oil control ring
11. Piston pin
12. Piston
13. Retaining ring
14. Connecting rod
15. Cylinder head
16. Head gasket
17. Exhaust valve
18. Intake valve
19. Spring cap
20. Valve spring
21. Spring retainer
22. Cylinder block
23. Crankshaft bushing
24. Oil seal
25. Magneto
26. Flywheel
27. Mounting plate
28. Fuel pump
29. Breather assy.
30. Carburetor
31. Camshaft assy.
32. Valve lifter
33. Rod bolt lock
34. Rod cap
35. Thrust washer
36. Bearing lock pin
37. Thrust washer
38. Ball bearing
39. Gasket
40. Crankcase cover
41. Bushing
42. Oil seal
43. Crankshaft

1. Governor shaft
2. Spool
3. Washer
4. Retaining ring
5. Gear & flyweight assy.
6. Bracket
7. Top compression ring
8. Second compression ring

HH80, VH80, HH100, VH100, HH120, (Both journals) . . 1.1865-1.1870 inches (30.14-30.15 mm)

VM80, HM80, HM100, VM100, TVM170, TVM195, TVM220
Flywheel end . . . . . 0.9985-0.9990 inch (25.36-25.38 mm)
Pto end . . . . 1.1870-1.1875 inches (30.15-30.16 mm)

Crankpin journal diameter—
HH80, VH80
HH100, HH120 . 1.3750-1.3755 inches (34.93-34.94 mm)

All other
models . . . . 1.1865-1.1870 inches (30.14-30.15 mm)

Connecting rods are available in a variety of sizes for reground crankshaft crankpin journals.

When reinstalling crankshaft, make certain timing marks on camshaft and crankshaft are properly aligned.

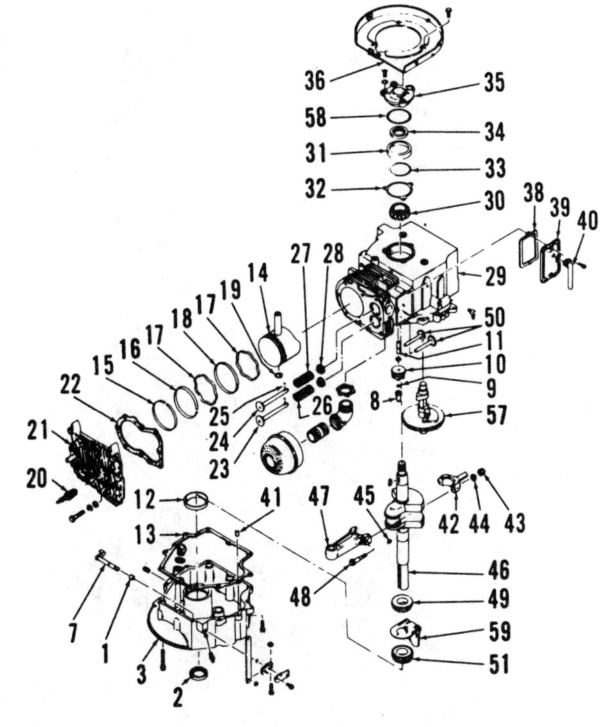

CAMSHAFT. Camshaft and camshaft gear are an integral part which rides on journals at each end of camshaft. Camshaft journal diameter is 0.6235-0.6240 inch (15.84-15.85 mm) and journal to bearing running clearance should be 0.003 inch (0.0762 mm).

Renew camshaft if gear teeth are worn or if journal or lobe surfaces are worn or damaged.

Camshaft equipped with compression release mechanism (Fig. T38) is easier to remove and install if crankshaft is rotated three teeth past aligned position which allows compression release mechanism to clear exhaust valve lifter. Compression release mechanism parts should work freely with no binding or sticking. Parts are not serviced separate from camshaft assembly.

Some models are equipped with an automatic timing advance mechanism (52 through 56–Fig. T43) which must work freely with no binding or sticking. Renew damaged parts as necessary. Make certain timing marks on camshaft gear and crankshaft gear are aligned after installation.

VALVE SYSTEM. Valve seats are machined directly into cylinder block assembly. Seats are ground at a 45° angle and should not exceed 3/64-inch (1.191 mm) in width.

Valve face is ground at a 45° angle and margin should not be less than 1/32-inch (0.794 mm). See Fig. T39. Valves are available with 1/32-inch (0.794 mm) oversize stem for use with oversize valve guide bore.

Valve guides are not renewable. If guides are excessively worn, ream to 0.3432-0.3442 inch (8.72-8.74 mm) and install valve with 1/32-inch (0.794 mm) oversize valve stem. Drill upper and

Bronze or aluminum bushing type main bearings are renewable and finish reamers are available from Tecumseh.

Standard crankshaft journal diameters are as follows:

Main journal diameter—
VM70, HM70,
HH70, VH70,
(Both journals) . . . . 0.9985-0.9990 inch (25.36-25.38 mm)

Fig. T42—Exploded view of VH80 or VH100 model vertical crankshaft engine.

12. Bearing cup
13. Gasket
14. Piston & pin assy.
15. Top compression ring
16. Second compression ring
17. Ring expander
18. Oil control ring
19. Retaining ring
20. Spark plug
21. Cylinder head
22. Head gasket
23. Exhaust valve
24. Intake valve
25. Pin
26. Exhaust valve spring
27. Intake valve spring
28. Spring cap
29. Cylinder block
30. Bearing cone
31. Bearing cup
32. Shim gasket
33. Steel washer (0.010 inch)
34. Oil seal
35. Bearing retainer cap
36. Blower air baffle
38. Gasket
39. Breather
40. Breather tube
42. Rod cap
43. Self locking nut
44. Washer
45. Crankshaft gear pin
46. Crankshaft
47. Connecting rod
48. Rod bolt
49. Crankshaft gear
50. Valve lifters
51. Bearing cone
57. Camshaft assy.
58. "O" ring
59. Oil slinger

1. Governor arm bushing
2. Oil seal
3. Mounting flange (engine base)
7. Governor arm
8. Thrust spool
9. Snap ring
10. Governor gear & weight assy.
11. Governor shaft

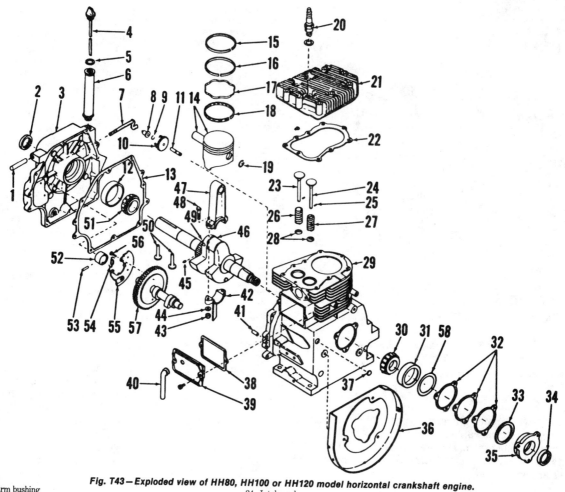

Fig. T43 — Exploded view of HH80, HH100 or HH120 model horizontal crankshaft engine.

1. Governor arm bushing
2. Oil Seal
3. Crankcase cover
4. Dipstick
5. Gasket
6. Oil filler tube
7. Governor arm
8. Thrust spool
9. Snap ring
10. Governor gear & weight assy.
11. Governor shaft
12. Bearing cup

13. Gasket
14. Piston & pin assy.
15. Top compression ring
16. Second compression ring
17. Oil ring expander
18. Oil control ring
19. Retaining ring
20. Spark plug
21. Cylinder head
22. Head gasket
23. Exhaust valve

24. Intake valve
25. Pin
26. Exhaust valve spring
27. Intake valve spring
28. Spring cap
29. Cylinder block
30. Bearing cone
31. Bearing cup
32. Shim gaskets
33. Steel washer (0.010 inch)
34. Oil seal
35. Bearing retainer cap

36. Blower air baffle
37. Plug
38. Gasket
39. Breather assy.
40. Breather tube
41. Dowel pin
42. Rod cap
43. Self locking nut
44. Washer
45. Crankshaft gear pin
46. Crankshaft

47. Connecting rod
48. Rod bolt
49. Crankshaft gear
50. Valve lifters
51. Bearing cone
52. Breaker cam
53. Push rod
54. Spring
55. Timing advance weight
56. Rivet
57. Camshaft assy.
58. "O" ring

lower valve spring caps as necessary for oversize valve stem.

Valve lifters should be identified before removal so they may be reinstalled in their original positions. Some models use lifters of different length. Short lifter is installed at intake position and longer lifter is installed at exhaust position.

Valve tappet gap (cold) is measured with piston at TDC on compression stroke. Correct clearance for all models is 0.010 inch (0.254 mm).

Adjust clearance by squarely grinding valve stem to increase clearance or grinding seat deeper to decrease clearance.

DYNA-STATIC BALANCER. Dyna-Static engine balancer used on some models consists of counterweighted

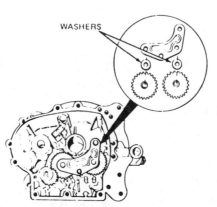

Fig. T44 — View showing Dyna-Static balancer gears installed in models so equipped. Note location of washers between gears retaining bracket.

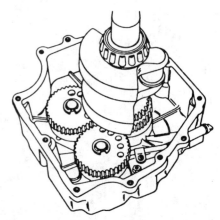

Fig. T45 — View showing Dyna-Static balancer gears installed in HH80, HH100 or HH120 models. Note gear retaining snap rings. Refer to Fig. T44 also.

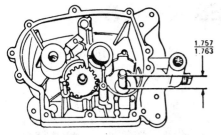

Fig. T46 — On HM80, VM80, HM100 and VM100 models, balancer gear shafts must be pressed into cover or engine base so a distance of 1.757-1.763 inches (44.52-44.78 mm) exists between shaft bore boss and edge of step cut as shown.

MEASURE FROM COVER BOSS TO RING GROOVE OUTER EDGE

Fig. T47 — On HH80, HH100 and HH120 models, press balancer gear shafts into cover until 1.7135-1.7185 inches (43.52-43.64 mm) exists between cover boss and outer edge of snap ring groove as shown.

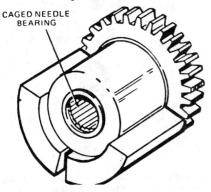

Fig. T48 — Using tool number 670210, press new needle bearing into HM80, VM80, HM100 or VM100 models balancer gears until bearing cage is flush to 0.015 inch (0.381 mm) below edge of bore

PRESS BEARINGS IN FLUSH TO .015 BELOW

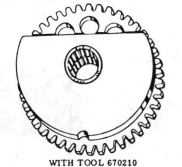

WITH TOOL 670210

Fig. T49 — Using tool number 670210, press new needle bearings into HH80, HH100 and HH120 models balancer gears until bearing cage is flush to 0.015 inch (0.381 mm) below edge of bore.

Fig. T50 — To time engine balancer gears, remove pipe plugs and insert alignment tool number 670240 through crankcase cover (HM80 and HM100) or engine base (VM80 and VM100) and into slots in balancer gears. Refer also to Fig. T52.

gears driven by crankshaft gear to counteract unbalance caused by counterweights on crankshaft.

Counterweight gears on medium frame models are held in position on shafts by a bracket bolted to crankcase or engine base (Fig. T44). Snap rings are used on heavy frame models to retain counterweight gears on shafts which are pressed into crankcase cover (Fig. T47).

Renewable balancer gear shafts are pressed into crankcase cover or engine base. On medium frame models, press shafts into cover or engine base until a distance of 1.757-1.763 inches (44.52-44.78 mm) exists between shaft bore boss and edge of step cut on shafts as shown in Fig. T46. Heavy frame model shafts should be pressed into cover until a distance of 1.7135-1.7185 inches (43.52-43.64 mm) exists between cover boss and outer edge of snap ring groove as shown in Fig. T47.

All balancer gears are equipped with renewable cage needle bearings. Using tool number 670210, press new bearings into gears until cage is flush to 0.015 inch (0.381 mm) below edge of bore.

When reassembling engine, balancer gears must be timed with crankshaft for correct operation. Refer to Figs. T50

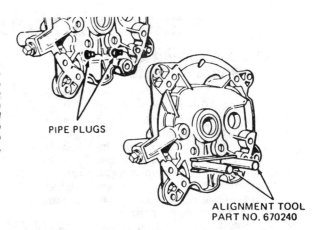

PIPE PLUGS

ALIGNMENT TOOL PART NO. 670240

and T51 and remove pipe plugs. Insert alignment tool number 670240 through crankcase cover or engine base of medium frame models and into timing slots in balancer gears. On heavy frame models use timing tool number 670239. On all models, rotate engine until piston is at TDC on compression stroke and install cover and gear assembly while tools retain gears in correct position. See Figs. T52 and T53.

When correctly assembled, piston should be on TDC and weights on balancer gears should be in directly opposite position. See Figs. T52 and T53.

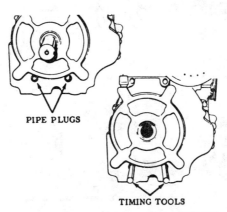

PIPE PLUGS

TIMING TOOLS

Fig. T51 — To time balancer gears on HH80, HH100 and HH120 models, remove pipe plugs and insert timing tools number 67039 through crankcase cover and into timing slots in balancer gears. Refer also to Fig. T53.

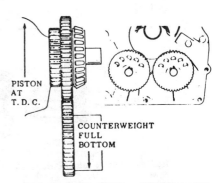

GEAR WEIGHTS

Fig. T52 — View showing correct balancer gear timing to crankshaft gear on HM80, VM80, HM100 and VM100 models. With piston at TDC, weights should be directly opposite.

PISTON AT T.D.C.

COUNTERWEIGHT FULL BOTTOM

Fig. T53 — On HH80, HH100 and HH120 models, balancer gears are correctly timed to crankshaft when piston is at TDC and weights are at full bottom position.

SERVICING TECUMSEH ACCESSORIES

12 VOLT STARTING AND CHARGING SYSTEMS

Engines with 3.5 horsepower (57 cc) or more may be equipped with a 12 volt direct current starting, generating and ignition unit system. The system includes the usual coil, condenser and breaker point for ignition, plus extra generating coils and flywheel magnets to generate alternating current. Also included is a silicon rectifier panel or regulator-rectifier for changing the generated alternating current to direct current, a series wound motor with a Bendix drive unit and a 12 volt wet cell storage battery.

Models LAV30, LAV35 and LAV40 may be equipped with a 12 volt starting motor with a right angle gear drive unit and a Nickel Cadmium battery pack. The "SAF-T-KEY" switch is located in the battery box cover. A battery charger which converts 110 volt ac house current to dc charging current is used to recharge the Nickel Cadmium battery pack.

Refer to the following paragraphs for service procedures on the units.

12 VOLT STARTER MOTOR (BENDIX DRIVE).
Refer to Fig. TE1 for exploded view of 12 volt starter motor and Bendix drive unit used on some engines. This motor should not be operated continuously for more than 10 seconds. Allow starter motor to cool one minute between each 10 second cranking period.

To perform a no-load test, remove starter motor and use a fully charged 6 volt battery. Maximum current draw should not exceed 25 amperes at 6 volts. Minimum rpm is 6500.

When assembling a starter motor, use 0.003 and 0.010 inch spacer washers (14 and 15) as required to obtain an armature end play of 0.005-0.015 inch (0.13-0.38 mm). Tighten nut on end of armature shaft to 100 in.-lbs. (11 N·m). Tighten the through-bolts (30) to 30-34 in.-lbs. (3-4 N·m).

12 VOLT STARTER MOTOR (SAF-T-KEY TYPE).
Refer to Fig. TE2 for exploded view of right angle drive gear unit and starter motor used on some Model LAV30, LAV35 and LAV40 models. Repair of this unit consists of renewing right angle gear drive parts or the starter motor assembly. Parts are not serviced separately for the motor (8). Bevel gears (9) are serviced only as a set.

When installing assembly on engine, adjust position of starter so there is a 1/16 inch (1.59 mm) clearance between crown of starter gear tooth and base of flywheel tooth.

NICKEL CADMIUM BATTERY.
The Nickel Cadmium battery pack (Fig. TE3) is a compact power supply for the "SAF-T-KEY" starter motor. To test battery pack, measure open circuit voltage across black and red wires. If voltage is 14.0 or above, battery is good and may be recharged. Charge battery only with charger provided with the system. This charger (Fig. TE4) converts 110 volt AC house current to DC charging current. Battery should be fully charged after 14-16 hours charging time. A fully charged battery should test 15.5-18.0 volts.

A further test of battery pack can be made by connecting a 1.4 ohm resistor across black and red wires (battery installed) for two minutes. Battery voltage at the end of two minutes must be 9.0 volts minimum. If battery passes this test, recharge battery for service.

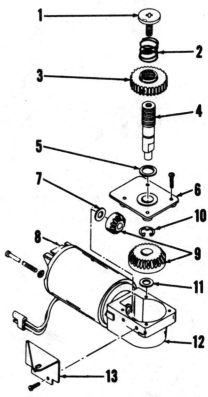

Fig. TE2—Twelve volt electric starter assembly used on some engines equipped with "SAF-T-KEY" Nickel Cadmium battery pack.

1. Screw L.H.	
2. Spring	8. Starter motor
3. Starter gear	9. Gear set
4. Shaft	10. Retaining ring
5. Thrust washer	11. Shim washer
6. Cover	12. Gear housing
7. Shim washer	13. Mounting bracket

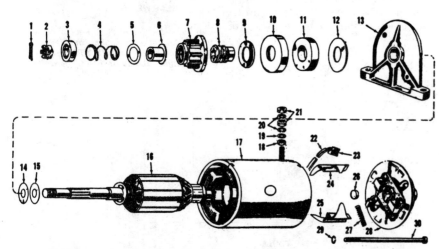

Fig. TE1—Twelve volt electric cranking motor used on some engines.

3. Pinion stop	14. Spacer washer, 0.003 in.	22. Insulation tubing
4. Antidrift spring	15. Spacer washer, 0.010 in.	23. Brush
5. Pinion washer		24. Insulation
6. Antidrift sleeve	16. Armature	25. Insulation
7. Pinion gear	18. Insulation washer	26. Thrust spacer
8. Screw shaft		28. Commutator end cap
9. Thrust washer		
10. Cushion		
11. Rubber cushion		
12. Thrust washer		
13. Drive end cap		

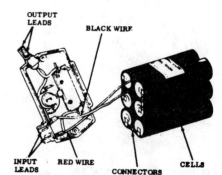

Fig. TE3—View of Nickel Cadmium battery pack connected to "SAF-T-KEY" starting switch.

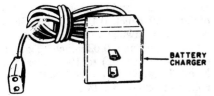

Fig. TE4—Recharge Nickel Cadmium battery pack only with the charger furnished with "SAF-T-KEY" starter system. Charger converts 110 volt ac to dc charging current.

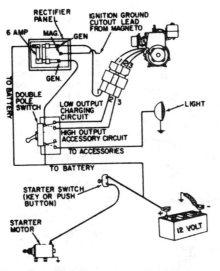

Fig. TE6—Wiring diagram of typical 7 ampere alternator and rectifier panel charging system. The double pole switch in one position reduces output to 3 amperes for charging or increases output to 7 amperes in other position to operate accessories.

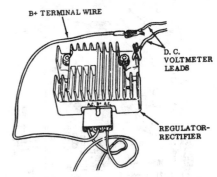

Fig. TE8—Connect dc voltmeter as shown when checking the regulator-rectifier.

ALTERNATOR CHARGING SYSTEMS.

Flywheel alternators are used on some engines for the charging system. The generated alternating current is converted to direct current by two rectifiers on the rectifier panel (Figs. TE5 and TE6) or the regulator-rectifier (Fig. TE7).

The system shown in Fig. TE5 has a maximum charging output of about 3 amperes at 3600 rpm. No current regulator is used on this low output system. The rectifier panel includes two diodes (rectifiers) and a 6 ampere fuse for overload protection.

The system shown in Fig. TE6 has a maximum output of 7 amperes. To prevent overcharging the battery, a double pole switch is used in low output position to reduce the output to 3 amperes for charging the battery. Move switch to high output position (7 amperes) when using accessories.

The system shown in Fig. TE7 has a maximum output of 7 amperes and uses a solid state regulator-rectifier which converts the generated alternating current to direct current for charging the battery. The regulator-rectifier also allows only the required amount of current flow for existing battery conditions. When battery is fully charged, current output is decreased to prevent overcharging the battery.

Testing. On models equipped with rectifier panel (Figs. TE5 or TE6), remove rectifiers and test them with either a continuity light or an ohmmeter. Rectifiers should show current flow in one direction only. Alternator output can be checked using an induction ammeter over the positive lead wire to battery.

On models equipped with the regulator-rectifier (Fig. TE7), check the system as follows: Disconnect B+ lead and connect a dc voltmeter as shown in Fig. TE8. With engine running near full throttle, voltage should be 14.0-14.7 volts. If voltage is above 14.7 or below 14.0 but above 0, the regulator-rectifier is defective. If voltmeter reading is 0 the regulator-rectifier or alternator coils may be defective. To test the alternator coils, connect an ac voltmeter to the ac leads as shown in Fig. TE9. With engine running at near full throttle, check ac voltage. If voltage is less than 20.0 volts, alternator is defective.

110 VOLT ELECTRIC STARTER

110 VOLT AC-DC STARTER. Some vertical crankshaft engines are available with a 110 volt ac electric starting system as shown in Fig. TE11. When switch button (6) is depressed, switch (21) is turned on and the motor rotates driven gear (23). As more pressure is applied, housing (11) moves down compressing springs (19). At this time, clutch facing (24) is forced tight between driven gear (23) and cone (30) and engine is cranked.

Electric motor (3) is serviced only as an assembly. All other parts shown are available separately. Shims (18) are used to adjust mesh of drive gear (5) to driven gear (23).

110 VOLT AC-DC STARTER. Some engines may be equipped with a 110 volt ac-dc starting motor (Fig. TE12). The rectifier assembly used with this starter converts 110 volt ac house current to approximately 100 volts dc. A thyrector (8) is used as a surge protector for the rectifiers.

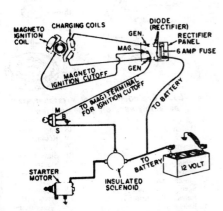

Fig. TE5—Wiring diagram of typical 3 ampere alternator and rectifier panel charging system.

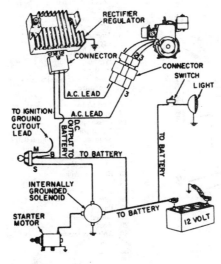

Fig. TE7—Wiring diagram of typical 7 or 10 ampere alternator and regulator-rectifier charging system.

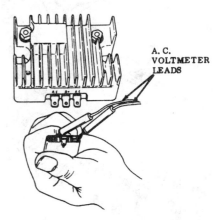

Fig. TE9—Connect ac voltmeter to ac leads as shown when checking alternator coils.

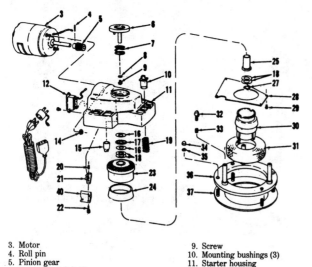

Fig. TE11—Exploded view of the 110 volt ac electric starter used on some engines.

16. Thrust races (2)
17. Thrust bearing
18. Thrust shims
19. Guide post springs (3)
20. Switch plunger spring
21. Switch
22. Screws (2)
23. Driven gear
24. Clutch facing
25. Bearing
27. Snap ring
28. Shield
29. Screws (4)
30. Lower cone
31. Screen
32. Cap studs (3)
33. Spring
34. Acorn nuts (4)
35. Lockwashers
36. Mounting ring
37. Reinforcing ring
40. Switch cover

3. Motor
4. Roll pin
5. Pinion gear
6. Starter switch button
7. Switch return spring
8. Washer
9. Screw
10. Mounting bushings (3)
11. Starter housing
12. Receptacle
14. Nuts (2)
15. Pinion bearing

necessary to obtain armature end play of 0.005-0.015 inch (0.13-0.38 mm). Shim washers are available in thicknesses of 0.005, 0.010 and 0.020 inch. Tighten elastic stop nut (32) to 100 in.-lbs. (11 N·m). Tighten nuts on the two through-bolts to 24-28 in.-lbs. (3 N·m).

To test the rectifier assembly, first use an ac voltmeter to check line voltage of the power supply, which should be approximately 115 volts. Connect the input cable (7) to the ac power supply. Connect a dc voltmeter to the two slotted terminals of the output cable (9). Move switch (3) to "ON" position and check the dc output voltage. Direct current output voltage should be a minimum of 100 volts. If a low voltage reading is obtained, rectifiers and/or thyrector could be faulty. Thyrector (8) can be checked after removal, by connecting the thyrector, a 7.5 watt ac light bulb and a 115 volt ac power supply in series. If light bulb glows, thyrector is faulty. Use an ohmmeter to check the rectifiers. Rectifiers must show continuity in one direction only.

To test starting motor, remove the dc output cable (9) from motor. Connect an ohmmeter between one of the flat receptacle terminals and motor housing. If a short is indicated, the motor is grounded and requires repair. Connect the ohmmeter between the two flat receptacle terminals. A resistance reading of 3-4 ohms must be obtained. If not, motor is faulty.

A no-load test can be performed with the starter removed. Do not operate the motor continuously for more than 15 seconds when testing. Maximum ampere draw should be 2 amperes and minimum rpm should be 8500.

Disassembly of starter motor is obvious after examination of the unit and reference to Fig. TE12. When reassembling, install shim washers (20) as

WIND-UP STARTERS

RATCHET STARTER. On models equipped with the ratchet starter, refer to Fig. TE13 and move release lever to "RELEASE" position to remove tension

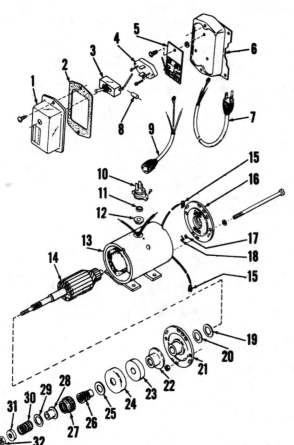

Fig. TE12 – Exploded view of 110 volt ac-dc starter motor and rectifier assembly.

1. Cover
2. Gasket
3. Switch
4. Rectifier bridge
5. Rectifier base
6. Mounting base
7. AC input cable
8. Thyrector
9. DC output cable
10. Receptacle
11. Washer
12. Insulator
13. Field coils & housing assy.
14. Armature
15. Brushes
16. End plate
17. Spring insulator
18. Brush spring
19. Nylon washer
20. Shim washers (0.005, 0.010 & 0.020 in.)
21. Drive end-plate
22. Thrust sleeve
23. Rubber cushion
24. Cup
25. Thrust washer
26. Screw shaft
27. Pinion gear
28. Spring sleeve
29. Washer
30. Antidrift spring
31. Pinion stop
32. Elastic stop nut

Fig. TE13 – Exploded view of a ratchet starter assembly used on some engines.

2. Handle
4. Clutch
5. Clutch spring
6. Bearing
7. Housing
8. Wind gear
9. Wave washer
10. Clutch washer
12. Spring & housing
13. Release dog spring
14. Release dog
15. Lock dog
16. Dog pivot retainers
17. Release gear
18. Spring cover
19. Retaining ring
20. Hub washer
21. Starter hub
22. Starter dog
23. Brake washer
24. Brake
25. Retainer
26. Screw L.H.
27. Centering pin
28. Hub & screen
29. Spacer washers
30. Lockwasher

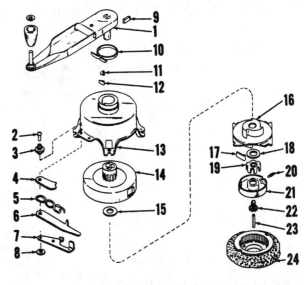

Fig. TE15 — Exploded view of remote release impulse starter used on some models. Refer also to Fig. TE16.

1. Cranking handle	9. Wind dog	17. Starter dog
2. Rivet	10. Brake band	18. Thrust washer
3. Pivot stud	11. Lock dog spring	19. Brake
4. Release lever	12. Lock dog	20. Spring
5. Release spring	13. Housing	21. Retainer
6. Trip lever	14. Spring & keeper	22. Screw L.H.
7. Trip release	15. Bearing washer	23. Centering pin
8. Washer	16. Hub	24. Hub & screen

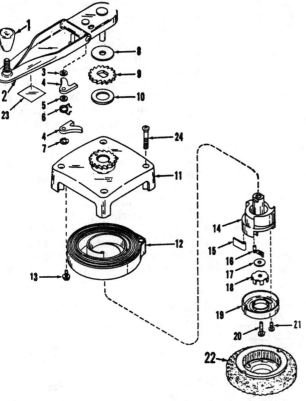

Fig. TE14 — Components of impulse mechanical starter. When handle (2) is used to wind spring (12) the energy put into the spring cranks the engine.

1. Handle knob	9. Ratchet	
2. Starter handle	10. Ratchet spacer	17. Brake washer
3. Dog washer	11. Housing assy.	18. Brake
4. Dog release	12. Spring & keeper	19. Retainer
5. Dog spacer	13. Keeper screw	20. Screw
6. Dog release spring	14. Spring hub assy.	21. Screw
7. Pivot retainer	15. Starter dog	22. Hub & screen
8. Ratchet	16. Spring & eyelet	24. Screw

from main spring. Remove starter assembly from engine. Remove left-hand thread screw (26), retainer hub (25), brake (24), washer (23) and six starter dogs (22). Note position of starter dogs in hub (21). Remove hub (21), washer (20), spring and housing (12), spring cover (18), release gear (17) and retaining ring (19) as an assembly. Remove retaining ring, then carefully separate these parts.

CAUTION: Do not remove main spring from housing (12). The spring and housing are serviced only as an assembly.

Remove snap rings (16), spacer washers (29), release dog (14), lock dog (15) and spring (13). Winding gear (8), clutch (4), clutch spring (5), bearing (6) and crank handle (2) can be removed after first removing the retaining screw and washers (10, 30 and 9).

Reassembly procedure is the reverse of disassembly. Centering pin (27) must align screw (26) with crankshaft center hole.

IMPULSE STARTER. To overhaul the impulse starter (Fig. TE14), first

release tension from main spring. Un-bolt and remove starter assembly from engine. Remove retaining screw (21), re-

Fig. TE16 — Exploded view of remote release impulse starter used on some models. Refer also to Fig. TE15.

1. Cranking handle	
2. Wind dog	
3. Brake band	
4. Lock dog spring	
5. Lock dog	
6. Housing	
7. Release lever	
8. Wave washer	
9. Release arm	
10. Hub	
11. Spacer washer	
12. Power spring	
13. Spring housing	
14. Shim	
15. Thrust washer	
16. Brake	
17. Starter dog	
18. Spring	
19. Retainer	
20. Shoulder nut	
21. Centering pin	
22. Hub & screen	

tainer (19), spring and eyelet (16) and starter dogs (15). Remove screw (20), brake (18), washer (17), cranking handle (2), bearing (8), ratchet (9) and spacer (10). Withdraw spring hub (14) and if spring and keeper assembly (12) are to be renewed, remove keeper screw (13). Carefully remove spring and keeper assembly.

CAUTION: Do not remove spring from keeper.

Remove pivot retainer (7), dog releases (4), spring (6), washer (3) and spacer (5) from handle (2).

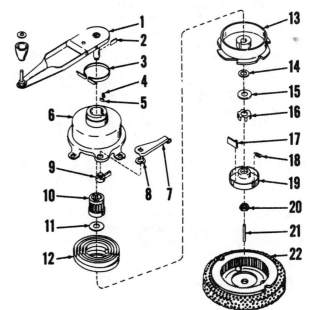

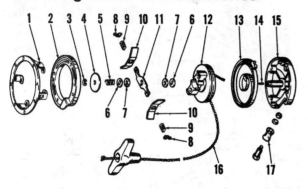

will increase about 1/3 when removed. To install new power spring in spring housing, clamp spring housing in a vise. Anchor outer end of spring in housing. Install spacer washer (11) and hub (10), engaging inner end of spring in hub notch. Insert crankpin handle shaft in hub and while pressing down on cranking handle, rotate crank to wind spring into housing. When all of the spring is in the housing, allow cranking handle and hub to unwind.

To remove release lever (7) and release arm (9), file off peened-over sides of release arm. Remove lever, wave washer and arm.

When reassembling, place release arm on wood block, install wave washer and release lever. Using a flat face punch, peen over edges of new release arm to secure release lever. The balance of

When reassembling, install spring hub (14) before placing spring and keeper in housing (11). Install ratchet (9) so teeth are opposite ratchet teeth on housing.

REMOTE RELEASE IMPULSE STARTER (SURE LOCK).

To disassemble the "Sure Lock" remote release impulse starter (Fig. TE15), first release tension from main spring. Unbolt and remove starter assembly from engine. Remove centering pin (23), then unscrew the left-hand threaded screw (22). Remove retainer (21), spring (20), brake (19), starter dog (17), thrust washer (18) and hub (16). Remove cranking handle (1), wind dog (9), brake band (10), lock dog spring (11) and lock dog (12). Carefully lift out spring and keeper assembly (14).

CAUTION: Do not attempt to remove spring from keeper. Spring and keeper are serviced only as an assembly.

Some models use pivot stud (3), release lever (4), release spring (5) and trip lever (6). Other models use rivet (2) and trip release (7). Removal of these parts is obvious after examination of the unit and reference to Fig. TE15.

Reassembly procedure is the reverse of disassembly. Press centering pin (23) into screw (22) about 1/3 of the pin length. Pin will align starter with center hole in crankshaft.

REMOTE RELEASE IMPULSE STARTER (SURE START).

Some models are equipped with the "Sure Start" remote release impulse starter shown in Fig. TE16. To disassemble the starter, first move release lever to remove power spring tension. Unbolt and remove starter assembly from engine. Remove centering pin (21), shoulder nut (20), retainer (19), spring (18), starter dog (17), thrust washer (15) and shim (14). Remove cranking handle (1), wind dog (2) and brake band (3). Rotate spring housing (13) spring while pushing down on top of hub (10). Withdraw spring housing, power spring and hub from housing (6). Lift hub from spring.

This is the only wind-up type starter spring which may be removed from the spring housing (keeper). To remove spring, grasp inner end of spring with pliers and pull outward. Spring diameter

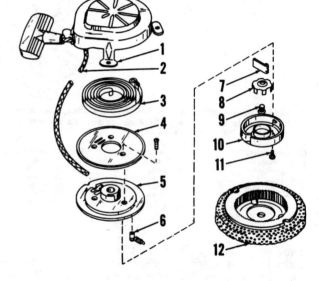

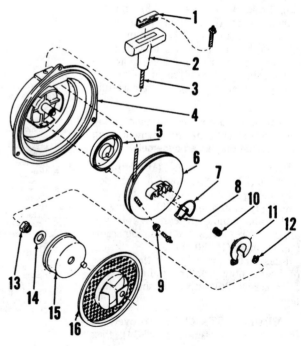

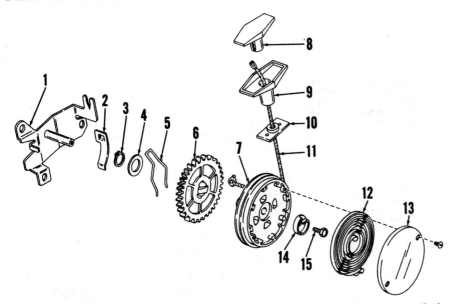

Fig. TE20—Exploded view of single-gear, side-mounted rewind starter used on some vertical crankshaft engines.

1. Mounting bracket
2. Rope clip
3. Snap ring
4. Washer
5. Brake
6. Gear
7. Pulley
8. Insert
9. Handle
10. Bracket
11. Rope
12. Rewind spring
13. Cover
14. Spring hub
15. Hub screw

reassembly is reverse of disassembly procedure. Install centering pin so end of pin extends below end of mounting legs. The pin will align the center of starter to center hole in end of crankshaft.

REWIND STARTERS

FRICTION SHOE TYPE. To disassemble the starter, refer to Fig. TE17 and hold starter rotor (12) securely with thumb and remove the four screws securing flanges (1 and 2) to cover (15). Remove flanges and release thumb pressure enough to allow spring to rotate pulley until spring (13) is unwound. Remove retaining ring (3), washer (4), spring (5), slotted washer (6) and fiber washer (7). Lift out friction shoe assembly (8, 9, 10 and 11), then remove second fiber washer and slotted washer. Withdraw rotor (12) with rope from cover and spring. Remove rewind spring from cover and unwind rope from rotor.

When reassembling, lubricate rewind spring, cover shaft and center bore in rotor with a light coat of Lubriplate or equivalent. Install rewind spring so windings are in same direction as removed spring. Install rope on rotor, then place rotor on cover shaft. Make certain inner and outer ends of spring are correctly hooked on cover and rotor. Preload the rewind spring by rotating the rotor two full turns. Hold rotor in preload position and install flanges (1 and 2). Check

sharp end of friction shoes (10) and sharpen or renew as necessary. Install washers (6 and 7), friction shoe assembly, spring (5), washer (4) and retaining ring (3). Make certain friction shoe assembly is installed properly for correct starter rotation. If properly installed, sharp ends of friction shoes will extend when rope is pulled.

Remove brass centering pin (14) from cover shaft, straighten pin if necessary, then reinsert pin 1/3 of its length into cover shaft. When installing starter on engine, centering pin will align starter with center hole in end of crankshaft.

DOG TYPE (FOUR-STROKE) To disassemble the dog type starter, refer to Fig. TE18 and release preload tension of rewind spring by pulling starter rope until notch in pulley half (5) is aligned with rope hole in cover (1). Use thumb pressure to prevent pulley from rotating. Engage rope in notch of pulley and slowly release thumb pressure to allow spring to unwind. Remove retainer screw (11), retainer (10) and spring (6). Remove brake screw (9), brake (8) and starter dog (7). Carefully remove pulley assembly with rope from spring and cover. Note direction of spring winding and carefully remove spring from cover. Unbolt and separate pulley halves (4 and 5) and remove rope.

To reassemble, reverse the disassembly procedure. Then, preload rewind spring by aligning notch in pulley with rope hole in cover. Engage rope in

notch and rotate pulley two full turns to properly preload the spring. Pull rope to fully extended position. Release handle and if spring is properly preloaded, the rope will fully rewind.

DOG TYPE (TWO-STROKE). Some engines are equipped with the dog type rewind starter shown in Fig. TE19. To disassemble the starter, first pull rope to fully extended position. Then, tie a slip knot in rope on outside of starter housing. Remove insert (1) and handle (2) from end of rope. Hold pulley with thumb pressure, untie slip knot and withdraw rope and grommet (9). Slowly release thumb pressure and allow pulley to rotate until spring is unwound. Remove retaining ring (12), brake (11), brake spring (10) and starter dog (8) with dog spring (7). Lift out pulley and hub assembly (6), then remove rewind spring (5).

Reassemble by reversing the disassembly procedure. When installing the rope, insert a suitable tool in the rope hole in pulley and rotate pulley six full turns until hole is aligned with rope hole in housing (4). Hold pulley in this position and install rope and grommet. Tie a slip knot in rope outside of housing and assemble handle, then insert on rope. Untie slip knot and allow rewind spring to wind the rope.

SIDE-MOUNTED TYPE (SINGLE GEAR). To disassemble the side-mounted starter (Fig. TE20), remove insert (8) and handle (9). Relieve spring tension by allowing spring cover (13) to rotate slowly. Rope will be drawn through bracket (10) and wrapped on pulley (7). Remove cover (13) and carefully remove rewind spring (12). Remove hub screw (15), spring hub (14), then withdraw pulley and gear assembly. Remove snap ring (3), washer (4), then withdraw pulley and gear assembly. Remove snap ring (3), washer (4), brake (5) and gear (6) from pulley.

Reassemble by reversing the disassembly procedure, keeping the following points in mind: Lubricate only the edges of rewind spring (12) and shaft on mounting bracket (1). Do not lubricate spiral in gear (6) or on pulley shaft. Rewind spring must be preloaded approximately 2½ turns.

When installing the starter assembly, adjust mounting bracket so side of teeth on gear (6) when in fully engaged position, has 1/16 inch (1.59 mm) clearance from base of flywheel gear teeth. Remove spark plug wire and test starter several times for gear engagement. A too-close adjustment could cause starter gear to hang up on flywheel gear when engine starts. This could damage the starter.

SIDE-MOUNTED TYPE (MULTI-PLE GEAR). Some engines may be equipped with the side-mounted starter shown in Fig. TE21. To disassemble this type starter, unbolt cover (11) and remove items (7, 8, 9, 10, 11 and 14) as an assembly. Pull rope out about 8 inches (203 mm), hold pulley and gear assembly (8) with thumb pressure and engage rope in notch in pulley. Slowly release thumb pressure and allow spring to unwind. Remove snap ring (7), pulley and gear (8), washer (9) and rewind spring (10) from cover. Rope (14) and rewind spring can now be removed if necessary. Gears (3 and 6) can be removed after first removing snap ring (12), driving out pin (13) and withdrawing shaft (1). Remove brake (4).

To reassemble, reverse the disassembly procedure. Preload the rewind spring two full turns.

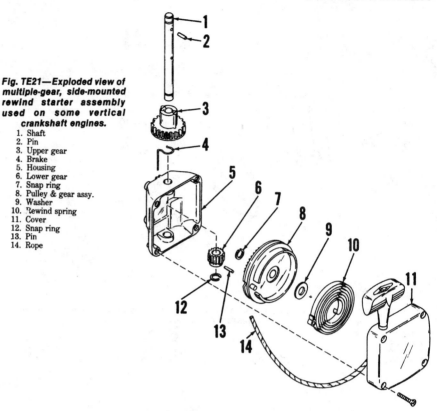

Fig. TE21—Exploded view of multiple-gear, side-mounted rewind starter assembly used on some vertical crankshaft engines.
1. Shaft
2. Pin
3. Upper gear
4. Brake
5. Housing
6. Lower gear
7. Snap ring
8. Pulley & gear assy.
9. Washer
10. Rewind spring
11. Cover
12. Snap ring
13. Pin
14. Rope

TECUMSEH CENTRAL WAREHOUSE DISTRIBUTORS
(Arranged Alphabetically by States)

These franchised firms carry extensive stocks of repair parts. Contact them for name of dealer in their area who will have replacement parts.

Charlie C. Jones Battery & Electric Co., Inc.
Phone: (602) 272-5621
2440 West McDowell Road
P.O. Box 6654
Phoenix, Arizona 85005

Pacific Power Equipment Company
Phone: (415) 692-1094
1565 Adrian Road
Burlingame, California 94010

Pacific Power Equipment Company
Phone: (303) 371-4081
500 Oakland Street
Denver, Colorado 80239

Spencer Engine Incorporated
Phone: (813) 253-6035
1114 West Cass Street
P.O. Box 2579
Tampa, Florida 33601

Sedco Incorporated
Phone: (404) 925-4706
1414 Red Plum Road NW
Norcross, Georgia 30093

Small Engine Clinic
Phone: (808) 488-0711
98019 Kam Highway
Honolulu, Hawaii 96701

Industrial Engine & Parts
Phone: (312) 927-4100
1133 West Pershing Road
Chicago, Illinois 60609

Medart Engines & Parts of Kansas
Phone: (913) 888-8828
15500 West 109th Street
Lenexa, Kansas 66219

Grayson Company of Louisiana
Phone: (318) 222-3211
100 Fannin
P.O. Box 206
Shreveport, Louisiana 71102

W. J. Connell Company
Phone: (617) 543-3600
65 Green Street
Route 106
Foxboro, Massachusetts 02035

Carl A. Anderson Inc. of Minnesota
Phone: (612) 452-2010
2737 South Lexington
Eagan, Minnesota 55121

Medart Engines & Parts
Phone: (314) 343-0505
100 Larkin William Industrial Ct.
Fenton, Missouri 63026

Original Equipment Incorporated
Phone: (406) 245-3081
905 Second Avenue North
Box 2135
Billings, Montana 59103

Carl A. Anderson Incorporated
Phone: (404) 339-4944
7410 "L" Street
P.O. Box 27139
Omaha, Nebraska 68127

E. J. Smith & Sons Company
Phone: (704) 394-3361
4250 Golf Acres Drive
P.O. Box 668887
Charlotte, North Carolina 28266

Uesco Warehouse Incorporated
Phone: (701) 237-0424
715 25th Street North
P.O. Box 2904
Fargo, North Dakota 58108

Gardner Engine & Parts Distribution
Phone: (614) 488-7951
1150 Chesapeake Avenue
Columbus, Ohio 43212

Mico Incorporated
Phone: (918) 627-1448
7450 East 46th Place
P.O. Box 470324
Tulsa, Oklahoma 74147

Brown & Wiser, Incorporated
Phone: (503) 692-0330
9991 South West Avery Street
Tualatin, Oregon 97062

Sullivan Brothers Incorporated
Phone: (215) 942-3686
Creek Road & Langoma Avenue
P.O. Box 140
Elverson, Pennsylvania 19520

Pitt Auto Electric Company
Phone: (412) 766-9112
2900 Stayton Street
Pittsburgh, Pennsylvania 15212

Locke Auto Electric Service
Incorporated
Phone: (605) 336-2780
231 North Dakota Avenue
P.O. Box 1165
Sioux Falls, South Dakota 57101

Medart Engines & Parts of Memphis
Phone: (901) 774-6371
674 Walnut Street
Memphis, Tennessee 38126

Engine Warehouse Incorporated
Phone: (713) 937-4000
7415 Empire Central Drive
Houston, Texas 77040

Frank Edwards Company
Phone: (801) 363-8851
100 South 300 West
P.O. Box 2158
Salt Lake City, Utah 84110

R B I Corporation
Phone: (804) 798-1541
101 Cedar Run Drive
Lake-Ridge Park
Ashland, Virginia 23005

BITCO Western Incorporated
Phone: (206) 682-4677
4030 1st Avenue South
P.O. Box 24707
Seattle, Washington 98124

Wisconsin Magneto Incorporated
Phone: (414) 445-2800
4727 North Teutonia Avenue
P.O. Box 09218
Milwaukee, Wisconsin 53209

CANADIAN DISTRIBUTORS

Suntester Equipment (Central) Ltd.
Phone: (403) 453-5791
13315 146th Street
Edmonton, Alberta, Canada T5L 4S8

Suntester Equipment (Central) Ltd.
Phone: (416) 624-6200
5466 Timberlea Boulevard
**Mississauga, Ontario, Canada
L4W 2T7**

GEAR TRANSMISSION SERVICE

PEERLESS

Series 700

GENERAL INFORMATION. The Peerless 700 series transmissions used in snowthrowers included in this service manual can have 3, 4, 5 or 6 forward speeds. All Series 700 transmissions are similar except for the number of spur gears on the countershaft and output shafts. For example, the four-speed unit will have four spur gears on each shaft, the five-speed unit will have five spur gears, etc. Spacers are used instead of spur gears in 3, 4 and some 5-speed transmissions. The output shaft drive sprocket can be located at either end of the output shaft.

Service procedures are the same for all models, but there is variation of internal design such as number of gears, thickness and number of spacer washers, production changes, etc. It is imperative that on transmission disassembly, the mechanic make note of thrust washer position, number of thrust washers used at each position and location of spacers. Although the following service procedures and illustrations are for a five-speed unit, follow same procedure for 3, 4 and 6-speed units.

DISASSEMBLY. To disassemble the transmission, place shift lever in neutral position, then remove the shift lever. Refer to Fig. PE1 and remove set screw (5), spring (6) and detent ball (7). Remove the six cap screws (1) and lift off transmission cover (2). Pull shifter assembly (35) upward (see Fig. PE2) and remove from case. Lift both the countershaft and output shaft and gear assemblies straight up out of case. Angle the shafts together (see Fig. PE3) to provide slack in reverse drive chain (21—Fig. PE1) and remove countershaft bushing (38), thrust washer (37) and sprocket (36). Remove reverse drive chain (21) from countershaft and reverse output sprocket (20).

NOTE: As disassembly progresses beyond this point, make note of number of thrust washers and washer thickness at each position and location and thickness of any spacers used in 3, 4 and 5-speed models.

Remove bushing (26), thrust washer (27) and gears (28 through 33) from

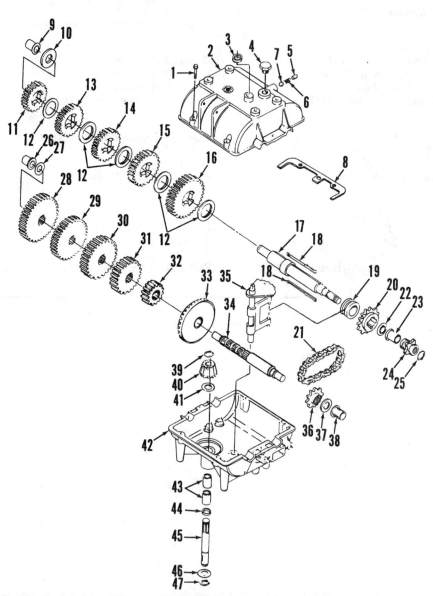

Fig. PE1—Exploded view of five-speed Series 700 Peerless transmission. Spacers will be used at gear locations (11 and 28) in four-speed transmissions and also at locations (13) and (29) on three-speed models. Some five-speed units may also have spacers between thrust washers (10 and 27) and gears at inboard ends of shafts. Threaded plug (4) opening has been provided in transmission cover for neutral start switch installation and is not a lubrication port. Only one part of the cover gasket (8) is shown. Shaft bushings (9, 23, 26 and 38) are alike but have different callouts as aid to service instructions.

1. Cap screws (6)	24. Output drive sprocket	36. Reverse drive sprocket
2. Transmission cover	25. Snap ring	37. Thrust washer
3. Shifter rod seal	26. Bushing	38. Bushing
4. Threaded plug	27. Thrust washer	39. Snap ring
5. Set screw	28. 5th speed drive gear	40. Bevel input gear
6. Spring	29. 4th speed drive gear	41. Thrust washer
7. Detent ball	30. 3rd speed drive gear	42. Transmission case
8. Cover gasket	31. 2nd speed drive gear	43. Needle bearings
9. Bushing	32. 1st speed drive gear	44. Square cut "O" ring
10. Thrust washer	33. Bevel drive gear	45. Input shaft
11. 5th speed output gear	34. Countershaft	46. Thrust washer
12. Shifting washers	35. Shift rod & fork assy.	47. Snap ring
13. 4th speed output gear	14. 3rd speed output gear	
14. 3rd speed output gear	15. 2nd speed output gear	
15. 2nd speed output gear	16. 1st speed output gear	
16. 1st speed output gear	17. Output shaft	
17. Output shaft	18. Shifting key	
18. Shifting key	19. Shift collar	
19. Shift collar	20. Reverse output sprocket	
20. Reverse output sprocket	21. Reverse drive chain	
21. Reverse drive chain	22. Thrust washer	
22. Thrust washer	23. Bushing	
23. Bushing		

countershaft (34) and remove the bushing (38), thrust washer (37) and countershaft reverse sprocket (36) from opposite end of shaft. These gears are splined to the shaft.

NOTE: In some transmissions, the countershaft bevel gear (33) and 1st speed spur gear (32) may be a combination bevel/spur gear.

On output shaft (17), output drive sprocket (24) may be at opposite end of shaft than from position shown in Fig. PE1, in which case the disassembly sequence will be changed. If assembly is same as shown in Fig. PE1, slide bushing (9), thick thrust washer (10), gears (11 through 16) and shifting washers (12) from end of shaft. Remove snap ring (25), output drive sprocket (24), bushing (23) and thrust washer (22) from opposite end of shaft, then slide output shaft reverse sprocket (20), shift collar (19) and shifter keys (18) from shaft.

Remove snap ring (39) and bevel input gear (40) from inner end of input shaft (45), then pull shaft to outside of transmission case. Remove square-cut "O" ring (44) from case and outer snap ring (47) and thrust washer (46) from input shaft (45).

REASSEMBLY. Thoroughly clean and inspect all parts and renew any excessively worn or damaged parts. Recommended lubricant is molybdenum disulfide E.P. lithium grease. All parts should be greased thoroughly as the transmission is reassembled.

If necessary to renew input shaft needle bearings (43), press on hardened (lettered) ends of bearing cages (lettered end facing away from other bearing) so cage is flush to 0.005 inch (0.127 mm) below flush with transmission case. Install snap ring (47) (sharp edge away from thrust washer) and thrust washer (46) on outer end of input shaft (45). Place new sealing ring (44) into transmission. Slide shaft into transmission case, then install thrust washer (41), bevel input gear (40) and snap ring (39). Be sure input shaft turns freely.

Install bevel gear (33) (or combination bevel gear/spur gear) and spur gears (28 through 32), smallest spur gear first as shown in Fig. PE2, onto countershaft (34—Fig. PE1). Install thrust washer (27)

and bushing (26). Lay the assembly aside in a clean space.

Install shift keys (18) and shift collar (19) onto output shaft (17) as shown in Fig. PE3; the thick side of the shift collar must be toward the shoulder on output shaft. Install the shifting washers (12) and spur gears (11 through 16) onto output shaft. The purpose of the shifting washers is to compress the shifting keys to allow the keys to slide into the gears. Therefore, the chamfer on the inside diameter of the washers must be toward the shifting keys. A running production change has been made in the shifting washers; refer to Fig. PE4. Early style and new style washers are interchangeable. The sides of the output shaft spur gears are different; the flat side of each gear is to be located facing the shifting keys or shoulder on the output shaft; refer to Fig. PE4. Install reverse output sprocket (20—Fig. PE1), thrust washer (22) and bushing (23). Install output drive sprocket (24) and snap ring (25).

Lay the two assembled shaft and gear units on a bench with ends pointing in correct direction and angled with reverse sprocket ends close together. Place the countershaft reverse drive sprocket (36) in the reverse drive chain (21), then slip the chain and sprocket onto the splined end of the countershaft and over the reverse sprocket on the output shaft. Put thrust washer (37) and bushing (38) on countershaft, then lift the two shafts as a unit and place them in the transmission case (42). Be sure the locator tangs on the shaft bushings engage the slots in the transmission case. Install the shifter rod and fork assembly in the case with the fork shifter pins engaging the slot in the shift collar.

Pack the transmission with 12 ounces (355 mL) of recommended grease. Place transmission cover (2) onto transmission case with new cover gaskets (8) and shifter rod seal (3). Install retaining cap screws (1) and tighten cap screws to a torque of 90-110 inch-pounds (10.2-11.2 N·m). Install detent ball (7), spring (6) and set screw (5). Turn the set screw in flush to two turns below flush until desired shift detent feel is reached. Turn the input shaft to check for binding; if no binding is noted, the transmission is now ready to be installed in the snowthrower.

Shifter Assembly

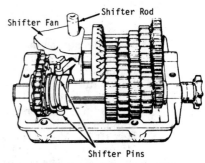

Fig. PE2—With cover removed, lift out shifter fan, rod and pin assembly from transmission case.

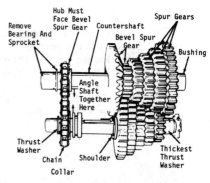

Fig. PE3—With shaft and gear asemblies removed, angle reverse sprocket end of shafts together to gain slack in chain for removal of countershaft bearing and sprocket.

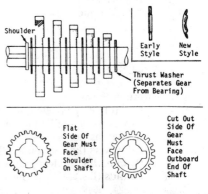

Fig. PE4—Drawing showing proper installation of output shaft gears and shifting washers. Early style and new style shifting washers are interchangeable.

METRIC/U.S. CONVERSION CHART

Linear Measurement
Inches (in.) = Millimeters (mm) × 0.03937
Millimeters (mm) = Inches (in.) × 25.4
Inches (in.) = Centimeters (cm) × 0.3937
Centimeters (cm) = Inches (in.) × 2.54
Inches (in.) = Meters (m) × 39.37
Meters (m) = Inches (in.) × 0.0254
Feet (ft.) = Meters (m) × 3.281
Meters (m) = Feet (ft.) × 0.3048
Yards (yd.) = Meters (m) × 1.0936
Meters (m) = Yards (yd.) × 0.9144

Volume
Cubic Centimeters (cc) = Liters (1) × 1000
Cubic Centimeters (cc) = Cubic Inches (in.³) × 16.388
Cubic Inches (in.³) = Cubic Centimeter (cc) × 0.061
Cubic Inches (in.³) = Liters (l) × 61.02
Liters (1) = Cubic Inches (in.³) × 0.0164
Liters (1) = Cubic Feet (ft.³) × 28.3168
Cubic Feet (ft.³) × 0.0283 = Cubic Meters (m³) × 35.336
Cubic Feet (ft.³) = Liters (l) × 28.3168

Liquid Measurement
Liters (l) = U.S. Quarts (qt.) × 0.9465
Liters (l) = U.S. Pints (pt.) × 0.4732
U.S. Pints (pt.) = Liters (l) × 2.113
U.S. Quarts (qt.) = Liters (l) × 1.056
Imperial (UK) Volume (pints, quarts, gallons) = U.S. Volume (pints, quarts, gallons) × 0.8326
U.S. Volume = Imperial Volume (UK) Volume × 1.201

Weight/Force
Newtons (N) = Kilograms (kg) × 9.8066
Newtons (N) = Pounds (lbs.) × 4.4482
Kilograms (kg) = Newtons (N) × 0.1019
Kilograms (kg) = Pounds (lbs.) × 0.4536
Pounds (lbs.) = Kilograms (kg) × 2.2046
Pounds (lbs.) = Newtons (N) × 0.2248

Tightening Torques
Newton-meters (N•m) = Foot-Pounds (ft.-lbs.) × 1.3558
Foot-pounds (ft.-lbs.) = Newton-meters (N•m) × 0.7376
Newton-meters (N•m) = Inch-pounds (in.-lbs.) × 0.113
Inch-pounds (in.-lbs.) = Newton-meters (N•m) × 8.8503
Newton-meters (N•m) = Kilogram-meters (kg-m) × 9.806
Kilogram-meters (kg-m) = Newton-meters (N•m) × 0.102
Kilogram-meters (kg-m) = Foot-pounds (ft.-lbs.) × 0.1383
Foot-pounds (ft.-lbs.) = Kilogram-meters (kg-m) × 7.233

METRIC/U.S. CONVERSION CHART (CONT.)

Horsepower/power

Kilowatts (kW) = Brake Horsepower (bhp) $\times$ 0.7457

Brake Horsepower (bhp) = Kilowatts (kW) 1.341

Temperature

Degrees Celcius (°C) = Degrees Farenheit (°F) − 32 ÷ 1.8

Degrees Farenheit (°F) = Degrees Celcius (°C) $\times$ 1.8 + 32

Pressure

Kilopascals (kPa) = Kilograms/Square Centimeter (kg/cm²) $\times$ 98.07

Kilopascals (kPa) = Pounds/Square Inch (psi) $\times$ 6.8965

Pounds/Square Inch (psi) = Kilopascals (kPa) $\times$ 0.145

Pounds/Square Inch (psi) = Kilograms/Square Centimeter (kg/cm²) $\times$ 14.223

Pounds/Square Inch (psi) = Bars $\times$ 14.5038

Pounds/Square Inch (psi) = Atmospheres (atm) $\times$ 14.6559

Bars = Pounds/Square Inch (psi) $\times$ 0.0689

Bars = Atmospheres (atm) $\times$ 0.9869

Atmospheres (atm) = Bars $\times$ 1.0132

Metric Prefixes

Tera (T) = 1,000,000,000,000

Giga (G) = 1,000,000,000

Mega (M) = 1,000,000

Kilo (k) = 1,000

Hecto (h) = 100

Deca (da) = 10

Deci (d) = 0.1

Centi (c) = 0.01

Milli (m) = 0.001

Micro (u) = 0.000,001

Nano (n) = 0.000,000,001

Pico (p) = 0.000,000,000,001

Femto (f) = 0.000,000,000,000,001

Atto (a) = 0.000,000,000,000,000,001

NOTES

NOTES

NOTES

NOTES

Genuine John Deere Service Literature

For ordering information, call John Deere at 1-800-522-7448.
All major credit cards accepted.

PARTS CATALOG

The parts catalog lists service parts available for your machine with exploded view illustrations to help you identify the correct parts. It is also useful in assembling and disassembling.

OPERATOR'S MANUAL

The operator's manual provides safety, operating, maintenance, and service information about John Deere machines.

The operator's manual and safety signs on your machine may also be available in other languages.

TECHNICAL AND SERVICE MANUALS

Technical and service manuals are service guides for your machine. Included in the manual are specifications, diagnosis, and adjustments. Also illustrations of assembly and disassembly procedures, hydraulic oil flows, and wiring diagrams.

Component technical manuals are required for some products. These supplemental manuals cover specific components.

FUNDAMENTALS OF SERVICE MANUALS

These basic manuals cover most makes and types of machines. FOS manuals tell you how to SERVICE machine systems. Each manual starts with basic theory and is fully illustrated with colorful diagrams and photographs. Both the "whys" and "hows" of adjustments and repairs are covered in this reference library.